THE ORIGIN OF SPECIES
A Variorum Text

The Origin of Species

A Variorum Text

Charles Darwin

Edited by
MORSE PECKHAM

PENN

University of Pennsylvania Press
Philadelphia

10 9 8 7 6 5 4 3 2 1

First paperback edition 2006

Published by
University of Pennsylvania Press
Philadelphia, Pennsylvania 19104-4112

Library of Congress Cataloging-in-Publication Data
Darwin, Charles, 1809–1882.
  The origin of species; a variorum text, edited by Morse Peckham.
    p. cm.
  Includes bibliographical references.
  ISBN-13: 978-0-8122-1954-8
  ISBN-10: 0-8122-1954-6 (pbk. : alk. paper)
  Note: With a reproduction of the t.p. of the original ed. published in 1859 under title: On the origin of species by means of natural selection.
  1. Evolution. 2. (Biology) Natural selection. I. Title II. Peckham, Morse.
QH365.O2 1959
575.0162   58010416

Dedicated
with profound intellectual gratitude
and
sincere friendship
to
LOREN EISELEY
and
IRVING HALLOWELL
of the
Department of Anthropology
The University of Pennsylvania

CONTENTS

INTRODUCTION

I. OCCASION AND PURPOSE OF THIS EDITION

"Much the greatest event that ever happened and much the best," Charles James Fox said of the French Revolution—to me, a remark even more pertinent to the publication of *On the Origin of Species*. Its greatness would justify the preparation of a variorum text, but there are sounder reasons. The scale on which Darwin carried out five revisions makes it impossible, without such a text, to comprehend the development of his book. Of the 3,878 sentences in the first edition, nearly 3,000, about 75 per cent, were rewritten from one to five times each. Over 1,500 sentences were added, and of the original sentences plus these, nearly 325 were dropped. Of the original and added sentences there are nearly 7,500 variants of all kinds.[1] In terms of net added sentences, the sixth edition is nearly a third as long again as the first.

But I have not undertaken this task because of the greatness of the book or even the scale of the revisions; I hope, rather, that the text will be a contribution to the history of biology, of ideas, and of modern culture. As a student of Victorian literature and its background, I became increasingly tantalized some years ago by the unsatisfactory discussion of the reception of the book, by confusing interpretations, by inadequate sketches of the development of Darwinism and the very different matter of evolutionary thought and its relation to the *Origin*. The larger outlines were fuzzily visible, but the illuminating detail was absent. Turning to the work itself, I quickly found part of the answer. Darwin complained often enough that few really understood the theory of Natural Selection, and were he to examine various cultural histories and even certain respected histories of science published since his death, he would still complain, perhaps quite bitterly. Yet the real reason for the fuzziness was not apparent. Modern editions, since 1898, have not included his tables of corrections, first introduced in the third edition, nor have any American editions I have examined, including the authorized Appleton edition of 1872.[2] By accident I came upon a passage taken from the 1859 edition. I was familiar with the passage, the heavily revised opening of the last chapter. At once I began to seek out, to examine, and to collect the earlier editions. The tables of changes surprised me. I was amazed at the extent and variety of the variations. Nothing I had read prepared me for such a situation, not even the *Life and Letters*. Nor did the lists of changes prepare me for what I was to discover when I did a little collation.

The cause of the unsatisfactory quality of such discussions as I had encountered now became apparent. Without a variorum text it is impossible to speak with accuracy on what Darwin said in the *Origin* at any given time. How could one know that a statement in the 1859 edition was not modified in 1860, 1861, 1866, 1869, and

1872? How many libraries even have a run of all authentic editions? In this country I know of none, though they may very well exist. Further, how could one relate to the *Origin* any reference in any scientific or non-scientific work of the time and know that the relation was valid for all editions? For example, part of the theological outcry can be traced to the fact that in the first edition there was much theological language which Darwin later excised. Without a variorum text, such a fact—and there are dozens of equal and greater importance—cannot even be known. It appeared to me unquestionable that it was impossible to write the early history of the development of Darwinian and evolutionary thought as affected by the *Origin* unless the student had a variorum text. And so, since biologists and even historians of science have more important things to do, and because the task seemed to call for the efforts of just such a harmless drudge as a student of literature, I undertook to create a variorum text.

Such a text will be useful, I hope, for some of the following purposes. To begin with, since Darwin in the *Origin* wrote, as he often said, an *"abstract* of an *abstract,"* he gave no references or authorities, except in the "Historical Sketch," which first appeared in the third edition, 1861. I have prepared a variorum *text.* The next step is a variorum *edition,* with the sources of his information tracked down and verified, authorities already published up to 1872 and authorities whom he consulted as revealed in the various volumes of his correspondence and in the letters, apparently a great body, which remain unpublished. Here is a task for a dozen maids with a dozen mops for more than a dozen years. Such an edition, once completed, would be a foundation for future studies. It would then be possible to organize in relation to the *Origin* Darwin's other works, his correspondence, and the correspondence of his contemporaries with whom he was in constant consultation by post: Hooker, Lyell, Huxley, Wallace, and others, of whom there were many. Much of this correspondence, especially to Darwin, remains to be discovered, made available, and published.

But this internal history of the development of the *Origin* is only part of the task, and the easier part. The external history, its relation to documents other than books by Darwin or letters to him or from him, is an even more complex task, and to my mind even more important. It is no exaggeration to say that we know more about the history of evolutionary thinking, both metaphysical and biological, before 1859 than we do after. It will now be possible to study this development in detail, since a variorum text will provide a fixed center for at least fifteen years after the publication of the first edition, until about 1875, and perhaps later. One may ask any number of questions on this subject. Did a given idea first appear in the *Origin* or did it appear elsewhere? What effect did the revisions have on the undertaking of various lines of investigation, or on the collection of facts? Did Darwin's most unusual conception of natural law have any effect on contemporary science, or on the philosophy of science, or on philosophy, or on the cultural milieu of generalized thinking? What was the specific relation of the *Origin* to the collapse of natural theology among scientists? Among non-scientists? Is there a discoverable relation between the triumph of the *Origin* and the revisions? Or, and this is possibly not a legitimate question, can one hazard that it would or would not have been so successful had he not undertaken such extensive revisions? What was the place, really, of the *Origin* in the development of metaphysical evolutionism?

The history of scholarship shows that the publication of all the variants of an important work has an immensely vitalizing impact on the field of study to which the work belongs. De Selincourt's edition of Wordsworth's *Prelude* was the specific cause of a reflorescence of Wordsworth study which has virtually revealed a new, or at least a decidedly different, man and poet. But heretofore such variorum texts have been principally found in the area of literature. Yet from the point of view of the history of science or the history of ideas or the history of culture, scientific works are just as important, and perhaps more so. Extensively revised texts are indeed even more common in science than in literature, especially since the immensely accelerated rate of scientific change began to show up about a century and a half ago It is my hope that other scholars will follow with editions of equally important works, such as Lyell's *Principles of Geology,* or Mill's *Logic,* or the quasi-scientific *Vestiges of Creation.* Above all I hope that the publication of the present edition will have a stimulating effect upon the fields of Darwinian study and the history of evolution that will lead to answers to the questions I have asked and to many more of even greater importance, particularly those questions which cannot even be proposed until Darwinian studies are much more advanced than they are at present.

II. Composition, Revision, and Publication

On the evening of July 1, 1858, the Linnaean Society, meeting in London, heard four communications, a letter from Sir Charles Lyell and Joseph Hooker to the society's secretary, J. J. Bennett, an "Extract from an unpublished Work on Species, by C. Darwin, Esq., consisting of a portion of a Chapter entitled, 'On the Variations of Organic Beings in a state of Nature; on the Natural Means of Selection; on the Comparison of Domestic Races and the True Species,' " an "Abstract of a letter from C. Darwin, Esq., to Prof. Asa Gray, Boston, U.S., dated Down, September 5th, 1857," and a paper "On the Tendency of Varieties to depart indefinitely from the Original Type. By Alfred Russel Wallace." [1]

Darwin was not present. He had remained at his home at Down, a village fifteen miles southeast of London, in the direction of Tunbridge Wells. He and family were in a state of panic. One daughter was ill from diphtheria, other members of the family of scarlet fever, of which an infant child had just died. The first nurse had been struck down by an ulcerated throat and quinsy, the second by scarlet fever; and since June 18th, when he had received Wallace's paper, he had been under the strain of having been anticipated in his theory of natural selection by the younger and almost unknown

Wallace, the anxiety of determining the proper thing to do, and his self-disgust at discovering that his devotion to science was not so great that he could not care about priority and fame. Incapable of decision and torn with anxiety and grief over his suffering family—scarlet fever and diphtheria were terrifying then—he had turned the whole Wallace matter over to Joseph Hooker, the botanist, and Sir Charles Lyell, the great geologist and for Darwin the Lord Chancellor of English science, men who had been long aware of what for years he had been engaged in developing. These men made the admirable decision to present the Darwin-Wallace material to the Linnaean Society, to the people who made a difference in such affairs, in a way that gave credit to Wallace's brilliance but simultaneously established Darwin's priority.[2]

Hooker wrote to Darwin shortly after the meeting, and Darwin replied on July 5th. There were now two questions, the publication of the communications to the Linnaean Society (Darwin had only to correct the proofs), and, far more important, a clear and sufficiently full statement by Darwin on his theory of natural selection and the origin of species, an unmistakable and convincing account of his revolutionary notions. What was to be done? Judging by later letters, both Lyell and Hooker urged him as forcefully as possible to print. "I could easily prepare an abstract of my whole work, but I can hardly see how it can be made scientific for a Journal, without giving facts, which would be impossible. Indeed a mere abstract cannot be very short. Could you give me any idea how many pages of the Journal could probably be spared me?" Or, perhaps, if it were not sufficiently scientific, it could be published as a pamphlet. At any rate he wanted advice.[3]

By now the children had been sent out of the house, presumably to Tunbridge Wells, from which on July 13th Darwin wrote to Hooker: he had received his suggestion for an article of thirty pages in the *Journal of the Linnaean Society,* and he did not care for it. "I will set to work at the abstract, though how on earth I shall make anything of an abstract in thirty pages of the Journal, I know not, but will try my best." Hooker kept at him, urging him once again "to print," and providing stimulating generalities and information; Darwin was still smarting, he admitted in the letter of July 13th, that he had not had "a grand enough soul not to care" about being forestalled; but by July 18th, at Sandown, on the Isle of Wight, he had turned his attention more fruitfully to his problem. The abstract would be longer, but it would still be "really impossible to do justice to the subject." His difficulty was this: The interrupted book was in itself an abstract. How could he write an abstract of an abstract, when the basic work had never been written and the abstract of that was only half done? Yet two days later, the day he received the proof-sheets of the communications to the Society, he began. The next day he wrote to Hooker that he would condense to the utmost, and if it turned out to be too long for the Society's resources, he would help pay for the printing.[4]

Yet after all it was a false start. His diary contained the entries: "July 20 to August 12, at Sandown, began Abstract of Species Book."

"September 16, Recommenced Abstract." He was happy to have started, and relieved that Hooker had consulted with George Busk, a member of the Society (in charge of the journal?), and could assure him that he need not hold the work down to thirty pages, since it could be published in parts. This plan involved presenting the abstract in the course of several evenings. "Variation under Domestication," probably what later became Chapter I of the book, forty-four manuscript pages, "would do," Darwin thought, for one evening. On August 11th he was still thinking of an abstract for the Society, but according to his diary, the next day stopped working, not to resume until September 16th, at Down. On October 6th, while thanking Hooker for making him write, he warned him that the work was growing to inordinate length, and on October 12th he announced that it would probably "run into a small volume, which will have to be published separately." [5]

Although Darwin always complained how slowly and badly he wrote, he made rapid progress. By November 14th he had finished the first six chapters, the next two could be abstracted from the unfinished book, but beyond this his "materials were less worked up." He expected to finish his "small book" in April. By December 24th he had written 330 folio pages and expected to write "150-200" more. The results would, he thought, be a duodecimo volume of about 400 pages, to be published separately. He underestimated by 100 pages, and the book was to be in traditional octavo size. But his forecast of finishing by April was correct. By March 2nd he had finished Chapter XII and by March 15th he had almost finished the last chapter but one. In another week he had done and was correcting for the press. "I hope in a month or six weeks to have proofsheets." [6]

But still there is no mention of a publisher. Months before, he had abandoned the notion of publication by the Linnaean Society, but he was a man of means, and aware of the *furor theologicus* raging in the land, and he could publish himself, as indeed he in a sense had once planned to, when he had offered to help the Society in the cost of the publication of the proposed pamphlet. But someone else thought differently. Sir Charles Lyell went to John Murray. It was a good choice. Since 1830 Murray had been the publisher of Sir Charles's great *Principles of Geology,* to the advantage of both parties, and in 1845 he had paid £150 for the copyright of the *Voyage of the Beagle,* which he published in his Home and Colonial Library. He had a varied and excellent list. Byron was one staple; the Murray guide books were another. Travel, history, memoirs, theology, piety, art, biography, the large and profitable series of school texts of the Rev. William Smith formed a large part of his list, but he had also brought out Layard on Nineveh, Sir David Brewster's popular works on astronomy, Dennis' great volumes on Etruria, Grote, Milman, the Admiralty publications, and, above all, the annual British Association reports. He was an amateur geologist, and like his father, John Murray II, fascinated by the scientific speculations of the day. He was interested at once in Lyell's suggestion. Sir Charles's authority alone would have been enough, but

Volume III of the Linnaean Society publications with the communications of July 1, 1858, had appeared; Darwin had written to Herbert Spencer that he was engaged on a work on "the changes of species"; gossip must have been circulating in the still rather small London world of literate and scientific men. Murray was ready to publish without seeing the MS. Lady Lyell, oddly enough, communicated the news to Darwin, perhaps in a letter to Mrs. Darwin, and Darwin, on March 28th, wrote to Lyell for further information; Lyell replied at once; Darwin was "much pleased" and on the 29th or 30th wrote to Murray. His book, he said, was a "popular abstract," and "*ought* to be popular"; he had been surprised at the "interest strangers and acquaintances" had shown. He hoped to be ready to go to press "early in May," and then "most earnestly" wished "to print at a rapid rate, for my health is much broken, and I want rest." Murray replied at once, offering to publish without seeing the MS and to pay Darwin "about two-thirds of the net proceeds." He emphasized the fact that he was so certain of the importance and success of the book that he was departing from his usual practice of always seeing the MS before a final decision. On April 2nd Darwin wrote to Hooker and replied to Murray, accepting his terms, but insisting that Murray was free to reject the MS after seeing some of it, if he felt it would not sell. To Hooker he expressed his conviction that the book would be popular among the scientific, semi-scientific, and non-scientific men. This is his typical modesty, but it was not quite consistent with his letter to Murray, to whom he stated that he felt it would be "interesting to all (and they are many) who care for the curious problems of the origin of all animal forms." What a way to refer to the question of questions! This is disingenuous; he was perfectly aware of the furor the work would cause. The direct interference of the Deity was, for most Victorians, even scientists, the only possible explanation for "the origin of all animal forms." This fantasy was precisely the last stronghold of British Natural Theology. Indeed he was nervous, and quite prepared for rejection, an attitude which gives some confirmation to the possibility that he was planning to publish himself. On April 5th, sooner than he had expected, he sent Murray the first three chapters.[7]

Murray was upset; he thought the idea impossible; and he turned to friends for advice. George Pollock, a lawyer, but not a scientist, thought that no living scientist could understand it—not much of an exaggeration—because it was too profound; and advised printing a thousand copies. Murray then turned to the Reverend Whitwell Elwin, the Rector of Booton, the editor of the *Quarterly*, still a chief Tory journal and still published by Murray. Elwin, currently also carrying on Croker's edition of Pope, was as incapable of understanding Darwin as he was of comprehending the finest poet in English and the most satisfactory human being among English authors. But he busied himself about the matter, went to see Lyell, since the latter had advised publication and he, Elwin, certainly did not, and wrote a letter to Murray to be sent on to Darwin. Lyell, he said, had suggested that Darwin publish his observations on pigeons with a brief statement on his theory. Later on a larger work could appear. "This appears to me an admirable suggestion. Everybody is interested in pigeons. The book would be reviewed in every journal in the kingdom and would soon be on every library table." This has the wild unreality of Darwin's own "curious problem." Lyell's action is odd, or Elwin misrepresented him. Elwin's letter was sent to Darwin, who dismissed both Lyell's and Elwin's suggestions as impracticable, but was grateful for the kindness of both men. Why? Again one thinks that the tone of this letter was determined by Darwin's readiness to publish himself, but that he wanted a commercial publisher because he thought the book would sell. At the same time he wanted no one to know his conviction of its possible popularity; "if it prove a dead failure, it would make me the more ridiculous." Nevertheless, Murray stuck by his decision, and April 9th told Darwin that the offer stood. Indeed, he required to see no more of the MS. The matter was settled, and by June 14th Darwin was reading proofs.[8]

In the meantime, however, trouble was beginning to develop. First, part of the MS, the section on Geographical Distribution, was lost on its way back from Hooker. Next, the whole MS, apparently, having been sent to Hooker, Mrs. Hooker found parts of it obscure. Darwin "trembled," and vowed to clarify his style in the proofs. He was to continue these efforts through six more editions and for twelve more years, for when he finally saw his book in print he found the style "incredibly bad," and the corrections were so heavy he wrote to Murray and offered to pay a substantial part of their cost. By June 21st he had corrected only 130 pages, and by the next day only 20 more. After another five weeks, five sheets were completely finished, and two more nearly ready, but for all he knew the book could be either good or bad. On September 10th the proofs were finished, leaving only some of the revises, index, and front matter. He told Lyell that he would have Murray send him copies of the corrected revises, through page 240. Thereafter he sent Lyell the corrected sheets, the last ones going on September 30th, except for the index, which was not in type. He dated the book Oct. 1st.[9]

It had been a long and dreary struggle. He was sure that 1,250 copies would be too many. The endless corrections, the despairing efforts to achieve clarity, the knowledge of what was involved, not only for him, but for science, and his wretched health, which limited his working to three hours in the morning—after twelve o'clock he could only suffer boredom and the frustration of delay—the last minute changes of fact and interpretation—all these had worn him out; as soon as he could travel, on October 2nd, he set out for Ilkley, in Yorkshire, to undergo a cure at a well-known hydropathic establishment. He stayed there until early in December; he was at Ilkley when the book was published.[10]

At first, then, he was still dissatisfied with the book. He was thinking of rewriting the larger, half-finished version, and he doubted if another edition of this little one would ever be called for. Yet if the October 15th *Athenaeum*, to which he had subscribed for years, was reaching him at Ilkley, or if he was buying it there, he should have been pleased at Murray's advertisement for his fall list. A full-page announcement, in two columns instead of the customary three, listed the thirty new books and editions Murray was about to bring out. Across the top of the page was the book of the season: *M'Clintock's Narrative of the Discovery of the Fate of Sir John Franklin.* Franklin was already part of the Victorian pantheon, and so, for years, had been Wellington, whose correspondence was the second title. Directly after Wellington came Darwin. It is amusing that even Samuel Smiles' marvelously successful *Self-Help* was only No. XXI. But in fact an even more effective kind of advertising had already started. On Thursday, September 18th, Sir Charles Lyell, as President of Section C, Geology, of the British Association, meeting at Aberdeen, had spoken provocatively of Mr. Charles Darwin's forthcoming work on the "difficult and mysterious subject" of the origin of species, a problem on which the author had "thrown a flood of light." And this in the presence of the Prince Consort, the Association President for that year. It was one of the few section meetings Albert attended. The *Athenaeum* for September 24th carried Lyell's remarks in detail.[11]

Murray continued building up for publication. Although the October 22nd issue of the *Athenaeum* listed only those of his new works which were ready, on October 29th he advertised the *Origin* again, and once more in a prominent position. On November 2nd, he sent Darwin a specimen copy, announced that the trade sale would be on November 22nd, and told him that he would absorb the corrections cost of £72/8/9. He must have felt very confident. Darwin protested, again apologizing for the heavy corrections. He was "infinitely pleased and proud at the appearance of my child," and in agreement about the price, which was to be 15s. November 11th and 12th he devoted to writing letters to individuals who were to receive copies: Agassiz, Gray, de Candolle, Falconer, Henslow, Lubbock, Wallace, Fox, Carpenter, and very likely others. On the 12th, Murray again advertised the work, as nearly ready, not so prominently on his list, but in considerably larger print, and again in an important position in the November 19th issue of the *Athenaeum*. But the issue carried an even finer advertisement, free. The editor, W. N. Dixon, devoted his lead article, five columns, to reviewing the book. It was so exquisitely calculated to arouse interest that were it not for the independent reputation of the paper, one would suspect that Murray had a hand in it. It was perfectly timed—the trade sale was to be the following Tuesday—and admirably hostile, raising all the most sensational points and deriving those conclusions bound to cause the highest anxiety and anger. "If a monkey has become a man—what may not a man become?" Thus early, Probably Arboreal intrudes himself.

Theologians will say—and they have a right to be heard—Why construct another elaborate theory to exclude Deity from renewed acts of creation? Why not at once admit that new species were introduced by the Creative energy of the Omnipotent? Why not accept direct interference, rather than evolutions of law, and needlessly indirect or remote action? Having introduced the author and his work we must leave them to the mercies of the Divinity Hall, the College, the Lecture Room, and the Museum.

And the whole is spiced, through a reference to Darwin's health (although this information came from the book itself), by the insinuation that the reviewer has Inside Information.[12]

Murray, if he had the divine élan of the true publisher, must have been delighted. Darwin was upset and even convinced that he had written "in a conceited and cocksure style" (nothing could be less true) and wrote to Hooker to find out the author. Hooker wrote a consoling reply; it made Darwin feel better and roused him against the anonymous reviewer. "But the manner in which he drags in immortality, and sets the priests at me, and leaves me to their mercies, is base. He would, on no account, burn me, but he will get the wood ready, and tell the black beasts how to catch me." And now, on Tuesday, November 22nd, came Mr. Murray's sale to the trade. Happily, a report on it was printed in "Our Weekly Gossip" in the November 26th *Athenaeum*.[13]

As was expected, the book on Franklin was the big seller, 7,600 copies, followed by *Self-Help*, 3,200 (wonderfully appropriate, this coincidental publication with the *Origin*), the life and correspondence of Sir Fowell Buxton, the abolitionist, 2,500 copies, a life of a Rev. Daniel Wilson, 2,200, and "Mr. Charles Darwin's work 'On the Origin of Species,' " 1,500. If this figure is correct, and there is no reason to doubt it, the statement so commonly made that the book sold out on the first day—a statement that many have interpreted as referring to retail sales from book shops—must be revised. The printing was 1,250 copies, of which 1,192 copies were for sale; it more than sold out. Murray wrote Darwin the good news at once, and said that a new edition must be prepared without delay. Darwin was confounded, probably by both statements, but prepared to set to work immediately, without his MS, since he planned to make few changes.[14]

The book that the trade bought so confidently, and the public so eagerly, is by no means a bad specimen of Victorian typography, the horrors of which have been much overrated. It was well printed, by W. Clowes and Sons, one of the four largest printers in London, a company still in business and still printing for John Murray, although their records, unfortunately, were lost in the blitz of World War II. The paper, bought from Spalding, was a heavy cream-colored stock which has scarcely discolored in nearly a hundred years. The binding, done by Edmonds and Remnants, was a royal green cloth, ribbed diagonally with wavy ribbing, not of the same quality as the paper, but reasonably sturdy for the times. It was blind-stamped with an elaborate paneled ornament on the sides and in excellent gilt on the spine. The exterior appearance is more characteristic of what we think of as Victorian book-making than the interior. The one plate was engraved by W. West, lithographers. The total manufacturing cost was £260/6/4. The 1,192 copies sold (5 went to Stationers' Hall for copyright purposes, 12 to Darwin, and 49 to reviewers) at the trade discount of 9/6. The retail price was 15/0. Darwin got £90 and Murray had left £57/4/2, instead of his one-third of £49/0/0. The reason for this discrepancy was that Darwin wished to be paid before the final casting of accounts, as indicated in the John Murray Letter Book.[15]

Darwin wrote Murray's good news to Lyell on November 24th, the date which he noted in his diary as the day of publication, and November 24th is always given as that date. But dates of publication are notoriously difficult to determine with exactness at this period, and this one is no exception. Why November 24th? The date of the trade sale was November 22nd. Deliveries could have been made to the major London book shops on the 23rd. On the 24th Darwin heard from Murray. At least that is the date given by Francis Darwin, and it seems reasonable, yet if Murray wrote on the 22nd, Darwin could have received the letter on the 23rd. In the letter to Lyell he speaks of having received a letter from him dated the 22nd, but he could have received it on the 24th or the 23rd. The language can be interpreted either way. On the other hand, in the *Athenaeum, The Saturday Review,* and *The Spectator* for November 26th, Murray had a one-column advertisement, dated "Albemarle-Street, November 1859," and headed "THIS DAY." The *Origin* is the first book on the list, despite the fact that *Self-Help*, the eighth, had outsold it on November 22nd. This would appear to establish the date of publication as November 26th. But if the November 26th weeklies were distributed on Thursdays, that might be the date intended in the advertisement, which is thus equivocal. Until contemporary evidence is found indicating when someone actually purchased it in a bookstore, the matter must remain in doubt, and even if a letter or diary records purchase on November 24th, there is still the possibility that it could have been bought on November 23rd. Even Darwin's evidence is not of great significance, for the November 24th date is entered in the diary under October 1st, and thus could have been made at any time after November 24th. Since the diary entries quoted by Francis Darwin have less the character of diary entries and more of summaries of work accomplished for the year, it seems quite likely that they were not written up until January, a possibility which makes it a little more likely that November 24th was the date Darwin heard from Murray. I am inclined to the position that Darwin remembered having heard from Murray on that day, and that we should think of the real date of publication as the 26th, the simplest interpretation of Murray's advertisement, at least until incontrovertible evidence indicates that someone actually bought or could not buy it on Wednesday, Thursday, or Friday. Alternatively, November 23rd, a possible date for deliveries to bookstores, is the next most attractive choice.[16]

If Darwin carried out the plans proposed in his letters, he left Ilkley on December 7th, stopped in London that night, called on Lyell the next day, and also went to see Murray. The latter visit was one of those rare occasions on which author and publisher confront each other with pleasure. Darwin could tell Murray that he was making progress in the preparation of the second edition, and Murray could give Darwin a very important piece of advice, which Lyell confirmed, or confirm what Lyell suggested, and which Hooker also had in mind—not to attempt to finish the unfinished book, or write the unwritten one of which the half-written was the theoretical abstract. "You have hit on the exact plan, which, on the advice of Lyell, Murray, etc., I mean to follow—viz., bring out separate volumes in detail—and I shall begin with domestic productions: . . ." It was an important decision, and a knowledge of it answers several questions: Why he did not write a big book on the origin of species; why he did not complete the half-written one; and, for our purposes, most important, why he spent such time and effort on revisions of the actual *Origin,* of which he was at first inclined to think so slightingly. Mankind owes much to Darwin's advisers, first for getting him to publish and second for persuading him to perfect the *Origin* and make it into a kind of Introduction to the rest of his works. It becomes analogous to Spencer's *First Principles,* the foundation and preface to his books on orchids, variation under domestication, man's ancestry, and, in one way or another, the balance of them, just as Spencer's book has a similar relation to his works on the principles of biology, psychology, sociology, and so on.[17]

By December 14th he had made further progress in his revision, "with a very few corrections," and Murray had decided to print 3,000 copies. A week later he speaks of the printing as having begun, refers to it as a reprint, but admits that he has made "a few important corrections." On December 26th came Huxley's famous review in the *Times.* It was thus not surprising to find an advertisement from Murray in the December 26th *Times* stating that the "5th thousand" was on sale. To be sure, anything over 4,000 is in the fifth thousand, but it must be noted that 1,250 and 3,000 add up to 4,250. Murray's advertisement may have been an anticipation; Darwin's diary notes that the second edition was published on January 7, 1860. Yet the Diary is not reliable; and the posting of the ledger entries as of December 30, 1859, suggests that Murray's regarded transactions involving this edition as completed by that time. All copies would not have to be bound to make enough available to fill back orders. I am inclined to believe that the *Times* date of December 26, 1859, is acceptable, although the date on the title-page is 1860.[18]

The new edition had the type, paper, and binding of the original. For textual purposes the fact that the same type was used is most important. In revising the work Darwin's determination to make only a few changes was carried out, at least in terms of the number of changes he made in later editions, particularly in the fourth, fifth, and sixth. Still, the total number in this edition is impressive enough: 9 sentences dropped; 483 rewritten or re-punctuated; 30 added. No chapter was untouched. The text, however, was not reset. The printers simply regarded Darwin's corrected copy of the first edition as still another revise. Instead of resetting the whole book, they reset only lines and paragraphs when necessary. Thus, frequently, it was not necessary to change the page lining, and the book had the same number of pages and gatherings as the first edition. Further, they took the occasion, either on their own or from Darwin's corrections, not only to correct most of the small number of typographical errors but also to reset a large number of lines which were either too closely or too loosely spaced. The appearance

19

of dozens of pages was much improved. Although this kind of re-setting is principally notable in the first edition, it is found as long as the original type, no matter how much revised, was used, that is, through the fifth edition, which is in consequence a remarkably correct, excellent, and uniform example of type-setting. Darwin received £313/12/5, almost exactly one third of the net. Again Darwin was paid ahead of the final reckoning. The first edition cost about 6/6 per copy; the second about 2/5. Hence the far greater profits did not depend merely on the increased size of the edition. The economic advantage of only partial resetting of the type for four successive editions and 9,750 copies is obvious. The price was reduced to 14/0, where it remained until the fourth edition. Deducting the copies needed to fill the orders remaining from the November 22nd trade sale and the copies sold at the sale in November, 1860, about 2,300 copies were sold in about ten months, a rate the book was not to reach again until the cheap edition was published in 1872.[19]

The first indication we have on future editions of the *Origin* appears in a letter of January 28, 1860, to Asa Gray, explaining that Darwin was "resolved to have the present book as it is (excepting errors, or here and there inserting short sentences) and use all my strength, *which is but little,* to bring out the first part (forming a separate volume, with index, &c) of the three volumes which will make my bigger work: . . ." Which three he means, aside from the book on variation under domestication, it is hard to say. His plans were to change as the years went by, but at least he had thoroughly determined to make the *Origin* his basic work. He adds, "I also intend to write a *short* preface with a brief history of the subject." And by the end of the month he had begun to collect materials.[20]

In the following November Murray sold 700 copies at his trade sale, but had less than half that number in stock. In the first year of its life, the book had sold about 3,800 copies. On November 22nd Darwin heard from Murray that "it was time to set to work at once on a new edition of the '*Origin.*'" He intended to correct, but not enlarge; "it will be only slightly altered." But in December he wrote to Murray that there will be many "corrections, or rather additions, which I have made in hopes of making my many rather stupid reviewers at least understand what is meant. I hope and think I shall improve the book considerably." By February 4, 1861, he was finishing the new edition, and on April 26th he referred to it as "just published."[21]

The third edition contained two important new features: a table of "Additions and Corrections, to the Second and Third Editions," by which is meant only the thirty-five passages he considered important enough to record, and "An Historical Sketch of the Recent Progress of Opinion on the Origin of Species." Altogether he dropped 33 sentences, altered 617, and added 266, together 14 per cent of the total number of variations, while the second edition had only 7 per cent. The text was 35 pages longer than in the two previous editions, and the "Historical Sketch" added six and a half pages in smaller type. Type and paper and binding were the same.

Sales must have slowed up considerably, in spite of the success of the book at the trade sale in November, 1860, for Murray printed only 2,000 copies, and this new stock lasted for nearly five years, a monthly sales rate of under 30. The cost of printing was about the same, because of the increased proportion of changes, but the paper and binding, of course, cost less. The unit manufacturing cost was 2/10. Darwin received £372/0/0 for the edition, and Murray made a profit of £239/19/6. Again there was a disparity in Murray's favor, presumably once more because of Darwin's desire to be paid without delay. After all, Murray had offered "about two-thirds" of the net.[22]

There is no indication as to when the stock was exhausted. After filling the orders left over from the previous November's trade sale Murray had fewer than 1,650 copies on hand. If these were finally sold out in the usual fall sale of 1865, he was in no great hurry for a new edition. According to his diary, Darwin began on the fourth edition on March 1, 1866, and finished it, except for the last revises, on May 10th. On July 5th he could write of the new edition as almost printed off. Murray delayed publication until the fall season. In the *Athenaeum* and in *The Saturday Review*, after several previous announcements, it was listed on December 15, 1866, as "Now Ready." [23]

The fourth edition was the most extensively revised yet, containing 21 per cent of the total number of variants. Darwin dropped 36 sentences, rewrote 1,073, and added 435, although in his new table of differences between the third and fourth editions he listed only 34 passages. He added two pages to the "Historical Sketch" and fifty-two pages to the text. An important structural change involved the addition of a number of new sub-headings within the chapters, and the change from the former place at the beginning of paragraphs to a position centered above paragraphs. The type, of course, was the same, but though Spalding provided the paper, it was whiter and took the ink less well, or perhaps the ink was unsatisfactory. The binder was again Edmonds and Remnants, but the binding was changed, still royal green, but simplified. Murray printed only 1,500 copies, a mere 250 more than the first edition. It was enough for two and one-half years at a monthly sales rate of 46; Murray was still advertising it in February, 1869, only a few months before the appearance of the fifth edition. The printing cost was higher as a result of the heavy revisions; between this and the smallness of the edition, the unit cost rose to 3/11. Darwin received only £250 and Murray £160, again more than one-third.[24]

The fifth edition was to be the most important yet published. We first hear of it on December 5, 1868, when Murray still had a large enough stock of the fourth to make it worth his while to advertise. On the other hand the sale of the fourth had been more rapid than that of the third, and Murray did not wait for the stock to be exhausted before informing Darwin it was time to set to work again. Darwin began on December 26, 1868, and continued until February 10, 1869. He was disgusted at how much he had to modify and felt he ought to add, but was determined not to add much. On January 31st he expected the new edition to be out in about two months, and by March 12th it was passing through the press. According to Francis Darwin it was published in May. However, the first advertisement I have been able to discover appeared in the *Times* for July 19th, under the heading of "New Books." The first advertisement in the *Athenaeum* appeared in the August 7th issue, and this paper reviewed the book, hostilely, on August 14th. Since the *Athenaeum's* reviews were usually prompt, appearing shortly after publication, the best publication date for the fifth edition is probably on or about August 7th.[25]

It was notable on several counts. For the first time Darwin used the famous phrase, taken from Spencer, "Survival of the Fittest," and it was the most extensively revised edition yet—indeed, if we except the bulk of the extra chapter added in the sixth edition, the most extensively revised of all. It contains 29 per cent of the total number of variants: 178 sentences dropped, 1,770 altered, and 227 added. Hence only two pages were added to the total length of the text. There are twenty-nine entries in the table of changes. Again the old type was only partially reset, but as yet showed no signs of wear. The paper, from Dickinson, rather than Spalding, was demy crown, a new size, and the book was imposed and gathered in 8's instead of 12's. It was the same width, but about a quarter-inch shorter, with a little more taken from the bottom margin than from the top. Hence the proportion of type to page is changed, and of margins to type-page. It looks more old-fashioned than the previous editions, less satisfying to modern taste. Victorian book-design was in fact entering its worst period at this time. The binder was the same, but a new blind-stamp was used on the sides, and the spine was new.

The edition was larger than the previous one. The basis was, of course, the increase in rate of sales shown by the fourth over the third. Murray was right in guessing that sales were on the rise. The fifth was to sell at the rate of nearly 60 a month. It cost less to print than the fourth, and the unit cost was of course less, 3/1, a decrease of 10d. from the fourth. Darwin got £315 and Murray £283/6/1, over 47 per cent of the net. The fifth edition returned to the original price of 15/0, already established after the original publication of the fourth at 14/0. Perhaps that is why there were still 57 on hand when the sixth edition was ready. But the sixth was to be an entirely new kind of edition, a popular edition to retail at 7/6.[26]

According to Francis Darwin, the new edition was begun in June, 1871, and finished January 10, 1872. By the middle of July, Darwin was struggling with his refutation of Mivart, the Roman Catholic biologist, about whom he had harsh and "mortified" things to say in his letters and whom Huxley was to trounce so thoroughly and deservedly the following November in the *Contemporary Review*. In August he expected to start correcting the sheets in a few weeks, and by late September he was satisfied with what he proposed to do to Mivart and had decided to devote a new chapter mainly to the cleverest enemy who had yet appeared, as well as the least fair. By late December he could write of being almost finished.

Francis Darwin may be right in stating that his father finished the revision on January 10th, but he is probably wrong about the date of publication, which he puts in the same month. In the *Athenaeum* and *The Saturday Review* Murray's first advertisements were on February 17th; the book was listed among the new works in the *Athenaeum* of March 2nd, and in *The Saturday Review* as "now ready" in the March 16th issue. Probably it was available sometime during the week of February 19th.[27]

The edition contained several important changes. The first word, "On," was dropped from the title. The considerations of objections were taken from Chapter IV and placed with new material, chiefly rebutting Mivart's attacks, in a new Chapter VII. Thus the old Chapters VII through XIV were renumbered VIII through XV. At the end was added a "Glossary of the Principal Scientific Terms used in the Present Volume," prepared by W. S. Dallas. This very useful addition was probably part of the plan to appeal to a popular market. Such a scheme was very likely Murray's, and my guess is that he suggested the glossary; it is more a publisher's idea than a scientist's. Including the new material on Mivart, the new edition had more variants than any of the previous five. Excluding that, it had fewer than the fifth but considerably more than the fourth. Darwin dropped 63 sentences, rewrote 1,669, and added 571. As in the fifth, hundreds of sentences were completely recast with only slight changes in meaning, the cumulative effect of which, however, was of great importance, as detailed studies of the text, if they are forthcoming, will show.[28]

The appearance of the book, aside from the old royal green color, was much different. It was completely reset, and thus a good many typographical errors were introduced which Darwin failed to catch, not surprising in view of the size of the type. It was the same font, but in a much smaller face, hitherto used only in the "Historical Sketch." The page was about a quarter-inch shorter and narrower than the fifth. The total length of the book was xxii + 458, 142 pages shorter than the fifth. The saving on paper was 6d. per copy. The unit cost was 2/8, at first glance not enough of a difference from the preceding edition, only 5d., to justify a halving of the price; but Murray knew what he was up to. The cost included a set of stereotype plates, £35/15/0, which he sold to Appleton in New York for £50/0/0. Hence the unit cost was reduced to 2/4. No doubt Francis Darwin expressed the opinion of the family and perhaps of his father when he regretted that this final edition should be so unpleasing and so discouraging, because of the smallness of the type, to a good many people who otherwise might have read it. But Murray had very good reasons for this cheap edition, in spite of the very small margin of profit on the first impression. The English literacy rate had gone up in the thirteen years since the first edition, and the passing of the Education Act of 1870 meant that it would go up even more. Popular Editions were becoming increasingly common in the trade. There was a large and growing new market. Darwinism was triumphing everywhere, and the controversy was by no means dying down. If anything, it was increasing in intensity and involving more people. The first impression of the new edition, 3,000 copies, lasted only a year. It was all sold out while 57 copies of the fifth were still on hand. Sales rose from fewer than 60 copies a month to nearly 250. Perhaps many people were kept from reading the book by its unattractive appearance and small type, but certainly they bought it. And this was only the beginning. It was reprinted in 1873, 1875, 1876, 1879, 1880. In 1878 (or 1876, *Cf.* App. III) it was odd. From 1872 to 1875 the title-page read "Sixth Edition, with Additions and Corrections." In 1876 to this line was added "to 1872." Collation reveals a number of typographical errors corrected, but only three changes in wording. If the new phrase is taken as modifying "Additions" as well as "Corrections," it is certainly a misleading exaggeration, but perhaps Murray's did not think of this interpretation.

The text of 1878, then, was the final text, and the careful publisher will make his reprints from this or any of the printings from these corrected plates up to 1891. In that year Murray began issuing an inexpensive one-volume reprint of an 1885 two-volume edition. Rather than present discursively the publishing history from 1875 to 1890, I present a summary of the information to be found in Appendix II. Thus there is no information about American or other foreign editions and translations. It is to be hoped that at some time full information of these editions will be made available to scholars and bibliographers.

Summary of the publishing history of the *Origin of Species,* English editions corrected by Darwin, 1859-1890.

Edition	Date	Printing
First	Nov. 26, 1859	1,250
Second	Dec. 26, 1859	3,000
Third	By Apr. 26, 1861	2,000
Fourth	Dec. 15, 1866	1,500
Fifth	Aug. 7, 1869	2,000
Sixth	Feb. 19, 1872	3,000
	1873	2,000
	1875	1,500
Sixth, corrected	1876	1,250
or	1878	2,000
	1880	2,000
	1882	2,000
	1882	2,000
	1883	2,000
	1885	2,000
	1886	2,000
	1887	3,500
	1889	2,000
	1890	2,000
		39,000

(Note: The dates as indicated above are for the most part not absolutely reliable. When the publisher's records disagree with the title-page dates, as in several of the 1880's issues, I have taken the former.)

By 1898 the total profits for Murray's were £2,709/19/8; for the Darwin family, £5,385/8/8, or £11/16/6 more than the stipulated two-thirds.

2/3

From 1872 to 1890 the average sale was 1,561 copies a year, a sales rate of 130 copies a month. It was that publisher's delight, a bread-and-butter book. In the first decade the rate was 65, in the second 98, and in the third 112. Forty-two per cent of the total was sold in the first fifteen years, 58 per cent in the second, to which should be added the several thousand of the expensive two-volume edition. The real triumph of Darwin's book came after his death. The profits of the American pirates must have been enormous. It would be interesting to know the sales in foreign languages. But now that the book is no longer protected by copyright, it would be as hopeless a task to search out all the reprints as it would be to discover those of its great—and almost as shattering—coeval, *The Rubáiyát of Omar Khayyám.*

III. Editorial Principles and Guide to the Text

The basic law for the preparation of a text is that any disturbance in the type means a high probability of a disturbance in the text. The fundamental questions about any document are: Is it authentic? Was the author responsible for any disturbances?

Ah, yes, but he didn't know

In the case of *The Origin of Species* the authenticity of the various versions is easily established. On the other hand, the preparation of the text, involving the enormous task of locating and recording nearly 7,500 departures from the original, is a different matter. After some years of consideration and consultation with various colleagues, the following plan was adopted.

It was decided, to begin with, that only published variants would be presented. Of non-published variants there are two classifications, MS, which, judging by Francis Darwin's remarks, have been destroyed,[1] and proofs. One set of proofs is in the Berg collection of the New York Public Library; another I saw briefly at a book-dealer's in London. It was corrected in two hands. The variants in both of these would probably be worth publishing, at least in a journal, but it is my opinion that such variants fall into the category of all MS, letters and journals, for example, and consequently have no proper place in an edition of this sort, the purpose of which is to make available what Darwin saw fit to present to the world. The growth of Darwin's ideas and the development of his information antecedent to the text and its internal history subsequent to publication are two entirely separate problems. Furthermore the use of material Darwin did not publish would necessarily be incomplete, while the text as offered here presents a sharply and clearly delimited and complete set of variants.

In making the edition, the first step was the decision to transcribe nothing, or at least as little as possible. (Actually fewer than 200 words have been inserted by hand.) The objective was to have the printers in 1958 set up their type from text proof-read by Darwin himself. This is an application of the basic law of the critical editing of printed material as defined above. I had a photographic copy made of the first edition, since, though it is not particularly rare, it has some value, and thus a photographic copy was less expensive than two copies of the first edition. After some years of search, with the aid of several obliging booksellers, who certainly made very little money from their efforts, I was able to locate two copies of each of the subsequent editions.

The first step of the collation was to compare the text of the second edition with that of the first, the third with the second, and so on, through the sixth. Whenever the text was the same, the text of the later edition was crossed out with a soft and easily erasable pencil line. When variants were discovered, they were left unlined; pick-up and drop words, as explained below, were also left unmarked. A partial collation of the 1878 sixth showed that, except for typographical errors, it was little changed. This first collation discovered about 90 per cent of the variants.

The next step was more complex and necessarily slower. Since it is, I believe, a novel method, it is worth describing in detail. In any case, it is my conviction that anyone who uses a text should know precisely the method, even to mechanical details, of collation used. First a sentence (or sentence equivalent, such as a chapter-title) was cut out of the photographic copy of the first edition, pasted on a piece of standard typing paper and numbered. Above the working space were arranged in order of publication the subsequent five editions. After some experimentation it was found most expeditious to remove a chapter at a time from each copy of the edition concerned, transfer the markings from the collated copy to the non-collated copy, interleave them to alternate face-up obverse with reverse, and number the pages, beginning with "1" and adding the letter of the edition, "*a*" to "*f*." One at a time the pages containing the sentence corresponding to the pasted sentence were placed immediately below the latter and compared. When a variant was confirmed or a variant missed in the first collation was discovered, the pencil lines with pick-up and drop words were checked, or the lines, if the variant had been overlooked, were erased to include the pick-up and drop words. The number of the sentence and the edition letter were written in the left-hand margin opposite the first word to be reproduced, and the sentence, or the variant with one marked-out line above and one below, when possible, was cut out and immediately pasted below the original sentence. In collating subsequent corresponding sentences, the text was read as far as the pick-up word in the variant, and followed through the variant to the drop word, when the eye returned to the original text. This device was followed because if, let us say, all of an *e* sentence is checked against all of a *c* sentence, an overlooked *c* variant from *a* or from *b* will not be discovered. Though this happens rarely, it can happen and, with the present editor, did. Consequently the last edition must be collated in the sentence-collation with particular care, since it is almost the last chance. This second stage discovered all but about 1.5 per cent of the variants.

In this process two rules are worth remembering. First, never handle more than one cut-out slip of paper at a time. Second, when one discovers that within a few minutes one has missed a variant to be picked up later, stop work. I found that an hour and a half of continuous work was my maximum, and that I could rarely work more than seven hours a day. Because of the first rule, new sentences must be pasted in the text in an order reversed from their chronological order of appearance; and on the principle of disturbing the text as little as possible, this must be the order of the final printed variorum text.

Finally, the whole text was checked sentence by sentence and variant by variant against a corrected copy of the sixth edition, that is, an 1878 sixth, first to discover the few significant variants in this text and second to make sure that every difference between the first edition and the sixth was accounted for by the variants discovered and recorded. The method is simple but time-consuming.

For these reasons the printed text in this edition has a most unconventional appearance. Every variant is printed directly beneath the original sentence, exactly as it appears in the MS, or perhaps the proper term is "paste-script." The reason for this device, which was suggested by Mr. Thomas Yoseloff, Director of the University of Pennsylvania Press, is that basic text and variants do not have to be set up separately, as would be the case if the variants were footnotes. The possibility of omitting a variant is reduced. Furthermore, the type suffers the minimum of manipulation when the galleys are divided into pages and placed in the forms. To be sure, this arrangement makes the text difficult to read, but I cannot imagine anyone reading such an edition in order to gain a general knowledge of the book. To make it easier for the eye to distinguish between text and variant, various typographical devices have been employed, in particular the use of roman numerals for the original and italic numerals for the variants.

It is obvious that the *Origin* presents a different problem from that of a literary text in which one is seeking to establish a text closest to what the author wrote or else to determine his final text, the Shakespeare or Wordsworth problems, for example. In this problem there is no question of what the author wrote, or of what was his final text. The object has been to make it possible for the student to read the variants against the original text and against each other as easily as possible. This is another reason for the unorthodox presentation of the text. The eye has to do the minimum amount of jumping and can most easily observe the differences if the variant is physically as close as possible to the original text and preceding variants. It is particularly important when, for reasons to be discussed below, a pair of pick-up and drop words encloses within the same sentence a series of separate variants connected by three or four words unchanged from the preceding text.

Every discovered variant has been included, no matter how minor, from a new passage of a number of sentences, in one case, nearly 400, to the addition or excision of a comma. For this there are two reasons. First, our libraries are stuffed with selective and eclectic texts, useless for precisely those reasons. It is impossible to predict whether or not a future scholar will find a variant significant, even though the editor may be convinced that it will never be of use to anyone. He must be as mechanical and as unthinking as possible. The fewer the judgments, the better the text. The second reason is more mechanical; if one is determined to find the most insignificant variant, the attention is so directed that more important variants are the more certainly discovered. Only obvious typographical errors have not been recorded. In case of doubt, the variant has been recorded. Another kind of variant may cause disagreement with my decisions. When is a rewritten sentence basically the same proposition and when is it a new proposition? In some dozens of cases this has been a difficult problem, but if the user of the text disagrees with me, he can do so because the necessary evidence is before him; and to provide the necessary evidence is all that I have tried to do.

The reader has a right to ask to what extent he may depend upon the correctness and completeness of the text. The Western world has had 500 years of experience with printed type. No one who is familiar with the history of book-making during that period will be anything but skeptical about any claim to perfection for a printed book, or any piece of printing more than a few words long. Consequently, I make no such claim. I believe I have discovered and recorded accurately all variants of more than insignificant semantic importance. I am also reasonably confident about changes in wording which Darwin believed to be stylistic improvements but which involve no apparent semantic distinction. About changes in spelling and punctuation I am less secure. However, in a text such as this, the kind of variant which may have escaped me has significance only if it occurs in considerable numbers, and the text is so long that a few such variants missed here and there could have little importance. I believe that the reader may use the text with a high degree of confidence if he is concerned with variants of two or more words, and with a little less confidence if the variant involves a single character, a letter, or a mark of punctuation.

Since I have not been able to find a model for this kind of edition, I have been forced to use a number of unusual typographical signs, or at least traditional signs in unusual ways. To explain these I give below examples of each kind of device, taken in order from the text, beginning with the "Introduction," followed by an explanation.

1   ON THE ORIGIN OF SPECIES

The number on the left in Roman type is the number assigned to each sentence, or sentence equivalent, such as chapter-title, summary of contents, chapter sub-division, etc. Each chapter has its own series. In the future it will be possible to refer to any sentence or variant in the *Origin* by reference to chapter, sentence-number, and this edition.

3:c of the organic beings inhabiting South
The italics indicate that this is a variant. The number is the sentence number; the letter is the edition letter: *a* for the first, *b* for the second, *c* for the third, *d* for the fourth, *e* for the fifth, and *f*

for the sixth, *g* for 1878 sixth. "of" is the pick-up word because it is the first time "of" appears in the sentence in *a;* "South" is the drop word because this is the last appearance of "South" in *a.* Frequently the second word in the variant is not the varying word, since it may be preceded by a word which is not making its first appearance in the sentence. With equal or greater frequency the pick-up and drop words will enclose two or more variants connected by words which do not vary. The purpose is to aid the reader in comprehending the continuity of variants. Frequently pick-up and drop words are different in grammatical form from the original, but only when they are not repeated in any form in that sentence. Different forms of "to be" and "to have" are never used as pick-up and drop words, unless there is not the faintest possibility of confusion. Punctuation is never used for pick-up or drop purposes, except for terminal purposes at the end of a sentence.

8   My work is now nearly finished; but as it will take me two or three more years to complete it, and as my health is far from strong, I have been urged to publish this Abstract.

8:f now (1859) nearly
The indentation for "My" indicates that in the original this sentence begins a new paragraph. If there is no contrary indication ("No ¶") this sentence begins a new paragraph in all subsequent editions.

26:[c] [See "Historical Sketch," 32-38]
The square brackets around "c" indicate that this sentence was omitted in third edition. The next pair of square brackets indicates that the enclosed words are the editor's. The numbers are sentence numbers.

41:e difficulties in accepting the theory/or how a simple/being or into an elaborately
The slash (/) separates the drop word of one variant from the pick-up word of another one in the same sentence.
    The next examples come from Chapter I.

3.1:d *[Center] Causes of Variability. [Space]*
The numerals plus letter indicate that this sentence was added after sentence 3 and before sentence 4 in the fourth edition. The square brackets around "Center" and "Space" indicate that these words are the editor's. They are in italics to emphasize their association with the variant numbers. "Center" means that "Causes of Variability" in *d* and all subsequent editions was centered on the page as a chapter sub-division. "Space" indicates that a blank line follows. Occasional irregularities of spacing for blank lines have not been noted. That there is a space after the chapter summary in the original and that this space is preserved in subsequent editions before the first sentence of the chapter or, beginning with *d*, before the sub-heading is indicated by a blank space before sentence 4 and after variant *3.1:d.*

9.1-3:e As far as I am . . . system. 2 With respect . . . conditions.
   3 The former . . . uniform.

"1-3" indicates three sentences added in the fifth edition between sentences 9 and 10. The first two have no variants; hence they are printed in paragraph form with the sub-number preceding the sentence which it identifies. The first new sentence is the first sentence in the variant series. The last sentence, in such situations, either has a variant or concludes a paragraph.

16.x-y:e generally, but erroneously, attributed to vitiated instincts.
   y Many
In the fifth edition sentence 16 is divided into two parts, of which the second begins with "Many," also a drop word.

42.x:c eyes are generally deaf.

42.y:c Colour
When sentences are divided and the first or *x* part has subsequent variants, all the variants of *x* are entered before the *y* part is given.

93.0.1:d

93.0.2:d

93.0.2.x-y:f

93.0.3:d

93:[c]

93.1:d

93.1:e

93.1:f
Between 92 and 94 a series of additions and excisions was executed in two stages. Three sentences were added in *d* after 92, of which the second was subdivided in *f*. In *c* 93 was omitted and a sentence which I rank as a different proposition was substituted for it in *d*. This was varied in the two subsequent editions.

156.x:e

156.x.x-y:f

156.y:e
In *e* 156 is divided into two parts. In *f* *156.x* is further subdivided.

166 + 167:f
In *f* 166 and 167 are combined into a single sentence.
   The next examples come from Chapter II.

27.1:d[¶] to *27.10:d*

27.10.1:f

27.11:d, etc.
In *d* thirteen sentences were added between 27 and 28, and in *f* one sentence was added between the tenth and eleventh of these added sentences. *27.1:d* begins a new paragraph.
   In Chapter IV after sentence 381 appear a large number of additions and variants, carried out in four strata. The third, *f*, includes Chapter VII, added in the sixth edition (see Part II of the Introduction, above, and Appendix II). Thus the first new sentence in Chapter VII is *VII.382.39.6.0.1:f*. This is cumbersome, but I believe clear, indicating four strata of variants.
   The final example comes from Chapter VI.

102.y.1.y:[e] = *102.y.3.1:e*

and

102.y.3.1:e = *102.y.1.y:[e]*

In the fifth edition a sentence after *102.y* is subdivided. Part *x* remains in its former order, after *102.y*. Part *y* is moved to a position after *102.y.3*.

Fortunately there are few such transferences; the order of the sentences is very rarely changed, in spite of all additions, variations, or excisions.

The various appendices include a statistical summary, a transcript of the publisher's records, a descriptive bibliography, and photographic reproductions of the principle binding variants. The original MS of this edition has been deposited in the University of Pennsylvania Library, together with a complete set of the first six editions and those of the subsequent impressions which I have been able to obtain and examine, as listed in Appendix II.

IV. Acknowledgments

My thanks are particularly due to the Committee on Research of the University of Pennsylvania for grants which aided me in the collection of materials for this edition of Darwin's *The Origin of Species;* and to the Officers and Trustees of the University of Pennsylvania for a semester's leave of absence to work on this edition. Mr. Peter J. Kroger of London has been extraordinarily helpful in locating the necessary copies of the various editions and imprints of the work. Mr. John Gray Murray of John Murray, Publishers, London, has been exceedingly kind in opening to me the archives of that publishing house and in granting permission to reproduce transcripts from the Murray ledgers.

Notes to the Introduction

I. *Occasion and Purpose of This Edition*

1. By "variant" is here meant "variant sentence." It may have one change in punctuation or a half-dozen verbal changes, or be a canceled sentence or a new one. I have not calculated the number of variants within sentences. There must be somewhere between 15,000 and 20,000.

2. An American will rarely happen on one of the original editions, or any English edition, so widely was the book published in the United States. Several university libraries I have examined have nothing but American editions; of these, few are early.

II. *Composition, Revision, and Publication*

1. Journal of the Linnaean Society, III, 1858.

2. To Hooker, July 5, 1858, LLCD, II, 126; to Hooker, June 29, 1858, LLCD, II, 120. The story has been told many times; hence I have not given much detail here. The best account I have seen is in Bert James Loewenberg, *Darwin, Wallace, and the Theory of*

Natural Selection, New Haven, 1957, unfortunately published privately, though charmingly, and limited to 385 copies.

3. To Hooker, July 5, 1858, LLCD, II, 126.

4. To Hooker, July 13, 1858, LLCD, II, 128; to Darwin, July 15, 1858, LLJH, II, 458; to Lyell, July 21, 1858, LLCD, II, 130; to Hooker, July 21, 1858, LLCD, II, 131.

5. Diary, LLCD, II, 131; to Hooker, July 30, 1858, LLCD, II, 132; to Hooker, Aug. 5, 1858, LLCD, II, 133; to Gray, Aug. 11, 1858, LLCD, I, 135; to Hooker, Oct. 6, 1858, LLCD, II, 137; to Hooker, Oct. 12, 1858, LLCD, II, 138. In the last letter there is mention of sending out either copies of the *Journal of the Linnaean Society* or off-prints.

6. To Hooker, Nov. 14, 1858, MLCD, I, 447; to Hooker, Mar. 12, 1859, LLCD, II, 147; to Hooker, Mar. 15, 1859, LLCD, II, 149; to Fox, Mar. 24, 1859, LLCD, II, 150. In the Mar. 2nd letter he actually says he has finished the chapter on geographical distribution, but this subject takes two chapters, XI and XII. Since the published Chapter XIII was almost finished on Mar. 15th, presumably the chapter on geographical distribution was divided at some later stage. Fox was a second cousin and an old friend. In a letter to Asa Gray, dated by F. Darwin as of April 4th, Darwin spoke of eleven chapters done and three to go. F. Darwin is probably in error, since this statement is consistent with the early March letters, but not, as we shall see, consistent with the early April letters. LLCD, II, 155.

7. To Hooker, Dec. 4, 1858, LLCD, II, 143; to Lyell, Mar. 28, 1859, LLCD, II, 152; JM, 168; Murray catalogue of Jan. 1860, bound with the U. of Pa. copy of the second edition; to Spencer, Nov. 25, 1858, LLCD, II, 141; to Lyell, Mar. 28, 1859, LLCD, II, 151; Mar. 11, 1859, LLCD, II, 152-3; to Murray, (no date), JM, 168-9, identified by Paston as Darwin's first letter to Murray, apparently correctly; to Darwin, Apr. 1, 1859, JM, 169; to Hooker, Apr. 2, 1859, LLCD, II, 153; to Murray, Apr. 2 (inferred), JM, 169; to Murray, Apr. 5, 1859, LLCD, II, 155.

8. JM, 170-3; to Hooker, Apr. or May, 1859, LLCD, II, 157; to Hooker, Apr. 11, 1859, LLCD, II, 156; to Murray, June 14, 1859, LLCD, II, 159. Paston does not give the dates for Elwin's letters or for his visit to Lyell. Murray presumably received the MS on Apr. 6th. One can sympathize with his bewilderment, considering how many people have read the whole book without understanding it; Murray did not even have the chapter on natural selection. It is possible, if Elwin was in London, as he must have been within a very short time to see Lyell, that his letter was sent to Darwin on Apr. 8th, that Darwin rejected his suggestion on the same day, and thus enabled Murray to reply on Apr. 9th. Alternatively, Murray could have determined to abide by his original offer in spite of Elwin and sent the letter along with his own of acceptance on the 9th. I am inclined to believe the latter. Pollock's judgment was weighty, and Murray was a man and a publisher of great integrity. For Elwin's attitude toward Darwin see his astounding letter on Darwin when *The Life and Letters of Charles Darwin* was pub-

lished. He there refers to Darwin's "inferior order" of mind. JM, 270-1.

9. To Hooker, Apr. 14, 1859, LLCD, II, 156; May 11, 1859, LLCD, II, 157; to Murray, June 14, 1859, LLCD, II, 159; to Lyell, June 21, 1859, LLCD, II, 160; to Hooker, June 22, 1859, LLCD, II, 160-1; to Murray, July 25, LLCD, II, 161; to Hooker, Sep. 11, 1859, LLCD, II, 165; Sep. 1, 1859, LLCD, II, 163; Sep. 11, 1859, LLCD, II, 165; to Lyell, Sep. 30, LLCD, II, 169-71; Diary, LLCD, II, 205.

10. For his health and departure to Yorkshire, see LLCD, II, 158-171, *passim*.

11. To Lyell, Oct. 23, 1859, LLCD, II, 175; to Hooker, Apr. 23, 1861, MLCD, I, 186; ATH, Oct. 15, 1859, No. 1668, p. 485; Sep. 24, 1859, No. 1665, p. 404.

12. ATH, Oct. 29, 1859, No. 1670, p. 553; LBJM; to Murray, LLCD, II, 178 (undated; probably Nov. 2 or 3, 1859); letters to other scientists, LLCD, II, 215-222; ATH, Nov. 12, 1859, No. 1672, p. 645; Nov. 19, 1859, No. 1673, p. 653, p. 660.

13. To Hooker, Sunday (Nov. 20?, 1859), LLCD, II, 222; to Darwin, Nov. 21, 1859, LLCD, II, 228 (also LLJH, I, 510); to Hooker, Nov., LLCD, II, 228; ATH, Nov. 26, 1859, No. 1674, p. 706.

14. Printing and sales records from ledgers at John Murray's. See Appendix II.

15. For the statement on Clowes see Charles Rosner, *Printer's Progress*, London and Cambridge, U.S.A., 1951, (p. 8 of Chapter I), "1851, Survey." (The book is not paginated nor are the chapters numbered.)

16. ATH, Nov. 26, 1859, No. 1674, p. 693; SPE, Nov. 26, 1859, No. 1639, p. 1219; SR, Nov. 26, 1859, VIII. No. 213, p. 656. Murray also ran advertisements in the London *Times* on the day of the sale, Nov. 22nd, p. 12.

17. To Murray, (no date, but from Ilkley), LLCD, II, 178; to Lyell, Dec. 2, LLCD, II, 238; to Huxley, MLCD, I, 131.

18. To Lubbock, Dec. 14, 1859, LLCD, II, 241; *Times*, Dec. 26, 1859, p. 10 (this issue carried the famous review by Huxley); a similar advertisement appeared, with the notation "NOW READY" in ATH, Dec. 31, 1859, No. 1679, p. 874; and in SR, Dec. 31, 1859, VIII, No. 218, p. 825; Diary, LLCD, II, 256.

19. For the Nov. 1860 sale, to Lyell, Nov. 24, 1860, LLCD, II, 352.

20. To Gray, Jan. 23, 1860, LLCD, II, 270; to Hooker, Jan. 31, 1860, LLCD, II, 274.

21. To Huxley, Nov. 22, 1860, LLCD, II, 351; to Bates, Nov. 22, 1860, MLCD, I, 177.

22. LLCD, II, 356; to Hooker, Feb. 4, 1861, LLCD, II, 361; to Davidson, Apr. 26, 1861, LLCD, II, 366.

23. Diary, LLCD, III, 42; to Wallace, July 5, 1866, LLCD, II, 46; ATH, Nov. 3, 1866, No. 2036, p. 557; Dec. 8, 1866, No. 2041, p. 739; Dec. 15, 1866, No. 2042, p. 808; SR, Dec. 15, 1866, XXII, No. 582, p. 752. The publication date and size of edition in MLCD, "Outline of Charles Darwin's Life," p. xxii, are in error.

24. In this edition alone an ordinary pencil eraser fades the print.

25. To Hooker, Dec. 5, 1868, MLCD, II, 375; Diary, LLCD, III,

106; to Hooker, Jan. 16, 1869, MLCD, II, 379; to Croll, Jan. 31, 1869, MLCD, II, 162; to Tait, Mar. 12, 1869, MLCD, II, 381; MLCD, I, xxii, for date of publication. LT, July 19, 1869, p. 12; ATH, Aug. 7, 1869, No. 2180, p. 164; ATH, Aug. 14, 1869, No. 2181, p. 210.

26. The fifth edition is the cause of complete mystery. Of all the editions, I have found the fifth the hardest to find. One I found in Boston, another was found for me in London, and a third in Dublin. Both the first and fourth are easier, the first for obvious reasons. But why should the fourth, in an edition of 1,500, be easier to come by than the fifth, with its 2,000 copies?

27. LLCD, III, 152; to Hooker, Sep. 16, 1871, MLCD, I, 332; "Mr. Darwin's Critics," *Darwiniana,* Vol. III of Huxley's Collected Essays; to Lubbock, Aug. 12, 1871, MLCD, I, 332; to Huxley, Sep. 21, 1871, LLCD, III, 148; to Häckel, Dec. 27, 1871, MLCD, I, 335; ATH, Feb. 17, 1872, No. 2312, p. 196; Mar. 2, No. 2314, p. 272; SR, Feb. 17, 1872, XXXII, No. 852, p. 234; Mar. 16, 1872, XXXII, No. 855.

28. Dallas, in 1868, was in charge of the Museum at York and was trying to get the situation of Assistant Secretary at the Geological Society. (MLCD, II, 353.) In 1867 he had prepared the index for the *Variation.* (LLCD, III, 74.) He translated Müller's *Für Darwin* and Krause's *Life of Erasmus Darwin,* to which Darwin, in 1879, prefixed a notice of his own. (LLCD, III, 364.)

III. *Editorial Principles and Guide to the Text*
   1. LLCD, I, 121.

ABBREVIATIONS

ATH: *The Athenaeum, Journal of Literature, Science, and the Fine Arts,* London, 1828-1921.

JM: George Paston, *At John Murray's, Records of a Literary Circle, 1843-1892,* London, 1932.

JMLB: Letter Book in the Archives of John Murray, Publishers, London.

LLCD: Francis Darwin, *The Life and Letters of Charles Darwin, Including an Autobiographical Chapter,* 3 vols., London, 1887.

LLJH: Leonard Huxley, *Life and Letters of Sir Joseph Dalton Hooker,* 2 vols., London, 1918.

LT: The London *Times.*

MLCD: Francis Darwin, *More Letters of Charles Darwin,* 2 vols., New York, 1903.

SPE: *The Spectator,* London, 1828-

SR: *The Saturday Review,* London, 1856-

THE ORIGIN OF SPECIES

By Charles Darwin

A Variorum Text

ON THE

ORIGIN OF SPECIES.

1:f THE ORIGIN OF SPECIES

1   "But with regard to the material world, we can at least go so far as this—we can perceive that events are brought about not by insulated interpositions of Divine power, exerted in each particular case, but by the establishment of general laws."

WHEWELL: Bridgewater Treatise.

1.1:b   "The only distinct meaning of the word 'natural' is *stated, fixed,* or *settled;* since what is natural as much requires and pre-supposes an intelligent agent to render it so, *i.e.,* to effect it continually or at stated times, as what is supernatural or mirac-ulous does to effect it for once."

BUTLER: Analogy of Revealed Religion.

2   "To conclude, therefore, let no man out of a weak conceit of sobriety, or an ill-applied moderation, think or maintain, that a man can search too far or be too well studied in the book of God's word, or in the book of God's works; divinity or philos-ophy; but rather let men endeavour an endless progress or proficience in both."

BACON: Advancement of Learning.

---

3   *Down, Bromley, Kent,*
October 1st, 1859. (1st Thousand).

3:c   *Down, Bromley, Kent,*
October 1st, 1859. (1st Thousand.)
Third Edit., March, 1861.

3:d   *Down, Bromley, Kent,*
November 24th, 1859. (1st Edition.)
Fourth Edition, June, 1866.

3:e   *Down, Beckenham, Kent,*
First Edition, November 24th, 1859.
Fifth Edition, May, 1869.

3:f   *Down, Beckenham, Kent,*
First Edition, November 24th, 1859.
Sixth Edition, Jan. 1872.

ON

THE ORIGIN OF SPECIES

BY MEANS OF NATURAL SELECTION,

OR THE

PRESERVATION OF FAVOURED RACES IN THE STRUGGLE FOR LIFE.

By CHARLES DARWIN, M.A.,

FELLOW OF THE ROYAL, GEOLOGICAL, LINNÆAN, ETC., SOCIETIES;
AUTHOR OF ' JOURNAL OF RESEARCHES DURING H. M. S. BEAGLE'S VOYAGE
ROUND THE WORLD.'

LONDON:

JOHN MURRAY, ALBEMARLE STREET.

1859.

[Note: This page used for listing Darwin's works, *c-f*]

LONDON: PRINTED BY W. CLOWES AND SONS, STAMFORD STREET,
AND CHARING CROSS.

CONTENTS

1

3   CHAPTER I.

4   Variation under Domestication.

43

CHAPTER II.

7   VARIATION UNDER NATURE.

9   # CHAPTER III.

10   STRUGGLE FOR EXISTENCE.

12   # CHAPTER IV.

13   NATURAL SELECTION.

15 CHAPTER V.

16 LAWS OF VARIATION.

18 CHAPTER VI.

19 DIFFICULTIES ON THEORY.

19:e DIFFICULTIES OF THE THEORY.

20.1:f

CHAPTER VII.

21

CHAPTER VII.

21:f VIII.

22

INSTINCT.

24

CHAPTER VIII.

24:f IX.

25

HYBRIDISM.

27   CHAPTER IX.

27:f X.

28   ON THE IMPERFECTION OF THE GEOLOGICAL RECORD.

30   CHAPTER X.

30:f XI.

31   ON THE GEOLOGICAL SUCCESSION OF ORGANIC BEINGS.

1   INSTRUCTION TO BINDER.

2 The Diagram to front page 117, and to face the latter part of
        the Volume.

1-2 [Note: In *a-d* on p. [x], after "Contents." In *e* on p. [xiv] after
    "Additions and corrections." In *f* on p. [x] after "Contents."]

2:c 123
2:d 130
2:e 132
2:f 90

1   ADDITIONS AND CORRECTIONS,

TO THE SECOND AND THIRD EDITIONS.

1: [Note: This section first appears in *c*]

1:d   TO THE FOURTH EDITION.

1:e   TO THE FIFTH EDITION.

1:f   TO THE SIXTH EDITION.

1.1:e NUMEROUS small corrections have been made in the present edition on various subjects, according as the evidence has become somewhat stronger or weaker. [No space or rule]

1.1:f made in the last and present editions

---

2 [The more important Additions and Corrections alone are here tabulated for the convenience of those interested in the subject, who possess only the earlier editions.

2:d subject, and who possess the third edition.

2:e The more important corrections and some additions are tabulated below for/possess the fourth edition.

2:f additions in the present volume are tabulated on the following page, for the convenience/possess the fifth edition.

3 Most of these additions have appeared in the German and in the Second American editions.

3:[d]

4 The paging of the first and second English editions is nearly the same.]

4:[d]

4.1-2:d The second edition was little more than a reprint of the first. 2 The third edition was largely corrected and added to.

4.2:e to, and the fourth much more largely.

4.2:f fourth and fifth still more

4.3:d As copies of the present work will be sent abroad, it may be of use to a few if I specify the state of the foreign editions: the second French and second German editions were from the third English, with some few of the additions given in the present edition.

4.3:e use if

4.3.x-y:f editions. *y* The third French

4.3.1:f A new fourth French edition has been translated by Colonel Moulinié; of which the first half is from the fifth English, and the latter half from the present edition.

4.4:d A new German edition is now preparing.

4.4:e A new third German edition, under the superintendence of Professor Victor Carus, was from the fourth English edition; and a fourth German edition will immediately appear, corrected up to the standard of the present edition.

4.4:f English edition; a fifth is now preparing by the same author from the present volume.

4.5:d The American second edition was from the English second, with some few of the additions given in the third.

4.5:e with a few/third English; as the work was unfortunately stereotyped, it remains very imperfect.

4.5:f The second American edition/third; and a third American edition has been printed from the fifth English edition.

4.6:d The Italian is from the third, and the Dutch from the second edition.]

4.6:e third, the Dutch, and two Russian editions, from the second English edition.

4.6:f Dutch and three Russian editions from the second English edition, and the Swedish from the fifth English edition.

Second Edition.	Third Edition.	Additions and Corrections.
Page	Page	
12	12	Case of correlation of colour and liability to poison given.
45	46	On the improbability of the occurrence of sudden and great modifications of form.
49	51	On the doctrine recently advanced by some naturalists that animals present no variations.
54	56	The meaning of the term "dominant species" made clearer.
73 95 }	76 100 }	The dependence of the fertility of clover on the visits of insects amplified.
81	84	Various objections to the term "Natural Selection" answered.
84	89	On the two assumptions of an indefinite and definite amount of variation.
102	107	Power of reversion exaggerated.
105	110	Mistaken idea of the efficiency of time alone in producing change.
112	117	On the accidental accumulation of similar variations effecting little without the aid of selection.
126	133 to 143	On the tendency of organisation to advance—On the present coexistence of high and lowly organised beings—Difficulty of conceiving how at the dawn of life the simplest beings were modified—Various objections considered—On convergence—On the indefinite multiplication of specific forms.
135	152	On mutilations being inherited.
137	154 to 157 }	The blindness of cave-animals amplified and corrected.
184	202	Case of ground-woodpecker amplified.
186 187 }	205 207 }	On the modification of the eye amplified.
192	211	On the ovigerous frena of cirripedes.
198	218	On the effect of little exercise and abundant food.
212	232	On variations in the habits of the hive-bee.
231	252	Thickness of walls of cell of hive-bee corrected.

5	Second Edition.	Third Edition.	Chief Additions and Corrections.
	Page	Page	
	235	256	On related variations in instinct and structure not necessarily simultaneous.
	269	291	On the fertility of crossed varieties a little amplified.
	285	308	The computation of time required for the denudation of the Weald omitted. I have been convinced of its inaccuracy in several respects by an excellent article in the 'Saturday Review,' Dec. 24, 1859.
	292	313	On deposition during subsidence partially modified—On the complete denudation of all the formations which must once have covered granitic areas.
	298	321	On the slight modifications which the latest tertiary forms have apparently undergone.
	299	323	On the nature of the links between past and present species, which geology has, and has not, revealed.
	303	328	On early transitional links.
	304	329	Case of fossil footsteps of birds in the United States, added in second edition in place of that of the whale, which is doubtful.
	319	346	On the fallacy of great size or strength saving an animal from extinction.
	336	363	On the degree of development of living forms compared with ancient forms, amplified.
	373	403	Additional facts on former glacial action in the Cordillera of S. America.
	391	422	On crossing keeping the birds of Madeira and Bermuda unmodified, added in second edition.
	439	471	Passage added on the authority of Von Baer on the embryonic similarity of Vertebrate animals.
	452	484	Distinction between nascent and rudimentary organs, indicated in the second edition.
	481	515	On the theological bearing of the general argument, added in second and present editions.
	484	519	On the probability of all organic beings having descended from one primordial form added to.

5:d	Third Edition.	Fourth Edition.	Chief Additions and Corrections.
	Page	Page	
	xiv	xiv	A notice is given of Dr. Wells' work, in which the doctrine of natural selection, as applied to man, was first clearly propounded.
	xvi	xvii	The account of Prof. Owen's views has been added to; and notices are given of the views of some other naturalists.
	18	18, 19	Facts are added on the multiple origin and antiquity of some of our cultivated plants and domesticated animals.
	48	50	On dimorphic and trimorphic animals and plants.

5:e	Fourth Edition.	Fifth Edition.	Chief Additions and Corrections.
	Page xvii	Page xvii	The account of Professor Owen's views on Species corrected.
	8	8	On the causes of Variability.
	44	45	Limit of variation under domestication.
	103	104	On the great importance of individual differences, and on the unimportance of single variations.
	117	120	Wagner Moritz on the importance of isolation in the formation of species.
	144	146	Fritz Müller on the lancelet not competing with higher fishes.
	148	150–157	The views of Bronn and Nägeli on characters which are important under a morphological point of view, but are unimportant physiologically, not being acted on through natural selection.
	157	165	Causes of Variability corrected.
	259	266–267	Additions on the instincts of the cuckoo.
	297	304	Additional cases of plants, which can be crossed by a distinct species, but are impotent with pollen from the same individual plant.
	324	331	Discussion on dimorphic and trimorphic plants corrected.
	327	333	Conclusions on the fertility of varieties when crossed, in contrast with species, corrected.
	330	336	Infertility of varieties of Verbascum when crossed.
	343	349	On subaerial denudation.
	344	352–354	On the rate of subaerial denudation, as measured by years, and on the probable rate of change of species.
	347	355	Absence of organic remains in certain great sedimentary formations.
	353	362	Intermediate varieties within the same formation.
	370	379	Age of the world in a habitable condition.
	371	380	On a monocotyledonous plant in the Cambrian formation.
	394	402	Gaudry, Professor, on the intermediate character of the fossil mammals of Attica; and Professor Huxley on the connecting forms between birds and reptiles.
	417	424	Günther, Dr., on the large proportion of fishes in common on opposite sides of the isthmus of Panama.
	432	439	Locusts transporting seeds.
	432	440	On a living seed in earth attached to a woodcock's leg.
	442	451–461	Mr. Croll on the alternate Glacial periods of the northern and southern hemispheres; and the bearing of this conclusion on geographical distribution.
	512	515	Professor Häckel on phylogeny, or the lines of descent of all organic beings.
	517	520	The whole discussion on embryology corrected, in small details.

5:e	Fourth Edition.	Fifth Edition.	Chief Additions and Corrections.
	Page 534	Page 536	Mr. G. H. Lewes on the functionless structure of the larva of Salamandra atra. The whole discussion on rudimentary organs has been slightly modified.
	537	539	Professor Weismann on the futility of the idea that rudiments complete the scheme of nature.
	571	573	On the question whether one or many forms of life first appeared.

5:f	Fifth Edition.	Sixth Edition.	Chief Additions and Corrections.
	Page 100	Page 68	Influence of fortuitous destruction on natural selection.
	158	101	On the convergence of specific forms.
	220	142	Account of the Ground-Woodpecker of La Plata modified.
	225	145	On the modification of the eye.
	230	149	Transitions through the acceleration or retardation of the period of reproduction.
	231	150	The account of the electric organ of fishes added to.
	233	151	Analogical resemblance between the eyes of Cephalopods and Vertebrates.
	234	153	Claparède on the analogical resemblance of the hair-claspers of the Acaridæ.
	248	162	The probable use of the rattle to the Rattle-snake.
	248	163	Helmholtz on the imperfection of the human eye.
	255	168	The first part of this new chapter consists of portions, in a much modified state, taken from chap. iv. of the former editions. The latter and larger part is new, and relates chiefly to the supposed incompetency of natural selection to account for the incipient stages of useful structures. There is also a discussion on the causes which prevent in many cases the acquisition through natural selection of useful structures. Lastly, reasons are given for disbelieving in great and sudden modifications. Gradations of character, often accompanied by changes of function, are likewise here incidentally considered.
	268	214	The statement with respect to young cuckoos ejecting their foster-brothers confirmed.
	270	215	On the cuckoo-like habits of the Molothrus.
	307	240	On fertile hybrid moths.
	319	248	The discussion on the fertility of hybrids not having been acquired through natural selection condensed and modified.
	326	252	On the causes of sterility of hybrids, added to and corrected.

6   POSTSCRIPT.

An admirable, and, to a certain extent, favourable Review of this work, including an able discussion on the Theological bearing of the belief in the descent of species, has now been separately published by Professor Asa Gray as a pamphlet, about 60 pages in length. It is entitled, 'Natural Selection not inconsistent with Natural Theology. A Free Examination of Darwin's Treatise on the Origin of Species, and of its American Reviewers. By ASA GRAY, M.D., Fisher Professor of Natural History in Harvard University. Reprinted from the Atlantic Monthly for July, August, and October, 1860. London: Trübner and Co., 60, Paternoster-row. Boston: Ticknor and Fields. 1861.' (Price 1s. 6d.) ·

6:[d]

An HISTORICAL SKETCH OF THE RECENT PROGRESS OF
OPINION ON THE ORIGIN OF SPECIES.

1: [Note: This section first appears in *c.*]

1:f AN HISTORICAL SKETCH OF THE PROGRESS OF OPINION ON THE
ORIGIN OF SPECIES, PREVIOUSLY TO THE PUBLICATION OF THE FIRST
EDITION OF THIS WORK.

2 I WILL here attempt to give a brief, but imperfect sketch of the
progress of opinion on the Origin of Species.

2:d here give

2:e brief sketch

3 The great majority of naturalists believe that species are immutable productions, and have been separately created.

3:e Until recently the great majority of naturalists believed that
species were immutable productions, and had been separately
created.

4 This view has been ably maintained by many authors.

5 Some few naturalists, on the other hand, believe that species
undergo modification, and that the existing forms of life have
descended by true generation from pre-existing forms.

5:d life are the descendants by true generation of pre-existing

5:e hand, have believed that species

6 Passing over authors from the classical period to that of Buffon,
with whose writings I am not familiar, Lamarck was the first
man whose conclusions on this subject excited much attention.

6.x:d over allusions to the subject in the classical writers,* the first
author who in modern times has treated it in a scientific spirit
was Buffon.

6.x.*.1-4:d * Aristotle, in his 'Physicæ Auscultationes' (lib. 2, cap.
8, s. 2), after remarking that rain does not fall in order to make
the corn grow, any more than it falls to spoil the farmer's corn
when threshed out of doors, applies the same argument to organisation; and adds (as translated by Mr. Clair Grece, who first
pointed out the passage to me), "So what hinders the different
parts [of the body] from having this merely accidental relation
in nature? as the teeth, for example, grow by necessity, the
front ones sharp, adapted for dividing, and the grinders flat,
and serviceable for masticating the food; since they were not
made for the sake of this, but it was the result of accident. 2 And
in like manner as to the other parts in which there appears to
exist an adaptation to an end. *3* Wheresoever, therefore, all
things together (that is all the parts of one whole) happened
like as if they were made for the sake of something, these were
preserved, having been appropriately constituted by an internal
spontaneity; and whatsoever things were not thus constituted,
perished, and still perish." *4* We here see the principle of

natural selection shadowed forth, but how little Aristotle fully comprehended the principle is shown by his remarks on the formation of the teeth.

6.x.1:d But as his opinions fluctuated greatly at different periods, and as he does not enter on the causes or means of the transformation of species, I need not here enter on details.

6.y:d[¶] Lamarck

7 This justly-celebrated naturalist first published his views in 1801, and he much enlarged them in 1809 in his 'Philosophie Zoologique,' and subsequently, in 1815, in his Introduction to his 'Hist. Nat. des Animaux sans Vertèbres.'

7:d 1801; he/1815, in the Introduction

8 In these works he upholds the doctrine that all species, including man, are descended from other species.

9 He first did the eminent service of arousing attention to the probability of all change in the organic as well as in the inorganic world being the result of law, and not of miraculous interposition.

9:d organic, as well as in the inorganic world, being

10 Lamarck seems to have been chiefly led to his conclusion on the gradual change of species, by the difficulty of distinguishing species and varieties, by the almost perfect gradation of forms in certain organic groups, and by the analogy of domestic productions.

10:d certain groups

11 With respect to the means of modification, he attributed something to the direct action of the physical conditions of life, something to the crossing of already existing forms, and much to use and disuse, that is, to the effects of habit.

12 To this latter agency he seems to attribute all the beautiful adaptations in nature;—such as the long neck of the giraffe for browsing on the branches of trees.

13 But he likewise believed in a law of progressive development; and as all the forms of life thus tended to progress, in order to account for the existence at the present day of very simple productions, he maintained that such forms were now spontaneously generated.*

13:d tend/day of simple productions, he maintains that such forms are now

13*.1-3 * I have taken the date of the first publication of Lamarck from Isid. Geoffroy Saint Hilaire's ('Hist. Nat. Générale,' tom. ii, p. 405, 1859) excellent history of opinion on this subject. 2 In this work a full account is given of Buffon's fluctuating conclusions on the same subject. 3 It is curious how largely my grandfather, Dr. Erasmus Darwin, anticipated the erroneous grounds of opinion, and the views of Lamarck, in his 'Zoonomia' (vol. i, p. 500-510), published in 1794.

13*.3:d anticipated the views and erroneous grounds of opinion of Lamarck in

13*.4 According to Isid. Geoffroy there is no doubt that Goethe was an extreme partisan of similar views, as shown in the Introduction to a work written in 1794 and 1795, but not published till long afterwards.

13*.4:d afterwards: he has pointedly remarked (Goethe als Naturforscher, von Dr. Karl Meding, s. 34) that the future question for naturalists will be how, for instance, cattle got their horns, and not for what they are used.

13*.5 It is rather a singular instance of the manner in which similar views arise at about the same period, that Goethe in Germany, Dr. Darwin in England, and Geoffroy Saint Hilaire (as we shall immediately see) in France, came to the same conclusion on the origin of species, in the years 1794-5.

13*.5:d same time, that

14 Geoffroy Saint Hilaire, as is stated in his 'Life,' written by his son, suspected, as early as 1795, that what we call species are various degenerations of the same type.

15 It was not until 1828 that he published his conviction that the same forms have not been perpetuated since the origin of all things.

16 Geoffroy seems to have relied chiefly on the conditions of life, or the *"monde ambiant,"* as the cause of change.

17 He was cautious in drawing conclusions, and did not believe that existing species are now undergoing modification; and, as his son adds, "C'est donc un problème à réserver entièrement à l'avenir, supposé même que l'avenir doive avoir prise sur lui."

17.1-7:d[¶] In 1813 Dr. W. C. Wells read before the Royal Society 'An Account of a White Female, part of whose Skin resembles that of a Negro;' but his paper was not published until his famous 'Two Essays upon Dew and Single Vision' appeared in 1818. 2 In this paper he distinctly recognises the principle of natural selection, and this is the first recognition which has been indicated; but he applies it only to the races of man, and to certain characters alone. 3 After remarking that negroes and mulattoes enjoy an immunity from certain tropical diseases, he observes, firstly, that all animals tend to vary in some degree, and, secondly, that agriculturists improve their domesticated animals by selection; and then, he adds, but what is done in this latter case "by art, seems to be done with equal efficacy, though more slowly, by nature, in the formation of varieties of mankind, fitted for the country which they inhabit. 4 Of the accidental varieties of man, which would occur among the first few and scattered inhabitants of the middle regions of Africa, some one would be better fitted than the others to bear the diseases of the country. 5 This race would consequently multiply, while the others would decrease; not only from their inability to sustain the attacks of disease, but from their incapacity of contending with their more vigorous neighbours. *6* The colour of this vigorous race I take for granted, from what has been already said, would be dark. *7* But the same disposition to form varieties still existing, a darker and a darker race would in the course of time occur; and as the darkest would be the best fitted for the climate, this would at length become the most prevalent, if not the only race, in the particular country in which it had originated."

17.7:f occur: and

17.8-9:d He then extends these same views to the white inhabitants of colder climates. *9* I am indebted to the Rev. Mr. Brace, of the United States, for having called my attention to the above passage in Dr. Wells' work.

17.9:e indebted to Mr. Rowley, of the United States, for having called my attention, through Mr. Brace, to the above

18   In England the Hon. and Rev. W. Herbert, afterwards Dean of Manchester, in the fourth volume of the 'Horticultural Transactions,' 1822, and in his work on the 'Amaryllidaceæ' (1837, p. 19, 339), declares that "horticultural experiments have established, beyond the possibility of refutation, that botanical species are only a higher and more permanent class of varieties."

18:d The Hon.

19 He extends the same view to animals.

20 The Dean believes that single species of each genus were created in an originally highly plastic condition, and that these have produced, chiefly by intercrossing, but likewise by variation, all our existing species.

21   In 1826, Professor Grant, in the concluding paragraph in his well-known paper (Edinburgh Philosophical Journal, vol. xiv. p. 283) on the Spongilla, clearly declares his belief that species are descended from other species, and that they become improved in the course of modification.

21:d In 1826 Professor/('Edinburgh Philosophical Journal,'

22 This same view was given in his 55th Lecture, published in the 'Lancet' in 1834.

23   In 1831 Mr. Patrick Matthew published his work on 'Naval Timber and Arboriculture,' in which he gives precisely the same view on the origin of species as that (presently to be alluded to) propounded by Mr. Wallace and myself in the 'Linnean Journal,' and as that enlarged on in the present volume.

23:d enlarged in

24 Unfortunately the view was given by Mr. Matthew very briefly in scattered passages in an Appendix to a work on a different subject, so that it remained unnoticed until Mr. Matthew himself drew attention to it in the 'Gardener's Chronicle,' on April 7th, 1860.

25 The differences of Mr. Matthew's view from mine are not of much importance: he seems to consider that the world was nearly depopulated at successive periods, and then re-stocked; and he gives, as an alternative, that new forms may be generated "without the presence of any mould or germ of former aggregates."

25:f gives as

26 I am not sure that I understand some passages; but it seems that he attributes much influence to the direct action of the conditions of life.

27 He clearly saw, however, the full force of the principle of natural selection.

28 In answer to a letter of mine (published in Gard. Chron., April 13th), fully acknowledging that Mr. Matthew had anticipated me, he with generous candour wrote a letter (Gard. Chron. May 12th) containing the following passage:—"To me the conception of this law of Nature came intuitively as a self-evident fact, almost without an effort of concentrated thought.

29 Mr. Darwin here seems to have more merit in the discovery than I have had; to me it did not appear a discovery.

30 He seems to have worked it out by inductive reason, slowly and with due caution to have made his way synthetically from fact to fact onwards; while with me it was by a general glance at the scheme of Nature that I estimated this select production of species as an *à priori* recognisable fact—an axiom requiring only to be pointed out to be admitted by unprejudiced minds of sufficient grasp."

28-30:[d]

30.1:d[¶] The celebrated geologist and naturalist, Von Buch, in his excellent 'Description Physique des Iles Canaries' (1836, p. 147), clearly expresses his belief that varieties slowly become changed into permanent species, which are no longer capable of intercrossing.

31 Rafinesque, in his 'New Flora of North America,' published in 1836, wrote (p. 6) as follows:—"All species might have been varieties once, and many varieties are gradually becoming species by assuming constant and peculiar characters:" but farther on (p. 18) he adds, "except the original types or ancestors of the genus."

31:f characters;" but

31.1:f In 1843-4 Professor Haldeman ('Boston Journal of Nat. Hist. U. States,' vol. iv. p. 468) has ably given the arguments for and against the hypothesis of the development and modification of species: he seems to lean towards the side of change.

32 The 'Vestiges of Creation' appeared in 1844.

33 In the tenth and much improved edition (1853) the anonymous

author says (p. 155):—"The proposition determined on after much consideration is, that the several series of animated beings, from the simplest and oldest up to the highest and most recent, are, under the providence of God, the results, *first*, of an impulse which has been imparted to the forms of life, advancing them, in definite times, by generation, through grades of organisation terminating in the highest dicotyledons and vertebrata, these grades being few in number, and generally marked by intervals of organic character, which we find to be a practical difficulty in ascertaining affinities; *second*, of another impulse connected with the vital forces, tending, in the course of generations, to modify organic structures in accordance with external circumstances, as food, the nature of the habitat, and the meteoric agencies, these being the 'adaptations' of the natural theologian."

34 The author apparently believes that organisation progresses by sudden leaps, but that the effects produced by the conditions of life are gradual.

35 He argues with much force on general grounds that species are not immutable productions.

36 But I cannot see how the two supposed "impulses" account in a scientific sense for the numerous and beautiful co-adaptations which we see throughout nature; I cannot see that we thus gain any insight how, for instance, a woodpecker has become adapted to its peculiar habits of life.

37 The work, from its powerful and brilliant style, though displaying in the earlier editions little accurate knowledge and a great want of scientific caution, immediately had a very wide circulation.

38 In my opinion it has done excellent service in calling in this country attention to the subject, in removing prejudice, and in thus preparing the ground for the reception of analogous views.

38:d service in this country in calling attention

39 In 1846 the veteran geologist M. J. d'Omalius d'Halloy published in an excellent, though short paper ('Bulletins de l'Acad. Roy. Bruxelles,' tom. xiii. p. 581), his opinion that it is more probable that new species have been produced by descent with modification, than that they have been separately created: the author first promulgated this opinion in 1831.

39.1:f excellent though

40 Professor Owen, in 1849 ('Nature of Limbs,' p. 86), wrote as follows:—"The archetypal idea was manifested in the flesh under diverse such modifications, upon this planet, long prior to the existence of those animal species that actually exemplify it.

41 To what natural laws or secondary causes the orderly succession and progression of such organic phenomena may have been committed, we, as yet, are ignorant."

42 In his Address to the British Association, in 1858, he speaks (p. li.) of "the axiom of the continuous operation of creative power, or of the ordained becoming of living things."

43 Farther on (p. xc.), after referring to geographical distribution, he adds, "These phenomena shake our confidence in the conclusion that the Apteryx of New Zealand and the Red Grouse of England were distinct creations in and for those islands respectively.

44 Always, also, it may be well to bear in mind that by the word 'creation' the zoologist means 'a process he knows not what.' "

45 He amplifies this idea by adding, that when such cases as that of the Red Grouse are "enumerated by the zoologist as evidence of distinct creation of the bird in and for such islands, he chiefly expresses that he knows not how the Red Grouse came to be there, and there exclusively; signifying also by this mode of expressing such ignorance his belief, that both the bird and the islands owed their origin to a great first Creative Cause."

45:e also, by this mode of expressing such ignorance, his belief that

45.1-7:d If we interpret these sentences given in the same Address, one by the other, it appears that this eminent philosopher felt in 1858 his confidence shaken that the Apteryx and the Red Grouse first appeared in their respective homes, 'he knew not how,' or by some process 'he knew not what.' 2 Since the publication in 1859 of my work on the 'Origin of Species,' but whether in consequence of it is doubtful, Professor Owen has clearly expressed his belief that species have not been separately created, and are not immutable productions; but he still ('Anatomy of the Vertebrates,' 1866) denies that we know the natural laws or secondary causes of the successive appearance of species; yet he at the same time admits that natural selection may have done something towards this end. 3 It is surprising that this admission should not have been made earlier, as Professor Owen now believes that he promulgated the theory of natural selection in a passage read before the Zoological Society in February, 1850 ('Transact.' vol. iv. p. 15); for in a letter to the 'London Review' (May 5, 1866, p. 516), commenting on some of the reviewer's criticisms, he says, "No naturalist can dissent from the truth of your perception of the essential identity of the passage cited with the basis of that [the so-called Darwinian] theory, the power, viz., of species to accommodate themselves, or bow to the influences of surrounding circumstances." 4 Further on in the same letter he speaks of himself as "the author of the same theory at the earlier date of 1850." 5 This belief in Professor Owen that he then gave to the world the theory of natural selection will surprise all those who are acquainted with the several passages in his works, reviews, and lectures, published since the 'Origin,' in which he strenuously opposes the theory; and it will please all those who are interested on this side of

65

the question, as it may be presumed that his opposition will now cease. *6* It should, however, be stated that the passage above referred to in the 'Zoological Transactions,' as I find on consulting it, applies exclusively to the extermination and preservation of animals, and in no way to their gradual modification, origination, or natural selection. *7* So far is this from being the case that Professor Owen actually begins the first of the two paragraphs (vol. iv. p. 15) with the following words:—"We have not a particle of evidence that any species of bird or beast that lived during the pliocene period has had its characters modified in any respect by the influence of time or of change of external circumstances."

45.2-7:[e]

45.8-13:e[¶] This Address was delivered after the papers, by Mr. Wallace and myself on the Origin of Species, presently to be referred to, had been read before the Linnean Society. *9* When the first edition of this work was published, I was so completely deceived, as were many others, by such expressions as "the continuous operation of creative power," that I included Professor Owen with other palæontologists as being firmly convinced of the immutability of species; but it appears ('Anat. of Vertebrates,' vol. iii. p. 796) that this was on my part a preposterous error. *10* In the last edition of this work I inferred, and the inference still seems to me perfectly just, from a passage beginning with the words "no doubt the type-form," &c. (Ibid. vol. i. p. xxxv.), that Professor Owen admitted that natural selection may have done something in the formation of new species; but this it appears (Ibid. vol. iii. p. 798) is inaccurate and without evidence. *11* I also gave some extracts from a correspondence between Professor Owen and the Editor of the 'London Review,' from which it appeared manifest to the Editor as well as to myself, that Professor Owen claimed to have promulgated the theory of natural selection before I had done so; and I expressed my surprise and satisfaction at this announcement; but as far as it is possible to understand certain recently published passages (Ibid. vol. iii. p. 798), I have either partially or wholly again fallen into error. *12* It is consolatory to me that others find Professor Owen's controversial writings as difficult to understand and to reconcile with each other, as I do. *13* As far as the mere enunciation of the principle of natural selection is concerned, it is quite immaterial whether or not Professor Owen preceded me, for both of us, as shown in this historical sketch, were long ago preceded by Dr. Wells and Mr. Matthew.

46   M. Isidore Geoffroy Saint Hilaire, in his Lectures delivered in 1850 (of which a Résumé appeared in the 'Revue et Mag. de Zoolog.,' Jan. 1851), briefly gives his reason for believing that specific characters "sont fixés, pour chaque espèce, tant qu'elle se perpétue au milieu des mêmes circonstances: ils se modifient, si les circonstances ambiantes viennent à changer."

47 "En résumé, *l'observation* des animaux sauvages démontre déjà la variabilité *limitée* des espèces.

48 Les *expériences* sur les animaux sauvages devenus domestiques, et sur les animaux domestiques redevenus sauvages, la démontrent plus clairement encore.

49 Ces mêmes expériences prouvent, de plus, que les différences produites peuvent être de *valeur générique.*"

50 In his 'Hist. Nat. Générale' (tom. ii. p. 430, 1859) he amplifies analogous conclusions.

51   From a circular lately issued it appears that Dr. Freke, in 1851 ('Dublin Medical Press,' p. 322), propounded the doctrine that all organic beings have descended from one primordial form.

52 His grounds of belief and treatment of the subject are wholly different from mine; but as Dr. Freke has now (1861) published his Essay on 'the Origin of Species by means of Organic Affinity,' the difficult attempt to give any idea of his views would be superfluous on my part.

53   Mr. Herbert Spencer, in an Essay (originally published in the 'Leader,' March 1852, and republished in his 'Essays' in 1858), has contrasted the theories of the Creation and the Development of organic beings with remarkable skill and force.

54 He argues from the analogy of domestic productions, from the changes which the embryos of many species undergo, from the difficulty of distinguishing species and varieties, and from the principle of general gradation, that species have been modified; and he attributes the modification to the change of circumstances.

55 The author (1855) has also treated Psychology on the principle of the necessary acquirement of each mental power and capacity by gradation.

56   In 1852 ('Revue Horticole,' p. 102) M. Naudin, a distinguished botanist, has expressly stated his belief that species are formed in an analogous manner as varieties are under cultivation; and the latter process he attributes to man's power of selection.

56:e In 1852 M. Naudin, a distinguished botanist, expressly stated, in an admirable paper on the Origin of Species ('Revue Horticole,' p. 102; since partly republished in the 'Nouvelles Archives du Muséum,' tom. i. p. 171), his

57 But he does not show how selection acts under nature.

58 He believes, like Dean Herbert, that species, when nascent, were more plastic than at present.

59 He lays weight on what he calls the principle of finality, "puissance mystérieuse, indéterminée; fatalité pour les uns; pour les

autres, volonté providentielle, dont l'action incessante sur les êtres vivants détermine, à toutes les époques de l'existence du monde, la forme, le volume, et la durée de chacun d'eux, en raison de sa destinée dans l'ordre de choses dont il fait partie.

60 C'est cette puissance qui harmonise chaque membre à l'ensemble en l'appropriant à la fonction qu'il doit remplir dans l'organisme général de la nature, fonction qui est pour lui sa raison d'être." *

60*.1-4 * From references in Bronn's 'Untersuchungen über die Entwickelungs-Gesetze' it appears that the celebrated botanist and palæontologist Unger published, in 1852, his belief that species undergo development and modification. 2 D'Alton, likewise, in Pander and d'Alton's work on Fossil Sloths, expressed, in 1821, a similar belief. 3 Similar views have, as is well known, been maintained by Oken in his mystical 'Natur-Philosophie.' 4 From other references in Godron's work 'Sur l'Espèce,' it seems that Bory St. Vincent, Burdach, Poiret, and Fries, have all admitted that new species are continually being produced.

60*.5 I may add, that of the thirty authors named in this Historical Sketch, who believe in the modification of species, or at least disbelieve in separate acts of creation, twenty-five have written on special branches of natural history: of these only three are simple geologists, nine are botanists, and thirteen zoologists; but several of the botanists and zoologists have written on palæontology or on geology.

60*.5:d of the thirty-four authors/creation, twenty-seven have written on special branches of natural history or geology.

61 In 1853, a celebrated geologist, Count Keyserling ('Bulletin de la Soc. Géolog.,' 2nd Ser., tom. x. p. 357), suggested that as new diseases, supposed to have been caused by some miasma, have arisen and spread over the world, so at certain periods the germs of existing species may have been chemically affected by circumambient molecules of a particular nature, and thus have given rise to new forms.

62 In this same year, 1853, Dr. Schaaffhausen published an excellent pamphlet ('Verhand. des Naturhist. Vereins der Preuss. Rheinlands,' &c.), in which he maintains the progressive development of organic forms on the earth.

63 He infers that many species have kept true for long periods, whereas a few have become modified.

64 The distinction of species he explains by the destruction of intermediate graduated forms.

65 "Thus living plants and animals are not separated from the extinct by new creations, but are to be regarded as their descendants through continued reproduction."

65.1:d[¶] A well-known French botanist, M. Lecoq, writes in 1854 ('Études sur Géograph. Bot.,' tom. i. p. 250), "On voit que nos recherches sur la fixité ou la variation de l'espèce, nous conduisent directement aux idées émises par deux hommes justement célèbres, Geoffroy Saint-Hilaire et Goethe."

65.1:f émises, par

65.2:d Some other passages scattered through M. Lecoq's large work, make it a little doubtful how far he extends his views on the modification of species.

66   The 'Philosophy of Creation' has been treated in a masterly manner by the Rev. Baden Powell, in his 'Essays on the Unity of Worlds,' 1855.

67 Nothing can be more striking than the manner in which he shows that the introduction of new species is "a regular, not a casual phenomenon," or, as Sir John Herschel expresses it, "a natural in contra-distinction to a miraculous process."

67:d contradistinction

68   The third volume of the 'Journal of the Linnean Society' contains papers, read July 1st, 1858, by Mr. Wallace and myself, in which, as stated in the introductory remarks to this volume, the theory of Natural Selection is promulgated by Mr. Wallace with admirable force and clearness.

68.1:d Von Baer, towards whom all zoologists feel so profound a respect, expressed about the year 1859 (see Prof. Rudolph Wagner, 'Zoologisch-Anthropologische Untersuchungen,' 1861, s. 51) his conviction, chiefly grounded on the laws of geographical distribution, that forms now perfectly distinct have descended from a single parent-form.

69   In June, 1859, Professor Huxley gave a lecture before the Royal Institution on the 'Persistent Types of Animal Life.'

70 Referring to such cases, he remarks, "It is difficult to comprehend the meaning of such facts as these, if we suppose that each species of animal and plant, or each great type of organisation, was formed and placed upon the surface of the globe at long intervals by a distinct act of creative power; and it is well to recollect that such an assumption is as unsupported by tradition or revelation as it is opposed to the general analogy of nature.

71 If, on the other hand, we view 'Persistent Types' in relation to that hypothesis which supposes the species living at any time to be the result of the gradual modification of pre-existing species —a hypothesis which, though unproven, and sadly damaged by some of its supporters, is yet the only one to which physiology lends any countenance; their existence would seem to show that the amount of modification which living beings have undergone during geological time is but very small in relation to the whole series of changes which they have suffered."

72   In December, 1859, Dr. Hooker published his 'Introduction to the Australian Flora.'

73 In the first part of this great work he admits the truth of the descent and modification of species, and supports this doctrine by many original observations.

74 The first edition of this work was published on November 24th, 1859, and the second edition on January 7th, 1860.

1 ON THE ORIGIN OF SPECIES.

1:f ORIGIN OF SPECIES.

2 INTRODUCTION.

3 WHEN on board H.M.S. 'Beagle,' as naturalist, I was much struck with certain facts in the distribution of the inhabitants of South America, and in the geological relations of the present to the past inhabitants of that continent.

3:c of the organic beings inhabiting South

4 These facts seemed to me to throw some light on the origin of species—that mystery of mysteries, as it has been called by one of our greatest philosophers.

4:c facts, as will be seen in the latter chapters of this volume, seemed to

5 On my return home, it occurred to me, in 1837, that something might perhaps be made out on this question by patiently accumulating and reflecting on all sorts of facts which could possibly have any bearing on it.

6 After five years' work I allowed myself to speculate on the subject, and drew up some short notes; these I enlarged in 1844 into a sketch of the conclusions, which then seemed to me probable: from that period to the present day I have steadily pursued the same object.

7 I hope that I may be excused for entering on these personal details, as I give them to show that I have not been hasty in coming to a decision.

8   My work is now nearly finished; but as it will take me two or three more years to complete it, and as my health is far from strong, I have been urged to publish this Abstract.

8:f now (1859) nearly/me many more years

9 I have more especially been induced to do this, as Mr. Wallace, who is now studying the natural history of the Malay archipelago, has arrived at almost exactly the same general conclusions that I have on the origin of species.

10 Last year he sent to me a memoir on this subject, with a request that I would forward it to Sir Charles Lyell, who sent it to the Linnean Society, and it is published in the third volume of the Journal of that Society.

10:b sent me

10:e In 1858 he

11 Sir C. Lyell and Dr. Hooker, who both knew of my work—the latter having read my sketch of 1844—honoured me by thinking it advisable to publish, with Mr. Wallace's excellent memoir, some brief extracts from my manuscripts.

12   This abstract, which I now publish, must necessarily be imperfect.

13 I cannot here give references and authorities for my several statements; and I must trust to the reader reposing some confidence in my accuracy.

14 No doubt errors will have crept in, though I hope I have always been cautious in trusting to good authorities alone.

15 I can here give only the general conclusions at which I have arrived, with a few facts in illustration, but which, I hope, in most cases will suffice.

16 No one can feel more sensible than I do of the necessity of hereafter publishing in detail all the facts, with references, on which my conclusions have been grounded; and I hope in a future work to do this.

17 For I am well aware that scarcely a single point is discussed in this volume on which facts cannot be adduced, often apparently leading to conclusions directly opposite to those at which I have arrived.

18 A fair result can be obtained only by fully stating and balancing the facts and arguments on both sides of each question; and this cannot possibly be here done.

18:e this is here impossible.

19   I much regret that want of space prevents my having the satisfaction of acknowledging the generous assistance which I have received from very many naturalists, some of them personally unknown to me.

20 I cannot, however, let this opportunity pass without expressing my deep obligations to Dr. Hooker, who for the last fifteen years has aided me in every possible way by his large stores of knowledge and his excellent judgment.

20:f who, for the last fifteen years, has

21   In considering the Origin of Species, it is quite conceivable that a naturalist, reflecting on the mutual affinities of organic beings, on their embryological relations, their geographical distribution, geological succession, and other such facts, might come to the conclusion that each species had not been independently created, but had descended, like varieties, from other species.

21:d that species

22 Nevertheless, such a conclusion, even if well founded, would be unsatisfactory, until it could be shown how the innumerable species inhabiting this world have been modified, so as to ac-

quire that perfection of structure and coadaptation which most justly excites our admiration.

22:c which justly

23 Naturalists continually refer to external conditions, such as climate, food, &c., as the only possible cause of variation.

24 In one very limited sense, as we shall hereafter see, this may be true; but it is preposterous to attribute to mere external conditions, the structure, for instance, of the woodpecker, with its feet, tail, beak, and tongue, so admirably adapted to catch insects under the bark of trees.

24:e In one limited

25 In the case of the misseltoe, which draws its nourishment from certain trees, which has seeds that must be transported by certain birds, and which has flowers with separate sexes absolutely requiring the agency of certain insects to bring pollen from one flower to the other, it is equally preposterous to account for the structure of this parasite, with its relations to several distinct organic beings, by the effects of external conditions, or of habit, or of the volition of the plant itself.

25:c mistletoe

26 The author of the 'Vestiges of Creation' would, I presume, say that, after a certain unknown number of generations, some bird had given birth to a woodpecker, and some plant to the misseltoe, and that these had been produced perfect as we now see them; but this assumption seems to me to be no explanation, for it leaves the case of the coadaptations of organic beings to each other and to their physical conditions of life, untouched and unexplained.

26:[c] [See "Historical Sketch," 32-38]

27 It is, therefore, of the highest importance to gain a clear insight into the means of modification and coadaptation.

28 At the commencement of my observations it seemed to me probable that a careful study of domesticated animals and of cultivated plants would offer the best chance of making out this obscure problem.

29 Nor have I been disappointed; in this and in all other perplexing cases I have invariably found that our knowledge, imperfect though it be, of variation under domestication, afforded the best and safest clue.

30 I may venture to express my conviction of the high value of such studies, although they have been very commonly neglected by naturalists.

31 From these considerations, I shall devote the first chapter of this Abstract to Variation under Domestication.

32 We shall thus see that a large amount of hereditary modification is at least possible; and, what is equally or more important, we

shall see how great is the power of man in accumulating by his Selection successive slight variations.

33 I will then pass on to the variability of species in a state of nature; but I shall, unfortunately, be compelled to treat this subject far too briefly, as it can be treated properly only by giving long catalogues of facts.

34 We shall, however, be enabled to discuss what circumstances are most favourable to variation.

35 In the next chapter the Struggle for Existence amongst all organic beings throughout the world, which inevitably follows from their high geometrical powers of increase, will be treated of.

35:b from the high geometrical ratio of their increase

35:f will be considered.

36 This is the doctrine of Malthus, applied to the whole animal and vegetable kingdoms.

37 As many more individuals of each species are born than can possibly survive; and as, consequently, there is a frequently recurring struggle for existence, it follows that any being, if it vary however slightly in any manner profitable to itself, under the complex and sometimes varying conditions of life, will have a better chance of surviving, and thus be *naturally selected.*

38 From the strong principle of inheritance, any selected variety will tend to propagate its new and modified form.

39 This fundamental subject of Natural Selection will be treated at some length in the fourth chapter; and we shall then see how Natural Selection almost inevitably causes much Extinction of the less improved forms of life, and induces what I have called Divergence of Character.

39:b life, and leads to what

40 In the next chapter I shall discuss the complex and little known laws of variation and of correlation of growth.

40:e variation.

41 In the four succeeding chapters, the most apparent and gravest difficulties on the theory will be given: namely, first, the difficulties of transitions, or in understanding how a simple being or a simple organ can be changed and perfected into a highly developed being or elaborately constructed organ; secondly, the subject of Instinct, or the mental powers of animals; thirdly, Hybridism, or the infertility of species and the fertility of varieties when intercrossed; and fourthly, the imperfection of the Geological Record.

41:e difficulties in accepting the theory/or how a simple being or a simple organ can/developed being or into an elaborately

41:f In the five succeeding

42 In the next chapter I shall consider the geological succession of

organic beings throughout time; in the eleventh and twelfth, their geographical distribution throughout space; in the thirteenth, their classification or mutual affinities, both when mature and in an embryonic condition.

42:f time; in the twelfth and thirteenth, their geographical distribution throughout space; in the fourteenth, their classification

43 In the last chapter I shall give a brief recapitulation of the whole work, and a few concluding remarks.

44   No one ought to feel surprise at much remaining as yet unexplained in regard to the origin of species and varieties, if he makes due allowance for our profound ignorance in regard to the mutual relations of all the beings which live around us.

45 Who can explain why one species ranges widely and is very numerous, and why another allied species has a narrow range and is rare?

46 Yet these relations are of the highest importance, for they determine the present welfare, and, as I believe, the future success and modification of every inhabitant of this world.

47 Still less do we know of the mutual relations of the innumerable inhabitants of the world during the many past geological epochs in its history.

48 Although much remains obscure, and will long remain obscure, I can entertain no doubt, after the most deliberate study and dispassionate judgment of which I am capable, that the view which most naturalists entertain, and which I formerly entertained—namely, that each species has been independently created—is erroneous.

48:f naturalists until recently entertained, and

49 I am fully convinced that species are not immutable; but that those belonging to what are called the same genera are lineal descendants of some other and generally extinct species, in the same manner as the acknowledged varieties of any one species are the descendants of that species.

50 Furthermore, I am convinced that Natural Selection has been the main but not exclusive means of modification.

50:e been the most important, but not the exclusive, means

CHAPTER I.

VARIATION UNDER DOMESTICATION.

3 Causes of Variability—Effects of Habit—Correlation of Growth—
Inheritance—Character of Domestic Varieties—Difficulty of distinguishing between Varieties and Species—Origin of Domestic
Varieties from one or more Species—Domestic Pigeons, their
Differences and Origin—Principle of Selection anciently followed, its Effects—Methodical and Unconscious Selection—Unknown Origin of our Domestic Productions—Circumstances
favourable to Man's power of Selection.

3:d Selection, anciently followed, their Effects

3:e Habit—Correlated Variation—Inheritance

3:f Habit and the use or disuse of Parts—Correlated

3.1:d [*Center*] *Causes of Variability*. [*Space*]

4 WHEN we look to the individuals of the same variety or sub-
variety of our older cultivated plants and animals, one of the
first points which strikes us, is, that they generally differ much
more from each other, than do the individuals of any one
species or variety in a state of nature.

4:b differ more from each other than

4:e WHEN we compare the individuals of the same/us is, that/gen-
erally differ from each other more than

4:f differ more from each other than

5 When we reflect on the vast diversity of the plants and animals
which have been cultivated, and which have varied during all
ages under the most different climates and treatment, I think
we are driven to conclude that this greater variability is simply
due to our domestic productions having been raised under conditions of life not so uniform as, and somewhat different from,
those to which the parent-species have been exposed under
nature.

5:e And if we reflect on/treatment, we/is due/parent-species had
been

6 There is, also, I think, some probability in the view propounded
by Andrew Knight, that this variability may be partly connected with excess of food.

6:b is also

6:f is, also, some

7 It seems pretty clear that organic beings must be exposed during
several generations to the new conditions of life to cause any

appreciable amount of variation; and that when the organisation has once begun to vary, it generally continues to vary for many generations.

7:d and that, when

7:e seems clear/to new conditions to cause/continues varying for

7:f any great amount

8 No case is on record of a variable being ceasing to be variable under cultivation.

8:e variable organism ceasing to vary under

9 Our oldest cultivated plants, such as wheat, still often yield new varieties: our oldest domesticated animals are still capable of rapid improvement or modification.

9:e still yield

9.1-3:e[¶] As far as I am able to judge, after long attending to the subject, the conditions of life appear to act in two ways,—directly on the whole organisation or on certain parts alone, and indirectly by affecting the reproductive system. *2* With respect to the direct action, we must bear in mind that in every case, as Professor Weismann has lately insisted, and as I have incidentally shown in my work on 'Variation under Domestication,' there are two factors: namely, the nature of the organism, and the nature of the conditions. *3* The former seems to be much the more important; for nearly similar variations sometimes arise under, as far as we can judge, dissimilar conditions; and, on the other hand, dissimilar variations under conditions which appear to be nearly uniform.

9.3:f hand, dissimilar variations arise under

9.4-9:e The effects on the offspring are either definite or indefinite. *5* They may be considered as definite when all or nearly all the offspring of individuals exposed to certain conditions during several generations are modified in the same manner. *6* It is extremely difficult to come to any conclusion in regard to the extent of the changes which have been thus definitely induced. *7* There can, however, be little doubt about many slight changes,—such as size from the amount of food, colour from the nature of the food, thickness of the skin and hair from climate, &c. *8* Each of the endless variations which we see in the plumage of our fowls must have had some efficient cause; and if the same cause were to act uniformly during a long series of generations on many individuals, all probably would be modified in the same manner. *9* Such facts as the complex and extraordinary out-growths which invariably follow from the insertion of a minute drop of poison by a gall-producing insect, show us what singular modifications might result in the case of plants from a chemical change in the nature of the sap.

9.10-12:e[¶] Indefinite variability is a much more common result of changed conditions than definite variability, and has probably played a more important part in the formation of our domestic races. *11* We see indefinite variability in the endless slight peculiarities which distinguish the individuals of the same species, and which cannot be accounted for by inheritance from either parent or from some more remote ancestor. *12* Even strongly-marked differences occasionally appear in the young of the same litter, and in seedlings from the same seed-capsules.

9.12:f seed-capsule.

9.13-14:e At long intervals of time, out of millions of individuals reared in the same country and fed on nearly the same food, deviations of structure so strongly pronounced as to deserve to be called monstrosities arise; but monstrosities cannot be separated by any distinct line from slighter variations. *14* All such changes of structure, whether extremely slight or strongly marked, which appear amongst many individuals living together, may be considered as the indefinite effects of the conditions of life on each individual organism, in nearly the same manner as a chill affects different men in an indefinite manner, according to their state of body or constitution, causing coughs or colds, rheumatism, or inflammations of various organs.

9.15:e[¶] With respect to what I have called the indirect action of changed conditions, namely, through the reproductive system being affected, we may infer that variability is thus induced, partly from the fact of this system being extremely sensitive to any change in the conditions, and partly from the similarity, as Kölreuter and others have remarked, between the variability which follows from the crossing of distinct species, and that which may be observed with all plants and animals when reared under new or unnatural conditions.

9.15:f observed with plants

9.16:e Many facts clearly show how eminently susceptible the reproductive system is to very slight changes in the surrounding conditions.

10   It has been disputed at what period of life the causes of variability, whatever they may be, generally act; whether during the early or late period of development of the embryo, or at the instant of conception.

10:[e]

11 Geoffroy St. Hilaire's experiments show that unnatural treatment of the embryo causes monstrosities; and monstrosities cannot be separated by any clear line of distinction from mere variations.

11:[e]

12 But I am strongly inclined to suspect that the most frequent cause of variability may be attributed to the male and female reproductive elements having been affected prior to the act of conception.

12:[e]

13 Several reasons make me believe in this; but the chief one is the remarkable effect which confinement or cultivation has on the functions of the reproductive system; this system appearing to be far more susceptible than any other part of the organisation, to the action of any change in the conditions of life.

13:b function

13:[e]

14 Nothing is more easy than to tame an animal, and few things more difficult than to get it to breed freely under confinement, even in the many cases when the male and female unite.

14:e even when

15 How many animals there are which will not breed, though living long under not very close confinement in their native country!

15:e though kept in an almost free state in

16 This is generally attributed to vitiated instincts; but how many cultivated plants display the utmost vigour, and yet rarely or never seed!

16.x-y:e generally, but erroneously, attributed to vitiated instincts. y Many

17 In some few such cases it has been found out that very trifling changes, such as a little more or less water at some particular period of growth, will determine whether or not the plant sets a seed.

17:b been discovered that

17:e few cases it has been discovered that a very trifling change, such/not a plant will produce seeds.

18 I cannot here enter on the copious details which I have collected on this curious subject; but to show how singular the laws are which determine the reproduction of animals under confinement, I may just mention that carnivorous animals, even from the tropics, breed in this country pretty freely under confinement, with the exception of the plantigrades or bear family; whereas, carnivorous birds, with the rarest exceptions, hardly ever lay fertile eggs.

18:b whereas carnivorous

18:c family, which seldom produce young; whereas

18:e here give the details which I have collected and elsewhere published on/may mention

19 Many exotic plants have pollen utterly worthless, in the same exact condition as in the most sterile hybrids.

19:e same condition

20 When, on the one hand, we see domesticated animals and plants, though often weak and sickly, yet breeding quite freely under confinement; and when, on the other hand, we see individuals, though taken young from a state of nature, perfectly tamed, long-lived, and healthy (of which I could give numerous

instances), yet having their reproductive system so seriously affected by unperceived causes as to fail in acting, we need not be surprised at this system, when it does act under confinement, acting not quite regularly, and producing offspring not perfectly like their parents or variable.

20:b parents.

20:e breeding freely/fail to act, we/does act under confinement, acting irregularly, and producing offspring somewhat unlike their parents.

20:f sickly, breeding/nature perfectly tamed, long-lived and healthy

21   Sterility has been said to be the bane of horticulture; but on this view we owe variability to the same cause which produces sterility; and variability is the source of all the choicest productions of the garden.

21:[e]

22 I may add, that as some organisms will breed most freely under the most unnatural conditions (for instance, the rabbit and ferret kept in hutches), showing that their reproductive system has not been thus affected; so will some animals and plants withstand domestication or cultivation, and vary very slightly—perhaps hardly more than in a state of nature.

22:b breed freely

22:e reproductive organs are not affected

22:f instance, rabbits and ferrets/not easily affected

23   A long list could easily be given of "sporting plants;" by this term gardeners mean a single bud or offset, which suddenly assumes a new and sometimes very different character from that of the rest of the plant.

23:e Some naturalists have maintained that all variations are connected with the act of sexual reproduction; but this is certainly an error; for I have given in another work a long list of "sporting plants," as they are called by gardeners:—that is, of plants which have suddenly produced a single bud with a new and sometimes widely different character from that of the other buds on the same plant.

24 Such buds can be propagated by grafting, &c., and sometimes by seed.

24:e These bud-variations, as they may be named, can be propagated by grafts, offsets, &c., and

24:f bud variations

25 These "sports" are extremely rare under nature, but far from rare under cultivation; and in this case we see that the treatment of the parent has affected a bud or offset, and not the ovules or pollen.

25:e They occur rarely under nature, but far from rarely under culture.

25:f but are far from rare under

26 But it is the opinion of most physiologists that there is no essential difference between a bud and an ovule in their earliest stages of formation; so that, in fact, "sports" support my view, that variability may be largely attributed to the ovules or pollen, or to both, having been affected by the treatment of the parent prior to the act of conception.

26:c support the view

26:[e]

27 These cases anyhow show that variation is not necessarily connected, as some authors have supposed, with the act of generation.

27:[e]

28   Seedlings from the same fruit, and the young of the same litter, sometimes differ considerably from each other, though both the young and the parents, as Müller has remarked, have apparently been exposed to exactly the same conditions of life; and this shows how unimportant the direct effects of the conditions of life are in comparison with the laws of reproduction, and of growth, and of inheritance; for had the action of the conditions been direct, if any of the young had varied, all would probably have varied in the same manner.

28:b reproduction, of growth

28:e[No ¶] As a single bud out of the many thousands produced year after year under uniform conditions on the same tree, has been known suddenly to assume a new character; and as buds on distinct trees, growing under different conditions, have sometimes yielded nearly the same variety—for instance, buds on peach-trees producing nectarines, and buds on common roses producing moss-roses—we clearly see that the nature of the conditions is of quite subordinate importance in comparison with the nature of the organism in determining each particular form of variation;—of not more importance than the nature of the spark by which a mass of combustible matter is ignited, has in determining the nature of the flames. *[Space]*

28:f thousands, produced year after year on the same tree under uniform conditions, has been/variation;—perhaps of not/spark, by

29 To judge how much, in the case of any variation, we should attribute to the direct action of heat, moisture, light, food, &c., is most difficult: my impression is, that with animals such agencies have produced very little direct effect, though apparently more in the case of plants.

29:[e]

30 Under this point of view, Mr. Buckman's recent experiments on plants seem extremely valuable.

30:b plants are extremely

30:[d]

31 When all or nearly all the individuals exposed to certain conditions are affected in the same way, the change at first appears to be directly due to such conditions; but in some cases it can be shown that quite opposite conditions produce similar changes of structure.

31:[e]

32 Nevertheless some slight amount of change may, I think, be attributed to the direct action of the conditions of life—as, in some cases, increased size from amount of food, colour from particular kinds of food and from light, and perhaps the thickness of fur from climate.

32:b food or from

32:c food, and perhaps

32:d climate. [Space]

32:[e]

32:1d [Center] Effects of Habit; Correlation of Growth; Inheritance. [Space]

32.1:f Effects of Habit and of the Use or Disuse of Parts; Correlated Variation; Inheritance.

33 Habit also has a decided influence, as in the period of flowering with plants when transported from one climate to another.

33:e Habits are inherited and have a decided influence; as in the period of the flowering of plants

33:f Changed habits produce an inherited effect, as

34 In animals it has a more marked effect; for instance, I find in the domestic duck that the bones of the wing weigh less and the bones of the leg more, in proportion to the whole skeleton, than do the same bones in the wild-duck; and I presume that this change may be safely attributed to the domestic duck flying much less, and walking more, than its wild parent.

34:e animals they have a more marked/wild-duck; and this change/ parents.

34:f With animals the increased use or disuse of parts has had a more marked influence; thus I find

35 The great and inherited development of the udders in cows and goats in countries where they are habitually milked, in comparison with the state of these organs in other countries, is another instance of the effect of use.

35:d is probably another instance of the effects

36 Not a single domestic animal can be named which has not in some country drooping ears; and the view suggested by some authors, that the drooping is due to the disuse of the muscles of the ear, from the animals not being much alarmed by danger, seems probable.

36:d view which has been suggested that the drooping/from the animals being seldom alarmed

36:e Not one of our domestic animals can

36:f seldom much alarmed, seems

37   There are many laws regulating variation, some few of which can be dimly seen, and will be hereafter briefly mentioned.

37:e Many laws regulate variation/will hereafter be briefly discussed.

38   I will here only allude to what may be called correlation of growth.

38:e called correlated variation.

39   Any change in the embryo or larva will almost certainly entail changes in the mature animal.

39:d will probably entail

39:e Important changes in the embryo

40   In monstrosities, the correlations between quite distinct parts are very curious; and many instances are given in Isidore Geoffroy St. Hilaire's great work on this subject.

41   Breeders believe that long limbs are almost always accompanied by an elongated head.

42   Some instances of correlation are quite whimsical: thus cats with blue eyes are invariably deaf; colour and constitutional peculiarities go together, of which many remarkable cases could be given amongst animals and plants.

42.x:c eyes are generally deaf.

42.x:d cats which are entirely white and have blue eyes

42.x:f deaf; but it has been lately stated by Mr. Tait that this is confined to the males.

42.y:c Colour

43   From the facts collected by Heusinger, it appears that white sheep and pigs are differently affected from coloured individuals by certain vegetable poisons.

43:c poisons: Professor Wyman has recently communicated to me a good illustration of this fact; on asking some farmers in Florida how it was that all their pigs were black, they informed him that the pigs ate the paint-root (Lachnanthes), which coloured their bones pink, and which caused the hoofs of all but the black varieties to drop off; and one of the "crackers" (*i.e.* Florida squatters) added, "we select the black members of a litter for raising, as they alone have a good chance of living."

43:d are injured by certain plants, whilst dark-coloured individuals escape: Professor

43:f in Virginia how/*i.e.* Virginia squatters

44   Hairless dogs have imperfect teeth; long-haired and coarse-

haired animals are apt to have, as is asserted, long or many horns; pigeons with feathered feet have skin between their outer toes; pigeons with short beaks have small feet, and those with long beaks large feet.

45 Hence, if man goes on selecting, and thus augmenting, any peculiarity, he will almost certainly unconsciously modify other parts of the structure, owing to the mysterious laws of the correlation of growth.

45:c certainly unintentionally modify

45:e certainly modify unintentionally other/laws of correlation.

46 The result of the various, quite unknown, or dimly seen laws of variation is infinitely complex and diversified.

46:e The results of the various, unknown, or but dimly understood laws of variation are infinitely

47 It is well worth while carefully to study the several treatises published on some of our old cultivated plants, as on the hyacinth, potato, even the dahlia, &c.; and it is really surprising to note the endless points in structure and constitution in which the varieties and sub-varieties differ slightly from each other.

47:d treatises on some

47:f points of structure

48 The whole organisation seems to have become plastic, and tends to depart in some small degree from that of the parental type.

48:e in a slight degree

48:f and departs in

49 Any variation which is not inherited is unimportant for us.

50 But the number and diversity of inheritable deviations of structure, both those of slight and those of considerable physiological importance, is endless.

50:c importance, are endless.

51 Dr. Prosper Lucas's treatise, in two large volumes, is the fullest and the best on this subject.

52 No breeder doubts how strong is the tendency to inheritance: like produces like is his fundamental belief: doubts have been thrown on this principle by theoretical writers alone.

52:e inheritance: that like produces/principle only by theoretical writers.

53 When a deviation appears not unfrequently, and we see it in the father and child, we cannot tell whether it may not be due to the same original cause acting on both; but when amongst individuals, apparently exposed to the same conditions, any very rare deviation, due to some extraordinary combination of circumstances, appears in the parent—say, once amongst several million individuals—and it reappears in the child, the mere

doctrine of chances almost compels us to attribute its reappearance to inheritance.

53:b When any deviation of structure often appears, and we see/ same cause having acted on

54 Every one must have heard of cases of albinism, prickly skin, hairy bodies, &c., appearing in several members of the same family.

55 If strange and rare deviations of structure are truly inherited, less strange and commoner deviations may be freely admitted to be inheritable.

56 Perhaps the correct way of viewing the whole subject, would be, to look at the inheritance of every character whatever as the rule, and non-inheritance as the anomaly.

56:f subject would

57   The laws governing inheritance are quite unknown; no one can say why the same peculiarity in different individuals of the same species, and in individuals of different species, is sometimes inherited and sometimes not so; why the child often reverts in certain characters to its grandfather or grandmother or other much more remote ancestor; why a peculiarity is often transmitted from one sex to both sexes, or to one sex alone, more commonly but not exclusively to the like sex.

57:b why a peculiarity/species, or in individuals

57.x-y:e are for the most part unknown. *y* No one can say why the same peculiarity in different individuals of the same species, or in different

57.y:f grandmother or more

58 It is a fact of some little importance to us, that peculiarities appearing in the males of our domestic breeds are often transmitted either exclusively, or in a much greater degree, to males alone.

58:d transmitted, either exclusively or

58:e to the males alone.

58:f some importance

59 A much more important rule, which I think may be trusted, is that, at whatever period of life a peculiarity first appears, it tends to appear in the offspring at a corresponding age, though sometimes earlier.

59:e to re-appear in the offspring

60 In many cases this could not be otherwise: thus the inherited peculiarities in the horns of cattle could appear only in the offspring when nearly mature; peculiarities in the silkworm are known to appear at the corresponding caterpillar or cocoon stage.

61 But hereditary diseases and some other facts make me believe that the rule has a wider extension, and that when there is no

apparent reason why a peculiarity should appear at any particular age, yet that it does tend to appear in the offspring at the same period at which it first appeared in the parent.

61:d that, **when**

62 I believe this rule to be of the highest importance in explaining the laws of embryology.

63 These remarks are of course confined to the first *appearance* of the peculiarity, and not to its primary cause, which may have acted on the ovules or male element; in nearly the same manner as in the crossed offspring from a short-horned cow by a long-horned bull, the greater length of horn, though appearing late in life, is clearly due to the male element.

63:c or on the male

63:f not to the primary cause which/as the increased length of the horns in the offspring from a short-horned cow by a long-horned bull, though

64   Having alluded to the subject of reversion, I may here refer to a statement often made by naturalists—namely, that our domestic varieties, when run wild, gradually but certainly revert in character to their aboriginal stocks.

64:e but invariably revert

65 Hence it has been argued that no deductions can be drawn from domestic races to species in a state of nature.

66 I have in vain endeavoured to discover on what decisive facts the above statement has so often and so boldly been made.

67 There would be great difficulty in proving its truth: we may safely conclude that very many of the most strongly-marked domestic varieties could not possibly live in a wild state.

68 In many cases we do not know what the aboriginal stock was, and so could not tell whether or not nearly perfect reversion had ensued.

69 It would be quite necessary, in order to prevent the effects of intercrossing, that only a single variety should be turned loose in its new home.

69:e be necessary/should have been turned

70 Nevertheless, as our varieties certainly do occasionally revert in some of their characters to ancestral forms, it seems to me not improbable, that if we could succeed in naturalising, or were to cultivate, during many generations, the several races, for instance, of the cabbage, in very poor soil (in which case, however, some effect would have to be attributed to the direct action of the poor soil), that they would to a large extent, or even wholly, revert to the wild aboriginal stock.

70:e attributed to the definite action

70:f improbable that if/attributed to the *definite* action/would, to a large

71 Whether or not the experiment would succeed, is not of great importance for our line of argument; for by the experiment itself the conditions of life are changed.

72 If it could be shown that our domestic varieties manifested a strong tendency to reversion,—that is, to lose their acquired characters, whilst kept under unchanged conditions, and whilst kept in a considerable body, so that free intercrossing might check, by blending together, any slight deviations of structure, in such case, I grant that we could deduce nothing from domestic varieties in regard to species.

72:b under the same conditions/deviations in their structure

73 But there is not a shadow of evidence in favour of this view: to assert that we could not breed our cart and race-horses, long and short-horned cattle, and poultry of various breeds, and esculent vegetables, for an almost infinite number of generations, would be opposed to all experience.

73:e for an unlimited number

74 I may add, that when under nature the conditions of life do change, variations and reversions of character probably do occur; but natural selection, as will hereafter be explained, will determine how far the new characters thus arising shall be preserved.

74:d that, when/preserved. [*Space*]

74:[e]

74.1:d Character of Domestic Varieties; difficulty of distinguishing between Varieties and Species; origin of Domestic Varieties from one or more Species. [*Space*]

75   When we look to the hereditary varieties or races of our domestic animals and plants, and compare them with species closely allied together, we generally perceive in each domestic race, as already remarked, less uniformity of character than in true species.

75:b with closely allied species, we

76 Domestic races of the same species, also, often have a somewhat monstrous character; by which I mean, that, although differing from each other, and from the other species of the same genus, in several trifling respects, they often differ in an extreme degree in some one part, both when compared one with another, and more especially when compared with all the species in nature to which they are nearest allied.

76:b and from other

76:d especially when compared with the species under nature

76:e races often have

77 With these exceptions (and with that of the perfect fertility of varieties when crossed,—a subject hereafter to be discussed), domestic races of the same species differ from each other in

the same manner as, only in most cases in a lesser degree than, do closely-allied species of the same genus in a state of nature.

77:d degree, in the same manner as do

77:f other in the same manner as do the closely-allied species of the same genus in a state of nature, but the differences in most cases are less in degree.

78 I think this must be admitted, when we find that there are hardly any domestic races, either amongst animals or plants, which have not been ranked by some competent judges as mere varieties, and by other competent judges as the descendants of aboriginally distinct species.

78:b by competent judges as mere

78:c competent judges as the descendants of aboriginally distinct species, and by other competent judges as mere varieties.

78:e This must be admitted as true, for the domestic races of many animals and plants have been ranked by some competent judges as the descendants

79 If any marked distinction existed between domestic races and species, this source of doubt could not so perpetually recur.

79:e If any well marked distinction existed between a domestic race and a species, this source of doubt would not

80 It has often been stated that domestic races do not differ from each other in characters of generic value.

81 I think it could be shown that this statement is hardly correct; but naturalists differ most widely in determining what characters are of generic value; all such valuations being at present empirical.

81:b differ widely

81:c it can be shown

81:e It can be/is not correct/differ much in determining

82 Moreover, on the view of the origin of genera which I shall presently give, we have no right to expect often to meet with generic differences in our domesticated productions.

82:e When it is explained how genera originate under nature, it will be seen that we/often to find a generic amount of difference in our domesticated races.

83 When we attempt to estimate the amount of structural difference between the domestic races of the same species, we are soon involved in doubt, from not knowing whether they have descended from one or several parent-species.

83:e In attempting

83:f between allied domestic races, we/they are descended from one or several parent species.

84 This point, if it could be cleared up, would be interesting; if, for instance, it could be shown that the greyhound, bloodhound, terrier, spaniel, and bull-dog, which we all know prop-

agate their kind so truly, were the offspring of any single species, then such facts would have great weight in making us doubt about the immutability of the many very closely allied and natural species—for instance, of the many foxes—inhabiting different quarters of the world.

84:b allied natural

84:f kind truly/many closely

85 I do not believe, as we shall presently see, that all our dogs have descended from any one wild species; but, in the case of some other domestic races, there is presumptive, or even strong, evidence in favour of this view.

85.x:b that the whole amount of difference between the several breeds of the dog has been produced under domestication; I believe that some small part of the difference is due to their being descended from distinct species.

85.x:c their having descended

85.x:f that a small/their being descended

85.y:b In the case of some other domesticated species, there is presumptive, or even strong evidence, that all the breeds have descended from a single wild stock.

85.y:e case of strongly marked races in some other domesticated species, there/all are descended

85.y:f presumptive or

86   It has often been assumed that man has chosen for domestication animals and plants having an extraordinary inherent tendency to vary, and likewise to withstand diverse climates.

87 I do not dispute that these capacities have added largely to the value of most of our domesticated productions; but how could a savage possibly know, when he first tamed an animal, whether it would vary in succeeding generations, and whether it would endure other climates?

88 Has the little variability of the ass or guinea-fowl, or the small power of endurance of warmth by the rein-deer, or of cold by the common camel, prevented their domestication?

88:c reindeer/camel prevented

88:d ass and goose, or the small/camel, prevented

89 I cannot doubt that if other animals and plants, equal in number to our domesticated productions, and belonging to equally diverse classes and countries, were taken from a state of nature, and could be made to breed for an equal number of generations under domestication, they would vary on an average as largely as the parent species of our existing domesticated productions have varied.

89:d would on an average vary as largely

90   In the case of most of our anciently domesticated animals and plants, I do not think it is possible to come to any definite

conclusion, whether they have descended from one or several species.

90:b several wild species.

90:c plants, it is not possible

90:e they are descended

91 The argument mainly relied on by those who believe in the multiple origin of our domestic animals is, that we find in the most ancient records, more especially on the monuments of Egypt, much diversity in the breeds; and that some of the breeds closely resemble, perhaps are identical with, those still existing.

91:d ancient times, on the monuments of Egypt, and in the lake-habitations of Switzerland, much/these ancient breeds closely resemble or are even identical with those

91:f resemble, or are even identical with, those still

92 Even if this latter fact were found more strictly and generally true than seems to me to be the case, what does it show, but that some of our breeds originated there, four or five thousand years ago?

92:[d]

93 But Mr. Horner's researches have rendered it in some degree probable that man sufficiently civilized to have manufactured pottery existed in the valley of the Nile thirteen or fourteen thousand years ago; and who will pretend to say how long before these ancient periods, savages, like those of Tierra del Fuego or Australia, who possess a semi-domestic dog, may not have existed in Egypt?

93:[c]

93.0.1:d But this only throws far backwards the history of civilization, and shows that animals were domesticated at a much earlier period than has hitherto been supposed.

93.0.2:d The lake-inhabitants of Switzerland cultivated several kinds of wheat and barley, the pea, the poppy for oil, and flax; and they possessed several domesticated animals; they had also commerce with other nations.

93.0.2.x-y:f animals. *y* They also carried on commerce

93.0.3:d All this clearly shows, as Heer has remarked, that they had at this early age progressed considerably in civilization; and this again implies a long-continued previous period of less advanced civilization, during which the domesticated animals, kept by the different tribes and in different districts, might have varied and given rise to distinct races.

93.0.3:f long continued

93.1:c Since the recent discoveries of flint tools or celts in the superficial deposits of France and England, few geologists will doubt that man, in a sufficiently civilized state to have manufactured weapons, existed at a period extremely remote as measured by years; and we know that at the present day there is hardly a tribe so barbarous as not to have domesticated at least the dog.

93.1:d Since the discovery of flint tools or celts in the superficial formations of France and England, all geologists believe that man in a barbarous condition existed at an enormously remote period; and we

93.1:e formations in many parts of the world, all geologists believe that barbarian man existed

93.1:f tools in the superficial formations of many/barbarous, as

94 The whole subject must, I think, remain vague; nevertheless, I may, without here entering on any details, state that, from geographical and other considerations, I think it highly probable that our domestic dogs have descended from several wild species.

94.x-y:c The origin of most of our domestic animals will probably for ever remain vague. *y* But I may here state, that looking to the domestic dogs of the whole world, I have, after a laborious collection of all known facts, come to the conclusion that several wild species of Canidæ have been tamed, and that their blood, more or less mingled, flows in the veins of our many domestic breeds.

94.y:d that, looking/mingled together, flows

94.y:e blood, in some cases mingled/our domestic

94.1:b Knowing, as we do, that savages are very fond of taming animals, it seems to me unlikely, in the case of the dog-genus, which is distributed in a wild state throughout the world, that since man first appeared one single species alone should have been domesticated.

94.1:[c]

95 In regard to sheep and goats I can form no opinion.

95:e no decided opinion.

96 I should think, from facts communicated to me by Mr. Blyth, on the habits, voice, and constitution, &c., of the humped Indian cattle, that these had descended from a different aboriginal stock from our European cattle; and several competent judges believe that these latter have had more than one wild parent.

96:c From/cattle, it is probable that these descended

96:d constitution, and structure of the humped/these are/European cattle; and some competent/have had several wild parents,— whether or not these deserve to be called species or races.

96:e Indian cattle, it is almost certain that they are/have had two or three wild progenitors,—whether

96:f species.

96.1:d This conclusion, as well as the specific distinction between the humped and common cattle, may indeed be looked at as almost established by the recent admirable researches of Professor Rütimeyer.

96.1:f well as that of the specific/may, indeed, be looked upon as established by the admirable

97 With respect to horses, from reasons which I cannot give here, I am doubtfully inclined to believe, in opposition to several authors, that all the races have descended from one wild stock.

97:c which I cannot give here, I am with much doubt inclined

97:e give, I am doubtfully inclined/races belong to the same species.

98 Mr. Blyth, whose opinion, from his large and varied stores of knowledge, I should value more than that of almost any one, thinks that all the breeds of poultry have proceeded from the common wild Indian fowl (Gallus bankiva).

98:d bankiva): having kept nearly all the English kinds alive, having bred and crossed them, and examined their skeletons, I have come to a similar conclusion,—the grounds of which will be given in a future work.

98:e Having kept nearly all the English breeds of the fowl alive, having bred and crossed them, and examined their skeletons, it appears to me almost certain that all are the descendants of the wild Indian fowl, Gallus bankiva; and this is the conclusion of Mr. Blyth, and of others who have studied this bird in India.

99 In regard to ducks and rabbits, the breeds of which differ considerably from each other in structure, I do not doubt that they all have descended from the common wild duck and rabbit.

99:b they have all descended

99:c structure, the evidence preponderates in favour of their having all

99:d evidence strongly preponderates

99:e rabbits, some breeds of which differ much from each other, the evidence is clear that they are all

100 The doctrine of the origin of our several domestic races from several aboriginal stocks, has been carried to an absurd extreme by some authors.

101 They believe that every race which breeds true, let the distinctive characters be ever so slight, has had its wild prototype.

102 At this rate there must have existed at least a score of species of wild cattle, as many sheep, and several goats in Europe alone, and several even within Great Britain.

103 One author believes that there formerly existed in Great Britain eleven wild species of sheep peculiar to it!

103:f existed eleven wild species of sheep peculiar to Great Britain!

104 When we bear in mind that Britain has now hardly one peculiar mammal, and France but few distinct from those of Germany and conversely, and so with Hungary, Spain, &c., but that each of these kingdoms possesses several peculiar breeds of cattle, sheep, &c., we must admit that many domestic breeds have originated in Europe; for whence could they have been

derived, as these several countries do not possess a number of peculiar species as distinct parent-stocks?

104:e for whence have they been derived, as these several countries could not possess so large a number of peculiar species for parent-stocks?

104:f now not one/Germany, and so/domestic breeds must have originated in Europe; for whence otherwise could they have been derived?

105 So it is in India.

106 Even in the case of the domestic dogs of the whole world, which I fully admit have probably descended from several wild species, I cannot doubt that there has been an immense amount of inherited variation.

107 Who can believe that animals closely resembling the Italian greyhound, the bloodhound, the bull-dog, or Blenheim spaniel, &c.—so unlike all wild Canidæ—ever existed freely in a state of nature?

106 + 7:c which I admit to have descended from several wild species, it cannot be doubted that there/variation; for who will believe/bull-dog, pug-dog, or

106 + 7:f domestic dog throughout the world, which I admit are descended/existed in

108 It has often been loosely said that all our races of dogs have been produced by the crossing of a few aboriginal species; but by crossing we can get only forms in some degree intermediate between their parents; and if we account for our several domestic races by this process, we must admit the former existence of the most extreme forms, as the Italian greyhound, bloodhound, bull-dog, &c., in the wild state.

108:b can only get forms

109 Moreover, the possibility of making distinct races by crossing has been greatly exaggerated.

110 There can be no doubt that a race may be modified by occasional crosses, if aided by the careful selection of those individual mongrels, which present any desired character; but that a race could be obtained nearly intermediate between two extremely different races or species, I can hardly believe.

110:c Many cases are on record, showing that a race may

110:d selection of the individuals which present the desired character; but to obtain a race nearly/species, would be very difficult.

110:f obtain a race intermediate between two quite distinct races, would

111 Sir J. Sebright expressly experimentised for this object, and failed.

111:c experimented

111:f experimented with this

112 The offspring from the first cross between two pure breeds is tolerably and sometimes (as I have found with pigeons) extremely uniform, and everything seems simple enough; but when these mongrels are crossed one with another for several generations, hardly two of them will be alike, and then the extreme difficulty, or rather utter hopelessness, of the task becomes apparent.

112:d hopelessness of

112:e alike; and then the extreme difficulty of the task

112:f pigeons) quite uniform in character, and everything/alike, and then the difficulty of the task becomes manifest.

113 Certainly, a breed intermediate between *two very distinct* breeds could not be got without extreme care and long-continued selection; nor can I find a single case on record of a permanent race having been thus formed.

113:e find a case

113:[f]

114   *On the Breeds of the Domestic Pigeon.—*

114:d [Center] *Breeds of the Domestic Pigeon, their Differences and Origin.* [Space]

115 Believing that it is always best to study some special group, I have, after deliberation, taken up domestic pigeons.

115:d[¶]

116 I have kept every breed which I could purchase or obtain, and have been most kindly favoured with skins from several quarters of the world, more especially by the Hon. W. Elliot from India, and by the Hon. C. Murray from Persia.

117 Many treatises in different languages have been published on pigeons, and some of them are very important, as being of considerable antiquity.

118 I have associated with several eminent fanciers, and have been permitted to join two of the London Pigeon Clubs.

119 The diversity of the breeds is something astonishing.

120 Compare the English carrier and the short-faced tumbler, and see the wonderful difference in their beaks, entailing corresponding differences in their skulls.

121 The carrier, more especially the male bird, is also remarkable from the wonderful development of the carunculated skin about the head, and this is accompanied by greatly elongated eyelids, very large external orifices to the nostrils, and a wide gape of mouth.

121:c head; and this

122 The short-faced tumbler has a beak in outline almost like that

95

of a finch; and the common tumbler has the singular and strictly inherited habit of flying at a great height in a compact flock, and tumbling in the air head over heels.

122:b singular inherited

123 The runt is a bird of great size, with long, massive beak and large feet; some of the sub-breeds of runts have very long necks, others very long wings and tails, others singularly short tails.

124 The barb is allied to the carrier, but, instead of a very long beak, has a very short and very broad one.

124:e of a long

125 The pouter has a much elongated body, wings, and legs; and its enormously developed crop, which it glories in inflating, may well excite astonishment and even laughter.

126 The turbit has a very short and conical beak, with a line of reversed feathers down the breast; and it has the habit of continually expanding slightly the upper part of the œsophagus.

126:f has a short and conical/expanding, slightly, the upper

127 The Jacobin has the feathers so much reversed along the back of the neck that they form a hood, and it has, proportionally to its size, much elongated wing and tail feathers.

127:c hood; and

127:f size, elongated

128 The trumpeter and laugher, as their names express, utter a very different coo from the other breeds.

129 The fantail has thirty or even forty tail-feathers, instead of twelve or fourteen, the normal number in all members of the great pigeon family; and these feathers are kept expanded, and are carried so erect that in good birds the head and tail touch; the oil-gland is quite aborted.

129:c fourteen—the normal/erect, that

129:f all the members of the great pigeon family: these

130 Several other less distinct breeds might have been specified.

130:b might be specified.

131 In the skeletons of the several breeds, the development of the bones of the face in length and breadth and curvature differs enormously.

132 The shape, as well as the breadth and length of the ramus of the lower jaw, varies in a highly remarkable manner.

133 The number of the caudal and sacral vertebræ vary; as does the number of the ribs, together with their relative breadth and the presence of processes.

133:c The caudal and sacral vertebræ vary in number; as

134 The size and shape of the apertures in the sternum are highly

96

variable; so is the degree of divergence and relative size of the two arms of the furcula.

135 The proportional width of the gape of mouth, the proportional length of the eyelids, of the orifice of the nostrils, of the tongue (not always in strict correlation with the length of beak), the size of the crop and of the upper part of the œsophagus; the development and abortion of the oil-gland; the number of the primary wing and caudal feathers; the relative length of wing and tail to each other and to the body; the relative length of leg and of the feet; the number of scutellæ on the toes, the development of skin between the toes, are all points of structure which are variable.

135:f body; the relative length of the leg and foot; the number

136 The period at which the perfect plumage is acquired varies, as does the state of the down with which the nestling birds are clothed when hatched.

137 The shape and size of the eggs vary.

138 The manner of flight differs remarkably; as does in some breeds the voice and disposition.

138:b flight, and in some breeds the voice and disposition, differ remarkably.

139 Lastly, in certain breeds, the males and females have come to differ to a slight degree from each other.

139:e differ in a slight

140   Altogether at least a score of pigeons might be chosen, which if shown to an ornithologist, and he were told that they were wild birds, would certainly, I think, be ranked by him as well-defined species.

140:c which, if/certainly be

141 Moreover, I do not believe that any ornithologist would place the English carrier, the short-faced tumbler, the runt, the barb, pouter, and fantail in the same genus; more especially as in each of these breeds several truly-inherited sub-breeds, or species as he might have called them, could be shown him.

141:d he would have

141:f would in this case place/species, as he would call

142   Great as the differences are between the breeds of pigeons, I am fully convinced that the common opinion of naturalists is correct, namely, that all have descended from the rock-pigeon (Columba livia), including under this term several geographical races or sub-species, which differ from each other in the most trifling respects.

142:e all are descended

142:f Great as are the differences between the breeds of the pigeon, I

143 As several of the reasons which have led me to this belief are in

97

some degree applicable in other cases, I will here briefly give them.

144 If the several breeds are not varieties, and have not proceeded from the rock-pigeon, they must have descended from at least seven or eight aboriginal stocks; for it is impossible to make the present domestic breeds by the crossing of any lesser number: how, for instance, could a pouter be produced by crossing two breeds unless one of the parent-stocks possessed the characteristic enormous crop?

145 The supposed aboriginal stocks must all have been rock-pigeons, that is, not breeding or willingly perching on trees.

145:f is, they did not breed or willingly perch

146 But besides C. livia, with its geographical sub-species, only two or three other species of rock-pigeons are known; and these have not any of the characters of the domestic breeds.

147 Hence the supposed aboriginal stocks must either still exist in the countries where they were originally domesticated, and yet be unknown to ornithologists; and this, considering their size, habits, and remarkable characters, seems very improbable; or they must have become extinct in the wild state.

147:e seems improbable

148 But birds breeding on precipices, and good fliers, are unlikely to be exterminated; and the common rock-pigeon, which has the same habits with the domestic breeds, has not been exterminated even on several of the smaller British islets, or on the shores of the Mediterranean.

149 Hence the supposed extermination of so many species having similar habits with the rock-pigeon seems to me a very rash assumption.

149:d to be a very

149:e seems a very

150 Moreover, the several above-named domesticated breeds have been transported to all parts of the world, and, therefore, some of them must have been carried back again into their native country; but not one has ever become wild or feral, though the dovecot-pigeon, which is the rock-pigeon in a very slightly altered state, has become feral in several places.

150:f one has become wild

151 Again, all recent experience shows that it is most difficult to get any wild animal to breed freely under domestication; yet on the hypothesis of the multiple origin of our pigeons, it must be assumed that at least seven or eight species were so thoroughly domesticated in ancient times by half-civilized man, as to be quite prolific under confinement.

151:c is difficult/yet, on

151:f get wild animals

152 An argument, as it seems to me, of great weight, and applicable in several other cases, is, that the above-specified breeds, though agreeing generally in constitution, habits, voice, colouring, and in most parts of their structure, with the wild rock-pigeon, yet are certainly highly abnormal in other parts of their structure: we may look in vain throughout the whole great family of Columbidæ for a beak like that of the English carrier, or that of the short-faced tumbler, or barb; for reversed feathers like those of the jacobin; for a crop like that of the pouter; for tail-feathers like those of the fantail.

152:b Jacobin

152:e argument of great weight/generally with the wild rock-pigeon in constitution, habits, voice, colouring, and in most parts of their structure, yet are certainly highly abnormal in other parts; we

153 Hence it must be assumed not only that half-civilized man succeeded in thoroughly domesticating several species, but that he intentionally or by chance picked out extraordinarily abnormal species; and further, that these very species have since all become extinct or unknown.

154 So many strange contingencies seem to me improbable in the highest degree.

154:c contingencies are improbable

155 Some facts in regard to the colouring of pigeons well deserve consideration.

156 The rock-pigeon is of a slaty-blue, and has a white rump (the Indian sub-species, C. intermedia of Strickland, having it bluish); the tail has a terminal dark bar, with the bases of the outer feathers externally edged with white; the wings have two black bars; some semi-domestic breeds and some apparently truly wild breeds have, besides the two black bars, the wings chequered with black.

156:c white croup (the Indian/breeds, and some apparently truly wild breeds, have

156.x:e slaty-blue, with white loins; the Indian sub-species, C. intermedia of Strickland, having this part bluish; the tail has a terminal dark bar, with the outer feathers externally edged at the base with white; the wings have two black bars.

156.x.x-y:f loins; but/Strickland, has this part bluish. *x.x* The tail/base with white. *x.y* The wings

156.y:e Some semi-domestic breeds, and some truly

157 These several marks do not occur together in any other species of the whole family.

158 Now, in every one of the domestic breeds, taking thoroughly well-bred birds, all the above marks, even to the white edging of the outer tail-feathers, sometimes concur perfectly developed.

159 Moreover, when two birds belonging to two distinct breeds are

crossed, neither of which is blue or has any of the above-specified marks, the mongrel offspring are very apt suddenly to acquire these characters; for instance, I crossed some uniformly white fantails with some uniformly black barbs, and they produced mottled brown and black birds; these I again crossed together, and one grandchild of the pure white fantail and pure black barb was of as beautiful a blue colour, with the white rump, double black wing-bar, and barred and white-edged tail-feathers, as any wild rock-pigeon!

159.x-y:c when birds belonging to two or more distinct breeds are crossed, none of which are blue or have any/characters. *y* To give one instance out of several which I have observed:—I crossed some white fantails, which breed very true, with some black barbs—and it so happens that blue varieties of barbs are so rare that I never heard of an instance in England; and the mongrels were black, brown, and mottled.

159.y.1:c I also crossed a barb with a spot, which is a white bird with a red tail and red spot on the forehead, and which notoriously breeds very true; the mongrels were dusky and mottled.

159.z:c I then crossed one of the mongrel barb-fantails with a mongrel barb-spot, and they produced a bird of as beautiful a blue colour, with the white croup, double

159.z:e white loins, double

160 We can understand these facts, on the well-known principle of reversion to ancestral characters, if all the domestic breeds have descended from the rock-pigeon.

160:c characters (confined, as far as I have seen, to colour alone), if

160:d characters, if

160:e breeds are **descended**

161 But if we deny this, we must make one of the two following highly improbable suppositions.

162 Either, firstly, that all the several imagined aboriginal stocks were coloured and marked like the rock-pigeon, although no other existing species is thus coloured and marked, so that in each separate breed there might be a tendency to revert to the very same colours and markings.

162:e first

163 Or, secondly, that each breed, even the purest, has within a dozen or, at most, within a score of generations, been crossed by the rock-pigeon: I say within a dozen or twenty generations, for we know of no fact countenancing the belief that the child ever reverts to some one ancestor, removed by a greater number of generations.

163:c dozen, or at most within a score, of generations, been

163:e for no instance is known of crossed descendants reverting to an ancestor of foreign blood, removed

164 In a breed which has been crossed only once with some distinct breed, the tendency to reversion to any character derived from such cross will naturally become less and less, as in each succeeding generation there will be less of the foreign blood; but when there has been no cross with a distinct breed, and there is a tendency in both parents to revert to a character, which has been lost during some former generation, this tendency, for all that we can see to the contrary, may be transmitted undiminished for an indefinite number of generations.

164:c parents to revert to a character which

164:e once, the tendency to revert to any character derived from such a cross will naturally/no cross, and there is a tendency in the breed to revert to a character which was lost

165 These two distinct cases are often confounded in treatises on inheritance.

165:c two quite distinct cases are often confounded by those who have written on inheritance.

165:e two distinct cases of reversion are often confounded together by

166 Lastly, the hybrids or mongrels from between all the domestic breeds of pigeons are perfectly fertile.

167 I can state this from my own observations, purposely made, on the most distinct breeds.

166 + 7:f all the breeds of the pigeon are perfectly fertile, as I can state from

168 Now, it is difficult, perhaps impossible, to bring forward one case of the hybrid offspring of two animals *clearly distinct* being themselves perfectly fertile.

168:f Now, hardly any cases have been ascertained with certainty of hybrids from two quite distinct species of animals being perfectly

169 Some authors believe that long-continued domestication eliminates this strong tendency to sterility: from the history of the dog I think there is some probability in this hypothesis, if applied to species closely related together, though it is unsupported by a single experiment.

169:d dog, and of some other domestic animals, there is great probability/related to each other, though

169.x-y:f strong tendency to sterility in species. *y* From/animals, this conclusion is probably quite correct, if/other.

170 But to extend the hypothesis so far as to suppose that species, aboriginally as distinct as carriers, tumblers, pouters, and fantails now are, should yield offspring perfectly fertile, *inter se,* seems to me rash in the extreme.

170:c fertile *inter*

170:f But to extend it so/*se,* would be rash

171   From these several reasons, namely, the improbability of man having formerly got seven or eight supposed species of pigeons to breed freely under domestication; these supposed species being quite unknown in a wild state, and their becoming nowhere feral; these species having very abnormal characters in certain respects, as compared with all other Columbidæ, though so like in most other respects to the rock-pigeon; the blue colour and various marks occasionally appearing in all the breeds, both when kept pure and when crossed; the mongrel offspring being perfectly fertile;—from these several reasons, taken together, I can feel no doubt that all our domestic breeds have descended from the Columba livia with its geographical sub-species.

171:c various black marks/together, we may safely conclude that

171:f namely,—the improbability of man having formerly made seven/domestication;—these supposed/their not having become anywhere feral;—these species presenting certain very abnormal characters, as compared with all other Columbidæ, though so like the rock-pigeon in most respects;—the occasional re-appearance of the blue colour and various black marks in all the breeds, both when kept pure and when crossed;—and lastly, the mongrel/breeds are descended from the rock-pigeon or Columba

172   In favour of this view, I may add, firstly, that C. livia, or the rock-pigeon, has been found capable of domestication in Europe and in India; and that it agrees in habits and in a great number of points of structure with all the domestic breeds.

172:f livia has

173 Secondly, although an English carrier or short-faced tumbler differs immensely in certain characters from the rock-pigeon, yet by comparing the several sub-breeds of these breeds, more especially those brought from distant countries, we can make an almost perfect series between the extremes of structure.

173:b these varieties, more

173:d or a short-faced/make in these two cases, and in some but not in all other cases, an almost

173:f Secondly, that, although/yet that, by comparing the several sub-breeds of these two races, more/make, between them and the rock-pigeon, an almost perfect series; so we can in some other cases, but not with all the breeds.

174 Thirdly, those characters which are mainly distinctive of each breed, for instance the wattle and length of beak of the carrier, the shortness of that of the tumbler, and the number of tail-feathers in the fantail, are in each breed eminently variable; and the explanation of this fact will be obvious when we come to treat of selection.

174:c Selection.

174:f breed are in each eminently variable, for instance the wattle and length of beak of the carrier, the shortness of that of the tumbler, and the number of tail-feathers in the fantail; and

175 Fourthly, pigeons have been watched, and tended with the utmost care, and loved by many people.

175:c watched and tended

176 They have been domesticated for thousands of years in several quarters of the world; the earliest known record of pigeons is in the fifth Ægyptian dynasty, about 3000 B.C., as was pointed out to me by Professor Lepsius; but Mr. Birch informs me that pigeons are given in a bill of fare in the previous dynasty.

177 In the time of the Romans, as we hear from Pliny, immense prices were given for pigeons; "nay, they are come to this pass, that they can reckon up their pedigree and race."

178 Pigeons were much valued by Akber Khan in India, about the year 1600; never less than 20,000 pigeons were taken with the court.

179 "The monarchs of Iran and Turan sent him some very rare birds"; and, continues the courtly historian, "His Majesty by crossing the breeds, which method was never practised before, has improved them astonishingly."

180 About this same period the Dutch were as eager about pigeons as were the old Romans.

181 The paramount importance of these considerations in explaining the immense amount of variation which pigeons have undergone, will be obvious when we treat of Selection.

182 We shall then, also, see how it is that the breeds so often have a somewhat monstrous character.

182:c the several breeds

183 It is also a most favourable circumstance for the production of distinct breeds, that male and female pigeons can be easily mated for life; and thus different breeds can be kept together in the same aviary.

184 I have discussed the probable origin of domestic pigeons at some, yet quite insufficient, length; because when I first kept pigeons and watched the several kinds, knowing well how true they bred, I felt fully as much difficulty in believing that they could ever have descended from a common parent, as any naturalist could in coming to a similar conclusion in regard to the many species of finches, or other large groups of birds, in nature.

184:c believing that since they were domesticated they could all have descended

184:d kinds, well knowing how truly they breed, I/since they had been domesticated they had all proceeded from

184:e other groups

103

185 One circumstance has struck me much; namely, that all the breeders of the various domestic animals and the cultivators of plants, with whom I have ever conversed, or whose treatises I have read, are firmly convinced that the several breeds to which each has attended, are descended from so many aboriginally distinct species.

185:c that nearly all/I have conversed

186 Ask, as I have asked, a celebrated raiser of Hereford cattle, whether his cattle might not have descended from long-horns, and he will laugh you to scorn.

186:d Long-horns, or both from a common parent-stock, and

187 I have never met a pigeon, or poultry, or duck, or rabbit fancier, who was not fully convinced that each main breed was descended from a distinct species.

188 Van Mons, in his treatise on pears and apples, shows how utterly he disbelieves that the several sorts, for instance a Ribston-pippin or Codlin-apple, could ever have proceeded from the seeds of the same tree.

189 Innumerable other examples could be given.

190 The explanation, I think, is simple: from long-continued study they are strongly impressed with the differences between the several races; and though they well know that each race varies slightly, for they win their prizes by selecting such slight differences, yet they ignore all general arguments, and refuse to sum up in their minds slight differences accumulated during many successive generations.

191 May not those naturalists who, knowing far less of the laws of inheritance than does the breeder, and knowing no more than he does of the intermediate links in the long lines of descent, yet admit that many of our domestic races have descended from the same parents—may they not learn a lesson of caution, when they deride the idea of species in a state of nature being lineal descendants of other species?

191:d species? [*Space*]
191:e races are descended

192   *Selection.—*
192:d [*Center*] *Principles of Selection anciently followed, and their Effects.* [*Space*]

193 Let us now briefly consider the steps by which domestic races have been produced, either from one or from several allied species.

193:d[¶]

194 Some little effect may, perhaps, be attributed to the direct action of the external conditions of life, and some little to habit; but he would be a bold man who would account by such

agencies for the differences of a dray and race horse, a grey-hound and bloodhound, a carrier and tumbler pigeon.

194:c may be attributed

194:e Some effect may be attributed to the direct and definite action/differences between a dray

194:f life, and some to habit

195 One of the most remarkable features in our domesticated races is that we see in them adaptation, not indeed to the animal's or plant's own good, but to man's use or fancy.

196 Some variations useful to him have probably arisen suddenly, or by one step; many botanists, for instance, believe that the fuller's teazle, with its hooks, which cannot be rivalled by any mechanical contrivance, is only a variety of the wild Dipsacus; and this amount of change may have suddenly arisen in a seed-ling.

197 So it has probably been with the turnspit dog; and this is known to have been the case with the ancon sheep.

198 But when we compare the dray-horse and race-horse, the drome-dary and camel, the various breeds of sheep fitted either for cul-tivated land or mountain pasture, with the wool of one breed good for one purpose, and that of another breed for another purpose; when we compare the many breeds of dogs, each good for man in very different ways; when we compare the game-cock, so pertinacious in battle, with other breeds so little quar-relsome, with "everlasting layers" which never desire to sit, and with the bantam so small and elegant; when we compare the host of agricultural, culinary, orchard, and flower-garden races of plants, most useful to man at different seasons and for dif-ferent purposes, or so beautiful in his eyes, we must, I think, look further than to mere variability.

198:f in different ways

199 We cannot suppose that all the breeds were suddenly produced as perfect and as useful as we now see them; indeed, in several cases, we know that this has not been their history.

199:d in many cases

200 The key is man's power of accumulative selection: nature gives successive variations; man adds them up in certain directions useful to him.

201 In this sense he may be said to make for himself useful breeds.

201:d to have made

202   The great power of this principle of selection is not hypo-thetical.

203 It is certain that several of our eminent breeders have, even within a single lifetime, modified to a large extent some breeds of cattle and sheep.

203:e extent the breeds

203:f extent their breeds

204 In order fully to realize what they have done, it is almost necessary to read several of the many treatises devoted to this subject, and to inspect the animals.

205 Breeders habitually speak of an animal's organisation as something quite plastic, which they can model almost as they please.

205:f something plastic

206 If I had space I could quote numerous passages to this effect from highly competent authorities.

207 Youatt, who was probably better acquainted with the works of agriculturists than almost any other individual, and who was himself a very good judge of an animal, speaks of the principle of selection as "that which enables the agriculturist, not only to modify the character of his flock, but to change it altogether.

207:f judge of animals

208 It is the magician's wand, by means of which he may summon into life whatever form and mould be pleases."

209 Lord Somerville, speaking of what breeders have done for sheep, says:—"It would seem as if they had chalked out upon a wall a form perfect in itself, and then had given it existence."

210 That most skilful breeder, Sir John Sebright, used to say, with respect to pigeons, that "he would produce any given feather in three years, but it would take him six years to obtain head and beak."

210[e]

211 In Saxony the importance of the principle of selection in regard to merino sheep is so fully recognised, that men follow it as a trade: the sheep are placed on a table and are studied, like a picture by a connoisseur; this is done three times at intervals of months, and the sheep are each time marked and classed, so that the very best may ultimately be selected for breeding.

212 What English breeders have actually effected is proved by the enormous prices given for animals with a good pedigree; and these have now been exported to almost every quarter of the world.

212:f have been

213 The improvement is by no means generally due to crossing different breeds; all the best breeders are strongly opposed to this practice, except sometimes amongst closely allied sub-breeds.

214 And when a cross has been made, the closest selection is far more indispensable even than in ordinary cases.

215 If selection consisted merely in separating some very distinct variety, and breeding from it, the principle would be so obvious as hardly to be worth notice; but its importance consists in the great effect produced by the accumulation in one direction,

during successive generations, of differences absolutely inappreciable by an uneducated eye—differences which I for one have vainly attempted to appreciate.

216 Not one man in a thousand has accuracy of eye and judgment sufficient to become an eminent breeder.

217 If gifted with these qualities, and he studies his subject for years, and devotes his lifetime to it with indomitable perseverance, he will succeed, and may make great improvements; if he wants any of these qualities, he will assuredly fail.

218 Few would readily believe in the natural capacity and years of practice requisite to become even a skilful pigeon-fancier.

219 The same principles are followed by horticulturists; but the variations are here often more abrupt.

220 No one supposes that our choicest productions have been produced by a single variation from the aboriginal stock.

221 We have proofs that this is not so in some cases, in which exact records have been kept; thus, to give a very trifling instance, the steadily-increasing size of the common gooseberry may be quoted.

221:c cases in

221:f this has not been so in several cases

222 We see an astonishing improvement in many florists' flowers, when the flowers of the present day are compared with drawings made only twenty or thirty years ago.

223 When a race of plants is once pretty well established, the seed-raisers do not pick out the best plants, but merely go over their seed-beds, and pull up the "rogues," as they call the plants that deviate from the proper standard.

224 With animals this kind of selection is, in fact, also followed; for hardly any one is so careless as to allow his worst animals to breed.

224:f fact, likewise followed; for hardly any one is so careless as to breed from his worst animals.

225 In regard to plants, there is another means of observing the accumulated effects of selection—namely, by comparing the diversity of flowers in the different varieties of the same species in the flower-garden; the diversity of leaves, pods, or tubers, or whatever part is valued, in the kitchen-garden, in comparison with the flowers of the same varieties; and the diversity of fruit of the same species in the orchard, in comparison with the leaves and flowers of the same set of varieties.

226 See how different the leaves of the cabbage are, and how extremely alike the flowers; how unlike the flowers of the heartsease are, and how alike the leaves; how much the fruit of the different kinds of gooseberries differ in size, colour, shape, and hairiness, and yet the flowers present very slight differences.

227 It is not that the varieties which differ largely in some one point do not differ at all in other points; this is hardly ever, perhaps never, the case.

227:c ever, I speak after careful observation, perhaps

227:d ever,—I speak after careful observation,—perhaps

228 The laws of correlation of growth, the importance of which should never be overlooked, will ensure some differences; but, as a general rule, I cannot doubt that the continued selection of slight variations, either in the leaves, the flowers, or the fruit, will produce races differing from each other chiefly in these characters.

228:e The law of correlated variation, the importance

228:f rule, it cannot be doubted

229 It may be objected that the principle of selection has been reduced to methodical practice for scarcely more than three-quarters of a century; it has certainly been more attended to of late years, and many treatises have been published on the subject; and the result, I may add, has been, in a corresponding degree, rapid and important.

229:b result has

230 But it is very far from true that the principle is a modern discovery.

231 I could give several references to the full acknowledgment of the importance of the principle in works of high antiquity.

231:f references to works of high antiquity, in which the full importance of the principle is acknowledged.

232 In rude and barbarous periods of English history choice animals were often imported, and laws were passed to prevent their exportation: the destruction of horses under a certain size was ordered, and this may be compared to the "roguing" of plants by nurserymen.

233 The principle of selection I find distinctly given in an ancient Chinese encyclopædia.

234 Explicit rules are laid down by some of the Roman classical writers.

235 From passages in Genesis, it is clear that the colour of domestic animals was at that early period attended to.

236 Savages now sometimes cross their dogs with wild canine animals, to improve the breed, and they formerly did so, as is attested by passages in Pliny.

237 The savages in South Africa match their draught cattle by colour, as do some of the Esquimaux their teams of dogs.

238 Livingstone shows how much good domestic breeds are valued by the negroes of the interior of Africa who have not associated with Europeans.

238:e Livingstone states that good/negroes in the interior

239 Some of these facts do not show actual selection, but they show that the breeding of domestic animals was carefully attended to in ancient times, and is now attended to by the lowest savages.

240 It would, indeed, have been a strange fact, had attention not been paid to breeding, for the inheritance of good and bad qualities is so obvious.

240:d obvious. [*Space*]

240.1:d [*Center*] *Unconscious Selection.* [*Space*]

241 At the present time, eminent breeders try by methodical selection, with a distinct object in view, to make a new strain or sub-breed, superior to anything existing in the country.

241:f anything of the kind in

242 But, for our purpose, a kind of Selection, which may be called Unconscious, and which results from every one trying to possess and breed from the best individual animals, is more important.

242:f purpose, a form of Selection

243 Thus, a man who intends keeping pointers naturally tries to get as good dogs as he can, and afterwards breeds from his own best dogs, but he has no wish or expectation of permanently altering the breed.

244 Nevertheless I cannot doubt that this process, continued during centuries, would improve and modify any breed, in the same way as Bakewell, Collins, &c., by this very same process, only carried on more methodically, did greatly modify, even during their own lifetimes, the forms and qualities of their cattle.

244:b Nevertheless we may infer that

244:f their lifetimes

245 Slow and insensible changes of this kind could never be recognised unless actual measurements or careful drawings of the breeds in question had been made long ago, which might serve for comparison.

245:f kind can never/question have been made long ago, which may serve

246 In some cases, however, unchanged or but little changed individuals of the same breed may be found in less civilised districts, where the breed has been less improved.

246:b unchanged, or

246:f breed exist in

247 There is reason to believe that King Charles's spaniel has been unconsciously modified to a large extent since the time of that monarch.

248 Some highly competent authorities are convinced that the setter is directly derived from the spaniel, and has probably been slowly altered from it.

249 It is known that the English pointer has been greatly changed within the last century, and in this case the change has, it is believed, been chiefly effected by crosses with the fox-hound; but what concerns us is, that the change has been effected unconsciously and gradually, and yet so effectually, that, though the old Spanish pointer certainly came from Spain, Mr. Borrow has not seen, as I am informed by him, any native dog in Spain like our pointer.

250 By a similar process of selection, and by careful training, the whole body of English racehorses have come to surpass in fleetness and size the parent Arab stock, so that the latter, by the regulations for the Goodwood Races, are favoured in the weights they carry.

250:f training, English racehorses/Arabs, so/weights which they

251 Lord Spencer and others have shown how the cattle of England have increased in weight and in early maturity, compared with the stock formerly kept in this country.

252 By comparing the accounts given in old pigeon treatises of carriers and tumblers with these breeds as now existing in Britain, India, and Persia, we can, I think, clearly trace the stages through which they have insensibly passed, and come to differ so greatly from the rock-pigeon.

252:f in various old treatises of the former and present state of carrier and tumbler pigeons in Britain, India, and Persia, we can trace

253 Youatt gives an excellent illustration of the effects of a course of selection, which may be considered as unconsciously followed, in so far that the breeders could never have expected or even have wished to have produced the result which ensued—namely, the production of two distinct strains.

253:d expected, or even have wished, to produce the result

253:f unconscious, in/even wished

254 The two flocks of Leicester sheep kept by Mr. Buckley and Mr. Burgess, as Mr. Youatt remarks, "have been purely bred from the original stock of Mr. Bakewell for upwards of fifty years.

255 There is not a suspicion existing in the mind of any one at all acquainted with the subject that the owner of either of them has deviated in any one instance from the pure blood of Mr. Bakewell's flock, and yet the difference between the sheep possessed by these two gentlemen is so great that they have the appearance of being quite different varieties."

255:e subject, that

256 If there exist savages so barbarous as never to think of the inherited character of the offspring of their domestic animals, yet any one animal particularly useful to them, for any special purpose, would be carefully preserved during famines and other accidents, to which savages are so liable, and such choice

animals would thus generally leave more offspring than the inferior ones; so that in this case there would be a kind of unconscious selection going on.

257 We see the value set on animals even by the barbarians of Tierra del Fuego, by their killing and devouring their old women, in times of dearth, as of less value than their dogs.

258 In plants the same gradual process of improvement, through the occasional preservation of the best individuals, whether or not sufficiently distinct to be ranked at their first appearance as distinct varieties, and whether or not two or more species or races have become blended together by crossing, may plainly be recognised in the increased size and beauty which we now see in the varieties of the heartsease, rose, pelargonium, dahlia, and other plants, when compared with the older varieties or with their parent-stocks.

259 No one would ever expect to get a first-rate heartsease or dahlia from the seed of a wild plant.

260 No one would expect to raise a first-rate melting pear from the seed of the wild pear, though he might succeed from a poor seedling growing wild, if it had come from a garden-stock.

261 The pear, though cultivated in classical times, appears, from Pliny's description, to have been a fruit of very inferior quality.

262 I have seen great surprise expressed in horticultural works at the wonderful skill of gardeners, in having produced such splendid results from such poor materials; but the art, I cannot doubt, has been simple, and, as far as the final result is concerned, has been followed almost unconsciously.

262:c art has been simple

263 It has consisted in always cultivating the best known variety, sowing its seeds, and, when a slightly better variety has chanced to appear, selecting it, and so onwards.

264 But the gardeners of the classical period, who cultivated the best pear they could procure, never thought what splendid fruit we should eat; though we owe our excellent fruit, in some small degree, to their having naturally chosen and preserved the best varieties they could anywhere find.

264:f pears which they

265 A large amount of change in our cultivated plants, thus slowly and unconsciously accumulated, explains, as I believe, the well-known fact, that in a vast number of cases we cannot recognise, and therefore do not know, the wild parent-stocks of the plants which have been longest cultivated in our flower and kitchen gardens.

265:e that in a number

265:f change, thus

266 If it has taken centuries or thousands of years to improve or

111

modify most of our plants up to their present standard of usefulness to man, we can understand how it is that neither Australia, the Cape of Good Hope, nor any other region inhabited by quite uncivilised man, has afforded us a single plant worth culture.

267 It is not that these countries, so rich in species, do not by a strange chance possess the aboriginal stocks of any useful plants, but that the native plants have not been improved by continued selection up to a standard of perfection comparable with that given to the plants in countries anciently civilised.

267:f with that acquired by the plants

268 In regard to the domestic animals kept by uncivilised man, it should not be overlooked that they almost always have to struggle for their own food, at least during certain seasons.

269 And in two countries very differently circumstanced, individuals of the same species, having slightly different constitutions or structure, would often succeed better in the one country than in the other, and thus by a process of "natural selection," as will hereafter be more fully explained, two sub-breeds might be formed.

269:b other; and

270 This, perhaps, partly explains what has been remarked by some authors, namely, that the varieties kept by savages have more of the character of species than the varieties kept in civilised countries.

270:f explains why the varieties kept by savages, as has been re-marked by some authors, have more of the character of true species

271 On the view here given of the all-important part which selection by man has played, it becomes at once obvious, how it is that our domestic races show adaptation in their structure or in their habits to man's wants or fancies.

271:f of the important

272 We can, I think, further understand the frequently abnormal character of our domestic races, and likewise their differences being so great in external characters and relatively so slight in internal parts or organs.

272:f external characters, and

273 Man can hardly select, or only with much difficulty, any deviation of structure excepting such as is externally visible; and indeed he rarely cares for what is internal.

274 He can never act by selection, excepting on variations which are first given to him in some slight degree by nature.

275 No man would ever try to make a fantail, till he saw a pigeon with a tail developed in some slight degree in an unusual manner, or a pouter till he saw a pigeon with a crop of somewhat

unusual size; and the more abnormal or unusual any character was when it first appeared, the more likely it would be to catch his attention.

275:d fantail till he saw a pigeon with a tail

276 But to use such an expression as trying to make a fantail, is, I have no doubt, in most cases, utterly incorrect.

277 The man who first selected a pigeon with a slightly larger tail, never dreamed what the descendants of that pigeon would become through long-continued, partly unconscious and partly methodical selection.

277:f methodical, selection.

278 Perhaps the parent bird of all fantails had only fourteen tail-feathers somewhat expanded, like the present Java fantail, or like individuals of other and distinct breeds, in which as many as seventeen tail-feathers have been counted.

278:c parent-bird

279 Perhaps the first pouter-pigeon did not inflate its crop much more than the turbit now does the upper part of its œsophagus, —a habit which is disregarded by all fanciers, as it is not one of the points of the breed.

280   Nor let it be thought that some great deviation of structure would be necessary to catch the fancier's eye: he perceives extremely small differences, and it is in human nature to value any novelty, however slight, in one's own possession.

281 Nor must the value which would formerly be set on any slight differences in the individuals of the same species, be judged of by the value which would now be set on them, after several breeds have once fairly been established.

281:f formerly have been set/by the value which is now set on them, after several breeds have fairly

282 Many slight differences might, and indeed do now, arise amongst pigeons, which are rejected as faults or deviations from the standard of perfection of each breed.

282:f It is known that with pigeons many slight variations now occasionally appear, but these are rejected/perfection in each

283 The common goose has not given rise to any marked varieties; hence the Thoulouse and the common breed, which differ only in colour, that most fleeting of characters, have lately been exhibited as distinct at our poultry-shows.

283:c Toulouse

284   I think these views further explain what has sometimes been noticed—namely, that we know nothing about the origin or history of any of our domestic breeds.

284:f These views appear to explain/know hardly anything about

285 But, in fact, a breed, like a dialect of a language, can hardly be said to have had a definite origin.

285:e have a distinct origin.

286 A man preserves and breeds from an individual with some slight deviation of structure, or takes more care than usual in matching his best animals and thus improves them, and the improved individuals slowly spread in the immediate neighbourhood.

286:d animals, and thus

287 But as yet they will hardly have a distinct name, and from being only slightly valued, their history will be disregarded.

287:f But they will as yet hardly/history will have been disregarded.

288 When further improved by the same slow and gradual process, they will spread more widely, and will get recognised as something distinct and valuable, and will then probably first receive a provincial name.

288:f will be recognised

289 In semi-civilised countries, with little free communication, the spreading and knowledge of any new sub-breed will be a slow process.

289:c sub-breed would be

289:e spreading of a new

290 As soon as the points of value of the new sub-breed are once fully acknowledged, the principle, as I have called it, of unconscious selection will always tend,—perhaps more at one period than at another, as the breed rises or falls in fashion,—perhaps more in one district than in another, according to the state of civilisation of the inhabitants,—slowly to add to the characteristic features of the breed, whatever they may be.

290:e value in a new strain are once acknowledged

290:f value are

291 But the chance will be infinitely small of any record having been preserved of such slow, varying, and insensible changes.

291:d changes. [*Space*]

291.1:d [*Center*] *Circumstances favourable to Man's Power of Selection.* [*Space*]

292   I must now say a few words on the circumstances, favourable, or the reverse, to man's power of selection.

292:f I will now

293 A high degree of variability is obviously favourable, as freely giving the materials for selection to work on; not that mere individual differences are not amply sufficient, with extreme care, to allow of the accumulation of a large amount of modification in almost any desired direction.

294 But as variations manifestly useful or pleasing to man appear only occasionally, the chance of their appearance will be much

increased by a large number of individuals being kept; and hence this comes to be of the highest importance to success.

294.x-y:f kept. *y* Hence, number is of the highest importance for success.

295 On this principle Marshall has remarked, with respect to the sheep of parts of Yorkshire, that "as they generally belong to poor people, and are mostly *in small lots,* they never can be improved."

295:d Marshall formerly remarked/that, "as

296:f Yorkshire, "as

296 On the other hand, nurserymen, from raising large stocks of the same plants, are generally far more successful than amateurs in getting new and valuable varieties.

296:f from keeping large stocks of the same plant, are/in raising new

297 The keeping of a large number of individuals of a species in any country requires that the species should be placed under favourable conditions of life, so as to breed freely in that country.

297:f A large number of individuals of an animal or plant can be reared only where the conditions for its propagation are favourable.

298 When the individuals of any species are scanty, all the individuals, whatever their quality may be, will generally be allowed to breed, and this will effectually prevent selection.

298:f individuals are scanty, all will be allowed to breed, whatever their quality may be, and

299 But probably the most important point of all, is, that the animal or plant should be so highly useful to man, or so much valued by him, that the closest attention should be paid to even the slightest deviation in the qualities or structure of each individual.

299:d all is that/attention is paid

299:f important element is that the animal or plant should be so highly valued by man/in its qualities or structure.

300 Unless such attention be paid nothing can be effected.

301 I have seen it gravely remarked, that it was most fortunate that the strawberry began to vary just when gardeners began to attend closely to this plant.

301:f attend to

302 No doubt the strawberry had always varied since it was cultivated, but the slight varieties had been neglected.

303 As soon, however, as gardeners picked out individual plants with slightly larger, earlier, or better fruit, and raised seedlings from them, and again picked out the best seedlings and bred from them, then, there appeared (aided by some crossing with

distinct species) those many admirable varieties of the strawberry which have been raised during the last thirty or forty years.

303:f then (with some aid by crossing distinct species) those many admirable varieties of the strawberry were raised which have appeared during the last half-century.

304   In the case of animals with separate sexes, facility in preventing crosses is an important element of success in the formation of new races,—at least, in a country which is already stocked with other races.

304:f With animals, facility/element in the formation

305 In this respect enclosure of the land plays a part.

306 Wandering savages or the inhabitants of open plains rarely possess more than one breed of the same species.

307 Pigeons can be mated for life, and this is a great convenience to the fancier, for thus many races may be kept true, though mingled in the same aviary; and this circumstance must have largely favoured the improvement and formation of new breeds.

307:c may be improved and kept

308 Pigeons, I may add, can be propagated in great numbers and at a very quick rate, and inferior birds may be freely rejected, as when killed they serve for food.

309 On the other hand, cats, from their nocturnal rambling habits, cannot be matched, and, although so much valued by women and children, we hardly ever see a distinct breed kept up; such breeds as we do sometimes see are almost always imported from some other country, often from islands.

309:c be easily matched/country.

309:f we rarely see a distinct breed long kept

310 Although I do not doubt that some domestic animals vary less than others, yet the rarity or absence of distinct breeds of the cat, the donkey, peacock, goose, &c., may be attributed in main part to selection not having been brought into play: in cats, from the difficulty in pairing them; in donkeys, from only a few being kept by poor people, and little attention paid to their breeding; in peacocks, from not being very easily reared and a large stock not kept; in geese, from being valuable only for two purposes, food and feathers, and more especially from no pleasure having been felt in the display of distinct breeds.

310:c their breeding; for recently in certain parts of Spain and of the United States this animal has been surprisingly modified and improved by careful selection: in peacocks/display of distinct breeds; but the goose seems to have a singularly inflexible organisation.

310:f goose, under the conditions to which it is exposed when domesticated, seems to have a singularly inflexible organisation, though it has varied to a slight extent, as I have elsewhere described.

310:1-8:e[¶] Some authors have maintained that the amount of variation in our domestic productions is soon reached, and can never afterwards be exceeded. *2* It would be somewhat rash to assert that the limit has been attained in any one case; for almost all our animals and plants have been greatly improved in many ways within a recent period; and this implies variation. *3* It would be equally rash to assert that characters now increased to their utmost limit, could not, after remaining fixed for many centuries, again vary under new conditions of life. *4* No doubt, as Mr. Wallace has remarked with much truth, a limit will be at last reached. *5* For instance there must be a limit to the fleetness of any terrestrial animal, as this will be determined by the friction to be overcome, the weight of body to be carried, and the power of contraction in the muscular fibres. *6* But what concerns us is that the domestic varieties of the same species differ from each other in almost every character, which man has attended to and selected, more than do the distinct species of the same genera. *7* Isidore Geoffroy St. Hilaire has proved this in regard to size, and so it is with colour and probably with the length of hair. *8* With respect to fleetness, which depends on many bodily characteristics, Eclipse was far fleeter, and a dray-horse is incomparably stronger than any two equine species.

310.8:f bodily characters, Eclipse/than any two natural species belonging to the same genus.

310.9:e So with plants, the seeds of the different varieties of the bean or maize differ more in size, than do the seeds of the distinct species in any one genus of the same two families.

310.9:f maize probably differ

310.10:e The same remark holds good in regard to the fruit of the several varieties of the plum, and still more so with the melon, as well as in endless other analogous cases.

310.10:f more strongly with the melon, as well as in many other

311 　　To sum up on the origin of our Domestic Races of animals and plants.

311:c domestic races

312 I believe that the conditions of life, from their action on the reproductive system, are so far of the highest importance as causing variability.

312:e Changed conditions of life are of the highest importance in causing variability, both directly by acting on the organisation, and indirectly by affecting the reproductive system.

312:f both by acting directly on

313 I do not believe that variability is an inherent and necessary contingency, under all circumstances, with all organic beings, as some authors have thought.

117

313:c It is not probable that variability

313:e contingent, under all circumstances.

314 The effects of variability are modified by various degrees of inheritance and of reversion.

314:e The greater or less force of inheritance and reversion determine whether variations shall endure.

315 Variability is governed by many unknown laws, more especially by that of correlation of growth.

315:e correlation.

315:f laws, of which correlated growth is probably the most important.

316 Something may be attributed to the direct action of the conditions of life.

316:e to the definite action of the conditions of life, but how much we do not know.

316:f Something, but how much we do not know, may/life.

317 Something must be attributed to use and disuse.

317:f Some, perhaps a great, effect may be attributed to the increased use or disuse of parts.

318 The final result is thus rendered infinitely complex.

319 In some cases, I do not doubt that the intercrossing of species, aboriginally distinct, has played an important part in the origin of our domestic productions.

319:c cases the intercrossing/has probably played/domestic breeds.

319:e intercrossing of aboriginally distinct species, has probably

319:f species appears to have played/our breeds.

320 When in any country several domestic breeds have once been established, their occasional intercrossing, with the aid of selection, has, no doubt, largely aided in the formation of new subbreeds; but the importance of the crossing of varieties has, I believe, been greatly exaggerated, both in regard to animals and to those plants which are propagated by seed.

320:c varieties has been

320:e When several breeds have once been formed in any country, their/aided in forming new sub-breeds; but the importance of crossing has been much exaggerated

320:f aided in the formation of new

321 In plants which are temporarily propagated by cuttings, buds, &c., the importance of the crossing both of distinct species and of varieties is immense; for the cultivator here quite disregards the extreme variability both of hybrids and mongrels, and the frequent sterility of hybrids; but the cases of plants not propagated by seed are of little importance to us, for their endurance is only temporary.

321:e With plants which/of crossing is immense; for the cultivator

may here disregard the extreme variability both of hybrids and
of mongrels/but plants

321:f mongrels, and the sterility

322 Over all these causes of Change I am convinced that the accumulative action of Selection, whether applied methodically and
more quickly, or unconsciously and more slowly, but more efficiently, is by far the predominant Power.

322:e Change the accumulative/methodically and quickly, or unconsciously and slowly, but

322:f Change, the accumulative/slowly but more efficiently, seems
to have been the predominant

CHAPTER II.

3 Variability—Individual differences—Doubtful species—Wide ranging, much diffused, and common species vary most—Species of
the larger genera in any country vary more than the species of
the smaller genera—Many of the species of the larger genera resemble varieties in being very closely, but unequally, related to
each other, and in having restricted ranges.

3:d in each country vary more frequently than

3:f species, vary most—Species of the larger genera in each

4 Before applying the principles arrived at in the last chapter to organic beings in a state of nature, we must briefly discuss whether
these latter are subject to any variation.

5 To treat this subject at all properly, a long catalogue of dry facts
should be given; but these I shall reserve for my future work.

5:d facts ought to be

5:f subject properly/for a future

6 Nor shall I here discuss the various definitions which have been
given of the term species.

7 No one definition has as yet satisfied all naturalists; yet every
naturalist knows vaguely what he means when he speaks of a
species.

7:f has satisfied

8 Generally the term includes the unknown element of a distinct
act of creation.

9 The term "variety" is almost equally difficult to define; but here
community of descent is almost universally implied, though it
can rarely be proved.

10 We have also what are called monstrosities; but they graduate
into varieties.

11 By a monstrosity I presume is meant some considerable deviation
of structure in one part, either injurious to or not useful to the
species, and not generally propagated.

11:c part, generally injurious to or not useful to the species.

11:d structure, generally

11:f injurious, or not

12 Some authors use the term "variation" in a technical sense, as
implying a modification directly due to the physical conditions
of life; and "variations" in this sense are supposed not to be

inherited: but who can say that the dwarfed condition of shells in the brackish waters of the Baltic, or dwarfed plants on Alpine summits, or the thicker fur of an animal from far northwards, would not in some cases be inherited for at least some few generations? and in this case I presume that the form would be called a variety.

12:f inherited; but/least a few

12.1:c[¶] It may perhaps be doubted whether monstrosities, or such sudden and great deviations of structure as we occasionally see in our domestic productions, more especially with plants, are ever permanently propagated in a state of nature.

12.1:d may be doubted whether sudden and great deviations of structure such as

12.1:f sudden and considerable deviations

12.2:c Monsters are very apt to be sterile; and almost every part of every organic being, at least with animals, is so beautifully related to its complex conditions of life that it seems as improbable that any part should have been suddenly produced perfect, as that a complex machine should have been invented by man in a perfect state.

12.2:d Almost/being is

12.3-6:c I have not, at least, been able to find good cases of species in a state of nature presenting modifications of structure resembling monstrosities observed in allied forms. *4* If such have occurred, their perpetuation will have been due to their beneficial nature, so that natural selection will have come into play. *5* Many cases are known of plants which regularly produce on different branches, or on the circumference and in the centre of umbels, &c., flowers of a widely different structure; and if the plant ceased to produce flowers of the one kind, a great change might perhaps suddenly be effected in the specific character; but then we do not at present know by what steps, or for what good, a plant produces two kinds of flowers. *6* With cultivated plants, in the few cases known of a variety habitually bearing flowers or fruit slightly different from each other, the production of the variety has been sudden.

12.3-6:[d]

12.6.1:d Under domestication monstrosities often occur which are comparable with normal structures.

12.6.1:e which resemble normal structures in widely different animals.

12.6.1:f monstrosities sometimes occur

12.6.2-3:d Thus pigs have often been born with a sort of proboscis like that of the tapir or elephant. *3* Now, if any wild species of the pig-genus had naturally possessed a proboscis, it might have been argued that this had suddenly appeared as a monstrosity; but I have as yet failed, after diligent search, to find, in nearly allied forms, cases of monstrosities and of normal structures resembling each other.

12.6.3:e this in like manner had suddenly/failed to find, after diligent search, cases of monstrosities resembling normal structures in nearly allied forms, and these alone would bear on the question.

12.6.2 + 3:f Thus pigs have occasionally been born with a sort of proboscis, and if any wild species of the same genus had naturally/this had appeared/alone bear

12.6.4:d If monstrous forms of this kind ever do appear in a state of nature and are capable of propagation (which is not always the case), as they occur rarely and singly, they must be crossed with the ordinary form, and their character will be transmitted in a modified state.

12.6.4:e singly, their preservation would depend on unusually favourable circumstances.

12.6.5:d If perpetuated in this crossed state, their preservation will be almost necessarily due to the modification being in some way beneficial to the animal under its then existing conditions of life; so that, even in this case, natural selection will come into play. [*Space*]

12.6.5:[e]

12.6.5.1:e They would, also, during the first and succeeding generations cross with the ordinary form, and thus they would almost inevitably lose their abnormal character.

12.6.5.1:f thus their abnormal character would almost inevitably be lost.

12.6.5.2:e But I shall have to return in a future chapter to the preservation and perpetuation of occasional variations. [*Space*]

12.6.5.2:f perpetuation of single or occasional

12.6.6:d [*Center*] *Individual Differences.* [*Space*]

13   Again, we have many slight differences which may be called individual differences, such as are known frequently to appear in the offspring from the same parents, or which may be presumed to have thus arisen, from being frequently observed in the individuals of the same species inhabiting the same confined locality.

13:d The many slight differences which frequently appear in the offspring from the same parents, or which may be presumed to have thus arisen, from being frequently observed in the individuals of the same species inhabiting the same confined locality, may be called individual differences.

13:f which appear/or which it may be presumed have thus arisen, from being observed

14 No one supposes that all the individuals of the same species are cast in the very same mould.

14:d in the same actual mould.

15 These individual differences are highly important for us, as they afford materials for natural selection to accumulate, in the same manner as man can accumulate in any given direction individual differences in his domesticated productions.

15:c us, for they are often inherited, as must be familiar to every one; and thus they afford/man accumulates

15:d selection to act on and accumulate

16 These individual differences generally affect what naturalists consider unimportant parts; but I could show by a long catalogue of facts, that parts which must be called important, whether viewed under a physiological or classificatory point of view, sometimes vary in the individuals of the same species.

17 I am convinced that the most experienced naturalist would be surprised at the number of the cases of variability, even in important parts of structure, which he could collect on good authority, as I have collected, during a course of years.

18 It should be remembered that systematists are far from pleased at finding variability in important characters, and that there are not many men who will laboriously examine internal and important organs, and compare them in many specimens of the same species.

18:d from being pleased

19 I should never have expected that the branching of the main nerves close to the great central ganglion of an insect would have been variable in the same species; I should have expected that changes of this nature could have been effected only by slow degrees: yet quite recently Mr. Lubbock has shown a degree of variability in these main nerves in Coccus, which may almost be compared to the irregular branching of the stem of a tree.

19:c It would never have been expected/species; it might have been thought that changes/degrees; yet

19:d yet recently Sir J. Lubbock

19:f yet Sir

20 This philosophical naturalist, I may add, has also quite recently shown that the muscles in the larvæ of certain insects are very far from uniform.

20:d also recently

20:e are far

20:f also shown

21 Authors sometimes argue in a circle when they state that important organs never vary; for these same authors practically rank that character as important (as some few naturalists have honestly confessed) which does not vary; and, under this point of view, no instance of an important part varying will ever be found: but under any other point of view many instances assuredly can be given.

21:d instance will ever be found of an important part varying: but

21:e rank those parts as important/which do not

22   There is one point connected with individual differences, which seems to me extremely perplexing: I refer to those genera which have sometimes been called "protean" or "polymorphic," in which the species present an inordinate amount of variation; and hardly two naturalists can agree which forms to rank as species and which as varieties.

22:c which is extremely

22:e have been/variation; and about which hardly two naturalists agree whether to rank them as species or as varieties.

22.x-y:f variation. *y* With respect to many of these forms, hardly

23 We may instance Rubus, Rosa, and Hieracium amongst plants, several genera of insects, and several genera of Brachiopod shells.

23:d insects, several genera of Brachiopod shells, and the Ruff (Machetes pugnax) amongst birds.

23:f insects and of Brachiopod shells.

24 In most polymorphic genera some of the species have fixed and definite characters.

25 Genera which are polymorphic in one country seem to be, with some few exceptions, polymorphic in other countries, and likewise, judging from Brachiopod shells, at former periods of time.

25:f with a few

26 These facts seem to be very perplexing, for they seem to show that this kind of variability is independent of the conditions of life.

26:c facts are very

27 I am inclined to suspect that we see in these polymorphic genera variations in points of structure which are of no service or disservice to the species, and which consequently have not been seized on and rendered definite by natural selection, as hereafter will be explained.

27:e suspect that we have at least in some of these polymorphic genera variations which are of no service or disservice to the species, and which

27:f that we see, at/genera, variations

27.1:d[¶] Individuals of the same species often present great differences of structure, not directly connected with variability, as in the two sexes, as in the two or three castes of sterile females or workers amongst insects, and as in the immature and larval states of all animals.

27.1:e structure, as in the two sexes of various animals, in the two or three castes of sterile females or workers amongst insects, and in the immature and larval states of many of the lower animals.

27.1:f present, as is known to every one, great differences of structure, independently of variation, as in the two sexes

27.2:d There are, however, other cases, namely of dimorphism and trimorphism, which might easily be, and have frequently been, confounded with variability, but which are really quite distinct.

27.2:e but which are quite

27.2:f There are, also, cases of dimorphism and trimorphism, both with animals and plants.

27.3:d I refer to the two or three distinct forms, which certain animals of either sex, and certain hermaphrodite plants, habitually present.

27.3:e three different forms

27.3:[f]

27.4:d Thus, Mr. Wallace, who has lately called special attention to the subject, has shown that the females of certain species of butterflies, in the Malayan archipelago, regularly appear under two or even three conspicuously distinct forms, not connected together by intermediate varieties.

27.4:f called attention

27.5:d The winged and frequently wingless states of so many Hemipterous insects may probably be included as a case of dimorphism, and not of mere variability.

27.5:[f]

27.6:d Fritz Müller, also, has recently described analogous but more extraordinary cases in the males of certain Brazilian Crustaceans: thus, the male of a Tanais regularly occurs under two widely different forms, not connected by any intermediate links; one of these forms has much stronger and differently shaped pincers for seizing the female, and the other, as if for compensation, has antennæ much more abundantly furnished with smelling-hairs, so as to have a better chance of finding the female.

27.6:e cases with the males of certain

27.6:f Müller has described/two distinct forms; one of these has strong and differently shaped pincers, and the other has antennæ much more abundantly furnished with smelling-hairs.

27.7:d Again, the males of another Crustacean, an Orchestia, occur under two distinct forms, with pincers differing much more from each other in structure, than do the pincers of most species of the same genus.

27.7:[f]

27.8:d With respect to plants I have recently shown that species in several widely distinct orders present two or even three forms, which are abruptly distinguished from each other in several important points, as in the size and colour of the pollen-grains; and these forms, though all hermaphrodites, differ from each other in their reproductive power, so that for full fertility, or indeed in some cases for any fertility, they must reciprocally impregnate each other.

27.8:e shown that in several widely distinct orders the species present

27.8:[f]

27.9:d Although the forms of the few dimorphic and trimorphic animals and plants which have been studied are not now connected together by intermediate links, it is probable that this will be found to occur in other cases; for Mr. Wallace observed a certain butterfly which presented in the same island a great range of varieties connected by intermediate links, and the extreme links of the chain closely resembled the two forms of an allied dimorphic species inhabiting another part of the Malay archipelago.

27.9.x-y:f Although in most of these cases, the two or three forms, both with animals and plants, are not now connected by intermediate gradations, it is probable that they were once thus connected. y Mr. Wallace, for instance, describes a certain butterfly which presents in the same island/resemble

27.10:d Thus also with ants, the several worker-castes are generally quite distinct; but in some cases, as we shall hereafter see, the castes are connected together by graduated varieties.

27.10:f by finely graduated varieties.

27.10.1:f So it is, as I have myself observed, with some dimorphic plants.

27.11:d It certainly at first appears a highly remarkable fact that the same female butterfly should have the power of producing at the same time three distinct female forms and a male; that a male Crustacean should generate two male forms and a female form, all widely different from each other; and that an hermaphrodite plant should produce from the same seed-capsule three distinct hermaphrodite forms, bearing three different kinds of females and three or even six different kinds of males.

27.11:f male; and that an hermaphrodite plant

27.12:d Nevertheless these cases are only exaggerations of the universal fact that every female animal produces males and females, which in some instances differ in so wonderful a manner from each other. [Space]

27.12:f of the common fact that the female produces offspring of two sexes which sometimes differ from each other in a wonderful manner.

27.13:d [Center] Doubtful Species. [Space]

28   Those forms which possess in some considerable degree the character of species, but which are so closely similar to some other forms, or are so closely linked to them by intermediate gradations, that naturalists do not like to rank them as distinct species, are in several respects the most important for us.

28:d The forms which possess/to other   ·

126

29 We have every reason to believe that many of these doubtful and closely-allied forms have permanently retained their characters in their own country for a long time; for as long, as far as we know, as have good and true species.

29:f closely allied/characters for a long time; for

30 Practically, when a naturalist can unite two forms together by others having intermediate characters, he treats the one as a variety of the other, ranking the most common, but sometimes the one first described, as the species, and the other as the variety.

30:f unite by means of intermediate links any two forms, he/other; ranking

31 But cases of great difficulty, which I will not here enumerate, sometimes occur in deciding whether or not to rank one form as a variety of another, even when they are closely connected by intermediate links; nor will the commonly-assumed hybrid nature of the intermediate links always remove the difficulty.

31:f sometimes arise in/nature of the intermediate forms always

32 In very many cases, however, one form is ranked as a variety of another, not because the intermediate links have actually been found, but because analogy leads the observer to suppose either that they do now somewhere exist, or may formerly have existed; and here a wide door for the entry of doubt and conjecture is opened.

33 Hence, in determining whether a form should be ranked as a species or a variety, the opinion of naturalists having sound judgment and wide experience seems the only guide to follow.

34 We must, however, in many cases, decide by a majority of naturalists, for few well-marked and well-known varieties can be named which have not been ranked as species by at least some competent judges.

35 That varieties of this doubtful nature are far from uncommon cannot be disputed.

36 Compare the several floras of Great Britain, of France or of the United States, drawn up by different botanists, and see what a surprising number of forms have been ranked by one botanist as good species, and by another as mere varieties.

36:c France, or

37 Mr. H. C. Watson, to whom I lie under deep obligation for assistance of all kinds, has marked for me 182 British plants, which are generally considered as varieties, but which have all been ranked by botanists as species; and in making this list he has omitted many trifling varieties, but which nevertheless have been ranked by some botanists as species, and he has entirely omitted several highly polymorphic genera.

38 Under genera, including the most polymorphic forms, Mr.

Babington gives 251 species, whereas Mr. Bentham gives only 112,—a difference of 139 doubtful forms!

39 Amongst animals which unite for each birth, and which are highly locomotive, doubtful forms, ranked by one zoologist as a species and by another as a variety, can rarely be found within the same country, but are common in separated areas.

40 How many of those birds and insects in North America and Europe, which differ very slightly from each other, have been ranked by one eminent naturalist as undoubted species, and by another as varieties, or, as they are often called, as geographical races!

40:f of the birds/called, geographical

40.1:d Mr. Wallace, in several valuable papers lately published on the various animals, especially on the Lepidoptera, inhabiting the islands of the great Malayan archipelago, shows that they may be classed under variable and under local forms, under geographical races or sub-species, and as true representative species.

40.1:e papers on the various/classed under four heads, namely, as variable forms, as local forms, as geographical races

40.2:d The variable forms vary much within the limits of the same island.

40.2:e The first or variable

40.3:d The local forms of the same species are moderately constant and distinct in each separate island; but when all such forms from the several islands are compared, the differences become so slight, numerous, and graduated, that it is impossible to define or describe many of them, though at the same time the extreme forms are sufficiently distinct.

40.3:e forms are/all the forms from the several islands are compared, the differences are seen to be so slight and graduated, that it is impossible to define or describe them, though

40.3:f all from the several islands are compared together, the differences

40.4:d The geographical races or sub-species are local forms completely fixed and isolated; but as they do not differ from each other by strongly marked and important characters, "there is no possible test but individual opinion to determine which of them shall be considered as species and which as varieties."

40.5:d Lastly, representative species fill the same place in the natural economy of each island as do the local forms and sub-species; but as they are distinguished from each other by a greater, though not by a definite, amount of difference than that between the local forms and sub-species, they are almost universally ranked by naturalists as true species.

40.5:e greater amount

40.6:d Nevertheless, no certain criterion can possibly be given by which variable forms, local forms, sub-species, and representative species can be recognised.

41 Many years ago, when comparing, and seeing others compare, the birds from the separate islands of the Galapagos Archipelago, both one with another, and with those from the American mainland, I was much struck how entirely vague and arbitrary is the distinction between species and varieties.

41:d[¶] from the closely neighbouring islands of the Galapagos

41:e archipelago

41:f archipelago, one

42 On the islets of the little Madeira group there are many insects which are characterized as varieties in Mr. Wollaston's admirable work, but which it cannot be doubted would be ranked as distinct species by many entomologists.

42:c but which certainly would

42:e but which would certainly be

43 Even Ireland has a few animals, now generally regarded as varieties, but which have been ranked as species by some zoologists.

44 Several most experienced ornithologists consider our British red grouse as only a strongly-marked race of a Norwegian species, whereas the greater number rank it as an undoubted species peculiar to Great Britain.

44:e Several experienced

45 A wide distance between the homes of two doubtful forms leads many naturalists to rank both as distinct species; but what distance, it has been well asked, will suffice? if that between America and Europe is ample, will that between the Continent and the Azores, or Madeira, or the Canaries, or Ireland, be sufficient?

45:d will that between Europe and the Azores, or Madeira, or the Canaries, or between the several islets in each of these small archipelagos, be sufficient?

45:e rank them as/suffice; if

45.0.1:d[¶] Mr. B. D. Walsh, a distinguished entomologist of the United States, has lately called attention to some cases, analogous with those of local forms and geographical races, yet very different from them.

45.0.2:d These cases he has fully described under the terms of Phytophagic varieties and Phytophagic species.

45.0.1 + 2:e lately described what he calls Phytophagic varieties

45.0.1 + 2:f has described

45.0.3:d Most vegetable-feeding insects live on one kind of plant or on one group of plants; but some feed indiscriminately on many widely distinct kinds, yet this induces no change in them.

45.0.3:e group of plants; some feed indiscriminately on many kinds, but do not in consequence vary.

45.0.4:d Mr. Walsh, however, has observed other cases in which either the larva or mature insect, or both states, are thus affected by slight, though constant, differences in colour, or size, or nature of their secretions.

45.0.4:e In several cases, however, insects found living on different plants have been observed by Mr. Walsh to present, either exclusively in their larval or mature state, or in both states, slight, though constant differences in colour, size, or in the nature

45.0.4:f plants, have been observed by Mr. Walsh to present in their larval

45.0.5-6:d In one case difference in food was accompanied by several slight but constant structural differences in the mature male alone. *6* In other cases both males and females are thus slightly affected.

45.0.5 + 6:e In some instances the males alone, in other instances both males and females, have been observed to be thus affected in a slight degree.

45.0.5 + 6:f observed thus to differ in a slight

45.0.7:d Lastly, differences of food apparently cause more marked and constant differences in colour or structure, or in both combined, in the larva and in the mature insect.

45.0.7:[e]

45.0.8:d Forms modified to this degree are ranked by all entomologists as distinct, though allied, species of the same genus.

45.0.8:e When the differences are rather more strongly marked, and when both sexes and all ages are affected, the forms would be ranked by all entomologists as species.

45.0.8:f forms are ranked by all entomologists as good species.

45.0.9:d The slighter differences, as in colour alone, and confined to the larva alone, or to the mature insect alone, are almost invariably looked at as mere varieties.

45.0.9:[e]

45.0.10:d But no man can draw the line for others, even if he can do so for himself, and determine with certainty which of the several phytophagic forms to call varieties and which to call species.

45.0.10:e But no observer can determine for others, even if he can do so for himself, which of these Phytophagic forms ought to be called species and which varieties.

45.0.10:f for another, even

45.0.11:d Mr. Walsh, who argues with much force that the different states have gradually passed into each other, is forced to assume that those forms, which it may be supposed would freely intercross, should be designated as varieties, whilst those which have probably lost this capacity for intercrossing should be called species.

45.0.11:e Mr. Walsh ranks the forms which it may be supposed would freely intercross together, as varieties; and those which appear to have lost this power, as species.

45.0.11:f intercross, as varieties

45.0.12:d As the difference in all these cases clearly depends on the insects having long fed on perfectly distinct plants, intermediate links between the several forms thus produced cannot be expected to be found; though formerly such must have existed, connecting the present divergent forms with their common progenitor.

45.0.12:e As the differences depend on the insects having long fed on distinct plants, it cannot be expected that intermediate links connecting the several forms should now be found.

45.0.13:d The naturalist thus loses his best guide in determining whether to rank such doubtful forms as varieties or species.

45.0.13:f rank doubtful

45.0.14:d This likewise necessarily occurs with closely allied organisms, of doubtful value, which inhabit separate continents or distant islands.

45.0.14:e organisms, which inhabit distinct continents or islands.

45.0.15:d But when an animal or plant ranges over the same continent or inhabits many islands in the same archipelago, and presents different forms in the different areas, there is always a chance, which is not rarely successful, that intermediate forms may be discovered which will link together the extreme states; and these are then degraded to the rank of varieties.

45.0.15:e When, on the other hand, an animal/continent, or/always a good chance that

45.0.15:f intermediate forms will be discovered which will link

45.1:c Some few naturalists maintain that animals never present varieties; but then these same naturalists rank the slightest differences as of specific value; and when even the same identical form is met with in two distant countries, or in two quite distinct geological formations, they go so far as to believe that two separate species are hidden under the same dress.

45.1:d[¶] Some/difference/even the identically same form is met with in two distant countries, or in two distinct

45.1:e differences as of specific value; and when the same identical form is met with in two distant countries, or in two geological formations, they believe that two distinct species

45.1:f difference

45.1.1:d The term species thus comes to be a mere useless mental abstraction, implying and assuming a separate act of creation.

45.1.1:f useless abstraction

46 It must be admitted that many forms, considered by highly-competent judges as varieties, have so perfectly the character of species that they are ranked by other highly competent judges as good and true species.

46:b highly-competent

46:c Finally, it cannot be disputed that many

46:d It cannot, however, be disputed/they have been ranked

46:e It is certain that many/varieties, so completely resemble species in character, that they have been thus ranked by other highly-competent judges.

46:f judges to be varieties, resemble species so completely in character

47 But to discuss whether they are rightly called species or varieties, before any definition of these terms has been generally accepted, is vainly to beat the air.

47:c whether such slightly different forms are

47:e whether they ought to be called

48   Many of the cases of strongly-marked varieties or doubtful species well deserve consideration; for several interesting lines of argument, from geographical distribution, analogical variation, hybridism, &c., have been brought to bear on the attempt to determine their rank.

48:d rank; but space does not here permit me to discuss them.

49 I will here give only a single instance,—the well-known one of the primrose and cowslip, or Primula veris and elatior.

49:b Primula vulgaris and veris.

49:[d]

50 These plants differ considerably in appearance; they have a different flavour and emit a different odour; they flower at slightly different periods; they grow in somewhat different stations; they ascend mountains to different heights; they have different geographical ranges; and lastly, according to very numerous experiments made during several years by that most careful observer Gärtner, they can be crossed only with much difficulty.

50:b flavour, and

50:[d]

51 We could hardly wish for better evidence of the two forms being specifically distinct.

51:[d]

52 On the other hand, they are united by many intermediate links, and it is very doubtful whether these links are hybrids; and there is, as it seems to me, an overwhelming amount of experimental evidence, showing that they descend from common parents, and consequently must be ranked as varieties.

52:c there is a large amount

52:[d]

53   Close investigation, in most cases, will bring naturalists to an agreement how to rank doubtful forms.

53:d[No ¶] in many cases

53:f will no doubt bring naturalists to agree how

54 Yet it must be confessed, that it is in the best-known countries that we find the greatest number of forms of doubtful value.

54:b confessed that it

54:d best known

54:f number of them.

55 I have been struck with the fact, that if any animal or plant in a state of nature be highly useful to man, or from any cause closely attract his attention, varieties of it will almost universally be found recorded.

55:f attracts

56 These varieties, moreover, will be often ranked by some authors as species.

56:d will often be ranked

57 Look at the common oak, how closely it has been studied; yet a German author makes more than a dozen species out of forms, which are very generally considered as varieties; and in this country the highest botanical authorities and practical men can be quoted to show that the sessile and pedunculated oaks are either good and distinct species or mere varieties.

57:c are almost universally considered

57:f considered by other botanists to be varieties; and in

57.1-3:d[¶] I may here allude to a remarkable memoir lately published by A. de Candolle, on the oaks of the whole world. *2* No one ever had more ample materials for the discrimination of the species, or could have worked on them with more zeal and sagacity. *3* He first gives in detail all the many points of structure which vary in the species, and estimates numerically the relative frequency of the variations.

57.3:f vary in the several species

57.4-5:d He specifies above a dozen characters which may be found varying even on the same branch, sometimes according to age or development, sometimes without any assignable reason. *5* Such characters of course are not of specific value, but they are, as Asa Gray has remarked in commenting on this memoir, such as generally enter into specific definitions.

57.5:f characters are not of course of specific

57.6-12:d De Candolle then goes on to say that he gives the rank of species to the forms that differ by characters never varying on the same tree, and never found connected by intermediate states. *7* After this discussion, the result of so much labour, he emphatically remarks: "They are mistaken, who repeat that the greater part of our species are clearly limited, and that the doubtful species are in a feeble minority. *8* This seemed to be true, so long as a genus was imperfectly known, and its species were founded upon a few specimens, that is to say, were provisional. *9* Just as we come to know them better, intermediate forms flow in, and doubts as to specific limits augment." *10* He also adds that it is the best known species which present the greatest number of spontaneous varieties and sub-varieties. *11* Thus Quercus robur has twenty-eight varieties, all of which, excepting six, are clustered round three sub-species, namely, Q. pedunculata, sessiliflora, and pubescèns. *12* The forms which connect these three sub-species are comparatively rare; and, as Asa Gray remarks, if these connecting forms, which are now rare, were to become extinct, the three sub-species would hold exactly the same relation to each other, as do the four or five provisionally admitted species which closely surround the typical Quercus robur.

57.12:e become wholly extinct

57.13-14:d Finally, De Candolle admits that out of the 300 species, which will be enumerated in his Prodromus as belonging to the oak family, at least two-thirds are provisional species, that is, are not known strictly to fulfil the definition above given of a true species. *14* For it should be added that De Candolle no longer believes that species are immutable creations, but concludes that the derivative theory of the succession of forms is the most natural one, "and the most accordant with the known facts in palæontology, geographical botany and zoology, of anatomical structure and classification;" but, he adds, direct proof is still wanting.

57.14:f It/theory is the most/classification."

58   When a young naturalist commences the study of a group of organisms quite unknown to him, he is at first much perplexed to determine what differences to consider as specific, and what as varieties; for he knows nothing of the amount and kind of variation to which the group is subject; and this shows, at least, how very generally there is some variation.

58:d and what as varietal; for

58:f perplexed in determining what differences

59   But if he confine his attention to one class within one country, he will soon make up his mind how to rank most of the doubtful forms.

60 His general tendency will be to make many species, for he will become impressed, just like the pigeon or poultry-fancier before alluded to, with the amount of difference in the forms which he is continually studying; and he has little general knowledge of analogical variation in other groups and in other countries, by which to correct his first impressions.

60:b poultry fancier

61 As he extends the range of his observations, he will meet with more cases of difficulty; for he will encounter a greater number of closely-allied forms.

62 But if his observations be widely extended, he will in the end generally be enabled to make up his own mind which to call

varieties and which species; but he will succeed in this at the expense of admitting much variation,—and the truth of this admission will often be disputed by other naturalists.

62:f mind; but

63 When, moreover, he comes to study allied forms brought from countries not now continuous, in which case he can hardly hope to find the intermediate links between his doubtful forms, he will have to trust almost entirely to analogy, and his difficulties will rise to a climax.

63:b difficulties rise

63:d case he cannot hope

63:f When he comes/find intermediate links, he will be compelled to trust

64 Certainly no clear line of demarcation has as yet been drawn between species and sub-species—that is, the forms which in the opinion of some naturalists come very near to, but do not quite arrive at the rank of species; or, again, between sub-species and well-marked varieties, or between lesser varieties and individual differences.

64:d at, the rank

65 These differences blend into each other in an insensible series; and a series impresses the mind with the idea of an actual passage.

65:f other by an insensible

66 Hence I look at individual differences, though of small interest to the systematist, as of high importance for us, as being the first step towards such slight varieties as are barely thought worth recording in works on natural history.

66:e as of the highest importance/steps

67 And I look at varieties which are in any degree more distinct and permanent, as steps leading to more strongly marked and more permanent varieties; and at these latter, as leading to sub-species, and to species.

67:e steps towards more strongly-marked and permanent varieties; and at the latter, as leading to sub-species or species.

67:f sub-species, and then to species.

68 The passage from one stage of difference to another and higher stage may be, in some cases, due merely to the long-continued action of different physical conditions in two different regions; but I have not much faith in this view; and I attribute the passage of a variety, from a state in which it differs very slightly from its parent to one in which it differs more, to the action of natural selection in accumulating (as will hereafter be more fully explained) differences of structure in certain definite directions.

68:e The passages from one stage of difference to another may, in some cases, be the simple result of the long-continued action of different physical conditions; but in most cases they may be attributed to the gradual accumulative action of natural selection, as hereafter to be more fully explained, on fluctuating variability.

68:f passage/may, in many cases, be the simple result of the nature of the organism and of the different physical conditions to which it has long been exposed; but with respect to the more important and adaptive characters, the passage from one stage of difference to another, may be safely attributed to the cumulative action of natural selection, hereafter to be explained, and to the effects of the increased use or disuse of parts.

69 Hence I believe a well-marked variety may be justly called an incipient species; but whether this belief be justifiable must be judged of by the general weight of the several facts and views given throughout this work.

69:b be called

69:e belief is justifiable

69:f A well-marked variety may therefore be called/judged by the weight/considerations to be given

70   It need not be supposed that all varieties or incipient species necessarily attain the rank of species.

70:f species attain

71 They may whilst in this incipient state become extinct, or they may endure as varieties for very long periods, as has been shown to be the case by Mr. Wollaston with the varieties of certain fossil land-shells in Madeira.

71:e may become/Madeira, and with plants by Gaston de Saporta.

72 If a variety were to flourish so as to exceed in numbers the parent species, it would then rank as the species, and the species as the variety; or it might come to supplant and exterminate the parent species; or both might co-exist, and both rank as independent species.

73 But we shall hereafter have to return to this subject.

73:e hereafter return

74   From these remarks it will be seen that I look at the term species, as one arbitrarily given for the sake of convenience to a set of individuals closely resembling each other, and that it does not essentially differ from the term variety, which is given to less distinct and more fluctuating forms.

74:d species as one arbitrarily given, for the sake of convenience, to a set

75 The term variety, again, in comparison with mere individual differences, is also applied arbitrarily, and for mere convenience sake.

75:b convenience'

75:d sake. [*Space*]

75:e arbitrarily, for convenience' sake.

75.1:d [*Center*] *Wide-ranging, much-diffused, and common Species vary most.* [*Space*]

76 Guided by theoretical considerations, I thought that some interesting results might be obtained in regard to the nature and relations of the species which vary most, by tabulating all the varieties in several well-worked floras.

77 At first this seemed a simple task; but Mr. H. C. Watson, to whom I am much indebted for valuable advice and assistance on this subject, soon convinced me that there were many difficulties, as did subsequently Dr. Hooker, even in stronger terms.

78 I shall reserve for my future work the discussion of these difficulties, and the tables themselves of the proportional numbers of the varying species.

78:f for a future

79 Dr. Hooker permits me to add, that after having carefully read my manuscript, and examined the tables, he thinks that the following statements are fairly well established.

80 The whole subject, however, treated as it necessarily here is with much brevity, is rather perplexing, and allusions cannot be avoided to the "struggle for existence," "divergence of character," and other questions, hereafter to be discussed.

81 Alph. De Candolle and others have shown that plants which have very wide ranges generally present varieties; and this might have been expected, as they become exposed to diverse physical conditions, and as they come into competition (which, as we shall hereafter see, is a far more important circumstance) with different sets of organic beings.

81:c is a more

81:e is an equally or more

81:f they are exposed

82 But my tables further show that, in any limited country, the species which are most common, that is abound most in individuals, and the species which are most widely diffused within their own country (and this is a different consideration from wide range, and to a certain extent from commonness), often give rise to varieties sufficiently well-marked to have been recorded in botanical works.

82:c oftenest

83 Hence it is the most flourishing, or, as they may be called, the dominant species,—those which range widely over the world, are the most diffused in their own country, and are the most numerous in individuals,—which oftenest produce well-marked varieties, or, as I consider them, incipient species.

83:c widely, are the most diffused

84 And this, perhaps, might have been anticipated; for, as varieties, in order to become in any degree permanent, necessarily have to struggle with the other inhabitants of the country, the species which are already dominant will be the most likely to yield offspring which, though in some slight degree modified, will still inherit those advantages that enabled their parents to become dominant over their compatriots.

84:b offspring, which/modified, still

84.1:c In these remarks on predominance, it should be understood that reference is made only to those forms which come into competition with each other, and more especially to the members of the same genus or class having nearly similar habits of life.

84.1:f only to the forms

84.2:c With respect to commonness or the number of individuals of any species, the comparison of course relates only to the members of the same group.

84.2:f to the number of individuals or commonness of species

84.3:c A plant may be said to be dominant if it be more numerous in individuals and more widely diffused than the other plants of the same country, not living under widely different conditions of life.

84.3:f One of the higher plants may be said/country, which live under nearly the same conditions.

84.4:c Such a plant is not the less dominant in the sense here used, because some conferva inhabiting the water or some parasitic fungus is infinitely more numerous in individuals and more widely diffused; if one kind of conferva or parasitic fungus exceeded its allies in the above respects, it would be a dominant form within its own class.

84.4:d [Space]

84.4.x-y:f A plant of this kind is not the less dominant because/individuals, and more widely diffused. *y* But if the conferva or parasitic fungus exceeds its allies in the above respects, it will then be dominant within

84.4.1:d [Center] *Species of the Larger Genera in each Country vary more frequently than the Species of the Smaller Genera.* [Space]

85   If the plants inhabiting a country and described in any Flora be divided into two equal masses, all those in the larger genera being placed on one side, and all those in the smaller genera on the other side, a somewhat larger number of the very common and much diffused or dominant species will be found on the side of the larger genera.

85:e genera (*i. e.,* those including many species) being

85:f country, as described in any Flora, be divided/other side, the former will be found to include a somewhat larger number of the very common and much diffused or dominant species.

86 This, again, might have been anticipated; for the mere fact of many species of the same genus inhabiting any country, shows that there is something in the organic or inorganic conditions of that country favourable to the genus; and, consequently, we might have expected to have found in the larger genera, or those including many species, a large proportional number of dominant species.

86:f This might have been

87 But so many causes tend to obscure this result, that I am surprised that my tables show even a small majority on the side of the larger genera.

88 I will here allude to only two causes of obscurity.

89 Fresh-water and salt-loving plants have generally very wide ranges and are much diffused, but this seems to be connected with the nature of the stations inhabited by them, and has little or no relation to the size of the genera to which the species belong.

89:f plants generally have very

90 Again, plants low in the scale of organisation are generally much more widely diffused than plants higher in the scale; and here again there is no close relation to the size of the genera.

91 The cause of lowly-organised plants ranging widely will be discussed in our chapter on geographical distribution.

91:c Geographical Distribution.

92 From looking at species as only strongly-marked and well-defined varieties, I was led to anticipate that the species of the larger genera in each country would oftener present varieties, than the species of the smaller genera; for wherever many closely related species (*i. e.,* species of the same genus) have been formed, many varieties or incipient species ought, as a general rule, to be now forming.

93 Where many large trees grow, we expect to find saplings.

94 Where many species of a genus have been formed through variation, circumstances have been favourable for variation; and hence we might expect that the circumstances would generally be still favourable to variation.

95 On the other hand, if we look at each species as a special act of creation, there is no apparent reason why more varieties should occur in a group having many species, than in one having few.

96 To test the truth of this anticipation I have arranged the plants of twelve countries, and the coleopterous insects of two districts, into two nearly equal masses, the species of the larger genera on one side, and those of the smaller genera on the other side, and it has invariably proved to be the case that a larger proportion of the species on the side of the larger genera present varieties, than on the side of the smaller genera.

96:e presented

97 Moreover, the species of the large genera which present many varieties, invariably present a larger average number of varieties than do the species of the small genera.

98 Both these results follow when another division is made, and when all the smallest genera, with from only one to four species, are absolutely excluded from the tables.

98:d all the least genera

98:f are altogether excluded

99 These facts are of plain signification on the view that species are only strongly marked and permanent varieties; for wherever many species of the same genus have been formed, or where, if we may use the expression, the manufactory of species has been active, we ought generally to find the manufactory still in action, more especially as we have every reason to believe the process of manufacturing new species to be a slow one.

99:d strongly-marked

100 And this certainly is the case, if varieties be looked at as incipient species; for my tables clearly show as a general rule that, wherever many species of a genus have been formed, the species of that genus present a number of varieties, that is of incipient species, beyond the average.

100:f certainly holds true, if

101 It is not that all large genera are now varying much, and are thus increasing in the number of their species, or that no small genera are now varying and increasing; for if this had been so, it would have been fatal to my theory; inasmuch as geology plainly tells us that small genera have in the lapse of time often increased greatly in size; and that large genera have often come to their maxima, declined, and disappeared.

102 All that we want to show is, that where many species of a genus have been formed, on an average many are still forming; and this holds good.

102:d that, where/this certainly holds good. [*Space*]

102.1:d [*Center*] *Many of the Species of the Larger Genera resemble Varieties in being very closely, but unequally, related to each other, and in having restricted ranges.* [*Space*]

102.1:e *Many of the Species included within the Larger*

103 There are other relations between the species of large genera and their recorded varieties which deserve notice.

104 We have seen that there is no infallible criterion by which to distinguish species and well-marked varieties; and in those cases in which intermediate links have not been found between doubtful forms, naturalists are compelled to come to a determination by the amount of difference between them, judging

by analogy whether or not the amount suffices to raise one or both to the rank of species.

104:f varieties; and when intermediate

105 Hence the amount of difference is one very important criterion in settling whether two forms should be ranked as species or varieties.

106 Now Fries has remarked in regard to plants, and Westwood in regard to insects, that in large genera the amount of difference between the species is often exceedingly small.

107 I have endeavoured to test this numerically by averages, and, as far as my imperfect results go, they always confirm the view.

107:b they confirm

108 I have also consulted some sagacious and most experienced observers, and, after deliberation, they concur in this view.

108:b and experienced

109 In this respect, therefore, the species of the larger genera resemble varieties, more than do the species of the smaller genera.

109:d genera, resemble

109:e resemble varieties

110 Or the case may be put in another way, and it may be said, that in the larger genera, in which a number of varieties or incipient species greater than the average are now manufacturing, many of the species already manufactured still to a certain extent resemble varieties, for they differ from each other by a less than usual amount of difference.

110:c than the usual

110:f by less

111   Moreover, the species of the large genera are related to each other, in the same manner as the varieties of any one species are related to each other.

112 No naturalist pretends that all the species of a genus are equally distinct from each other; they may generally be divided into sub-genera, or sections, or lesser groups.

113 As Fries has well remarked, little groups of species are generally clustered like satellites around certain other species.

113:f around other

114 And what are varieties but groups of forms, unequally related to each other, and clustered round certain forms—that is, round their parent-species?

115 Undoubtedly there is one most important point of difference between varieties and species; namely, that the amount of difference between varieties, when compared with each other or with their parent-species, is much less than that between the species of the same genus.

116 But when we come to discuss the principle, as I call it, of Divergence of Character, we shall see how this may be explained, and how the lesser differences between varieties will tend to increase into the greater differences between species.

117   There is one other point which seems to me worth notice.

117:c which is worth

118 Varieties generally have much restricted ranges: this statement is indeed scarcely more than a truism, for if a variety were found to have a wider range than that of its supposed parent-species, their denominations ought to be reversed.

118:d for, if/denominations would be reversed.

119 But there is also reason to believe, that those species which are very closely allied to other species, and in so far resemble varieties, often have much restricted ranges.

119:c believe that

119:f that the species

120 For instance, Mr. H. C. Watson has marked for me in the well-sifted London Catalogue of plants (4th edition) 63 plants which are therein ranked as species, but which he considers as so closely allied to other species as to be of doubtful value: these 63 reputed species range on an average over 6.9 of the provinces into which Mr. Watson has divided Great Britain.

121 Now, in this same catalogue, 53 acknowledged varieties are recorded, and these range over 7.7 provinces; whereas, the species to which these varieties belong range over 14.3 provinces.

122 So that the acknowledged varieties have very nearly the same restricted average range, as have those very closely allied forms, marked for me by Mr. Watson as doubtful species, but which are almost universally ranked by British botanists as good and true species.

122.1:d [Center] Summary. [Space]

123   Finally, then, varieties have the same general characters as species, for they cannot be distinguished from species,—except, firstly, by the discovery of intermediate linking forms, and the occurrence of such links cannot affect the actual characters of the forms which they connect; and except, secondly, by a certain amount of difference, for two forms, if differing very little, are generally ranked as varieties, notwithstanding that intermediate linking forms have not been discovered; but the amount of difference considered necessary to give to two forms the rank of species is quite indefinite.

123:b secondly by

123:c varieties cannot be/intermediate forms linking them together, and the occurrence of such links does not affect the character of the forms

123:e except, first, by the discovery of intermediate linking forms;

and, secondly, by a certain indefinite amount of difference between them; for two/that they cannot be closely connected; but/give to any two forms the rank of species cannot be defined.

123:f Finally, varieties cannot be

124 In genera having more than the average number of species in any country, the species of these genera have more than the average number of varieties.

125 In large genera the species are apt to be closely, but unequally, allied together, forming little clusters round certain species.

125:b unequally allied

125:c certain other species.

125:d unequally, allied

125:f round other

126 Species very closely allied to other species apparently have restricted ranges.

127 In all these several respects the species of large genera present a strong analogy with varieties.

127:f these respects

128 And we can clearly understand these analogies, if species have once existed as varieties, and have thus originated: whereas, these analogies are utterly inexplicable if each species has been independently created.

128:e species once existed as varieties, and thus/inexplicable if species are independent creations.

129 We have, also, seen that it is the most flourishing and dominant species of the larger genera which on an average vary most; and varieties, as we shall hereafter see, tend to become converted into new and distinct species.

129:b flourishing or dominant

129:c genera within each class which

129:d average yield the greatest number of varieties; and varieties, as

130 The larger genera thus tend to become larger; and throughout nature the forms of life which are now dominant tend to become still more dominant by leaving many modified and dominant descendants.

130:d Thus the larger genera tend to become larger

131 But by steps hereafter to be explained, the larger genera also tend to break up into smaller genera.

132 And thus, the forms of life throughout the universe become divided into groups subordinate to groups.

CHAPTER III.

3 Bears on natural selection—The term used in a wide sense—Geometrical powers of increase—Rapid increase of naturalised animals and plants—Nature of the checks to increase—Competition universal—Effects of climate—Protection from the number of individuals—Complex relations of all animals and plants throughout nature—Struggle for life most severe between individuals and varieties of the same species; often severe between species of the same genus—The relation of organism to organism the most important of all relations.

3:c Its bearing on natural/Geometrical ratio of increase—Rapid

4 BEFORE entering on the subject of this chapter, I must make a few preliminary remarks, to show how the struggle for existence bears on Natural Selection.

5 It has been seen in the last chapter that amongst organic beings in a state of nature there is some individual variability; indeed I am not aware that this has ever been disputed.

6 It is immaterial for us whether a multitude of doubtful forms be called species or sub-species or varieties; what rank, for instance, the two or three hundred doubtful forms of British plants are entitled to hold, if the existence of any well-marked varieties be admitted.

7 But the mere existence of individual variability and of some few well-marked varieties, though necessary as the foundation for the work, helps us but little in understanding how species arise in nature.

8 How have all those exquisite adaptations of one part of the organization to another part, and to the conditions of life, and of one distinct organic being to another being, been perfected?

8:c life, and of one organic being to

9 We see these beautiful co-adaptations most plainly in the woodpecker and missletoe; and only a little less plainly in the humblest parasite which clings to the hairs of a quadruped or feathers of a bird; in the structure of the beetle which dives through the water; in the plumed seed which is wafted by the gentlest breeze; in short, we see beautiful adaptations everywhere and in every part of the organic world.

9:c and the mistletoe

10 Again, it may be asked, how is it that varieties, which I have called incipient species, become ultimately converted into good

and distinct species, which in most cases obviously differ from each other far more than do the varieties of the same species?

11 How do those groups of species, which constitute what are called distinct genera, and which differ from each other more than do the species of the same genus, arise?

12 All these results, as we shall more fully see in the next chapter, follow inevitably from the struggle for life.

12:b follow from

13 Owing to this struggle for life, any variation, however slight and from whatever cause proceeding, if it be in any degree profitable to an individual of any species, in its infinitely complex relations to other organic beings and to external nature, will tend to the preservation of that individual, and will generally be inherited by its offspring.

13:c beings and to its physical conditions of life, will tend

13:e struggle, variations, however slight, and from whatever cause proceeding, if they be in any degree profitable to the individuals of a species, in their infinitely complex relations to other organic beings and to their physical conditions of life, will tend to the preservation of such individuals, and will generally be inherited by the offspring.

14 The offspring, also, will thus have a better chance of surviving, for, of the many individuals of any species which are periodically born, but a small number can survive.

15 I have called this principle, by which each slight variation, if useful, is preserved, by the term of Natural Selection, in order to mark its relation to man's power of selection.

15.1:e But the expression often used by Mr. Herbert Spencer of the Survival of the Fittest is more accurate, and is sometimes equally convenient.

16 We have seen that man by selection can certainly produce great results, and can adapt organic beings to his own uses, through the accumulation of slight but useful variations, given to him by the hand of Nature.

17 But Natural Selection, as we shall hereafter see, is a power incessantly ready for action, and is as immeasurably superior to man's feeble efforts, as the works of Nature are to those of Art.

18 We will now discuss in a little more detail the struggle for existence.

19 In my future work this subject shall be treated, as it well deserves, at much greater length.

19:e at greater length.

20 The elder De Candolle and Lyell have largely and philosophically shown that all organic beings are exposed to severe competition.

21 In regard to plants, no one has treated this subject with more

spirit and ability than W. Herbert, Dean of Manchester, evidently the result of his great horticultural knowledge.

22 Nothing is easer than to admit in words the truth of the universal struggle for life, or more difficult—at least I have found it so—than constantly to bear this conclusion in mind.

23 Yet unless it be thoroughly engrained in the mind, I am convinced that the whole economy of nature, with every fact on distribution, rarity, abundance, extinction, and variation, will be dimly seen or quite misunderstood.

23:c mind, the whole

24 We behold the face of nature bright with gladness, we often see superabundance of food; we do not see, or we forget, that the birds which are idly singing round us mostly live on insects or seeds, and are thus constantly destroying life; or we forget how largely these songsters, or their eggs, or their nestlings, are destroyed by birds and beasts of prey; we do not always bear in mind, that though food may be now superabundant, it is not so at all seasons of each recurring year.

24:d mind, that, though/year. [Space]

24:f not see or we forget, that the birds

24.1:d [Center] The Term, Struggle for Existence, used in a large sense. [Space]

25   I should premise that I use the term Struggle for Existence in a large and metaphorical sense, including dependence of one being on another, and including (which is more important) not only the life of the individual, but success in leaving progeny.

25:d use this term in

25:f sense including

26 Two canine animals in a time of dearth, may be truly said to struggle with each other which shall get food and live.

26:d animals, in

27 But a plant on the edge of a desert is said to struggle for life against the drought, though more properly it should be said to be dependent on the moisture.

28 A plant which annually produces a thousand seeds, of which on an average only one comes to maturity, may be more truly said to struggle with the plants of the same and other kinds which already clothe the ground.

28:f which only one on an average comes

29 The missletoe is dependent on the apple and a few other trees, but can only in a far-fetched sense be said to struggle with these trees, for if too many of these parasites grow on the same tree, it will languish and die.

29:c mistletoe

29:d for, if

29:f languishes and dies.

30 But several seedling missletoes, growing close together on the same branch, may more truly be said to struggle with each other.

30:c mistletoes

31 As the missletoe is disseminated by birds, its existence depends on birds; and it may metaphorically be said to struggle with other fruit-bearing plants, in order to tempt birds to devour and thus disseminate its seeds rather than those of other plants.

31:c mistletoe

31:f on them; and it may/plants, in tempting the birds to devour and thus disseminate its seeds.

32 In these several senses, which pass into each other, I use for convenience sake the general term of struggle for existence.

32:b convenience'

32:d existence. [*Space*]

32:f Struggle for Existence.

32.1:d [*Center*] *Geometrical Ratio of Increase.* [*Space*]

33 A struggle for existence inevitably follows from the high rate at which all organic beings tend to increase.

34 Every being, which during its natural lifetime produces several eggs or seeds, must suffer destruction during some period of its life, and during some season or occasional year, otherwise, on the principle of geometrical increase, its numbers would quickly become so inordinately great that no country could support the product.

35 Hence, as more individuals are produced than can possibly survive, there must in every case be a struggle for existence, either one individual with another of the same species, or with the individuals of distinct species, or with the physical conditions of life.

36 It is the doctrine of Malthus applied with manifold force to the whole animal and vegetable kingdoms; for in this case there can be no artificial increase of food, and no prudential restraint from marriage.

37 Although some species may be now increasing, more or less rapidly, in numbers, all cannot do so, for the world would not hold them.

38 There is no exception to the rule that every organic being naturally increases at so high a rate, that if not destroyed, the earth would soon be covered by the progeny of a single pair.

38:d that, if

39 Even slow-breeding man has doubled in twenty-five years, and at this rate, in a few thousand years, there would literally not be standing room for his progeny.

39:c standing-room

39:f rate, in less than a thousand

40 Linnæus has calculated that if an annual plant produced only two seeds—and there is no plant so unproductive as this—and their seedlings next year produced two, and so on, then in twenty years there would be a million plants.

40:c no plant nearly so unproductive

40:f no plant so

41 The elephant is reckoned to be the slowest breeder of all known animals, and I have taken some pains to estimate its probable minimum rate of natural increase: it will be under the mark to assume that it breeds when thirty years old, and goes on breeding till ninety years old, bringing forth three pair of young in this interval; if this be so, at the end of the fifth century there would be alive fifteen million elephants, descended from the first pair.

41:b reckoned the slowest

41:d that it begins breeding when

41:e it will be safest to assume

41:f increase; it/forth six young in the interval, and surviving till one hundred years old; if this be so, after a period of from 740 to 750 years there would be nearly nineteen million elephants alive, descended

42   But we have better evidence on this subject than mere theoretical calculations, namely, the numerous recorded cases of the astonishingly rapid increase of various animals in a state of nature, when circumstances have been favourable to them during two or three following seasons.

43 Still more striking is the evidence from our domestic animals of many kinds which have run wild in several parts of the world: if the statements of the rate of increase of slow-breeding cattle and horses in South-America, and latterly in Australia, had not been well authenticated, they would have been quite incredible.

43:b South America

43:c have been incredible.

43:f world; if

44 So it is with plants: cases could be given of introduced plants which have become common throughout whole islands in a period of less than ten years.

45 Several of the plants now most numerous over the wide plains of La Plata, clothing square leagues of surface almost to the exclusion of all other plants, have been introduced from Europe; and there are plants which now range in India, as I hear from Dr. Falconer, from Cape Comorin to the Himalaya, which have been imported from America since its discovery.

45:b plants, such as the cardoon and a tall thistle, now most

45:d cardoon, and a tall

45:f cardoon and a tall thistle, which are now the commonest over/
exclusion of every other plant, have been introduced

46 In such cases, and endless instances could be given, no one supposes that the fertility of these animals or plants has been suddenly and temporarily increased in any sensible degree.

46:f endless others could be given, no one supposes, that the fertility
of the animals

47 The obvious explanation is that the conditions of life have been
very favourable, and that there has consequently been less destruction of the old and young, and that nearly all the young
have been enabled to breed.

47:f been highly favourable

48 In such cases the geometrical ratio of increase, the result of which
never fails to be surprising, simply explains the extraordinarily
rapid increase and wide diffusion of naturalised productions in
their new homes.

48:f Their geometrical/explains their extraordinarily rapid increase
and wide diffusion in

49 In a state of nature almost every plant produces seed, and
amongst animals there are very few which do not annually pair.

49:f every full-grown plant annually produces

50 Hence we may confidently assert, that all plants and animals are
tending to increase at a geometrical ratio, that all would most
rapidly stock every station in which they could any how exist,
and that the geometrical tendency to increase must be checked
by destruction at some period of life.

50:d ratio,—that all/exist,—and that

50:f exist,—and that this geometrical

51 Our familiarity with the larger domestic animals tends, I think,
to mislead us: we see no great destruction falling on them, and
we forget that thousands are annually slaughtered for food,
and that in a state of nature an equal number would have
somehow to be disposed of.

51:f them, but we do not keep in mind that thousands

52 The only difference between organisms which annually produce eggs or seeds by the thousand, and those which produce
extremely few, is, that the slow-breeders would require a few
more years to people, under favourable conditions, a whole
district, let it be ever so large.

53 The condor lays a couple of eggs and the ostrich a score, and
yet in the same country the condor may be the more numerous
of the two: the Fulmar petrel lays but one egg, yet it is believed
to be the most numerous bird in the world.

54 One fly deposits hundreds of eggs, and another, like the hippo-
bosca, a single one; but this difference does not determine how

many individuals of the two species can be supported in a district.

55 A large number of eggs is of some importance to those species, which depend on a rapidly fluctuating amount of food, for it allows them rapidly to increase in number.

55:b species which

55:f on a fluctuating

56 But the real importance of a large number of eggs or seeds is to make up for much destruction at some period of life; and this period in the great majority of cases is an early one.

57 If an animal can in any way protect its own eggs or young, a small number may be produced, and yet the average stock be fully kept up; but if many eggs or young are destroyed, many must be produced, or the species will become extinct.

58 It would suffice to keep up the full number of a tree, which lived on an average for a thousand years, if a single seed were produced once in a thousand years, supposing that this seed were never destroyed, and could be ensured to germinate in a fitting place.

59 So that in all cases, the average number of any animal or plant depends only indirectly on the number of its eggs or seeds.

59:d that, in

60 In looking at Nature, it is most necessary to keep the foregoing considerations always in mind—never to forget that every single organic being around us may be said to be striving to the utmost to increase in numbers; that each lives by a struggle at some period of its life; that heavy destruction inevitably falls either on the young or old, during each generation or at recurrent intervals.

60:f being may

61 Lighten any check, mitigate the destruction ever so little, and the number of the species will almost instantaneously increase to any amount.

61:d amount. [*Space*]

62 The face of Nature may be compared to a yielding surface, with ten thousand sharp wedges packed close together and driven inwards by incessant blows, sometimes one wedge being struck, and then another with greater force.

62:[b]

62.1:d [*Center*] Nature of the Checks to Increase. [*Space*]

63 What checks the natural tendency of each species to increase in number is most obscure.

63:b The causes which check the natural/number are most

63:d increase are

64 Look at the most vigorous species; by as much as it swarms in

numbers, by so much will its tendency to increase be still further increased.

64:f will it tend to increase still further.

65 We know not exactly what the checks are in even one single instance.

65:f are even in a single

66 Nor will this surprise any one who reflects how ignorant we are on this head, even in regard to mankind, so incomparably better known than any other animal.

66:f mankind, although so

67 This subject has been ably treated by several authors, and I shall, in my future work, discuss some of the checks at considerable length, more especially in regard to the feral animals of South America.

67:f subject of the checks to increase has/I hope in a future work to discuss it at

68 Here I will make only a few remarks, just to recall to the reader's mind some of the chief points.

69 Eggs or very young animals seem generally to suffer most, but this is not invariably the case.

70 With plants there is a vast destruction of seeds, but, from some observations which I have made, I believe that it is the seedlings which suffer most from germinating in ground already thickly stocked with other plants.

70:f made it appears that the seedlings suffer

71 Seedlings, also, are destroyed in vast numbers by various enemies; for instance, on a piece of ground three feet long and two wide, dug and cleared, and where there could be no choking from other plants, I marked all the seedlings of our native weeds as they came up, and out of the 357 no less than 295 were destroyed, chiefly by slugs and insects.

71:f 295, were

72 If turf which has long been mown, and the case would be the same with turf closely browsed by quadrupeds, be let to grow, the more vigorous plants gradually kill the less vigorous, though fully grown, plants: thus out of twenty species growing on a little plot of turf (three feet by four) nine species perished from the other species being allowed to grow up freely.

72:c fully grown plants

72:f plot of mown turf/perished, from

73 The amount of food for each species of course gives the extreme limit to which each can increase; but very frequently it is not the obtaining food, but the serving as prey to other animals, which determines the average numbers of a species.

74 Thus, there seems to be little doubt that the stock of partridges,

grouse, and hares on any large estate depends chiefly on the destruction of vermin.

75 If not one head of game were shot during the next twenty years in England, and, at the same time, if no vermin were destroyed, there would, in all probability, be less game than at present, although hundreds of thousands of game animals are now annually killed.

75:f annually shot.

76 On the other hand, in some cases, as with the elephant and rhinoceros, none are destroyed by beasts of prey: even the tiger in India most rarely dares to attack a young elephant protected by its dam.

76:c elephant, none/prey; for even

77 Climate plays an important part in determining the average numbers of a species, and periodical seasons of extreme cold or drought, I believe to be the most effective of all checks.

77:c drought seem to

78 I estimated that the winter of 1854-55 destroyed four-fifths of the birds in my own grounds; and this is a tremendous destruction, when we remember that ten per cent. is an extraordinarily severe mortality from epidemics with man.

78:c estimated (chiefly from the greatly reduced numbers of nests in the spring) that the winter

79 The action of climate seems at first sight to be quite independent of the struggle for existence; but in so far as climate chiefly acts in reducing food, it brings on the most severe struggle between the individuals, whether of the same or of distinct species, which subsist on the same kind of food.

80 Even when climate, for instance extreme cold, acts directly, it will be the least vigorous, or those which have got least food through the advancing winter, which will suffer most.

80:f vigorous individuals, or

81 When we travel from south to north, or from a damp region to a dry, we invariably see some species gradually getting rarer and rarer, and finally disappearing; and the change of climate being conspicuous, we are tempted to attribute the whole effect to its direct action.

82 But this is a very false view: we forget that each species, even where it most abounds, is constantly suffering enormous destruction at some period of its life, from enemies or from competitors for the same place and food; and if these enemies or competitors be in the least degree favoured by any slight change of climate, they will increase in numbers, and, as each area is already fully stocked with inhabitants, the other species will decrease.

82:b is a false

82:f numbers; and as/species must decrease.

83 When we travel southward and see a species decreasing in numbers, we may feel sure that the cause lies quite as much in other species being favoured, as in this one being hurt.

84 So it is when we travel northward, but in a somewhat lesser degree, for the number of species of all kinds, and therefore of competitors, decreases northwards; hence in going northward, or in ascending a mountain, we far oftener meet with stunted forms, due to the *directly* injurious action of climate, than we do in proceeding southwards or in descending a mountain.

85 When we reach the Arctic regions, or snow-capped summits, or absolute deserts, the struggle for life is almost exclusively with the elements.

86   That climate acts in main part indirectly by favouring other species, we may clearly see in the prodigious number of plants in our gardens which can perfectly well endure our climate, but which never become naturalised, for they cannot compete with our native plants, nor resist destruction by our native animals.

86:b plants nor

86:f we clearly/plants which in our gardens can

87   When a species, owing to highly favourable circumstances, increases inordinately in numbers in a small tract, epidemics— at least, this seems generally to occur with our game animals— often ensue: and here we have a limiting check independent of the struggle for life.

88 But even some of these so-called epidemics appear to be due to parasitic worms, which have from some cause, possibly in part through facility of diffusion amongst the crowded animals, been disproportionably favoured: and here comes in a sort of struggle between the parasite and its prey.

89   On the other hand, in many cases, a large stock of individuals of the same species, relatively to the numbers of its enemies, is absolutely necessary for its preservation.

90 Thus we can easily raise plenty of corn and rape-seed, &c., in our fields, because the seeds are in great excess compared with the number of birds which feed on them; nor can the birds, though having a super-abundance of food at this one season, increase in number proportionally to the supply of seed, as their numbers are checked during winter: but any one who has tried, knows how troublesome it is to get seed from a few wheat or other such plants in a garden; I have in this case lost every single seed.

90:b garden: I

90:f winter; but

91 This view of the necessity of a large stock of the same species for

its preservation, explains, I believe, some singular facts in nature, such as that of very rare plants being sometimes extremely abundant in the few spots where they do occur; and that of some social plants being social, that is, abounding in individuals, even on the extreme confines of their range.

91:f abundant, in the few spots where they do exist; and/extreme verge of

92 For in such cases, we may believe, that a plant could exist only where the conditions of its life were so favourable that many could exist together, and thus save each other from utter destruction.

92:b save the species from

93 I should add that the good effects of frequent intercrossing, and the ill effects of close interbreeding, probably come into play in some of these cases; but on this intricate subject I will not here enlarge.

93:d enlarge. [Space]

93:f of intercrossing/interbreeding, no doubt come into play in many of these cases; but I will not here enlarge on this subject.

93.1:d [Center] *Complex Relations of all Animals and Plants to each other in the Struggle for Existence.* [Space]

94 Many cases are on record showing how complex and unexpected are the checks and relations between organic beings, which have to struggle together in the same country.

95 I will give only a single instance, which, though a simple one, has interested me.

95:f one, interested

96 In Staffordshire, on the estate of a relation where I had ample means of investigation, there was a large and extremely barren heath, which had never been touched by the hand of man; but several hundred acres of exactly the same nature had been enclosed twenty-five years previously and planted with Scotch fir.

96:b relation, where

97 The change in the native vegetation of the planted part of the heath was most remarkable, more than is generally seen in passing from one quite different soil to another: not only the proportional numbers of the heath-plants were wholly changed, but twelve species of plants (not counting grasses and carices) flourished in the plantations, which could not be found on the heath.

97:f another; not

98 The effect on the insects must have been still greater, for six insectivorous birds were very common in the plantations, which were not to be seen on the heath; and the heath was frequented by two or three distinct insectivorous birds.

99 Here we see how potent has been the effect of the introduction

of a single tree, nothing whatever else having been done, with the exception that the land had been enclosed, so that cattle could not enter.

99:e exception of the land having been enclosed

100 But how important an element enclosure is, I plainly saw near Farnham, in Surrey.

101 Here there are extensive heaths, with a few clumps of old Scotch firs on the distant hill-tops: within the last ten years large spaces have been enclosed, and self-sown firs are now springing up in multitudes, so close together that all cannot live.

102 When I ascertained that these young trees had not been sown or planted, I was so much surprised at their numbers that I went to several points of view, whence I could examine hundreds of acres of the unenclosed heath, and literally I could not see a single Scotch fir, except the old planted clumps.

103 But on looking closely between the stems of the heath, I found a multitude of seedlings and little trees, which had been perpetually browsed down by the cattle.

103:f trees which

104 In one square yard, at a point some hundred yards distant from one of the old clumps, I counted thirty-two little trees; and one of them, judging from the rings of growth, had during twenty-six years tried to raise its head above the stems of the heath, and had failed.

104:b them, with twenty-six rings of growth, had during many years
104:f had, during

105 No wonder that, as soon as the land was enclosed, it became thickly clothed with vigorously growing young firs.

106 Yet the heath was so extremely barren and so extensive that no one would ever have imagined that cattle would have so closely and effectually searched it for food.

107 Here we see that cattle absolutely determine the existence of the Scotch fir; but in several parts of the world insects determine the existence of cattle.

108 Perhaps Paraguay offers the most curious instance of this; for here neither cattle nor horses nor dogs have ever run wild, though they swarm southward and northward in a feral state; and Azara and Rengger have shown that this is caused by the greater number in Paraguay of a certain fly, which lays its eggs in the navels of these animals when first born.

109 The increase of these flies, numerous as they are, must be habitually checked by some means, probably by birds.

109:c probably by other parasitic insects.

110 Hence, if certain insectivorous birds (whose numbers are prob-

ably regulated by hawks or beasts of prey) were to increase in Paraguay, the flies would decrease—then cattle and horses would become feral, and this would certainly greatly alter (as indeed I have observed in parts of South America) the vegetation: this again would largely affect the insects; and this, as we just have seen in Staffordshire, the insectivorous birds, and so onwards in ever-increasing circles of complexity.

110:c birds were to decrease in Paraguay, the parasitic insects would probably increase; and this would lessen the number of the navel-frequenting flies—then cattle/we have just seen

111 We began this series by insectivorous birds, and we have ended with them.

111:[f]

112 Not that in nature the relations can ever be as simple as this.

112:f that under nature the relations will ever

113 Battle within battle must ever be recurring with varying success; and yet in the long-run the forces are so nicely balanced, that the face of nature remains uniform for long periods of time, though assuredly the merest trifle would often give the victory to one organic being over another.

113:f must be continually recurring/remains for long periods of time uniform, though

114 Nevertheless so profound is our ignorance, and so high our presumption, that we marvel when we hear of the extinction of an organic being; and as we do not see the cause, we invoke cataclysms to desolate the world, or invent laws on the duration of the forms of life!

114:f life! [*Space*]

115 I am tempted to give one more instance showing how plants and animals, most remote in the scale of nature, are bound together by a web of complex relations.

115:f animals, remote

116 I shall hereafter have occasion to show that the exotic Lobelia fulgens, in this part of England, is never visited by insects, and consequently, from its peculiar structure, never can set a seed.

116:e never sets a seed.

116:f fulgens is never visited in my garden by

117 Many of our orchidaceous plants absolutely require the visits of moths to remove their pollen-masses and thus to fertilise them.

117:d Nearly all our/visits of insects to remove

118 I have, also, reason to believe that humble-bees are indispensable to the fertilisation of the heartsease (Viola tricolor), for other bees do not visit this flower.

118:c I find from experiments that humble-bees are almost indispensable

119 From experiments which I have tried, I have found that the visits of bees, if not indispensable, are at least highly beneficial to the fertilisation of our clovers; but humble-bees alone visit the common red clover (Trifolium pratense), as other bees cannot reach the nectar.

119:b have lately tried/of bees are necessary for the fertilisation of some kinds of clover; but humble-bees alone visit the red

119.x:c I have also found that the visits of bees are necessary for the fertilisation of some kinds of clover: for instance, 20 heads of Dutch clover (Trifolium repens) yielded 2,290 seeds; but 20 other heads protected from bees produced not one.

119.x:e seeds, but

119.x.1:c Again, 100 heads of red clover (T. pratense) produced 2,700 seeds, but the same number of protected heads produced not a single seed.

119.y:c Humble-bees alone visit red clover, as

119.1:c It has been suggested that moths may serve to fertilise the clovers; but I doubt this in the case of the red clover, from their weight being apparently not sufficient to depress the wing-petals.

119.1:e doubt whether they could do so in the case of the red clover, from their weight not being sufficient

120 Hence I have very little doubt, that if the whole genus of humble-bees became extinct or very rare in England, the heartsease and red clover would become very rare, or wholly disappear.

120:c Hence we may infer as highly probable that

120:d that, if

121 The number of humble-bees in any district depends in a great degree on the number of field-mice, which destroy their combs and nests; and Mr. H. Newman, who has long attended to the habits of humble-bees, believes that "more than two-thirds of them are thus destroyed all over England."

121:d nests; and Col. Newman

121:f great measure on

122 Now the number of mice is largely dependent, as every one knows, on the number of cats; and Mr. Newman says, "Near villages and small towns I have found the nests of humble-bees more numerous than elsewhere, which I attribute to the number of cats that destroy the mice."

122:d and Col. Newman

123 Hence it is quite credible that the presence of a feline animal in large numbers in a district might determine, through the intervention first of mice and then of bees, the frequency of certain flowers in that district!

124 In the case of every species, many different checks, acting at different periods of life, and during different seasons or years, probably come into play; some one check or some few being

generally the most potent, but all concurring in determining the average number or even the existence of the species.

124:b concur

124:f potent; but all will concur in

125 In some cases it can be shown that widely-different checks act on the same species in different districts.

126 When we look at the plants and bushes clothing an entangled bank, we are tempted to attribute their proportional numbers and kinds to what we call chance.

127 But how false a view is this!

128 Every one has heard that when an American forest is cut down, a very different vegetation springs up; but it has been observed that the trees now growing on the ancient Indian mounds, in the Southern United States, display the same beautiful diversity and proportion of kinds as in the surrounding virgin forests.

128:b observed that ancient Indian ruins in the Southern United States, which must formerly have been cleared of trees, now display

129 What a struggle between the several kinds of trees must here have gone on during long centuries, each annually scattering its seeds by the thousand; what war between insect and insect— between insects, snails, and other animals with birds and beasts of prey—all striving to increase, and all feeding on each other or on the trees or their seeds and seedlings, or on the other plants which first clothed the ground and thus checked the growth of the trees!

129:d other, or on the trees, their seeds

129:e struggle must have gone on during long centuries between the several kinds of trees, each/increase, all

130 Throw up a handful of feathers, and all must fall to the ground according to definite laws; but how simple is this problem compared to the action and reaction of the innumerable plants and animals which have determined, in the course of centuries, the proportional numbers and kinds of trees now growing on the old Indian ruins!

130:c is the problem where each shall fall compared to that of the action

130:f all fall

131 The dependency of one organic being on another, as of a parasite on its prey, lies generally between beings remote in the scale of nature.

132 This is often the case with those which may strictly be said to struggle with each other for existence, as in the case of locusts and grass-feeding quadrupeds.

132:e This is likewise sometimes the case

133 But the struggle almost invariably will be most severe between

the individuals of the same species, for they frequent the same districts, require the same food, and are exposed to the same dangers.

133:e struggle will almost invariably be

134 In the case of varieties of the same species, the struggle will generally be almost equally severe, and we sometimes see the contest soon decided: for instance, if several varieties of wheat be sown together, and the mixed seed be resown, some of the varieties which best suit the soil or climate, or are naturally the most fertile, will beat the others and so yield more seed, and will consequently in a few years quite supplant the other varieties.

134:f years supplant

135 To keep up a mixed stock of even such extremely close varieties as the variously coloured sweet-peas, they must be each year harvested separately, and the seed then mixed in due proportion, otherwise the weaker kinds will steadily decrease in numbers and disappear.

136 So again with the varieties of sheep: it has been asserted that certain mountain-varieties will starve out other mountain-varieties, so that they cannot be kept together.

137 The same result has followed from keeping together different varieties of the medicinal leech.

138 It may even be doubted whether the varieties of any one of our domestic plants or animals have so exactly the same strength, habits, and constitution, that the original proportions of a mixed stock could be kept up for half a dozen generations, if they were allowed to struggle together, like beings in a state of nature, and if the seed or young were not annually sorted.

138:b half-a-dozen

138:d sorted. [*Space*]

138:e any of our

138:f stock (crossing being prevented) could/together, in the same manner as beings/annually preserved in due proportion.

138.1:d [*Center*] Struggle for Life most severe between Individuals *and Varieties of the same Species*. [*Space*]

139 As species of the same genus have usually, though by no means invariably, some similarity in habits and constitution, and always in structure, the struggle will generally be more severe between species of the same genus, when they come into competition with each other, than between species of distinct genera.

139:d invariably, much similarity

139:f As the species of the same genus usually have, though/between them, if they/between the species

140 We see this in the recent extension over parts of the United

States of one species of swallow having caused the decrease of another species.

141 The recent increase of the missel-thrush in parts of Scotland has caused the decrease of the song-thrush.

142 How frequently we hear of one species of rat taking the place of another species under the most different climates!

143 In Russia the small Asiatic cockroach has everywhere driven before it its great congener.

143.1:d In Australia the imported hive-bee is rapidly exterminating the small, stingless native bee.

144 One species of charlock will supplant another, and so in other cases.

144:d charlock has been known to supplant another species; and

145 We can dimly see why the competition should be most severe between allied forms, which fill nearly the same place in the economy of nature; but probably in no one case could we precisely say why one species has been victorious over another in the great battle of life.

146 A corollary of the highest importance may be deduced from the foregoing remarks, namely, that the structure of every organic being is related, in the most essential yet often hidden manner, to that of all other organic beings, with which it comes into competition for food or residence, or from which it has to escape, or on which it preys.

147 This is obvious in the structure of the teeth and talons of the tiger; and in that of the legs and claws of the parasite which clings to the hair on the tiger's body.

148 But in the beautifully plumed seed of the dandelion, and in the flattened and fringed legs of the water-beetle, the relation seems at first confined to the elements of air and water.

149 Yet the advantage of plumed seeds no doubt stands in the closest relation to the land being already thickly clothed by other plants; so that the seeds may be widely distributed and fall on unoccupied ground.

149:d clothed with other

150 In the water-beetle, the structure of its legs, so well adapted for diving, allows it to compete with other aquatic insects, to hunt for its own prey, and to escape serving as prey to other animals.

151 The store of nutriment laid up within the seeds of many plants seems at first sight to have no sort of relation to other plants.

152 But from the strong growth of young plants produced from such seeds (as peas and beans), when sown in the midst of long grass, I suspect that the chief use of the nutriment in the seed is to favour the growth of the young seedling, whilst struggling with other plants growing vigorously all around.

152:c grass, it may be suspected that

152:f seeds, as peas and beans, when/growth of the seedlings, whilst

153   Look at a plant in the midst of its range, why does it not double or quadruple its numbers?

154 We know that it can perfectly well withstand a little more heat or cold, dampness or dryness, for elsewhere it ranges into slightly hotter or colder, damper or drier districts.

155 In this case we can clearly see that if we wished in imagination to give the plant the power of increasing in number, we should have to give it some advantage over its competitors, or over the animals which preyed on it.

155:f wish/prey

156 On the confines of its geographical range, a change of constitution with respect to climate would clearly be an advantage to our plant; but we have reason to believe that only a few plants or animals range so far, that they are destroyed by the rigour of the climate alone.

156:f destroyed exclusively by the rigour of the climate.

157 Not until we reach the extreme confines of life, in the arctic regions or on the borders of an utter desert, will competition cease.

157:b Arctic

158 The land may be extremely cold or dry, yet there will be competition between some few species, or between the individuals of the same species, for the warmest or dampest spots.

159   Hence, also, we can see that when a plant or animal is placed in a new country amongst new competitors, though the climate may be exactly the same as in its former home, yet the conditions of its life will generally be changed in an essential manner.

159:f Hence we/competitors, the conditions of its life will generally be changed in an essential manner, although the climate may be exactly the same as in its former home.

160 If we wished to increase its average numbers in its new home, we should have to modify it in a different way to what we should have done in its native country; for we should have to give it some advantage over a different set of competitors or enemies.

160:c what we should have to do in

160:f If its average numbers are to increase in its new/what we should have had to do

161   It is good thus to try in our imagination to give any form some advantage over another.

161:e in imagination

161:f any one species an advantage

162 Probably in no single instance should we know what to do, so as to succeed.

162:f do.

163 It will convince us of our ignorance on the mutual relations of all organic beings; a conviction as necessary, as it seems to be difficult to acquire.

163:d seems difficult

163:f This ought to convince/it is difficult to

164 All that we can do, is to keep steadily in mind that each organic being is striving to increase at a geometrical ratio; that each at some period of its life, during some season of the year, during each generation or at intervals, has to struggle for life, and to suffer great destruction.

164:c increase in a geometrical

164:f life and

165 When we reflect on this struggle, we may console ourselves with the full belief, that the war of nature is not incessant, that no fear is felt, that death is generally prompt, and that the vigorous, the healthy, and the happy survive and multiply.

CHAPTER IV.

2:e Natural Selection; or the Survival of the Fittest.

3 Natural Selection—its power compared with man's selection—its power on characters of trifling importance—its power at all ages and on both sexes—Sexual Selection—On the generality of intercrosses between individuals of the same species—Circumstances favourable and unfavourable to Natural Selection, namely, intercrossing, isolation, number of individuals—Slow action—Extinction caused by Natural Selection—Divergence of Character, related to the diversity of inhabitants of any small area, and to naturalisation—Action of Natural Selection, through Divergence of Character and Extinction, on the descendants from a common parent—Explains the Grouping of all organic beings.

3:c beings—Advance in organisation—Low forms preserved—Objections considered—Indefinite multiplication of species—Summary.

3:d by Natural Selection—Divergence of Character related

3:e unfavourable to the results of Natural Selection, namely/considered—Uniformity of certain characters due to their unimportance and to their not having been acted on by Natural Selection—Indefinite

3:f by Natural Selection—Divergence of Character, related/through Divergence of Character, and Extinction, on/grouping/preserved—Convergence of character—Indefinite

4 How will the struggle for existence, discussed too briefly in the last chapter, act in regard to variation?

4:e existence, briefly discussed in the last

5 Can the principle of selection, which we have seen is so potent in the hands of man, apply in nature?

5:f apply under nature?

6 I think we shall see that it can act most effectually.

6:f most efficiently.

7 Let it be borne in mind in what an endless number of strange peculiarities our domestic productions, and, in a lesser degree, those under nature, vary; and how strong the hereditary tendency is.

7:e Let the endless number of peculiar variations in our domestic productions, and, in a lesser degree, in those under nature, be borne in mind; as well as the strength of the hereditary tendency.

7:f of slight variations and individual differences occurring in our

8 Under domestication, it may be truly said that the whole organisation becomes in some degree plastic.

8.1:c But the variability, which we almost universally meet with in our domestic productions, is not directly produced, as Hooker and Asa Gray have well remarked, by man; he can neither originate varieties, nor prevent their occurrence; he can only preserve and accumulate such as do occur; unintentionally he exposes organic beings to new and changing conditions of life, and variability ensues; but similar changes of conditions might and do occur under nature.

8.1.x-y:f occur. *y* Unintentionally

9 Let it be borne in mind how infinitely complex and close-fitting are the mutual relations of all organic beings to each other and to their physical conditions of life.

9:c it also be/life; and consequently what infinitely varied diversities of structure may be of use to each being under changing conditions of life.

10 Can it, then, be thought improbable, seeing that variations useful to man have undoubtedly occurred, that other variations useful in some way to each being in the great and complex battle of life, should sometimes occur in the course of thousands of generations?

10:f should occur in the course of many successive generations?

11 If such do occur, can we doubt (remembering that many more individuals are born than can possibly survive) that individuals having any advantage, however slight, over others, would have the best chance of surviving and of procreating their kind?

12 On the other hand, we may feel sure that any variation in the least degree injurious would be rigidly destroyed.

13 This preservation of favourable variations and the rejection of injurious variations, I call Natural Selection.

13:e and the destruction of injurious variations, I call Natural Selection, or the Survival of the Fittest.

13:f favourable individual differences and variations, and the destruction of those which are injurious, I have called Natural

14 Variations neither useful nor injurious would not be affected by natural selection, and would be left a fluctuating element, as perhaps we see in the species called polymorphic.

14:e left either a fluctuating element, as perhaps we see in certain polymorphic species, or would ultimately become fixed, owing to the nature of the organism and the nature of the conditions.

14.1-2:c[¶] Several writers have misapprehended or objected to the term Natural Selection. *2* Some have even imagined that natural selection induces variability, whereas it implies only the preservation of such variations as occur and are beneficial to the being under its conditions of life.

14.2:f as arise and

14.3-4:c No one objects to agriculturists speaking of the potent effects of man's selection; and in this case the individual differences given by nature, which man for some object selects, must of necessity first occur. *4* Others have objected that the term selection implies conscious choice in the animals which become modified; and it has even been urged that as plants have no volition, natural selection is not applicable to them!

14.4:d urged that, as

14.5:c In the literal sense of the word, no doubt, natural selection is a misnomer; but who ever objected to chemists speaking of the elective affinities of the various elements?—and yet an acid cannot strictly be said to elect the base with which it will in preference combine.

14.5:e selection is a false term; but/it in preference combines.

14.6-9:c It has been said that I speak of natural selection as an active power or Deity; but who objects to an author speaking of the attraction of gravity as ruling the movements of the planets? *7* Every one knows what is meant and is implied by such metaphorical expressions; and they are almost necessary for brevity. *8* So again it is difficult to avoid personifying the word Nature; but I mean by Nature, only the aggregate action and product of many natural laws, and by laws the sequence of events as ascertained by us. *9* With a little familiarity such superficial objections will be forgotten.

15   We shall best understand the probable course of natural selection by taking the case of a country undergoing some physical change, for instance, of climate.

15:d some slight physical

16 The proportional numbers of its inhabitants would almost immediately undergo a change, and some species might become extinct.

16:f inhabitants will almost/species will probably become

17 We may conclude, from what we have seen of the intimate and complex manner in which the inhabitants of each country are bound together, that any change in the numerical proportions of some of the inhabitants, independently of the change of climate itself, would most seriously affect many of the others.

17:b would seriously affect

17:e affect the others.

17:f proportions of the inhabitants

18 If the country were open on its borders, new forms would certainly immigrate, and this also would seriously disturb the relations of some of the former inhabitants.

18:d would often seriously

18:f this would likewise seriously

19 Let it be remembered how powerful the influence of a single introduced tree or mammal has been shown to be.

20 But in the case of an island, or of a country partly surrounded by barriers, into which new and better adapted forms could not freely enter, we should then have places in the economy of nature which would assuredly be better filled up, if some of the original inhabitants were in some manner modified; for, had the area been open to immigration, these same places would have been seized on by intruders.

20:e seized by intruders.

20:f seized on by

21 In such case, every slight modification, which in the course of ages chanced to arise, and which in any way favoured the individuals of any of the species, by better adapting them to their altered conditions, would tend to be preserved; and natural selection would thus have free scope for the work of improvement.

21:e such cases, slight modifications, which in any way favoured the individuals of any species

22 We have reason to believe, as stated in the first chapter, that a change in the conditions of life, by specially acting on the reproductive system, causes or increases variability; and in the foregoing case the conditions of life are supposed to have undergone a change, and this would manifestly be favourable to natural selection, by giving a better chance of profitable variations occurring; and unless profitable variations do occur, natural selection can do nothing.

22:e life cause or excite a tendency to vary; and in the foregoing case the conditions are supposed to have changed, and this/ unless such do occur

22.x-y:f have good reason to believe, as shown in the first/life give a tendency to increased variability; and in the foregoing cases the conditions have changed, and this/selection, by affording a better chance of the occurrence of profitable variations. *y* Unless such occur

22.1:e Under the term of "variations," it must never be forgotten that mere individual differences are always included.

22.1:f are included.

23 Not that, as I believe, any extreme amount of variability is necessary; as man can certainly produce great results by adding up in any given direction mere individual differences, so could Nature, but far more easily, from having incomparably longer time at her disposal.

23:c Not that any extreme/could natural selection, but/time for action.

23:e As/produce a great result with his domestic animals and plants by adding up in any given direction individual

23:f can produce

24 Nor do I believe that any great physical change, as of climate, or any unusual degree of isolation to check immigration, is actually necessary to produce new and unoccupied places for natural selection to fill up by modifying and improving some of the varying inhabitants.

24:f is necessary in order that new and unoccupied places should be left, for natural selection to fill up by improving

25 For as all the inhabitants of each country are struggling together with nicely balanced forces, extremely slight modifications in the structure or habits of one inhabitant would often give it an advantage over others; and still further modifications of the same kind would often still further increase the advantage.

25:c advantage, as long as the being continued under the same conditions of life and profited by similar means of subsistence and defence.

25:e one species would often give/long as the species continued

26 No country can be named in which all the native inhabitants are now so perfectly adapted to each other and to the physical conditions under which they live, that none of them could anyhow be improved; for in all countries, the natives have been so far conquered by naturalised productions, that they have allowed foreigners to take firm possession of the land.

26:f allowed some foreigners

26:e could be still better adapted or improved

27 And as foreigners have thus everywhere beaten some of the natives, we may safely conclude that the natives might have been modified with advantage, so as to have better resisted such intruders.

27:e thus in every country beaten/resisted the intruders.

28 As man can produce and certainly has produced a great result by his methodical and unconscious means of selection, what may not nature effect?

28:b Nature

28:c not natural selection effect?

28:f produce, and certainly has produced, a great

29 Man can act only on external and visible characters: nature cares nothing for appearances, except in so far as they may be useful to any being.

29:b Nature

29:c characters: Nature (if I may be allowed thus to personify the natural preservation of varying and favoured individuals during the struggle for existence) cares/they are useful

29:e characters: Nature, if I may be allowed to personify the natural preservation or survival of the fittest, cares

30 She can act on every internal organ, on every shade of constitutional difference, on the whole machinery of life.

31 Man selects only for his own good; Nature only for that of the being which she tends.

31:f good: Nature

32 Every selected character is fully exercised by her; and the being is placed under well-suited conditions of life.

32:e her, as is implied by the fact of their selection.

33 Man keeps the natives of many climates in the same country; he seldom exercises each selected character in some peculiar and fitting manner; he feeds a long and a short beaked pigeon on the same food; he does not exercise a long-backed or long-legged quadruped in any peculiar manner; he exposes sheep with long and short wool to the same climate.

34 He does not allow the most vigorous males to struggle for the females.

35 He does not rigidly destroy all inferior animals, but protects during each varying season, as far as lies in his power, all his productions.

36 He often begins his selection by some half-monstrous form; or at least by some modification prominent enough to catch his eye, or to be plainly useful to him.

36:e catch the eye or

37 Under nature, the slightest difference of structure or constitution may well turn the nicely-balanced scale in the struggle for life, and so be preserved.

37:e differences

38 How fleeting are the wishes and efforts of man! how short his time! and consequently how poor will his products be, compared with those accumulated by nature during whole geological periods.

38:b Nature

38:e will be his results, compared

39 Can we wonder, then, that nature's productions should be far "truer" in character than man's productions; that they should be infinitely better adapted to the most complex conditions of life, and should plainly bear the stamp of far higher workmanship?

39:b Nature's

40 It may be said that natural selection is daily and hourly scrutinising, throughout the world, every variation, even the slightest; rejecting that which is bad, preserving and adding up all that is good; silently and insensibly working, whenever

and wherever opportunity offers, at the improvement of each organic being in relation to its organic and inorganic conditions of life.

40:b may metaphorically be

40:e world, the slightest variations; rejecting those that are bad, preserving and adding up all that are good

40:f whenever and wherever opportunity offers

41 We see nothing of these slow changes in progress, until the hand of time has marked the long lapse of ages, and then so imperfect is our view into long past geological ages, that we only see that the forms of life are now different from what they formerly were.

41:d long-past

41:e marked the lapse/that we see only that

41.1:c[¶] In order that any great amount of modification should thus in the course of time be produced, it is necessary to believe that when a variety has once arisen, it again varies, after perhaps a long interval of time; and that its varieties, if favourable, are again preserved, and so onwards.

41.1:e modification in any part should be effected, a variety when once formed must again, perhaps after a long interval of time, vary or present individual differences of the same favourable nature, and these must be again preserved, and so onwards step by step.

41.1:f modification should be effected in a species, a variety/nature as before; and these

41.2:c That varieties more or less different from the parent-stock occasionally arise, few will deny; but that the process of variation should be thus indefinitely prolonged is an assumption, the truth of which must be judged of by how far the hypothesis accords with and explains the general phenomena of nature.

41.2.x-y:e Seeing that individual differences of all kinds perpetually recur, this can hardly be considered as an unwarrantable assumption. *y* But whether all this has actually taken place must be judged by

41.2.x-y:f differences of the same kind/But whether it is true, we can judge only by seeing how

41.3:c On the other hand, the ordinary belief that the amount of possible variation is a strictly limited quantity is likewise a simple assumption.

41.3:e quantity is a simple

41.3:f quantity is likewise a simple

42 Although natural selection can act only through and for the good of each being, yet characters and structures, which we are apt to consider as of very trifling importance, may thus be acted on.

43 When we see leaf-eating insects green, and bark-feeders mottled-

grey; the alpine ptarmigan white in winter, the red-grouse the colour of heather, and the black-grouse that of peaty earth, we must believe that these tints are of service to these birds and insects in preserving them from danger.

43:e heather, we

44 Grouse, if not destroyed at some period of their lives, would increase in countless numbers; they are known to suffer largely from birds of prey; and hawks are guided by eyesight to their prey,—so much so, that on parts of the Continent persons are warned not to keep white pigeons, as being the most liable to destruction.

44:b eyesight to their prey—so much

45 Hence I can see no reason to doubt that natural selection might be most effective in giving the proper colour to each kind of grouse, and in keeping that colour, when once acquired, true and constant.

45:e Hence natural

45:f be effective

46 Nor ought we to think that the occasional destruction of an animal of any particular colour would produce little effect: we should remember how essential it is in a flock of white sheep to destroy every lamb with the faintest trace of black.

46:f destroy a lamb

46.1:c We have seen how in Florida the colour of the hogs, when feeding on the "paint root," determines whether they shall live or die.

46.1:e how the colour of the hogs, when feeding on the "paint-root" in Florida, determines

46.1:f hogs, which feed on the "paint-root" in Virginia, determines

47 In plants the down on the fruit and the colour of the flesh are considered by botanists as characters of the most trifling importance: yet we hear from an excellent horticulturist, Downing, that in the United States smooth-skinned fruits suffer far more from a beetle, a curculio, than those with down; that purple plums suffer far more from a certain disease than yellow plums; whereas another disease attacks yellow-fleshed peaches far more than those with other coloured flesh.

47:f In plants, the down on/Curculio

48 If, with all the aids of art, these slight differences make a great difference in cultivating the several varieties, assuredly, in a state of nature, where the trees would have to struggle with other trees and with a host of enemies, such differences would effectually settle which variety, whether a smooth or downy, a yellow or purple fleshed fruit, should succeed.

49 In looking at many small points of difference between species, which, as far as our ignorance permits us to judge, seem to be

quite unimportant, we must not forget that climate, food, &c., probably produce some slight and direct effect.

49:b seem quite

49:e &c., may have produced some direct effect.

49:f &c., have no doubt produced

50 It is, however, far more necessary to bear in mind that there are many unknown laws of correlation of growth, which, when one part of the organisation is modified through variation, and the modifications are accumulated by natural selection for the good of the being, will cause other modifications, often of the most unexpected nature.

50:c variation and/being, cause

50:e is also necessary to bear in mind that, owing to the law of correlation, when one part varies, and the variations are accumulated through natural selection, other modifications, often of the most unexpected nature, will ensue.

51 As we see that those variations which under domestication appear at any particular period of life, tend to reappear in the offspring at the same period;—for instance, in the seeds of the many varieties of our culinary and agricultural plants; in the caterpillar and cocoon stages of the varieties of the silkworm; in the eggs of poultry, and in the colour of the down of their chickens; in the horns of our sheep and cattle when nearly adult;—so in a state of nature, natural selection will be enabled to act on and modify organic beings at any age, by the accumulation of profitable variations at that age, and by their inheritance at a corresponding age.

51:b accumulation of variations profitable at that

51:c instance, in the shape, size, and flavour of the seeds

51:f which, under

52 If it profit a plant to have its seeds more and more widely disseminated by the wind, I can see no greater difficulty in this being effected through natural selection, than in the cotton-planter increasing and improving by selection the down in the pods on his cotton-trees.

53 Natural selection may modify and adapt the larva of an insect to a score of contingencies, wholly different from those which concern the mature insect.

54 These modifications will no doubt affect, through the laws of correlation, the structure of the adult; and probably in the case of those insects which live only for a few hours, and which never feed, a large part of their structure is merely the correlated result of successive changes in the structure of their larvæ.

54:d will probably affect

53 + 4:e mature insect; and these modifications may affect, through correlation, the structure of the adult.

171

55 So, conversely, modifications in the adult will probably often affect the structure of the larva; but in all cases natural selection will ensure that modifications consequent on other modifications at a different period of life, shall not be in the least degree injurious: for if they became so, they would cause the extinction of the species.

55:e adult may affect/ensure that they shall not be injurious: for if they were so, the species would become extinct.

56 Natural selection will modify the structure of the young in relation to the parent, and of the parent in relation to the young.

57 In social animals it will adapt the structure of each individual for the benefit of the community; if each in consequence profits by the selected change.

57:e benefit of the whole community; if this in consequence profits

57:f community; if the community profits

58 What natural selection cannot do, is to modify the structure of one species, without giving it any advantage, for the good of another species; and though statements to this effect may be found in works of natural history, I cannot find one case which will bear investigation.

59 A structure used only once in an animal's whole life, if of high importance to it, might be modified to any extent by natural selection; for instance, the great jaws possessed by certain insects, and used exclusively for opening the cocoon—or the hard tip to the beak of nestling birds, used for breaking the egg.

59:b insects, used exclusively

59:e animal's life

59:f beak of unhatched birds

60 It has been asserted, that of the best short-beaked tumbler-pigeons more perish in the egg than are able to get out of it; so that fanciers assist in the act of hatching.

60:e tumbler-pigeons a greater number perish

61 Now, if nature had to make the beak of a full-grown pigeon very short for the bird's own advantage, the process of modification would be very slow, and there would be simultaneously the most rigorous selection of the young birds within the egg, which had the most powerful and hardest beaks, for all with weak beaks would inevitably perish: or, more delicate and more easily broken shells might be selected, the thickness of the shell being known to vary like every other structure.

61:e selection of all the young

61.1-6:f[¶] It may be well here to remark that with all beings there must be much fortuitous destruction, which can have little or no influence on the course of natural selection. 2 For instance a vast number of eggs or seeds are annually devoured, and these could be modified through natural selection only if they varied in some manner which protected them from their enemies. *3* Yet many of these eggs or seeds would perhaps, if not destroyed, have yielded individuals better adapted to their conditions of life than any of those which happened to survive. *4* So again a vast number of mature animals and plants, whether or not they be the best adapted to their conditions, must be annually destroyed by accidental causes, which would not be in the least degree mitigated by certain changes of structure or constitution which would in other ways be beneficial to the species. *5* But let the destruction of the adults be ever so heavy, if the number which can exist in any district be not wholly kept down by such causes,—or again let the destruction of eggs or seeds be so great that only a hundredth or a thousandth part are developed,—yet of those which do survive, the best adapted individuals, supposing that there is any variability in a favourable direction, will tend to propagate their kind in larger numbers than the less well adapted. *6* If the numbers be wholly kept down by the causes just indicated, as will often have been the case, natural selection will be powerless in certain beneficial directions; but this is no valid objection to its efficiency at other times and in other ways; for we are far from having any reason to suppose that many species ever undergo modification and improvement at the same time in the same area.

62   *Sexual Selection.—*
62:d [Center] Sexual Selection. [Space]

63 Inasmuch as peculiarities often appear under domestication in one sex and become hereditarily attached to that sex, the same fact probably occurs under nature, and if so, natural selection will be able to modify one sex in its functional relations to the other sex, or in relation to wholly different habits of life in the two sexes, as is sometimes the case with insects.

63:d[¶]

63.x-y:f that sex, so no doubt it will be under nature. *y* Thus it is rendered possible for tne two sexes to be modified through natural selection in relation to different habits of life, as is sometimes the case; or for one sex to be modified in relation to the other sex, as commonly occurs.

64 And this leads me to say a few words on what I call Sexual Selection.

64:f This/what I have called

65 This depends, not on a struggle for existence, but on a struggle between the males for possession of the females; the result is not death to the unsuccessful competitor, but few or no offspring.

65.x-y:f This form of selection depends, not on a struggle for existence in relation to other organic beings or to external conditions, but on a struggle between the individuals of one sex, generally the males, for the possession of the other sex. *y* The result

66 Sexual selection is, therefore, less rigorous than natural selection.

67 Generally, the most vigorous males, those which are best fitted for their places in nature, will leave most progeny.

68 But in many cases, victory will depend not on general vigour, but on having special weapons, confined to the male sex.

68:b victory depends not

68:f not so much on general vigour, as on having

69 A hornless stag or spurless cock would have a poor chance of leaving offspring.

69:e leaving numerous offspring.

70 Sexual selection by always allowing the victor to breed might surely give indomitable courage, length to the spur, and strength to the wing to strike in the spurred leg, as well as the brutal cockfighter, who knows well that he can improve his breed by careful selection of the best cocks.

70:e as in the case of the brutal cockfighter, who knows well how to improve his breed by the careful

70:f selection, by always/leg, in nearly the same manner as does the brutal cockfighter by the careful selection of his best

71 How low in the scale of nature this law of battle descends, I know not; male alligators have been described as fighting, bellowing, and whirling round, like Indians in a war-dance, for the possession of the females; male salmons have been seen fighting all day long; male stag-beetles often bear wounds from the huge mandibles of other males.

71:b nature the law

71:c stag-beetles sometimes bear

71:d other males; the males of certain hymenopterous insects have been frequently seen by that inimitable observer M. Fabre, fighting for a particular female, who sits by an apparently unconcerned beholder of the struggle, and then retires with the conqueror.

71:f salmons have been observed fighting all/female who sits by, an apparently

72 The war is, perhaps, severest between the males of polygamous animals, and these seem oftenest provided with special weapons.

73 The males of carnivorous animals are already well armed; though to them and to others, special means of defence may be given through means of sexual selection, as the mane to the lion, the shoulder-pad to the boar, and the hooked jaw to the male salmon; for the shield may be as important for victory, as the sword or spear.

73:e lion, and the hooked

74   Amongst birds, the contest is often of a more peaceful character.

75 All those who have attended to the subject, believe that there is the severest rivalry between the males of many species to attract by singing the females.

75:*f* attract, by singing, the females.

76 The rock-thrush of Guiana, birds of Paradise, and some others, congregate; and successive males display their gorgeous plumage and perform strange antics before the females, which standing by as spectators, at last choose the most attractive partner.

76:*b* which, standing

76:*f* paradise/display with the most elaborate care, and show off in the best manner their gorgeous plumage; they likewise perform

77 Those who have closely attended to birds in confinement well know that they often take individual preferences and dislikes: thus Sir R. Heron has described how one pied peacock was eminently attractive to all his hen birds.

77:*f* how a pied

78 It may appear childish to attribute any effect to such apparently weak means: I cannot here enter on the details necessary to support this view; but if man can in a short time give elegant carriage and beauty to his bantams, according to his standard of beauty, I can see no good reason to doubt that female birds, by selecting, during thousands of generations, the most melodious or beautiful males, according to their standard of beauty, might produce a marked effect.

78:*e* [birds.] I cannot here enter on the necessary details; but

78:*f* give beauty and an elegant carriage to his

79 I strongly suspect that some well-known laws with respect to the plumage of male and female birds, in comparison with the plumage of the young, can be explained on the view of plumage having been chiefly modified by sexual selection, acting when the birds have come to the breeding age or during the breeding season; the modifications thus produced being inherited at corresponding ages or seasons, either by the males alone, or by the males and females; but I have not space here to enter on this subject.

79:*b* laws, with

79:*e* Some/explained through the action of sexual selection on variations occurring at different ages, and being transmitted to the males alone or to both sexes at a corresponding age; but

79:*f* can partly be explained/ages, and transmitted/at corresponding ages; but

80   Thus it is, as I believe, that when the males and females of any animal have the same general habits of life, but differ in structure, colour, or ornament, such differences have been mainly caused by sexual selection; that is, individual males

175

have had, in successive generations, some slight advantage over other males, in their weapons, means of defence, or charms; and have transmitted these advantages to their male offspring.

80:e that is, by individual males having had, in successive/charms; and having transmitted

80:f charms, which they have transmitted to their male offspring alone.

81 Yet, i would not wish to attribute all such sexual differences to this agency: for we see peculiarities arising and becoming attached to the male sex in our domestic animals (as the wattle in male carriers, horn-like protuberances in the cocks of certain fowls, &c.), which we cannot believe to be either useful to the males in battle, or attractive to the females.

81:e (as the greater development of the wattle in male carrier-pigeons, horn-like protuberances in certain fowls, &c.), which are in no way useful.

81:f all sexual differences to this agency: for we see in our domestic animals peculiarities arising and becoming attached to the male sex, which apparently have not been augmented through selection by man.

82 We see analogous cases under nature, for instance, the tuft of hair on the breast of the turkey-cock, which can hardly be either useful or ornamental to this bird;—indeed, had the tuft appeared under domestication, it would have been called a monstrosity.

82:d nature,—for

82:e which cannot be useful and can hardly be ornamental;—indeed

82:f The tuft of hair on the breast of the wild turkey-cock cannot be of any use, and it is doubtful whether it can be ornamental in the eyes of the female bird;—indeed

83   *Illustrations of the action of Natural Selection.—*

83:d [*Center*] *Illustrations of the Action of Natural Selection.* [*Space*]

83:e *Illustrations of the Action of Natural Selection, or the Survival of the Fittest.*

84 In order to make it clear how, as I believe, natural selection acts, I must beg permission to give one or two imaginary illustrations.

84:d[¶]

85 Let us take the case of a wolf, which preys on various animals, securing some by craft, some by strength, and some by fleetness; and let us suppose that the fleetest prey, a deer for instance, had from any change in the country increased in numbers, or that other prey had decreased in numbers, during that season of the year when the wolf is hardest pressed for food.

85:e when the wolf was hardest

86 I can under such circumstances see no reason to doubt that the swiftest and slimmest wolves would have the best chance of surviving, and so be preserved or selected,—provided always that they retained strength to master their prey at this or at some other period of the year, when they might be compelled to prey on other animals.

86:c Under such circumstances the swiftest

86:f when they were compelled

87 I can see no more reason to doubt this, than that man can improve the fleetness of his greyhounds by careful and methodical selection, or by that unconscious selection which results from each man trying to keep the best dogs without any thought of modifying the breed.

87:e methodical selection, or by unconscious

87:f doubt that this would be the result, than that man should be able to improve the fleetness of his greyhounds by careful and methodical selection, or by that kind of unconscious selection which follows from

88 Even without any change in the proportional numbers of the animals on which our wolf preyed, a cub might be born with an innate tendency to pursue certain kinds of prey.

88:[e]

89 Nor can this be thought very improbable; for we often observe great differences in the natural tendencies of our domestic animals; one cat, for instance, taking to catch rats, another mice; one cat, according to Mr. St. John, bringing home winged game, another hares or rabbits, and another hunting on marshy ground and almost nightly catching woodcocks or snipes.

89:[e]

90 The tendency to catch rats rather than mice is known to be inherited.

90:[e]

91 Now, if any slight innate change of habit or of structure benefited an individual wolf, it would have the best chance of surviving and of leaving offspring.

91:[e]

92 Some of its young would probably inherit the same habits or structure, and by the repetition of this process, a new variety might be formed which would either supplant or coexist with the parent-form of wolf.

92:b parent form

92:d supplant, or coexist with, the parent

92:[e]

93 Or, again, the wolves inhabiting a mountainous district, and those frequenting the lowlands, would naturally be forced to hunt different prey; and from the continued preservation of

the individuals best fitted for the two sites, two varieties might slowly be formed.

93:c varieties would slowly

93:[e]

94 These varieties would cross and blend where they met; but to this subject of intercrossing we shall soon have to return.

94:[e]

95 I may add, that, according to Mr. Pierce, there are two varieties of the wolf inhabiting the Catskill Mountains in the United States, one with a light greyhound-like form, which pursues deer, and the other more bulky, with shorter legs, which more frequently attacks the shepherd's flocks.

95.1-3:e[¶] It should be observed that, in the above illustration, I speak of the slimmest individual wolves, and not of any single strongly-marked variation having been preserved. *2* In former editions of this work I sometimes spoke as if this latter alternative had frequently occurred. *3* I saw the great importance of individual differences, and this led me fully to discuss the results of unconscious selection by man, which depends on the preservation of the better adapted or more valuable individuals, and on the destruction of the worst.

95.3:f preservation of all the more or less valuable

95.4-6:e I saw, also, that the preservation in a state of nature of any occasional deviation of structure, such as a monstrosity, would be a rare event; and that, if preserved, it would generally be lost by subsequent intercrossing with ordinary individuals. *5* Nevertheless, until reading an able and valuable article in the 'North British Review' (1867), I did not appreciate how rarely single variations, whether slight or strongly-marked, could be perpetuated. *6* The author takes the case of a pair of animals, which produce during their lifetime two hundred offspring, of which, from various causes of destruction, only two on an average survive to procreate their kind.

95.6:f animals, producing

95.7-11:e This is rather an extreme estimate for most of the higher animals, but by no means so for many of the lower organisms. *8* He then shows that if a single individual were born, which varied in some manner, giving it twice as good a chance of life as that of the other individuals, yet the chances would be strongly against its survival. *9* Supposing it to survive and to breed, and that half its young inherited the favourable variation; still, as the Reviewer goes on to show, the young would have only a slightly better chance of surviving and breeding; and this chance would go on decreasing in the succeeding generations. *10* The justice of these remarks cannot, I think, be disputed. *11* If, for instance, a bird of some kind could procure its food more easily by having its beak curved, and if one were born with its beak strongly curved, and which consequently flourished, nevertheless there would be a very poor chance of this one individual perpetuating its kind to the exclusion of the common form; but there can hardly be a doubt, judging by what we see taking place under domestication, that this result would follow from the preservation during many generations of a large number of individuals with more or less curved beaks, and from the destruction of a still larger number with the straightest beaks.

95.11:f less strongly curved

95.12:e[¶] It should not, however, be overlooked that certain variations, which no one would rank as mere individual differences, frequently recur owing to a similar organisation being similarly acted on,—of which fact numerous instances could be given with our domestic productions.

95.12:f certain rather strongly marked variations

95.13:e In such cases, if a varying individual did not actually transmit to its offspring its newly-acquired character, it would undoubtedly transmit, as long as the existing conditions remained the same, a still stronger tendency to vary in the same manner.

95.13:f if the varying individual/undoubtedly transmit to them, as long

95.14:e The conditions might indeed act in so energetic and definite a manner as to lead to the same modification in all the individuals of the species without the aid of selection.

95.14:f There can also be little doubt that the tendency to vary in the same manner has often been so strong that all the individuals of the same species have been similarly modified without the aid of any form of selection.

95.15:e But we may suppose that the conditions sufficed to affect only a third, or fourth, or tenth part of the individuals; and several such cases could be given; for instance, it has been estimated by Graba that in the Faroe Islands about one-fifth of the guillemots, which all breed together, consist of a well-marked variety; and this was formerly ranked as a distinct species under the name of Uria lacrymans.

95.15.x-y:f Or only a third, fifth, or tenth part of the individuals may have been thus affected, of which fact several instances could be given. *y* Thus Graba estimates that about one-fifth of the guillemots in the Faroe Islands consist of a variety so well marked, that it was formerly ranked as a distinct species under the name of Uria lacrymans.

95.16:e Now, in such cases, if the variation were of a beneficial nature, the original form would soon be supplanted by the modified form, through the survival of the fittest.

95.16:f In cases of this kind, if

95.17:e[¶] With reference to the effects of intercrossing and of competition, it should be borne in mind that most animals and plants keep to their proper homes, and do not needlessly wander about; we see this even with migratory birds, which almost always return to the same district.

95.17:f To the effects of intercrossing in eliminating variations of all kinds, I shall have to recur; but it may be here remarked that/same spot.

95.18-19:e Consequently each newly-formed variety would generally be at first local, as seems to be the common rule with varieties in a state of nature; so that similarly modified individuals would soon exist in a small body together, and would often breed together. *19* If the new variety was successful in its battle for life, it would slowly spread from a central spot, competing with and conquering the unchanged individuals on the margins of an ever-increasing circle.

95.19:f variety were successful/central district, competing

95.20-22:e But to the subject of intercrossing we shall have to return. *21* It may be objected by those who have not attended to natural history, that the long-continued accumulation of individual differences could not give rise to parts or organs which seem to us, and are often called, new. *22* But, as we shall hereafter find, it is difficult to advance any good instance of a really new organ; even so complex and perfect an organ as the eye can be shown to graduate downwards into mere tissue sensitive to diffused light.

95.20-22:[f]

96 Let us now take a more complex case.

96:e It may be worth while to give another and more complex illustration of the action of natural selection.

97 Certain plants excrete a sweet juice, apparently for the sake of eliminating something injurious from their sap: this is effected by glands at the base of the stipules in some Leguminosæ, and at the back of the leaf of the common laurel.

97:e excrete sweet/effected, for instance, by/backs of the leaves

97:f from the sap

98 This juice, though small in quantity, is greedily sought by insects.

98:e insects; but their visits do not in any way benefit the plant.

99 Let us now suppose a little sweet juice or nectar to be excreted by the inner bases of the petals of a flower.

99:e Now, let us suppose that the juice or nectar was excreted from the inside of the flowers of a certain number of plants of any species.

100 In this case insects in seeking the nectar would get dusted with pollen, and would certainly often transport the pollen from one flower to the stigma of another flower.

100:e Insects/transported it from one flower to another.

100:f would often

101 The flowers of two distinct individuals of the same species

would thus get crossed; and the act of crossing, we have good reason to believe (as will hereafter be more fully alluded to), would produce very vigorous seedlings, which consequently would have the best chance of flourishing and surviving.

101:e crossing, as we have good reason to believe, would produce vigorous

101:f crossing, as can be fully proved, gives rise to vigorous

102 Some of these seedlings would probably inherit the nectar-excreting power.

102:c would almost certainly inherit

102:[e]

103 Those individual flowers which had the largest glands or nectaries, and which excreted most nectar, would be oftenest visited by insects, and would be oftenest crossed; and so in the long-run would gain the upper hand.

103:e The plants which produced flowers with the largest glands or nectaries, excreting most nectar, would oftenest be visited by insects, and would oftenest be crossed; and so in the long-run would gain the upper hand and form a local variety.

104 Those flowers, also, which had their stamens and pistils placed, in relation to the size and habits of the particular insects which visited them, so as to favour in any degree the transportal of their pollen from flower to flower, would likewise be favoured or selected.

104:d insects, which

104:e The flowers/insect which/transportal of the pollen, would likewise be favoured.

105 We might have taken the case of insects visiting flowers for the sake of collecting pollen instead of nectar; and as pollen is formed for the sole object of fertilisation, its destruction appears a simple loss to the plant; yet if a little pollen were carried, at first occasionally and then habitually, by the pollen-devouring insects from flower to flower, and a cross thus effected, although nine-tenths of the pollen were destroyed, it might still be a great gain to the plant; and those individuals which produced more and more pollen, and had larger and larger anthers, would be selected.

105:e sole purpose of fertilisation, its destruction appears to be a simple/gain to the plant; and the individuals/had larger anthers

105:f gain to the plant to be thus robbed; and the individuals

106　When our plant, by this process of the continued preservation or natural selection of more and more attractive flowers, had been rendered highly attractive to insects, they would, unintentionally on their part, regularly carry pollen from flower to flower; and that they can most effectually do this, I could easily show by many striking instances.

106:d can effectually

106:e by the above process long continued, had been rendered/that they do this effectually, I could easily show by many striking facts.

107 I will give only one—not as a very striking case, but as likewise illustrating one step in the separation of the sexes of plants, presently to be alluded to.

107:e one, as likewise/plants.

108 Some holly-trees bear only male flowers, which have four stamens producing rather a small quantity of pollen, and a rudimentary pistil; other holly-trees bear only female flowers; these have a full-sized pistil, and four stamens with shrivelled anthers, in which not a grain of pollen can be detected.

108:b producing a rather small

109 Having found a female tree exactly sixty yards from a male tree, I put the stigmas of twenty flowers, taken from different branches, under the microscope, and on all, without exception, there were pollen-grains, and on some a profusion of pollen.

109:e were a few pollen-grains, and on some a profusion.

110 As the wind had set for several days from the female to the male tree, the pollen could not thus have been carried.

111 The weather had been cold and boisterous, and therefore not favourable to bees, nevertheless every female flower which I examined had been effectually fertilised by the bees, accidentally dusted with pollen, having flown from tree to tree in search of nectar.

111:e by the bees, which had flown from tree to tree in search of nectar.

112 But to return to our imaginary case: as soon as the plant had been rendered so highly attractive to insects that pollen was regularly carried from flower to flower, another process might commence.

113 No naturalist doubts the advantage of what has been called the "physiological division of labour;" hence we may believe that it would be advantageous to a plant to produce stamens alone in one flower or on one whole plant, and pistils alone in another flower or on another plant.

114 In plants under culture and placed under new conditions of life, sometimes the male organs and sometimes the female organs become more or less impotent; now if we suppose this to occur in ever so slight a degree under nature, then as pollen is already carried regularly from flower to flower, and as a more complete separation of the sexes of our plant would be advantageous on the principle of the division of labour, individuals with this tendency more and more increased, would be continually favoured or selected, until at last a complete separation of the sexes would be effected.

114:d then, as pollen

114:e last a complete separation of the sexes might be

114.1:d It would take up too much space to show the various steps, through dimorphism and other means, by which the separation of the sexes in plants of various kinds is apparently now in progress; but I may add that some of the species of holly in North America are, according to Asa Gray, in an intermediate condition, or, as he expresses himself, the flowers are more or less diœciously polygamous.

114.1:e expresses it, are

114.1:f America, are, according to Asa Gray, in an exactly intermediate

115   Let us now turn to the nectar-feeding insects in our imaginary case: we may suppose the plant of which we have been slowly increasing the nectar by continued selection, to be a common plant; and that certain insects depended in main part on its nectar for food.

115:e insects: we may suppose the plant, of

116   I could give many facts, showing how anxious bees are to save time; for instance, their habit of cutting holes and sucking the nectar at the bases of certain flowers, which they can, with a very little more trouble, enter by the mouth.

116:d time: for

116:f facts showing/flowers, which with a very little more trouble, they can enter

117   Bearing such facts in mind, I can see no reason to doubt that an accidental deviation in the size and form of the body, or in the curvature and length of the proboscis, &c., far too slight to be appreciated by us, might profit a bee or other insect, so that an individual so characterised would be able to obtain its food more quickly, and so have a better chance of living and leaving descendants.

117:d mind, it may be believed that an accidental

118   Its descendants would probably inherit a tendency to a similar slight deviation of structure.

117 + 8:e that under certain circumstances individual differences in the curvature or length of the proboscis, &c., too/so that certain individuals would be able to obtain their food more quickly than others; and thus the communities to which they belonged would flourish and throw off many swarms inheriting the same peculiarities.

119   The tubes of the corollas of the common red and incarnate clovers (Trifolium pratense and incarnatum) do not on a hasty glance appear to differ in length; yet the hive-bee can easily suck the nectar out of the incarnate clover, but not out of the common red clover, which is visited by humble-bees alone; so

that whole fields of the red clover offer in vain an abundant supply of precious nectar to the hive-bee.

119:d corolla

119:e fields of the red clover in vain offer an abundant

119:f fields of the red clover offer in vain an abundant

119.0.1:d That this nectar is much liked by the hive-bee is certain; for I have repeatedly seen, but only in the autumn, many hive-bees sucking the flowers through holes in the base of the tube which had been bitten by humble-bees.

119.0.1:e holes bitten in the base of the tube by humble-bees.

119.1:c The difference in the length of the corolla which determines the visits of the hive-bee must be very trifling; for I have been informed, that when the red clover has been mown, the flowers of the second crop are somewhat smaller, and that these are abundantly visited by hive-bees.

119.1:d corolla in the two kinds of clover, which determines the visits of the hive-bee, must be very trifling; for I have been assured that when red

119.1:e these are visited by many hive-bees.

119.1.1:d I do not know whether this statement is accurate; nor whether another published statement can be trusted, namely, that the Ligurian bee, which is generally considered a mere variety and which freely crosses with the common hive-bee, is able to reach and suck the nectar of the common red clover.

119.1.1:f variety of the common hive-bee, and which freely crosses with it, is able to reach and suck the nectar of the red clover.

120 Thus it might be a great advantage to the hive-bee to have a slightly longer or differently constructed proboscis.

120:d Thus, in a country where this kind of clover abounded, it

121 On the other hand, I have found by experiment that the fertility of clover greatly depends on bees visiting and moving parts of the corolla, so as to push the pollen on to the stigmatic surface.

121:b clover depends

121:c hand, the fertility of clover, as previously stated, depends on bees moving the petals, so

122 Hence, again, if humble-bees were to become rare in any country, it might be a great advantage to the red clover to have a shorter or more deeply divided tube to its corolla, so that the hive-bee could visit its flowers.

121 + 2:d hand, as the fertility of this clover absolutely depends on bees moving the petals, if humble-bees/advantage to the plant to have a shorter or more deeply divided corolla, so that hive-bees should be induced to suck its flowers.

121 + 2:e bees visiting the flowers, if

121 + 2:f should be enabled to suck

123 Thus I can understand how a flower and a bee might slowly become, either simultaneously or one after the other, modified and adapted in the most perfect manner to each other, by the continued preservation of individuals presenting mutual and slightly favourable deviations of structure.

123:d presenting slight deviations of structure mutually favourable to each other.

123:e adapted to each other in the most perfect manner, by the continued preservation of all the individuals which presented slight

124　I am well aware that this doctrine of natural selection, exemplified in the above imaginary instances, is open to the same objections which were at first urged against Sir Charles Lyell's noble views on "the modern changes of the earth, as illustrative of geology;" but we now very seldom hear the action, for instance, of the coast-waves, called a trifling and insignificant cause, when applied to the excavation of gigantic valleys or to the formation of the longest lines of inland cliffs.

124:b now seldom

124:e hear the agencies still at work, spoken as trifling or insignificant, when applied to the excavation of the deepest valleys or to the formation of long lines

124:f agencies which we see still at work, spoken of as trifling or insignificant, when used in explaining the excavation of the deepest valleys or the formation

125 Natural selection can act only by the preservation and accumulation of infinitesimally small inherited modifications, each profitable to the preserved being; and as modern geology has almost banished such views as the excavation of a great valley by a single diluvial wave, so will natural selection, if it be a true principle, banish the belief of the continued creation of new organic beings, or of any great and sudden modification in their structure.

125:c of small

125:d sudden modifications

125:e selection acts/sudden modification

125:f will natural selection banish the belief

126　*On the Intercrossing of Individuals.—*

126:b [Center] *On the Intercrossing of Individuals.* [Space]

127 I must here introduce a short digression.

127:d[¶]

128 In the case of animals and plants with separated sexes, it is of course obvious that two individuals must always unite for each birth; but in the case of hermaphrodites this is far from obvious.

128:b always (with the exception of the curious and not well-understood cases of parthenogenesis) unite

185

129 Nevertheless I am strongly inclined to believe that with all hermaphrodites two individuals, either occasionally or habitually, concur for the reproduction of their kind.

129:e Nevertheless there is a reason to

130 This view, I may add, was first suggested by Andrew Knight.

130:b view was

130:f view was long ago doubtfully suggested by Sprengel, Knight and Kölreuter.

131 We shall presently see its importance; but I must here treat the subject with extreme brevity, though I have the materials prepared for an ample discussion.

132 All vertebrate animals, all insects, and some other large groups of animals, pair for each birth.

133 Modern research has much diminished the number of supposed hermaphrodites, and of real hermaphrodites a large number pair; that is, two individuals regularly unite for reproduction, which is all that concerns us.

134 But still there are many hermaphrodite animals which certainly do not habitually pair, and a vast majority of plants are hermaphrodites.

135 What reason, it may be asked, is there for supposing in these cases that two individuals ever concur in reproduction?

136 As it is impossible here to enter on details, I must trust to some general considerations alone.

137 In the first place, I have collected so large a body of facts, showing, in accordance with the almost universal belief of breeders, that with animals and plants a cross between different varieties, or between individuals of the same variety but of another strain, gives vigour and fertility to the offspring; and on the other hand, that *close* interbreeding diminishes vigour and fertility; that these facts alone incline me to believe that it is a general law of nature (utterly ignorant though we be of the meaning of the law) that no organic being self-fertilises itself for an eternity of generations; but that a cross with another individual is occasionally—perhaps at very long intervals—indispensable.

137:d being fertilises itself

137:e nature that no organic being fertilises itself for a perpetuity of generations/at long intervals of time—indispensable.

137:f facts, and made so many experiments, showing

138 On the belief that this is a law of nature, we can, I think, understand several large classes of facts, such as the following, which on any other view are inexplicable.

139 Every hybridizer knows how unfavourable exposure to wet is to the fertilisation of a flower, yet what a multitude of flowers have their anthers and stigmas fully exposed to the weather! but if an

186

occasional cross be indispensable, the fullest freedom for the entrance of pollen from another individual will explain this state of exposure, more especially as the plant's own anthers and pistil generally stand so close together that self-fertilisation seems almost inevitable.

139.x-y:e weather! *y* If an occasional cross be indispensable, notwithstanding that the plant's own anthers and pistil stand so near each other as almost to ensure self-fertilisation, the fullest freedom for the entrance of pollen from another individual will explain the above state of exposure of the organs.

140 Many flowers, on the other hand, have their organs of fructification closely enclosed, as in the great papilionaceous or pea-family; but in several, perhaps in all, such flowers, there is a very curious adaptation between the structure of the flower and the manner in which bees suck the nectar; for, in doing this, they either push the flower's own pollen on the stigma, or bring pollen from another flower.

140:c but in most of these flowers there is a very curious adaptation between their structure and the manner

140:e is a curious

140:f but these almost invariably present beautiful and curious adaptations in relation to the visits of insects.

141 So necessary are the visits of bees to papilionaceous flowers, that I have found, by experiments published elsewhere, that their fertility is greatly diminished if these visits be prevented.

141:c to many papilionaceous

141:f flowers, that their

142 Now, it is scarcely possible that bees should fly from flower to flower, and not carry pollen from one to the other, to the great good, as I believe, of the plant.

142:f possible for insects to fly/good of the plant.

143 Bees will act like a camel-hair pencil, and it is quite sufficient just to touch the anthers of one flower and then the stigma of another with the same brush to ensure fertilisation; but it must not be supposed that bees would thus produce a multitude of hybrids between distinct species; for if you bring on the same brush a plant's own pollen and pollen from another species, the former will have such a prepotent effect, that it will invariably and completely destroy, as has been shown by Gärtner, any influence from the foreign pollen.

143:f Insects act like a camel-hair pencil, and it is sufficient, to ensure fertilisation, just to touch with the same brush the anthers of one flower and then the stigma of another; but/if a plant's own pollen and that from another species are placed on the same stigma, the former is so prepotent that it invariably and completely destroys, as has been shown by Gärtner, the influence of the foreign

144   When the stamens of a flower suddenly spring towards the pistil, or slowly move one after the other towards it, the contrivance seems adapted solely to ensure self-fertilisation; and no doubt it is useful for this end: but, the agency of insects is often required to cause the stamens to spring forward, as Kölreuter has shown to be the case with the barberry; and curiously in this very genus, which seems to have a special contrivance for self-fertilisation, it is well known that if very closely-allied forms or varieties are planted near each other, it is hardly possible to raise pure seedlings, so largely do they naturally cross.

144:b barberry; and in/if closely-allied

144:d but the/that, if

145 In many other cases, far from there ɔeing any aids for self-fertilisation, there are special contrivances, as I could show from the writings of C. C. Sprengel and from my own observations, which effectually prevent the stigma receiving pollen from its own flower: for instance, in Lobelia fulgens, there is a really beautiful and elaborate contrivance by which every one of the infinitely numerous pollen-granules are swept out of the conjoined anthers of each flower, before the stigma of that individual flower is ready to receive them; and as this flower is never visited, at least in my garden, by insects, it never sets a seed, though by placing pollen from one flower on the stigma of another, I raised plenty of seedlings; and whilst other species of Lobelia growing close by, which is visited by bees, seeds freely.

145:c by which all the infinitely

145.x-y:f In numerous other cases, far from self-fertilisation being favoured, there are special contrivances which effectually prevent the stigma receiving pollen from its own flower, as I could show from the works of Sprengel and others, as well as from my own observations: for/seedlings. *y* Another species of Lobelia, which is visited by bees, seeds freely in my garden.

146 In very many other cases, though there be no special mechanical contrivance to prevent the stigma of a flower receiving its own pollen, yet, as C. C. Sprengel has shown, and as I can confirm, either the anthers burst before the stigma is ready for fertilisation, or the stigma is ready before the pollen of that flower is ready, so that these plants have in fact separated sexes, and must habitually be crossed.

146:f there is no/stigma receiving pollen from the same flower, yet, as Sprengel, and more recently Hildebrand, and others, have shown/these so-named dichogamous plants

146.1:d So it is with the reciprocally dimorphic and trimorphic plants previously alluded to.

147 How strange are these facts!

148 How strange that the pollen and stigmatic surface of the same flower, though placed so close together, as if for the very pur-

188

pose of self-fertilisation, should in so many cases be mutually useless to each other!

148:f should be in so many cases mutually

149 How simply are these facts explained on the view of an occasional cross with a distinct individual being advantageous or indispensable!

150 If several varieties of the cabbage, radish, onion, and of some other plants, be allowed to seed near each other, a large majority, as I have found, of the seedlings thus raised will turn out mongrels: for instance, I raised 233 seedling cabbages from some plants of different varieties growing near each other, and of these only 78 were true to their kind, and some even of these were not perfectly true.

150:f majority of the seedlings thus raised turn out, as I have found, mongrels

151 Yet the pistil of each cabbage-flower is surrounded not only by its own six stamens, but by those of the many other flowers on the same plant.

151:c plant; and the pollen of each flower readily gets on its own stigma without insect-agency; for I have found that a plant carefully protected produced the full number of pods.

151:f that plants carefully protected from insects produce

152 How, then, comes it that such a vast number of the seedlings are mongrelized?

153 I suspect that it must arise from the pollen of a distinct *variety* having a prepotent effect over a flower's own pollen; and that this is part of the general law of good being derived from the intercrossing of distinct individuals of the same species.

153:f It must

154 When distinct *species* are crossed the case is directly the reverse, for a plant's own pollen is always prepotent over foreign pollen; but to this subject we shall return in a future chapter.

154:c pollen is almost always

154:f is reversed, for

155 In the case of a gigantic tree covered with innumerable flowers, it may be objected that pollen could seldom be carried from tree to tree, and at most only from flower to flower on the same tree, and that flowers on the same tree can be considered as distinct individuals only in a limited sense.

155:e of a large tree

155:f same tree; and flowers

156 I believe this objection to be valid, but that nature has largely provided against it by giving to trees a strong tendency to bear flowers with separated sexes.

157 When the sexes are separated, although the male and female flowers may be produced on the same tree, we can see that

189

pollen must be regularly carried from flower to flower; and this will give a better chance of pollen being occasionally carried from tree to tree.

157:f tree, pollen

158 That trees belonging to all Orders have their sexes more often separated than other plants, I find to be the case in this country; and at my request Dr. Hooker tabulated the trees of New Zealand, and Dr. Asa Gray those of the United States, and the result was as I anticipated.

159 On the other hand, Dr. Hooker has recently informed me that he finds that the rule does not hold in Australia; and I have made these few remarks on the sexes of trees simply to call attention to the subject.

159.x-y:f Hooker informs me that the rule does not hold good in Australia; but if most of the Australian trees are dichogamous, the same result would follow as if they bore flowers with separated sexes. *y* I have made these few remarks on trees

160 Turning for a very brief space to animals: on the land there are some hermaphrodites, as land-mollusca and earth-worms; but these all pair.

160:f for a brief space to animals: various terrestrial species are hermaphrodites, such as the land-mollusca

161 As yet I have not found a single case of a terrestrial animal which fertilises itself.

161:f single terrestrial animal which can fertilise itself.

162 We can understand this remarkable fact, which offers so strong a contrast with terrestrial plants, on the view of an occasional cross being indispensable, by considering the medium in which terrestrial animals live, and the nature of the fertilising element; for we know of no means, analogous to the action of insects and of the wind in the case of plants, by which an occasional cross could be effected with terrestrial animals without the concurrence of two individuals.

162:f This remarkable/plants, is intelligible on the view of an occasional cross being indispensable; for owing to the nature of the fertilising element there are no means, analogous to the action of insects and of the wind with plants

163 Of aquatic animals, there are many self-fertilising hermaphrodites; but here currents in the water offer an obvious means for an occasional cross.

163:f here the currents of water

164 And, as in the case of flowers, I have as yet failed, after consultation with one of the highest authorities, namely, Professor Huxley, to discover a single case of an hermaphrodite animal with the organs of reproduction so perfectly enclosed within the body, that access from without and the occasional influence

of a distinct individual can be shown to be physically impossible.

164:f As in/single hermaphrodite/enclosed that access from without, and the occasional influence of a distinct individual, can

165 Cirripedes long appeared to me to present a case of very great difficulty under this point of view; but I have been enabled, by a fortunate chance, elsewhere to prove that two individuals, though both are self-fertilising hermaphrodites, do sometimes cross.

165:f present, under this point of view, a case of great difficulty; but I have been enabled, by a fortunate chance, to

166 It must have struck most naturalists as a strange anomaly that, in the case of both animals and plants, species of the same family and even of the same genus, though agreeing closely with each other in almost their whole organisation, yet are not rarely, some of them hermaphrodites, and some of them unisexual.

166:f that, both with animals and plants, some species/organisation, are hermaphrodites, and some unisexual.

167 But if, in fact, all hermaphrodites do occasionally intercross with other individuals, the difference between hermaphrodites and unisexual species, as far as function is concerned, becomes very small.

167:f intercross, the difference between them and unisexual species is, as far as function is concerned, very

168 From these several considerations and from the many special facts which I have collected, but which I am not here able to give, I am strongly inclined to suspect that, both in the vegetable and animal kingdoms, an occasional intercross with a distinct individual is a law of nature.

168:d suspect, **that**

168:f am unable here to give, it appears that with animals and plants an occasional intercross between distinct individuals is a very general, if not universal, law of nature.

169 I am well aware that there are, on this view, many cases of difficulty, some of which I am trying to investigate.

169:[f]

170 Finally then, we may conclude that in many organic beings, a cross between two individuals is an obvious necessity for each birth; in many others it occurs perhaps only at long intervals; but in none, as I suspect, can self-fertilisation go on for perpetuity.

170:[f]

171 *Circumstances favourable to Natural Selection.—*
171:d [*Center*] *Selection.* [*Space*]

171:e Circumstances favourable for the production of new forms through Natural Selection.

172 This is an extremely intricate subject.

172:d[¶]

173 A large amount of inheritable and diversified variability is favourable, but I believe mere individual differences suffice for the work.

173:c amount of variability will evidently be favourable for the work of natural selection, but mere individual differences probably suffice.

173:e A great amount of variability, under which term individual differences are always included, will evidently be favourable.

174 A large number of individuals, by giving a better chance for the appearance within any given period of profitable variations, will compensate for a lesser amount of variability in each individual, and is, I believe, an extremely important element of success.

174:e appearance of profitable variations within any given period, will compensate

174:f chance within any given period for the appearance of profitable variations, will/believe, a highly important

175 Though nature grants vast periods of time for the work of natural selection, she does not grant an indefinite period; for as all organic beings are striving, it may be said, to seize on each place in the economy of nature, if any one species does not become modified and improved in a corresponding degree with its competitors, it will soon be exterminated.

175:c Nature/striving to seize

175:e grants long periods/will be

175.1-2:c Unless favourable variations be inherited by some at least of the offspring, nothing can be effected by natural selection. 2 Non-inheritance of any new character is, in fact, the same thing as reversion to the character of the grandparents or more remote ancestors; and no doubt the tendency to reversion may often have checked or prevented the action of natural selection; but its importance has been greatly exaggerated by some writers.

175.2:[e]

175.3:c If the tendency to reversion has not prevented man from creating innumerable hereditary races in the animal and vegetable kingdoms, why should it have stopped the process of natural selection?

175.3:e The tendency to reversion may often check or prevent the work; but as this tendency has not prevented man from forming by selection numerous domestic races, why should it prevail against natural selection?

176　In man's methodical selection, a breeder selects for some definite object, and free intercrossing will wholly stop his work.

176:e In the case of methodical

176:f and if the individuals be allowed freely to intercross, his work will completely fail.

177　But when many men, without intending to alter the breed, have a nearly common standard of perfection, and all try to get and breed from the best animals, much improvement and modification surely but slowly follow from this unconscious process of selection, notwithstanding a large amount of crossing with inferior animals.

177:f try to procure and breed from the best animals, improvement surely but slowly follows from this unconscious process of selection, notwithstanding that there is no separation of selected individuals.

178　Thus it will be in nature; for within a confined area, with some place in its polity not so perfectly occupied as might be, natural selection will always tend to preserve all the individuals varying in the right direction, though in different degrees, so as better to fill up the unoccupied place.

178:e not perfectly occupied, natural

178:f be under nature/place in the natural polity not perfectly occupied, all/degrees, will tend to be preserved.

179　But if the area be large, its several districts will almost certainly present different conditions of life; and then if natural selection be modifying and improving a species in the several districts, there will be intercrossing with the other individuals of the same species on the confines of each.

179:d then, if

179:e be very large/then, if the same species undergoes modification in different parts, the newly-formed varieties will intercross on the confines of each district.

179:f be large/in different districts, the newly-formed

180　And in this case the effects of intercrossing can hardly be counterbalanced by natural selection always tending to modify all the individuals in each district in exactly the same manner to the conditions of each; for in a continuous area, the conditions will generally graduate away insensibly from one district to another.

180:b area, the physical conditions at least will

180:c physical conditions will

180:e But we shall see in the seventh chapter that intermediate varieties, inhabiting an intermediate district, whether the result of the crossing of other varieties, or originally formed with an intermediate character, will in the long run generally be supplanted by one of the varieties on either hand.

180:f in the sixth chapter that intermediate varieties, inhabiting intermediate districts, will in the long run generally be supplanted by one of the adjoining varieties.

181 The intercrossing will most affect those animals which unite for each birth, which wander much, and which do not breed at a very quick rate.

181:e Intercrossing will affect those animals most which unite for each birth and wander

181:f will chiefly affect those animals which

182 Hence in animals of this nature, for instance in birds, varieties will generally be confined to separated countries; and this I believe to be the case.

182:c and this I find to

182:e Hence with animals of this nature, for instance birds

183 In hermaphrodite organisms which cross only occasionally, and likewise in animals which unite for each birth, but which wander little and which can increase at a very rapid rate, a new and improved variety might be quickly formed on any one spot, and might there maintain itself in a body, so that whatever intercrossing took place would be chiefly between the individuals of the same new variety.

183:e With hermaphrodite/likewise with animals/little and can/ body and afterwards spread, so that the crossing would be chiefly between the individuals of the new variety living together in the same place.

183:f at a rapid/spread, so that the individuals of the new variety would chiefly cross together.

184 A local variety when once thus formed might subsequently slowly spread to other districts.

184:[e]

185 On the above principle, nurserymen always prefer getting seed from a large body of plants of the same variety, as the chance of intercrossing with other varieties is thus lessened.

185:e On this principle, nurserymen always prefer saving seed from a large body of plants, as the chance of intercrossing is

186 Even in the case of slow-breeding animals, which unite for each birth, we must not overrate the effects of intercrosses in retarding natural selection; for I can bring a considerable catalogue of facts, showing that within the same area, varieties of the same animal can long remain distinct, from haunting different stations, from breeding at slightly different seasons, or from varieties of the same kind preferring to pair together.

186:e case of animals which breed slowly and unite for each birth, we must not assume that the effects of natural selection will always be immediately overpowered by free intercrossing; for I can bring a considerable body of facts, showing that within the same area, varieties of the same animal may long

186:f Even with animals which unite for each birth, and which do not propagate rapidly, we must not assume that free intercrossing would always eliminate the effects of natural selection; for I can bring forward a considerable body of facts showing that within the same area, two varieties of the same animal may long remain distinct, from haunting different stations, from breeding at slightly different seasons, or from the individuals of each variety preferring

187   Intercrossing plays a very important part in nature in keeping the individuals of the same species, or of the same variety, true and uniform in character.

187:f nature by keeping

188 It will obviously thus act far more efficiently with those animals which unite for each birth; but I have already attempted to show that we have reason to believe that occasional intercrosses take place with all animals and with all plants.

188:e but as already stated we

188:f but, as already stated, we/all animals and plants.

189 Even if these take place only at long intervals, I am convinced that the young thus produced will gain so much in vigour and fertility over the offspring from long-continued self-fertilisation, that they will have a better chance of surviving and propagating their kind; and thus, in the long run, the influence of intercrosses, even at rare intervals, will be great.

189:c intervals, the young

189:e intervals of time, the young

189:f influence of crosses, even

190 If there exist organic beings which never intercross, uniformity of character can be retained amongst them, as long as their conditions of life remain the same, only through the principle of inheritance, and through natural selection destroying any which depart from the proper type; but if their conditions of life change and they undergo modification, uniformity of character can be given to their modified offspring, solely by natural selection preserving the same favourable variations.

190:e inheritance and/preserving similar favourable

190.x-y:f With respect to organic beings extremely low in the scale, which do not propagate sexually, nor conjugate, and which cannot possibly intercross, uniformity of character can be retained by them under the same conditions of life, only through the principle of inheritance, and through natural selection which will destroy any individuals departing from the proper type. *y* If the conditions of life change and the form undergoes modification, uniformity of character can be given to the modified

191   Isolation, also, is an important element in the process of natural selection.

191:e in the changes effected through natural

191:f in the modification of species through natural

192 In a confined or isolated area, if not very large, the organic and inorganic conditions of life will generally be in a great degree uniform; so that natural selection will tend to modify all the individuals of a varying species throughout the area in the same manner in relation to the same conditions.

192:e generally be almost uniform/all the varying individuals of the same species in the same manner.

193 Intercrosses, also, with the individuals of the same species, which otherwise would have inhabited the surrounding and differently circumstanced districts, will be prevented.

193:e Intercrossing with the inhabitants of the surrounding districts will, also, be

193:f be thus prevented.

193.1:e Moritz Wagner has lately published an interesting essay on this subject, and has shown that the service rendered by isolation in preventing crosses between newly formed varieties is probably greater even than I have supposed.

193.1:f than I supposed.

193.2:e But from reasons already assigned I can by no means agree with this naturalist, that migration and isolation are necessary for the formation of new species.

193.2:f necessary elements for

194 But isolation probably acts more efficiently in checking the immigration of better adapted organisms, after any physical change, such as of climate or elevation of the land, &c.; and thus new places in the natural economy of the country are left open for the old inhabitants to struggle for, and become adapted to, through modifications in their structure and constitution.

194:d struggle for and

194:e The importance of isolation is likewise great in preventing, after any physical change in the conditions, such as of climate, elevation of the land, &c., the immigration of better adapted organisms; and thus new places in the natural economy of the district are left/to.

194:f district will be left open to be filled up by the modification of the old inhabitants.

195 Lastly, isolation, by checking immigration and consequently competition, will give time for any new variety to be slowly improved; and this may sometimes be of importance in the production of new species.

195:e isolation will give time for a new

195:f be improved at a slow rate; and this may sometimes be of much importance.

196 If, however, an isolated area be very small, either from being

surrounded by barriers, or from having very peculiar physical conditions, the total number of the individuals supported on it will necessarily be very small; and fewness of individuals will greatly retard the production of new species through natural selection, by decreasing the chance of the appearance of favourable variations.

196:e of the inhabitants will be small; and this will retard the production of new species through natural selection, by decreasing the chances of the appearance of favourable individual differences.

196:f chances of favourable variations arising.

196.1:c[¶] The mere lapse of time by itself does nothing either for or against natural selection.

196.1:f nothing, either

196.2:c I state this because it has been erroneously asserted that the element of time is assumed by me to play an all-important part in natural selection, as if all species were necessarily undergoing slow modification from some innate law.

196.2:e time has been assumed by me to play an all-important part in modifying species, as if all were necessarily undergoing change through the action of some innate law.

196.2:f species, as if all the forms of life were necessarily undergoing change through some

196.3:c Lapse of time is only so far highly important, as it gives a better chance of beneficial variations arising, being selected, accumulated, and fixed, in relation to the slowly changing organic and inorganic conditions of life.

196.3:e far important, and its importance in this respect is great, that it/selected, increased, and fixed

196.3:f arising, and of their being selected, accumulated, and fixed.

196.4:c It likewise favours the *direct* action of new or changed physical conditions of life.

196.4:e favours the definite action of the conditions

196.4:f likewise tends to increase the direct action of the physical conditions of life, in relation to the constitution of each organism.

197   If we turn to nature to test the truth of these remarks, and look at any small isolated area, such as an oceanic island, although the total number of the species inhabiting it, will be found to be small, as we shall see in our chapter on geographical distribution; yet of these species a very large proportion are endemic,—that is, have been produced there, and nowhere else.

197:c Geographical Distribution

197:d it will

197:e although the number of the different species inhabiting it is small/yet of the species/else in the world.

197:f number of species inhabiting/yet of these species

198 Hence an oceanic island at first sight seems to have been highly favourable for the production of new species.

199 But we may thus greatly deceive ourselves, for to ascertain whether a small isolated area, or a large open area like a continent, has been most favourable for the production of new organic forms, we ought to make the comparison within equal times; and this we are incapable of doing.

199:e thus deceive

200 Although I do not doubt that isolation is of considerable importance in the production of new species, on the whole I am inclined to believe that largeness of area is of more importance, more especially in the production of species, which will prove capable of enduring for a long period, and of spreading widely.

200:c Although isolation/importance, especially

200:e isolation is of great importance in the production of new/ area is still more important, especially for the production of species, which shall prove

201 Throughout a great and open area, not only will there be a better chance of favourable variations arising from the large number of individuals of the same species there supported, but the conditions of life are infinitely complex from the large number of already existing species; and if some of these many species become modified and improved, others will have to be improved in a corresponding degree or they will be exterminated.

201:d are much more complex/degree, or

201:f variations, arising

202 Each new form, also, as soon as it has been much improved, will be able to spread over the open and continuous area, and will thus come into competition with many others.

202:f many other forms.

203 Hence more new places will be formed, and the competition to fill them will be more severe, on a large than on a small and isolated area.

203:[d]

204 Moreover, great areas, though now continuous, owing to oscillations of level, will often have recently existed in a broken condition, so that the good effects of isolation will generally, to a certain extent, have concurred.

204:d to former oscillations of level, will often have existed

204:f continuous, will often, owing to former oscillations of level, have existed in a broken condition; so

205 Finally, I conclude that, although small isolated areas probably have been in some respects highly favourable for the production of new species, yet that the course of modification will generally have been more rapid on large areas; and what is more important, that the new forms produced on large areas, which al-

ready have been victorious over many competitors, will be those that will spread most widely, will give rise to most new varieties and species, and will thus play an important part in the changing history of the organic world.

205:c play the most important

205.x-y:f areas have been in some/widely, and will give rise to the greatest number of new varieties and species. *y* They will thus play a more important

206 　We can, perhaps, on these views, understand some facts which will be again alluded to in our chapter on geographical distribution; for instance, that the productions of the smaller continent of Australia have formerly yielded, and apparently are now yielding, before those of the larger Europæo-Asiatic area.

206:c Geographical Distribution

206:f In accordance with this view, we can, perhaps, understand/ instance, the fact of the productions of the smaller continent of Australia now yielding before

207 Thus, also, it is that continental productions have everywhere become so largely naturalised on islands.

208 On a small island, the race for life will have been less severe, and there will have been less modification and less extermination.

209 Hence, perhaps, it comes that the flora of Madeira, according to Oswald Heer, resembles the extinct tertiary flora of Europe.

209:e it is that

209:f Hence, we can understand how it is that the flora of Madeira, according to Oswald Heer, resembles to a certain extent the extinct

210 All fresh-water basins, taken together, make a small area compared with that of the sea or of the land; and, consequently, the competition between fresh-water productions will have been less severe than elsewhere; new forms will have been more slowly formed, and old forms more slowly exterminated.

210.x-y:f land. *y* Consequently/forms will have been then more slowly produced, and

211 And it is in fresh water that we find seven genera of Ganoid fishes, remnants of a once preponderant order: and in fresh water we find some of the most anomalous forms now known in the world, as the Ornithorhynchus and Lepidosiren, which, like fossils, connect to a certain extent orders now widely separated in the natural scale.

211:e extent orders at present widely

211:f fresh-water basins that/widely sundered in

212 These anomalous forms may almost be called living fossils; they have endured to the present day, from having inhabited a

confined area, and from having thus been exposed to less severe competition.

212:d may be/and thus having been

212:e and having been exposed to less varied and therefore less

212:f and from having been exposed to less varied, and therefore less severe, competition.

213 To sum up the circumstances favourable and unfavourable to natural selection, as far as the extreme intricacy of the subject permits.

213:e unfavourable for the production of new species through natural

213:f up, as far as the extreme intricacy of the subject permits, the circumstances favourable and unfavourable for the production of new species through natural selection.

214 I conclude, looking to the future, that for terrestrial productions a large continental area, which will probably undergo many oscillations of level, and which consequently will exist for long periods in a broken condition, will be the most favourable for the production of many new forms of life, likely to endure long and to spread widely.

214:b condition, is the most

214:c I conclude that for terrestrial productions a large continental area, which has undergone many oscillations/condition, has been the most/life, fitted to endure

214:e level, will have been the most/endure for a long time and to

215 For the area will first have existed as a continent, and the inhabitants, at this period numerous in individuals and kinds, will have been subjected to very severe competition.

215:b area first existed

215:d to severe

215:e Whilst the area existed as a continent, the inhabitants will have been numerous in individuals and kinds, and will have been subjected

216 When converted by subsidence into large separate islands, there will still exist many individuals of the same species on each island: intercrossing on the confines of the range of each species will thus be checked: after physical changes of any kind, immigration will be prevented, so that new places in the polity of each island will have to be filled up by modifications of the old inhabitants; and time will be allowed for the varieties in each to become well modified and perfected.

216:c still have existed many/thus have been checked/immigration will have been prevented, so that new places in the polity of each island will have had to be filled/time will have been allowed

216:e range of each new species will have been checked

216:f up by the modification

217 When, by renewed elevation, the islands shall be re-converted into a continental area, there will again be severe competition: the most favoured or improved varieties will be enabled to spread: there will be much extinction of the less improved forms, and the relative proportional numbers of the various inhabitants of the renewed continent will again be changed; and again there will be a fair field for natural selection to improve still further the inhabitants, and thus produce new species.

217:c islands were re-converted/again have been severe/varieties will have been enabled to spread: there will have been much/ continent will again have been changed; and again there will have been a fair field

217:e inhabitants of the reunited continent

217:f been very severe/thus to produce

218 That natural selection will always act with extreme slowness, I fully admit.

218:d selection always acts

218:e selection acts

218:f selection generally acts

219 Its action depends on there being places in the polity of nature, which can be better occupied by some of the inhabitants of the country undergoing modification of some kind.

219:d better filled through some

219:e The result depends/modifications

219:f It can act only when there are places in the natural polity of a district which can be better occupied by the modification of some of its existing inhabitants.

220 The existence of such places will often depend on physical changes, which are generally very slow, and on the immigration of better adapted forms having been checked.

220:e forms being checked.

220:f The occurrence of such/which generally take place very slowly, and/being prevented.

221 But the action of natural selection will probably still oftener depend on some of the inhabitants becoming slowly modified; the mutual relations of many of the other inhabitants being thus disturbed.

221:c some few of the inhabitants becoming

221:e But the effects of natural

221:f As some few of the old inhabitants become modified, the mutual relations of others will often be disturbed; and this will create new places, ready to be filled up by better adapted forms; but all this will take place very slowly.

222 Nothing can be effected, unless favourable variations occur, and variation itself is apparently always a very slow process.

222:c always a slow

222:e Although all the individuals of the same species differ more or less from each other, differences of the right nature, better adapted to the then existing conditions, may not soon occur.

222:f differ in some slight degree from each other, it would often be long before differences of the right nature in various parts of the organisation might occur.

223 The process will often be greatly retarded by free intercrossing.

223:e The results will

223:f The result would often

224 Many will exclaim that these several causes are amply sufficient wholly to stop the action of natural selection.

224:e sufficient to neutralise the power of

225 I do not believe so.

226 On the other hand, I do believe that natural selection will always act very slowly, often only at long intervals of time, and generally on only a very few of the inhabitants of the same region at the same time.

226:b selection always acts very

226:d slowly, generally at only long

226:e But I/selection generally acts very slowly in effecting changes, at long intervals of time, and only on a few/region.

226:f selection will generally act very slowly, only at long

227 I further believe, that this very slow, intermittent action of natural selection accords perfectly well with what geology tells us of the rate and manner at which the inhabitants of this world have changed.

227:c this slow/inhabitants of the world

227:e that these slow, intermittent results of natural selection accord perfectly with

227:f results accord well with

228 Slow though the process of selection may be, if feeble man can do much by his powers of artificial selection, I can see no limit to the amount of change, to the beauty and infinite complexity of the coadaptations between all organic beings, one with another and with their physical conditions of life, which may be effected in the long course of time by nature's power of selection.

228:e much by artificial/power of selection or the survival of the fittest.

228:f beauty and complexity/which may have been effected in the long course of time through nature's power of selection, that is by the survival

229 *Extinction.—*

229:d [*Center*] *Extinction caused by Natural Selection.* [*Space*]

230 This subject will be more fully discussed in our chapter on Geology; but it must be here alluded to from being intimately connected with natural selection.

230:d[¶]

230:f must here be alluded

231 Natural selection acts solely through the preservation of variations in some way advantageous, which consequently endure.

232 But as from the high geometrical powers of increase of all organic beings, each area is already fully stocked with inhabitants, it follows that as each selected and favoured form increases in number, so will the less favoured forms decrease and become rare.

232:b geometrical ratio of increase

232:c From/already stocked with the full number of its existing inhabitants, and as most areas are already stocked with a great diversity of forms, it/number, so generally will

232:d that, as

232:e Owing to the high

232:f geometrical rate of increase of all organic beings, each area is already fully stocked with inhabitants; and it follows from this, that as the favoured forms increase in number, so, generally, will the less favoured decrease

233 Rarity, as geology tells us, is the precursor to extinction.

234 We can, also, see that any form represented by few individuals will, during fluctuations in the seasons or in the number of its enemies, run a good chance of utter extinction.

234:f We can see that any form which is represented by few individuals will run a good chance of utter extinction, during great fluctuations in the nature of the seasons, or from a temporary increase in the number of its enemies.

235 But we may go further than this; for as new forms are continually and slowly being produced, unless we believe that the number of specific forms goes on perpetually and almost indefinitely increasing, numbers inevitably must become extinct.

235:d increasing, many inevitably

235:f for, as new forms are produced, unless we admit that specific forms can go on indefinitely increasing in number, many old forms must become extinct.

236 That the number of specific forms has not indefinitely increased, geology shows us plainly; and indeed we can see reason why they should not have thus increased, for the number of places in the polity of nature is not indefinitely great,—not that we have any means of knowing that any one region has as yet got its maximum of species.

236:c geology tells us plainly; and we shall presently attempt to show why it is that the number of species throughout the world has not become immeasurably great.

236:f geology plainly tells us; and

237 Probably no region is as yet fully stocked, for at the Cape of Good Hope, where more species of plants are crowded together than in any other quarter of the world, some foreign plants have become naturalised, without causing, as far as we know, the extinction of any natives.

237:[c]

238 Furthermore, the species which are most numerous in individuals will have the best chance of producing within any given period favourable variations.

238:c We have seen that the species which are most numerous in individuals have

238:e producing favourable variations within any given period.

239 We have evidence of this, in the facts given in the second chapter, showing that it is the common species which afford the greatest number of recorded varieties, or incipient species.

239:c facts stated in/which offer the greatest

239:f common and diffused or dominant species which/varieties.

240 Hence, rare species will be less quickly modified or improved within any given period, and they will consequently be beaten in the race for life by the modified descendants of the commoner species.

240:f period; they/by the modified and improved descendants

241 From these several considerations I think it inevitably follows, that as new species in the course of time are formed through natural selection, others will become rarer and rarer, and finally extinct.

242 The forms which stand in closest competition with those undergoing modification and improvement, will naturally suffer most.

243 And we have seen in the chapter on the Struggle for Existence that it is the most closely-allied forms,—varieties of the same species, and species of the same genus or of related genera,— which, from having nearly the same structure, constitution, and habits, generally come into the severest competition with each other.

244 Consequently, each new variety or species, during the progress of its formation, will generally press hardest on its nearest kindred, and tend to exterminate them.

243 + 4:f other; consequently

245 We see the same process of extermination amongst our domesticated productions, through the selection of improved forms by man.

246 Many curious instances could be given showing how quickly new breeds of cattle, sheep, and other animals, and varieties of flowers, take the place of older and inferior kinds.

247 In Yorkshire, it is historically known that the ancient black cattle were displaced by the long-horns, and that these "were swept away by the short-horns" (I quote the words of an agricultural writer) "as if by some murderous pestilence."

248   *Divergence of Character.*—
248:d [Center] Divergence of Character. [Space]

249 The principle, which I have designated by this term, is of high importance on my theory, and explains, as I believe, several important facts.
249:c importance, and
249:d[¶]

250 In the first place, varieties, even strongly marked ones, though having somewhat of the character of species—as is shown by the hopeless doubts in many cases how to rank them—yet certainly differ from each other far less than do good and distinct species.
250:f differ far less from each other than

251 Nevertheless, according to my view, varieties are species in the process of formation, or are, as I have called them, incipient species.

252 How, then, does the lesser difference between varieties become augmented into the greater difference between species?

253 That this does habitually happen, we must infer from most of the innumerable species throughout nature presenting well-marked differences; whereas varieties, the supposed prototypes and parents of future well-marked species, present slight and ill-defined differences.

254 Mere chance, as we may call it, might cause one variety to differ in some character from its parents, and the offspring of this variety again to differ from its parent in the very same character and in a greater degree; but this alone would never account for so habitual and large an amount of difference as that between varieties of the same species and species of the same genus.
254:c between well-marked varieties
254:e large a degree of difference as that between the species of the same genus.

255   As has always been my practice, let us seek light on this head from our domestic productions.
255:e practice, I have sought light

256 We shall here find something analogous.
256.1:c It will be admitted that the production of races so different as short-horn and Hereford cattle, race and cart horses, the

several breeds of pigeons, &c., could never have been effected by the mere chance accumulation of variations of a similar character during many successive generations.

256.1:f accumulation of similar variations during

257 A fancier is struck by a pigeon having a slightly shorter beak; another fancier is struck by a pigeon having a rather longer beak; and on the acknowledged principle that "fanciers do not and will not admire a medium standard, but like extremes," they both go on (as has actually occurred with tumbler-pigeons) choosing and breeding from birds with longer and longer beaks, or with shorter and shorter beaks.

257:c In practice, a fancier is, for instance, struck by a pigeon having a slightly/with the sub-breeds of the tumbler-pigeon

258 Again, we may suppose that at an early period one man preferred swifter horses; another stronger and more bulky horses.

258:f period of history, the men of one nation or district required swifter horses, whilst those of another required stronger and bulkier horses.

259 The early differences would be very slight; in the course of time, from the continued selection of swifter horses by some breeders, and of stronger ones by others, the differences would become greater, and would be noted as forming two sub-breeds; finally, after the lapse of centuries, the sub-breeds would become converted into two well-established and distinct breeds.

259.x-y:f slight; but, in the course/horses in the one case, and of stronger ones in the other, the differences/sub-breeds. y Ultimately, after the lapse of centuries, these sub-breeds

260 As the differences slowly become greater, the inferior animals with intermediate characters, being neither very swift nor very strong, will have been neglected, and will have tended to disappear.

260:e strong, would have been

260:f differences became/would not have been used for breeding, and will thus have tended to disappear.

261 Here, then, we see in man's productions the action of what may be called the principle of divergence, causing differences, at first barely appreciable, steadily to increase, and the breeds to diverge in character both from each other and from their common parent.

261:f character, both

262 But how, it may be asked, can any analogous principle apply in nature?

263 I believe it can and does apply most efficiently, from the simple circumstance that the more diversified the descendants from any one species become in structure, constitution, and habits, by so much will they be better enabled to seize on many and

widely diversified places in the polity of nature, and so be enabled to increase in numbers.

263:c efficiently (though it was a long time before I saw how), from

264 We can clearly see this in the case of animals with simple habits.

264:c clearly discern this

265 Take the case of a carnivorous quadruped, of which the number that can be supported in any country has long ago arrived at its full average.

266 If its natural powers of increase be allowed to act, it can succeed in increasing (the country not undergoing any change in its conditions) only by its varying descendants seizing on places at present occupied by other animals: some of them, for instance, being enabled to feed on new kinds of prey, either dead or alive; some inhabiting new stations, climbing trees, frequenting water, and some perhaps becoming less carnivorous.

266:f power

267 The more diversified in habits and structure the descendants of our carnivorous animal became, the more places they would be enabled to occupy.

267:f become, the more places they will be

268 What applies to one animal will apply throughout all time to all animals—that is, if they vary—for otherwise natural selection can do nothing.

268:e can effect nothing.

269 So it will be with plants.

270 It has been experimentally proved, that if a plot of ground be sown with one species of grass, and a similar plot be sown with several distinct genera of grasses, a greater number of plants and a greater weight of dry herbage can thus be raised.

270:e herbage can be raised by the latter process.

270:f raised in the latter than in the former case.

271 The same has been found to hold good when first one variety and then several mixed varieties of wheat have been sown on equal spaces of ground.

271:c when one variety and several

272 Hence, if any one species of grass were to go on varying, and those varieties were continually selected which differed from each other in at all the same manner as distinct species and genera of grasses differ from each other, a greater number of individual plants of this species of grass, including its modified descendants, would succeed in living on the same piece of ground.

272:f and the varieties/in the same manner, though in a very slight degree, as do the distinct species and genera of grasses, a greater number of individual plants of this species, including

273 And we well know that each species and each variety of grass is annually sowing almost countless seeds; and thus, as it may be said, is striving its utmost to increase its numbers.

273:f And we know/seeds; and is thus striving, as it may be said, to the utmost to increase in number.

274 Consequently, I cannot doubt that in the course of many thousands of generations, the most distinct varieties of any one species of grass would always have the best chance of succeeding and of increasing in numbers, and thus of supplanting the less distinct varieties; and varieties, when rendered very distinct from each other, take the rank of species.

274:d thousand generations

274:e Consequently, in the course

274:f would have

275 The truth of the principle, that the greatest amount of life can be supported by great diversification of structure, is seen under many natural circumstances.

275:c principle that

276 In an extremely small area, especially if freely open to immigration, and where the contest between individual and individual must be severe, we always find great diversity in its inhabitants.

276:f be very severe

277 For instance, I found that a piece of turf, three feet by four in size, which had been exposed for many years to exactly the same conditions, supported twenty species of plants, and these belonged to eighteen genera and to eight orders, which shows how much these plants differed from each other.

278 So it is with the plants and insects on small and uniform islets; and so in small ponds of fresh water.

278:e islets; also in

278:f islets: also

279 Farmers find that they can raise most food by a rotation of plants belonging to the most different orders: nature follows what may be called a simultaneous rotation.

280 Most of the animals and plants which live close round any small piece of ground, could live on it (supposing it not to be in any way peculiar in its nature), and may be said to be striving to the utmost to live there; but, it is seen, that where they come into the closest competition with each other, the advantages of diversification of structure, with the accompanying differences of habit and constitution, determine that the inhabitants, which thus jostle each other most closely, shall, as a general rule, belong to what we call different genera and orders.

280:f supposing its nature not to be in any way peculiar), and may/ competition, the advantages

281 The same principle is seen in the naturalisation of plants through man's agency in foreign lands.

282 It might have been expected that the plants which have succeeded in becoming naturalised in any land would generally have been closely allied to the indigenes; for these are commonly looked at as specially created and adapted for their own country.

282:e plants which would succeed in becoming

283 It might, also, perhaps have been expected that naturalised plants would have belonged to a few groups more especially adapted to certain stations in their new homes.

284 But the case is very different; and Alph. De Candolle has well remarked in his great and admirable work, that floras gain by naturalisation, proportionally with the number of the native genera and species, far more in new genera than in new species.

284:f remarked, in his

285 To give a single instance: in the last edition of Dr. Asa Gray's 'Manual of the Flora of the Northern United States,' 260 naturalised plants are enumerated, and these belong to 162 genera.

286 We thus see that these naturalised plants are of a highly diversified nature.

287 They differ, moreover, to a large extent from the indigenes, for out of the 162 genera, no less than 100 genera are not there indigenous, and thus a large proportional addition is made to the genera of these States.

287:c 162 naturalised genera, no/made to the endemic genera of the United States.

287:d made to the genera naturally living in the United States.

287:f made to the genera now living

288 By considering the nature of the plants or animals which have struggled successfully with the indigenes of any country, and have there become naturalised, we can gain some crude idea in what manner some of the natives would have had to be modified, in order to have gained an advantage over the other natives; and we may, I think, at least safely infer that diversification of structure, amounting to new generic differences, would have been profitable to them.

288:b we may gain some crude/would have to be modified, in order to gain/ may at/differences, would be profitable to them.

288:f which have in any country struggled successfully with the indigenes, and have there/over their compatriots; and we may at least infer

289 The advantage of diversification in the inhabitants of the same region is, in fact, the same as that of the physiological

division of labour in the organs of the same individual body—a subject so well elucidated by Milne Edwards.

289:f diversification of structure in the inhabitants

290 No physiologist doubts that a stomach by being adapted to digest vegetable matter alone, or flesh alone, draws most nutriment from these substances.

290:b stomach adapted to digest

291 So in the general economy of any land, the more widely and perfectly the animals and plants are diversified for different habits of life, so will a greater number of individuals be capable of there supporting themselves.

292 A set of animals, with their organisation but little diversified, could hardly compete with a set more perfectly diversified in structure.

293 It may be doubted, for instance, whether the Australian marsupials, which are divided into groups differing but little from each other, and feebly representing, as Mr. Waterhouse and others have remarked, our carnivorous, ruminant, and rodent mammals, could successfully compete with these well-pronounced orders.

293:f these well-developed orders.

294 In the Australian mammals, we see the process of diversification in an early and incomplete stage of development.

294:d development. [*Space*]

294.1:d [*Center*] *The probable Action of Natural Selection, through Divergence of Character and Extinction, on the Descendants of a Common Ancestor.* [*Space*]

294.1:e The Probable Results of the Action of Natural Selection through Divergence of Character and Extinction, in the Descendants of a Common Ancestor.

294.1:f The Probable Effects of the Action/Extinction, on the Descendants

295 After the foregoing discussion, which ought to have been much amplified, we may, I think, assume that the modified descendants of any one species will succeed by so much the better as they become more diversified in structure, and are thus enabled to encroach on places occupied by other beings.

295:c may assume

295:e which has been much compressed, we

295:f succeed so

296 Now let us see how this principle of great benefit being derived from divergence of character, combined with the principles of natural selection and of extinction, will tend to act.

296:b of benefit

296:d extinction, tends

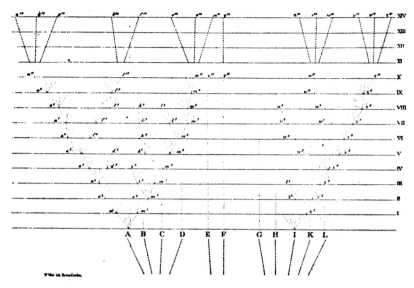

297   The accompanying diagram will aid us in understanding this rather perplexing subject.

298 Let A to L represent the species of a genus large in its own country; these species are supposed to resemble each other in unequal degrees, as is so generally the case in nature, and as is represented in the diagram by the letters standing at unequal distances.

299 I have said a large genus, because we have seen in the second chapter, that on an average more of the species of large genera vary than of small genera; and the varying species of the large genera present a greater number of varieties.

299:f because as we saw in the second chapter, on an average more species vary in large genera than in small genera; and

300 We have, also, seen that the species, which are the commonest and the most widely-diffused, vary more than rare species with restricted ranges.

300:d than do the rare and restricted species.

301 Let (A) be a common, widely-diffused, and varying species, belonging to a genus large in its own country.

302 The little fan of diverging dotted lines of unequal lengths proceeding from (A), may represent its varying offspring.

302:d The branching and diverging

303 The variations are supposed to be extremely slight, but of the most diversified nature; they are not supposed all to appear simultaneously, but often after long intervals of time; nor are they all supposed to endure for equal periods.

211

304 Only those variations which are in some way profitable will be preserved or naturally selected.

305 And here the importance of the principle of benefit being derived from divergence of character comes in; for this will generally lead to the most different or divergent variations (represented by the outer dotted lines) being preserved and accumulated by natural selection.

306 When a dotted line reaches one of the horizontal lines, and is there marked by a small numbered letter, a sufficient amount of variation is supposed to have been accumulated to have formed a fairly well-marked variety, such as would be thought worthy of record in a systematic work.

306:f accumulated to form it into a fairly

307 The intervals between the horizontal lines in the diagram, may represent each a thousand generations; but it would have been better if each had represented ten thousand generations.

307:e generations, or ten thousand.

307:f thousand or more generations.

308 After a thousand generations, species (A) is supposed to have produced two fairly well-marked varieties, namely a^1 and m^1.

309 These two varieties will generally continue to be exposed to the same conditions which made their parents variable, and the tendency to variability is in itself hereditary, consequently they will tend to vary, and generally to vary in nearly the same manner as their parents varied.

309:f generally still be/hereditary; consequently they will likewise tend to vary, and commonly in nearly the same manner as did their parents.

310 Moreover, these two varieties, being only slightly modified forms, will tend to inherit those advantages which made their common parent (A) more numerous than most of the other inhabitants of the same country; they will likewise partake of those more general advantages which made the genus to which the parent-species belonged, a large genus in its own country.

310:b their parent

310:f they will also partake

311 And these circumstances we know to be favourable to the production of new varieties.

311:f And all these circumstances are favourable

312 If, then, these two varieties be variable, the most divergent of their variations will generally be preserved during the next thousand generations.

313 And after this interval, variety a^1 is supposed in the diagram to have produced variety a^2, which will, owing to the principle of divergence, differ more from (A) than did variety a^1.

314 Variety m^1 is supposed to have produced two varieties, namely m^2 and s^2, differing from each other, and more considerably from their common parent (A).

315 We may continue the process by similar steps for any length of time; some of the varieties, after each thousand generations, producing only a single variety, but in a more and more modified condition, some producing two or three varieties, and some failing to produce any.

316 Thus the varieties or modified descendants, proceeding from the common parent (A), will generally go on increasing in number and diverging in character.

316:f descendants of the common

317 In the diagram the process is represented up to the ten-thousandth generation, and under a condensed and simplified form up to the fourteen-thousandth generation.

318 But I must here remark that I do not suppose that the process ever goes on so regularly as is represented in the diagram, though in itself made somewhat irregular.

318:d irregular, nor that it goes on continuously; it is far more probable that each form remains for long periods unaltered, and then again undergoes modification.

319 I am far from thinking that the most divergent varieties will invariably prevail and multiply: a medium form may often long endure, and may or may not produce more than one modified descendant; for natural selection will always act according to the nature of the places which are either unoccupied or not perfectly occupied by other beings; and this will depend on infinitely complex relations.

319:c varieties invariably

319:d Nor do I suppose that the most divergent varieties are invariably preserved: a medium

320 But as a general rule, the more diversified in structure the descendants from any one species can be rendered, the more places they will be enabled to seize on, and the more their modified progeny will be increased.

319:c progeny will increase.

321 In our diagram the line of succession is broken at regular intervals by small numbered letters marking the successive forms which have become sufficiently distinct to be recorded as varieties.

322 But these breaks are imaginary, and might have been inserted anywhere, after intervals long enough to have allowed the accumulation of a considerable amount of divergent variation.

323 As all the modified descendants from a common and widely-diffused species, belonging to a large genus, will tend to partake of the same advantages which made their parent successful

in life, they will generally go on multiplying in number as well as diverging in character: this is represented in the diagram by the several divergent branches proceeding from (A).

324 The modified offspring from the later and more highly improved branches in the lines of descent, will, it is probable, often take the place of, and so destroy, the earlier and less improved branches: this is represented in the diagram by some of the lower branches not reaching to the upper horizontal lines.

325 In some cases I do not doubt that the process of modification will be confined to a single line of descent, and the number of the descendants will not be increased; although the amount of divergent modification may have been increased in the successive generations.

325:f cases no doubt the process/number of modified descendants/been augmented.

326 This case would be represented in the diagram, if all the lines proceeding from (A) were removed, excepting that from a^1 to a^{10}.

327 In the same way, for instance, the English race-horse and English pointer have apparently both gone on slowly diverging in character from their original stocks, without either having given off any fresh branches or races.

327:f way the English race-horse

328　After ten thousand generations, species (A) is supposed to have produced three forms, a^{10}, f^{10}, and m^{10}, which, from having diverged in character during the successive generations, will have come to differ largely, but perhaps unequally, from each other and from their common parent.

329 If we suppose the amount of change between each horizontal line in our diagram to be excessively small, these three forms may still be only well-marked varieties; or they may have arrived at the doubtful category of sub-species; but we have only to suppose the steps in the process of modification to be more numerous or greater in amount, to convert these three forms into well-defined species: thus the diagram illustrates the steps by which the small differences distinguishing varieties are increased into the larger differences distinguishing species.

329:e varieties; but/into doubtful or at last well-defined

329.x-y:f well-defined species. *y* Thus

330 By continuing the same process for a greater number of generations (as shown in the diagram in a condensed and simplified manner), we get eight species, marked by the letters between a^{14} and m^{14}, all descended from (A).

331 Thus, as I believe, species are multiplied and genera are formed.

332 In a large genus it is probable that more than one species would vary.

333 In the diagram I have assumed that a second species (I) has produced, by analogous steps, after ten thousand generations, either two well-marked varieties (w^{10} and z^{10}) or two species, according to the amount of change supposed to be represented between the horizontal lines.

334 After fourteen thousand generations, six new species, marked by the letters n^{14} to z^{14}, are supposed to have been produced.

335 In each genus, the species, which are already extremely different in character, will generally tend to produce the greatest number of modified descendants; for these will have the best chance of filling new and widely different places in the polity of nature: hence in the diagram I have chosen the extreme species (A), and the nearly extreme species (I), as those which have largely varied, and have given rise to new varieties and species.

335:c The species of a genus which are/for they will

335:e In any genus, the species which are already very different in character from each other, will generally

335:f for these will have the best chance of seizing on new and widely

336 The other nine species (marked by capital letters) of our original genus, may for a long period continue transmitting unaltered descendants; and this is shown in the diagram by the dotted lines not prolonged far upwards from want of space.

336:b continue to transmit

336:e for long but unequal periods/lines unequally prolonged upwards.

337 But during the process of modification, represented in the diagram, another of our principles, namely that of extinction, will have played an important part.

338 As in each fully stocked country natural selection necessarily acts by the selected form having some advantage in the struggle for life over other forms, there will be a constant tendency in the improved descendants of any one species to supplant and exterminate in each stage of descent their predecessors and their original parent.

338:e original progenitor.

339 For it should be remembered that the competition will generally be most severe between those forms which are most nearly related to each other in habits, constitution, and structure.

340 Hence all the intermediate forms between the earlier and later states, that is between the less and more improved state of a species, as well as the original parent-species itself, will generally tend to become extinct.

340:c improved states

340:f improved states of the same species, as well as

341 So it probably will be with many whole collateral lines of descent, which will be conquered by later and improved lines of descent.

341:f improved lines.

342 If, however, the modified offspring of a species get into some distinct country, or become quickly adapted to some quite new station, in which child and parent do not come into competition, both may continue to exist.

342:e which offspring and progenitor do

343 If then our diagram be assumed to represent a considerable amount of modification, species (A) and all the earlier varieties will have become extinct, having been replaced by eight new species (a^{14} to m^{14}); and (I) will have been replaced by six (n^{14} to z^{14}) new species.

343:c If, then, our

343:f extinct, being replaced by eight new species (a^{14} to m^{14}); and species (I) will be replaced

344 But we may go further than this.

345 The original species of our genus were supposed to resemble each other in unequal degrees, as is so generally the case in nature; species (A) being more nearly related to B, C, and D, than to the other species; and species (I) more to G, H, K, L, than to the others.

346 These two species (A) and (I), were also supposed to be very common and widely diffused species, so that they must originally have had some advantage over most of the other species of the genus.

346:d (I) were

347 Their modified descendants, fourteen in number at the fourteen-thousandth generation, will probably have inherited some of the same advantages: they have also been modified and improved in a diversified manner at each stage of descent, so as to have become adapted to many related places in the natural economy of their country.

348 It seems, therefore, to me extremely probable that they will have taken the places of, and thus exterminated, not only their parents (A) and (I), but likewise some of the original species which were most nearly related to their parents.

348:c therefore, extremely

349 Hence very few of the original species will have transmitted offspring to the fourteen-thousandth generation.

350 We may suppose that only one (F), of the two species which

were least closely related to the other nine original species, has transmitted descendants to this late stage of descent.

350:e species (E and F) which

351   The new species in our diagram descended from the original eleven species, will now be fifteen in number.

352 Owing to the divergent tendency of natural selection, the extreme amount of difference in character between species a^{14} and z^{14} will be much greater than that between the most different of the original eleven species.

352:e most distinct of

353 The new species, moreover, will be allied to each other in a widely different manner.

354 Of the eight descendants from (A) the three marked a^{14}, q^{14}, p^{14}, will be nearly related from having recently branched off from a^{10}; b^{14} and f^{14}, from having diverged at an earlier period from a^{5}, will be in some degree distinct from the three first-named species; and lastly, o^{14}, e^{14}, and m^{14}, will be nearly related one to the other, but from having diverged at the first commencement of the process of modification, will be widely different from the other five species, and may constitute a sub-genus or even a distinct genus.

354:c b^{14}, and f^{14}

354:d but, from

354:e or a distinct genus.

355   The six descendants from (I) will form two sub-genera or even genera.

355:e or genera.

356 But as the original species (I) differed largely from (A), standing nearly at the extreme points of the original genus, the six descendants from (I) will, owing to inheritance, differ considerably from the eight descendants from (A); the two groups, moreover, are supposed to have gone on diverging in different directions.

356:b inheritance alone, differ

356:e extreme end of

357 The intermediate species, also (and this is a very important consideration), which connected the original species (A) and (I), have all become, excepting (F), extinct, and have left no descendants.

358 Hence the six new species descended from (I), and the eight descended from (A), will have to be ranked as very distinct genera, or even as distinct sub-families.

359   Thus it is, as I believe, that two or more genera are produced by descent, with modification, from two or more species of the same genus.

359:b descent with

360 And the two or more parent-species are supposed to have descended from some one species of an earlier genus.

360:f to be descended

361 In our diagram, this is indicated by the broken lines, beneath the capital letters, converging in sub-branches downwards towards a single point; this point representing a single species, the supposed single parent of our several new sub-genera and genera.

361:e supposed progenitor of

361:f point; this point represents a species, the supposed

362 It is worth while to reflect for a moment on the character of the new species F^{14}, which is supposed not to have diverged much in character, but to have retained the form of (F), either unaltered or altered only in a slight degree.

363 In this case, its affinities to the other fourteen new species will be of a curious and circuitous nature.

364 Having descended from a form which stood between the two parent-species (A) and (I), now supposed to be extinct and unknown, it will be in some degree intermediate in character between the two groups descended from these species.

364:f Being descended/these two species.

365 But as these two groups have gone on diverging in character from the type of their parents, the new species (F^{14}) will not be directly intermediate between them, but rather between types of the two groups; and every naturalist will be able to bring some such case before his mind.

365:f able to call such cases

366 In the diagram, each horizontal line has hitherto been supposed to represent a thousand generations, but each may represent a million or hundred million generations, and likewise a section of the successive strata of the earth's crust including extinct remains.

366:e million or several million generations; it may also represent a section

366:f represent a million or more generations

367 We shall, when we come to our chapter on Geology, have to refer again to this subject, and I think we shall then see that the diagram throws light on the affinities of extinct beings, which, though generally belonging to the same orders, or families, or genera, with those now living, yet are often, in some degree, intermediate in character between existing groups; and we can understand this fact, for the extinct species lived at very ancient epochs when the branching lines of descent had diverged less.

367:f orders, families/lived at various remote epochs

368　I see no reason to limit the process of modification, as now explained, to the formation of genera alone.

369　If, in our diagram, we suppose the amount of change represented by each successive group of diverging dotted lines to be very great, the forms marked a^{14} to p^{14}, those marked b^{14} and f^{14}, and those marked o^{14} to m^{14}, will form three very distinct genera.

369:e be great

369:f the diagram

370　We shall also have two very distinct genera descended from (I); and as these latter two genera, both from continued divergence of character and from inheritance from a different parent, will differ widely from the three genera descended from (A), the two little groups of genera will form two distinct families, or even orders, according to the amount of divergent modification supposed to be represented in the diagram.

370.x-y:e (I), differing widely from the descendants of (A). *y* These two groups of genera will thus form two distinct families, or orders

371　And the two new families, or orders, will have descended from two species of the original genus; and these two species are supposed to have descended from one species of a still more ancient and unknown genus.

371:e orders, are descended from two species of the original genus, and these are supposed to be descended from some still more ancient and unknown form.

372　We have seen that in each country it is the species of the larger genera which oftenest present varieties or incipient species.

372:e species belonging to the larger

373　This, indeed, might have been expected; for as natural selection acts through one form having some advantage over other forms in the struggle for existence, it will chiefly act on those which already have some advantage; and the largeness of any group shows that its species have inherited from a common ancestor some advantage in common.

373:f for, as

374　Hence, the struggle for the production of new and modified descendants, will mainly lie between the larger groups, which are all trying to increase in number.

374:c descendants will/groups which

375　One large group will slowly conquer another large group, reduce its numbers, and thus lessen its chance of further variation and improvement.

376　Within the same large group, the later and more highly perfected sub-groups, from branching out and seizing on many new places in the polity of Nature, will constantly tend to supplant and destroy the earlier and less improved sub-groups.

377 Small and broken groups and sub-groups will finally tend to disappear.

377:b finally disappear.

378 Looking to the future, we can predict that the groups of organic beings which are now large and triumphant, and which are least broken up, that is, which as yet have suffered least extinction, will for a long period continue to increase.

378:f that is, which have as yet suffered least extinction, will, for a long period, continue

379 But which groups will ultimately prevail, no man can predict; for we well know that many groups, formerly most extensively developed, have now become extinct.

379:f we know

380 Looking still more remotely to the future, we may predict that, owing to the continued and steady increase of the larger groups, a multitude of smaller groups will become utterly extinct, and leave no modified descendants; and consequently that of the species living at any one period, extremely few will transmit descendants to a remote futurity.

380:d consequently that, of

380:e that owing

381 I shall have to return to this subject in the chapter on Classification, but I may add that on this view of extremely few of the more ancient species having transmitted descendants, and on the view of all the descendants of the same species making a class, we can understand how it is that there exist but very few classes in each main division of the animal and vegetable kingdoms.

381:e that as according to this view extremely few of the more ancient species have transmitted descendants, and as all the descendants of the same species form a class, we can understand how it is that there exist so few

381:f as, according to this view, extremely/transmitted descendants to the present day, and, as/exists

382 Although extremely few of the most ancient species may now have living and modified descendants, yet at the most remote geological period, the earth may have been as well peopled with many species of many genera, families, orders, and classes, as at the present day.

382:e Although few of the most ancient species have left modified descendants, yet, at remote geological periods, the earth may have been almost as well peopled with species

382:f present time.

382.1:c[¶] *On the degree to which Organisation tends to advance.—*

382.1:d [Center] On the Degree/advance. [Space]

382.2:c Natural selection acts, as we have seen, exclusively by the preservation and accumulation of variations, which are beneficial under the organic and inorganic conditions of life to which each creature is at each successive period exposed.

382.2:d[¶] which have been beneficial/creature has been exposed at each successive period of time.

382.2:e Natural Selection exclusively acts by the preservation and accumulation of variations, which are beneficial/creature is exposed at each successive period of life.

382.2:f Selection acts exclusively by/at all periods of life.

382.3:c The ultimate result will be that each creature will tend to become more and more improved in relation to its conditions of life.

382.3:d result is that each creature tends

382.3:e relation to their conditions.

382.3:f relation to its conditions.

382.4:c This improvement will, I think, inevitably lead to the gradual advancement of the organisation of the greater number of living beings throughout the world.

382.4:d improvement inevitably leads to

382.5-7:c But here we enter on a very intricate subject, for naturalists have not defined to each other's satisfaction what is meant by an advance in organisation. *6* Amongst the vertebrata the degree of intellect and an approach in structure to man clearly come into play. *7* It might be thought that the amount of change which the various parts and organs undergo in their development from the embryo to maturity would suffice as a standard of comparison; but there are cases, as with certain parasitic crustaceans, in which several parts of the structure become less perfect, so that the mature animal cannot be called higher than its larva.

382.7:d organs pass through in their

382.8:c Von Baer's standard seems the most widely applicable and the best, namely, the amount of differentiation of the different parts (in the adult state, as I should be inclined to add) and their specialisation for different functions; or, as Milne Edwards would express it, the completeness of the division of physiological labour.

382.8:d parts of the same organic being (and, as I should be inclined to add, in the adult state), and their

382.8:e being, in the adult state as I should be inclined to add, and their

382.9:c But we shall see how obscure a subject this is if we look, for instance, to fish, amongst which some naturalists rank those as highest which, like the sharks, approach nearest to reptiles; whilst other naturalists rank the common bony or teleostean fishes as the highest, inasmuch as they are most strictly fish-like, and differ most from the other vertebrate classes.

382.9:e obscure this subject is if

382.9:f to fishes, amongst/nearest to amphibians; whilst

382.10:c Still more plainly we see the obscurity of the subject by turning to plants, with which the standard of intellect is of course quite excluded; and here some botanists rank those plants as highest which have every organ, as sepals, petals, stamens, and pistils, fully developed in each flower; whereas other botanists, probably with more truth, look at the plants which have their several organs much modified and somewhat reduced in number as being of the highest rank.

382.10:d plants, amongst which the standard

382.10:e modified and reduced in number as the highest.

382.10:f We see still more plainly the obscurity

382.11:c[¶] If we look at the differentiation and specialisation of the several organs of each being when adult (and this will include the advancement of the brain for intellectual purposes) as the best standard of highness of organisation, natural selection clearly leads towards highness; for all physiologists admit that the specialisation of organs, inasmuch as they perform in this state their functions better, is an advantage to each being; and hence the accumulation of variations tending towards specialisation is within the scope of natural selection.

382.11:e If we take, as the standard of high organisation, the amount of differentiation and specialisation of the several organs in each being when/purposes), natural selection clearly

382.11:f take as the standard of high/towards this standard: for/as in this state they perform their

382.12:c On the other hand, we can see, bearing in mind that all organic beings are striving to increase at a high ratio and to seize on every ill-occupied place in the economy of nature, that it is quite possible for natural selection gradually to fit an organic being to a situation in which several organs would be superfluous and useless: in such cases there might be retrogression in the scale of organisation.

382.12:d superfluous or useless

382.12:f every unoccupied or less well occupied place/fit a being/ there would be

382.13:c Whether organisation on the whole has actually advanced from the remotest geological periods to the present day will be more conveniently discussed in our chapter on Geological Succession.

382.14-16:c[¶] But it may be objected that if all organic beings thus tend to rise in the scale, how is it that throughout the world a multitude of the lowest forms still exist; and how is it that in each great class some forms are far more highly developed than others? *15* Why have not the more highly developed forms everywhere supplanted and exterminated the lower? *16* Lamarck, who believed in an innate and inevitable tendency towards perfection in all organic beings, seems to have felt this difficulty so strongly, that he was led to suppose that new and simple forms were continually being produced by spontaneous generation.

382.16:e forms are continually

382.17:c I need hardly say that Science in her present state does not countenance the belief that living creatures are now ever produced from inorganic matter.

382.17:e Science, however, under her present aspect does not countenance the belief, whatever the future may reveal, that living creatures are now being generated.

382.17:f Science has not as yet proved the truth of this belief, whatever the future may reveal.

382.18:c On my theory the present existence of lowly organised productions offers no difficulty; for natural selection includes no necessary and universal law of advancement or development —it only takes advantage of such variations as arise and are beneficial to each creature under its complex relations of life.

382.18:e On our theory the continued existence of lowly organisms offers no difficulty; for natural selection, or the survival of the fittest, does not necessarily include progressive development

382.19-20:c And it may be asked what advantage, as far as we can see, would it be to an infusorian animalcule—to an intestinal worm—or even to an earth-worm, to be highly organised? *20* If it were no advantage, these forms would be left by natural selection unimproved or but little improved; and might remain for indefinite ages in their present little advanced condition.

382.20:f left, by natural selection, unimproved/present lowly condition.

382.21-2:c And geology tells us that some of the lowest forms, as the infusoria and rhizopods, have remained for an enormous period in nearly their present state. *22* But to suppose that most of the many now existing low forms have not in the least advanced since the first dawn of life would be rash; for every naturalist who has dissected some of the beings now ranked as very low in the scale, must have been struck with their really wondrous and beautiful organisation.

382.22:f be extremely rash

382.23:c[¶] Nearly the same remarks are applicable if we look to the great existing differences in the grades of organisation which occur within almost every great group; for instance, to the co-existence of mammals and fish in the vertebrata,—to the co-existence of man and the ornithorhynchus amongst mammalia—or of the shark and amphioxus, which latter fish in the extreme simplicity of its structure closely approaches the invertebrate classes.

382.23:d shark and lancelet (Branchiostoma), which

382.23:e look to the different grades of organisation within each great group; for instance, in the vertebrata, to the co-existence of mammals and fish—amongst mammalia, to the co-existence of man and the ornithorhynchus—amongst fishes, to the co-existence of the shark and lancelet (Branchiostoma), which

382.23:f within the same great/shark and the lancelet (Amphioxus), which

382.24:c But mammals and fish hardly come into competition with each other; the advancement of certain mammals or of the whole class to the highest grade of organisation would not lead to their taking the place of, and thus exterminating, fishes.

382.24:e advancement of the whole class of mammals, or of certain members in this class, to the highest grade would

382.24:f place of fishes.

382.25:c Physiologists believe that the brain must be bathed by warm blood to be highly active, and this requires aërial respiration; so that warm-blooded mammals when inhabiting the water live under some disadvantages compared with fishes.

382.25:e disadvantages in comparison with

382.25:f water lie under a disadvantage in having to come continually to the surface to breathe.

382.26:c In this latter class, members of the shark family would not, it is probable, tend to supplant the amphioxus; the struggle for existence in the case of the amphioxus apparently will lie with members of the invertebrate classes.

382.26:d supplant the lancelet; the struggle for existence in the case of the lancelet apparently

382.26:e not tend to supplant the lancelet; for the lancelet, as I hear from Fritz Müller, has as sole companion and competitor on the barren sandy shore of South Brazil, an anomalous annelid.

382.26:f With fishes, members

382.27-8:c The three lowest orders of mammals, namely, marsupials, edentata, and rodents, co-exist in South America in the same region with numerous monkeys, and probably interfere little with each other. *28* Although organisation, on the whole, may have advanced and be advancing throughout the world, yet the scale will still present all degrees of perfection; for the high advancement of certain whole classes, or of certain members of each class, does not at all necessarily lead to the extinction of those groups with which they do not enter into close competition.

382.28:e be still advancing throughout the world, yet the scale will always present many degrees

382.29:c In some cases, as we shall hereafter see, lowly organised forms seem to have been preserved to the present day from inhabiting peculiar or isolated stations, where they have been subjected to less severe competition, and where they have existed in scanty numbers, which, as already explained, retards the chance of favourable variations arising.

382.29:e inhabiting confined or peculiar stations/and where their scanty numbers have retarded

382.29:f day, from

382.30:c[¶] Finally, I believe that lowly organised forms now exist in numbers throughout the world, and in nearly every class, from various causes.

382.30:e exist throughout the world, from various causes.

382.31:c In some cases favourable variations may never have arisen for natural selection to act on and accumulate.

382.31:e cases variations or individual differences of a favourable nature may

382.32-3:c In no case, probably, has time sufficed for the utmost possible amount of development. *33* In some few cases there may have been what we must call retrogression of organisation.

382.33:e there has been

382.34:c But the main cause lies in the circumstance that under very simple conditions of life a high organisation would be of no service,—possibly would be of actual disservice, as being of a more delicate nature, and more liable to be put out of order and thus injured.

382.34:e lies in the fact that/order and injured.

382.35:c[¶] A difficulty, diametrically opposite to this which we have just been considering, has been advanced, namely, looking to the dawn of life, when all organic beings, as we may imagine, presented the simplest structure, how could the first steps in advancement or in the differentiation and specialisation of parts have arisen?

382.35:d namely, that, looking

382.35:e Looking to the first dawn/may believe, presented the simplest structure, how, it has been asked, could the first steps in the advancement or differentiation of parts

382.35.1:d Mr. Herbert Spencer would probably answer that as soon as the most simple unicellular organism came by growth or division to be compounded of several cells, or became attached to any supporting surface, his law would come into action, namely, "that homologous units of any order become differentiated in proportion as their relations to incident forces become different."

382.35.1:e soon as simple/law "that/different" would come into action.

382.35.1:f that, as soon

382.36:c I can make no sufficient answer; and can only say that as we have no facts to guide us, all speculation on the subject would be baseless and useless.

382.36:d But as we have no facts to guide us, all speculation on the subject is useless.

382.36:f us, speculation on the subject is almost useless.

382.37:c It is, however, an error to suppose that there would be no struggle for existence, and, consequently, no natural selection, until many forms had been produced: variations in a single species inhabiting an isolated station might be beneficial, and through their preservation either the whole mass of individuals might become modified, or two distinct forms might arise.

382.37:e beneficial, and thus the whole mass of individuals might be modified

382.38:c But I must recur to what was stated towards the close of the Introduction, where I say that no one ought to feel surprise at much remaining as yet unexplained on the origin of species, if due allowance be made for our profound ignorance on the mutual relations of the inhabitants of the world during the many past epochs in its history. [*Space*]

382.38:d But, as I remarked towards the close of the Introduction, no

382.38:e if we make due allowance for/during the past

382.38:f world at the present time, and still more so during past ages.

VII.382.38.0.0.1:f [*Center*] CHAPTER VII. [*Space*]

382.38.1:d [*Center*] *Various Objections considered.* [*Space*]

VII.382.38.1:f MISCELLANEOUS OBJECTIONS TO THE THEORY OF NATURAL SELECTION. [*Space*]

VII.382.38.1.0.1:f Longevity—Modifications not necessarily simultaneous—Modifications apparently of no direct service—Progressive development—Characters of small functional importance, the most constant—Supposed incompetence of natural selection to account for the incipient stages of useful structures—Causes which interfere with the acquisition through natural selection of useful structures—Gradations of structure with changed functions—Widely different organs in members of the same class, developed from one and the same source—Reasons for disbelieving in great and abrupt modifications. [*Space*]

382.39:c[¶] I will here notice a few miscellaneous objections which have been advanced against my views, as some of the previous discussions may perhaps thus be made clearer.

382.39:d clearer; but it would be useless to discuss all of them, as they have been made by writers who have not taken the trouble to understand my views.

VII.382.39:f I WILL devote this chapter to the consideration of various miscellaneous/understand the subject.

382.39.1:d Thus a distinguished German naturalist has recently asserted that the weakest part of my theory is, that I consider all organic beings as imperfect: what I have really said is, that all are not as perfect in relation to the conditions under which they live, as they might be; and this is shown to be the case by so many native forms yielding their places in many quarters of the world to intruding and naturalised foreigners.

382.39.1:e to their conditions as/forms in many quarters of the world yielding their places to intruding foreigners which have become naturalised.

VII.382.39.1:f has asserted/perfect as they might have been in relation to their conditions; and this/world having yielded their places to intruding foreigners.

382.39.2:d Nor can all organic beings, even if at any one time perfectly adapted to their conditions of life, remain so, when these conditions slowly change; and no one will dispute that the physical conditions of each country, as well as the number and kind of its inhabitants, are liable to change.

382.39.2:e if they were at/when the conditions slowly change, unless they likewise change; and no/numbers and kinds

VII.382.39.2:f can organic/life, have remained so, when their conditions changed, unless they themselves likewise changed; and no/inhabitants, have undergone many mutations.

382.39.3-6:d Thus again, a French author, in opposition to the whole tenor of this volume, assumes that, according to my view, species undergo great and abrupt changes, and then he triumphantly asks how this is possible, seeing that such modified forms would be crossed by the many which have remained unchanged. *4* No doubt the small changes or variations which do occur are incessantly checked and retarded by intercrossing; but the frequent existence of varieties in the same country with the parent species shows that crossing does not necessarily prevent their formation; and in the still more frequent cases of local forms or geographical races, crossing cannot come into play. *5* It should also be borne in mind that the offspring from a cross between a modified and unmodified species tends partially to inherit the characters of both parents, and natural selection assuredly will preserve even slight approaches to any change of structure which is beneficial. *6* Moreover such crossed offspring, from partaking of the same constitution with the modified parent, and from being still exposed to the same conditions, will be far more liable than other individuals of the same species again to vary or be modified in a similar manner.

382.39.3-6:[e]

VII.382.39.6.0.1-4:f[¶] A critic has lately insisted, with some parade of mathematical accuracy, that longevity is a great advantage to all species, so that he who believes in natural selection "must arrange his genealogical tree" in such a manner that all the descendants have longer lives than their progenitors! *2* Cannot our critic conceive that a biennial plant or one of the lower animals might range into a cold climate and perish there every winter; and yet, owing to advantages gained through natural selection, survive from year to year by means of its seeds or ova?

3 Mr. E. Ray Lankester has recently discussed this subject, and he concludes, as far as its extreme complexity allows him to form a judgment, that longevity is generally related to the standard of each species in the scale of organisation, as well as to the amount of expenditure in reproduction and in general activity. *4* And these conditions have, it is probable, been largely determined through natural selection.

382.40:c It has been argued that as none of the animals and plants of Egypt, of which we know anything, have changed during the last 3000 years, so probably none have been modified in other parts of the world.

382.40:d[¶] modified in any other part of the world.

VII.382.40:f last three or four thousand years, so probably have none in any part

VII.382.40.0.1:f But, as Mr. G. H. Lewes has remarked, this line of argument proves too much, for the ancient domestic races figured on the Egyptian monuments, or embalmed, are closely similar or even identical with those now living; yet all naturalists admit that such races have been produced through the modification of their original types.

382.41:c The many animals which have remained unchanged since the commencement of the glacial period would have been an incomparably stronger case, for these have been exposed to great changes of climate and have migrated over great distances; whereas, in Egypt, during the last 3000 years, the conditions of life, as far as we know, have remained absolutely uniform.

VII.382.41:f period, would/last several thousand years, the conditions

382.42:c The fact of little or no modification having been effected since the glacial period would be of some avail against those who believe in the existence of an innate and necessary law of development, but is powerless against the doctrine of natural selection, which only implies that variations occasionally occurring in single species are under favourable conditions preserved.

382.42:d which implies only that variations occasionally occur in single species, and that these when favourable are preserved; but this will occur only at long intervals of time after changes in the conditions of each country.

382.42:e believe in an innate/selection or the survival of the fittest, which implies only that variations or individual differences of a favourable nature occasionally arise in a few species, and are then preserved.

VII.382.42:f would have been of some/implies that when variations or individual differences of a beneficial nature happen to arise, these will be preserved; but this will be effected only under certain favourable circumstances.

382.43:c As Mr. Fawcett has well asked, what would be thought of a man who argued that because he could show that Mont Blanc and the other Alpine peaks had exactly the same height 3000 years ago as at present, consequently that these mountains had never been slowly upraised, and that the height of other mountains in other parts of the world had not recently been increased by slow degrees?

382.43:d that, because

382.43:[e]

382.44:c[¶] It has been objected, if natural selection be so powerful, why has not this or that organ been recently modified and improved?

382.44:e powerful an agent, why

VII.382.44:[f]

382.45:c Why has not the proboscis of the hive-bee been lengthened so as to reach the nectar in the flower of the red-clover?

382.45:d nectar of

VII.382.45:[f]

382.46-7:c Why has not the ostrich acquired the power of flight? *47* But granting that these organs have happened to vary in the right direction, granting that there has been time sufficient for the slow work of natural selection, checked as it will be by intercrossing and the tendency to reversion, who will pretend that he knows the natural history of any one organic being sufficiently well to say whether any particular change would be to its advantage?

382.47:d these parts and organs have varied in the right direction—granting/say that any particular change would on the whole be

382.47:e selection, the effects being often checked as they will be by intercrossing and the tendency to reversion, who will pretend that he knows the life-history of any one/would be on the whole to its advantage?

VII.382.46-7:[f]

382.48-9:c Can we feel sure that a long proboscis would not be a disadvantage to the hive-bee in sucking the innumerable small flowers which it frequents? *49* Can we feel sure that a long proboscis would not, by correlation of growth, almost necessarily give increased size to other parts of the mouth, perhaps interfering with the delicate cell-constructing work?

382.49:e correlation, almost

VII.382.48-9:[f]

382.50:c In the case of the ostrich a moment's reflection will show that an enormous supply of food would be necessary in this bird of the desert, to supply force to move its huge body through the air.

382.50:d ostrich, a moment's

382.50:e show what an enormous supply of food would be necessary to supply force for this bird of the desert to move

VII.382.50:[f]

382.51:c But such ill-considered objections are hardly worth notice.

VII.382.51:[f]

382.52:c[¶] The celebrated palæontologist, Professor Bronn, in his German translation of this work, has advanced various good objections to my views, and other remarks in its favour.

382.53:c Of the objections, some seem to me unimportant, some few are owing to misapprehension, and some are incidentally noticed in various parts of this volume.

382.53:[e]

382.54:c On the erroneous supposition that all the species of a region are believed by me to be changing at the same time, he justly asks how it is that all the forms of life do not present a fluctuating and inextricably confused body? but it is sufficient for us if some few forms at any one time are variable, and few will dispute that this is the case.

382.54:[e]

382.55:c He asks, how can it be on the principle of natural selection that a variety should live in abundance side by side with the parent species; for the variety during its formation is supposed to have supplanted the intermediate forms between itself and the parent species, and yet it has not supplanted the parent species itself, for both are supposed now to live side by side?

382.55:d He asks, how, on the principle of natural selection, can a variety live in abundance side by side with the parent-species; for a variety during its formation is supposed to supplant the intermediate forms between itself and the parent-species, and yet it has not supplanted even the parent-species, for both now live together?

382.52 + 5:e[¶] The celebrated palæontologist, Bronn, at the close of his German translation of this work, asks, how, on the principle of natural selection, can a variety live side by side with the parent-species?

382.56:c If the variety and parent species have become fitted to slightly different habits of life, they might live together; though in the case of animals which freely cross and move about, varieties seem to be almost always confined to distinct localities.

382.56:d parent-species have become fitted for slightly

382.56:e If both have/life or conditions, they/cross and wander much about

382.57:c But is it the case that varieties of plants and of the lower animals are often found in abundance side by side with the parent forms?

382.57:d parent-forms

382.57:[e]

382.58:c Laying aside the polymorphic species in which the innumerable variations that occur seem neither advantageous nor disadvantageous to the species, and have not been fixed; laying aside also temporary variations, such as albinism, &c., my impression is that varieties and the supposed parent species are generally found, inhabiting either distinct stations, high land or low land, dry or moist districts, or distinct regions.

382.58:d aside polymorphic species, in which innumerable variations seem/found inhabiting distinct stations

382.58:e But if we put on one side polymorphic species, in which the variability seems to be of a peculiar nature, and all mere temporary variations, such as size, albinism, &c., the more permanent varieties are generally found, as far as I can judge, inhabiting

VII.382.56 + 8:f together; and if we lay on one side/can discover, inhabiting distinct stations,—such as high land or low land, dry or moist districts.

VII.382.56 + 8.o.o.1:f Moreover, in the case of animals which wander much about and cross freely, their varieties seem to be generally confined to distinct regions.

382.59:c[¶] Again, Professor Bronn truly remarks, that distinct species do not differ from each other in single characters alone, but in many; and he asks, how it comes that natural selection should always have simultaneously affected many parts of the organisation?

382.59:e[No ¶] Bronn also insists that distinct species never differ from each other only in single characters, but in many parts; and he asks, how it comes that natural selection should invariably have affected simultaneously many

VII.382.59:f[¶] it always comes that many parts of the organisation should have been modified at the same time through variation and natural selection?

382.60:c Probably the whole amount of difference has not been simultaneously effected; and the unknown laws of correlation will certainly account for, but not strictly explain, much simultaneous modification.

382.60.x-y:e But there is not the least necessity for believing that all the parts have been simultaneously modified; they may have been gained one after the other, and from being transmitted together, they appear to us as if simultaneously formed. *y* Correlation, however, will account for various parts changing, when any one part changes.

VII.382.60.x-y:f But there is no necessity for supposing that all the parts of any being have been simultaneously modified. *y* The most striking modifications, excellently adapted for some purpose, might, as was formerly remarked, be acquired by successive variations, if slight, first in one part and then in another; and as they would be transmitted all together, they would appear to us as if they had been simultaneously developed.

382.61:c Anyhow, we see in our domestic varieties the very same fact: though our domestic races may differ much in some one organ from the other races of the same species, yet the other parts of the organisation will always be found in some degree different.

382.61:d yet the remaining parts

382.61:e We have evidence of this in our domestic races, which though they may differ greatly in some one selected character, always differ to a certain extent in other characters.

VII.382.61:[f]

VII.382.61.0.0.1-5:f The best answer, however, to the above objection is afforded by those domestic races which have been modified, chiefly through man's power of selection, for some special purpose. *2* Look at the race and dray horse, or at the greyhound and mastiff. *3* Their whole frames and even their mental characteristics have been modified; but if we could trace each step in the history of their transformation,—and the latter steps can be traced,—we should not see great and simultaneous changes, but first one part and then another slightly modified and improved. *4* Even when selection has been applied by man to some one character alone,—of which our cultivated plants offer the best instances,—it will invariably be found that although this one part, whether it be the flower, fruit, or leaves, has been greatly changed, almost all the other parts have been slightly modified. *5* This may be attributed partly to the principle of correlated growth, and partly to so-called spontaneous variation.

382.62:c Professor Bronn likewise asks with striking effect how, for instance in the mouse or hare genus, natural selection will account for the several species (descended, I may remark, from a parent of unknown character) having longer or shorter tails, longer or shorter ears, and fur of different colours; how will it account for one species of plant having pointed and another species obtuse leaves?

382.62:d genus, can natural selection account for the several/colours; how can it account for one species of plant having pointed and another species blunt leaves?

382.62:e[¶] Bronn, again, asks how natural selection can account for differences between species, which appear to be of no service to these species, such as the length of the ears or tail, or the folds of the enamel in the teeth, of the several species of hares and mice?

VII.382.62.x-y:f A much more serious objection has been urged by Bronn, and recently by Broca, namely, that many characters appear to be of no service whatever to their possessors, and therefore cannot have been influenced through natural selection. *y* Bronn adduces the length of the ears and tails in the different species of hares and mice,—the complex folds of enamel in the teeth of many animals, and a multitude of analogous cases.

382.63-5:c I can give no definite answer to such questions; but I might ask in return, were these differences, on the doctrine of independent creation, formed for no purpose? *64* If of use, or if due to correlation of growth, they could assuredly be formed through the natural preservation of such useful or correlated variations. *65* I believe in the doctrine of descent with modification, notwithstanding that this or that particular change of structure cannot be accounted for, because this doctrine groups together and explains, as we shall see in the latter chapters, many general phenomena of nature.

382.65:d in the later chapters

382.63-5:[e]

382.65.0.1:e With respect to plants, this subject has been recently discussed by Nägeli in an admirable essay.

VII.382.65.0.1:f been discussed

382.65.0.2:e He admits that natural selection has effected much, but he urges that the families of plants differ chiefly from each other in morphological characters, which seem quite unimportant for the welfare of the species.

VII.382.65.0.2:f but he insists that/which appear to be quite

382.65.0.3:e He consequently believes in an innate tendency towards perfection or progressive development.

VII.382.65.0.3:f towards progressive and more perfect development.

382.65.0.4:e He specifies the arrangement of the cells in the tissues, and of the leaves on the axis, as cases in which natural selection would fail to act.

VII.382.65.0.4:f selection could not have acted.

382.65.0.5:e To these may be added the numerical divisions in the parts of the flower, the position of the ovules, the shape of the seed, when not of any use for dissemination, &c.

382.65.0.6-9:e Professor Weismann, in discussing Nägeli's essay, accounts for such differences by the nature of the varying organism under the action of certain conditions; and this is the same with what I have called the direct and definite action of the conditions of life, causing all or nearly all the individuals of the same species to vary in the same manner. *7* When we remember such cases as the formation of the more complex galls, and certain monstrosities, which cannot be accounted for by reversion, cohesion, &c., and sudden strongly-marked deviations of structure, such as the appearance of a moss-rose on a common rose, we must admit that the organisation of the individual is capable through its own laws of growth, under certain conditions, of undergoing great modifications, independently of the gradual accumulation of slight inherited modifications. *8* Various morphological differences probably come under this head, to which we shall recur; but many differences may at the present time be of high service, or may formerly have been so, although we are not able to perceive their use; and these will have been acted on by natural selection. *9* A still larger number of morphological differences may certainly be looked at as the necessary result—through pressure, the withdrawal or excess of nutriment, an early-formed part affecting a part subsequently developed, correlation, &c.—of other adaptive changes, through which all species must have passed during their long course of descent and modification.

382.65.0.10-12:e[¶] No one will maintain that we as yet know the uses of all the parts of any one plant, or the functions of each cell in any one organ. *11* Five or six years ago, endless peculiarities of structure in the flowers of orchids, great ridges and crests, and the relative positions of the various parts would have been considered as useless morphological differences; but now we know that they are of great service, and must have been under the dominion of natural selection. *12* No one at present can explain why the leaves in a spire diverge from each other at certain angles; but we can see that their arrangement is related to their standing at equal distances from the leaves on all sides; and we may reasonably expect that the angles will hereafter be shown to follow from some such cause, as the addition of new leaves to the crowded spire in the bud, as inevitably as the angles of a bee's cell follow from the manner in which the insects work together.

VII.382.65.0.6-12:[f]

VII.382.65.0.12.1-7:f[¶] There is much force in the above objection. *2* Nevertheless, we ought, in the first place, to be extremely cautious in pretending to decide what structures now are, or have formerly been, of use to each species. *3* In the second place, it should always be borne in mind that when one part is modified, so will be other parts, through certain dimly seen causes, such as an increased or diminished flow of nutriment to a part, mutual pressure, an early developed part affecting one subsequently developed, and so forth,—as well as through other causes which lead to the many mysterious cases of correlation, which we do not in the least understand. *4* These agencies may be all grouped together, for the sake of brevity, under the expression of the laws of growth. *5* In the third place, we have to allow for the direct and definite action of changed conditions of life, and for so-called spontaneous variations, in which the nature of the conditions apparently plays a quite subordinate part. *6* Bud-variations, such as the appearance of a moss-rose on a common rose, or of a nectarine on a peach-tree, offer good instances of spontaneous variations; but even in these cases, if we bear in mind the power of a minute drop of poison in producing complex galls, we ought not to feel too sure that the above variations are not the effect of some local change in the nature of the sap, due to some change in the conditions. *7* There must be some efficient cause for each slight individual difference, as well as for more strongly marked variations which occasionally arise; and if the unknown cause were to act persistently, it is almost certain that all the individuals of the species would be similarly modified.

VII.382.65.0.12.8-10:f[¶] In the earlier editions of this work I under-rated, as it now seems probable, the frequency and importance of modifications due to spontaneous variability. *9* But it is impossible to attribute to this cause the innumerable structures which are so well adapted to the habits of life of each species. *10* I can no more believe in this, than that the well-adapted form of a race-horse or greyhound, which before the principle of selection by man was well understood, excited so much surprise in the minds of the older naturalists, can thus be explained.

VII.382.65.0.12.11-4:f[¶] It may be worth while to illustrate some of the foregoing remarks. *12* With respect to the assumed inutility of various parts and organs, it is hardly necessary to observe that even in the higher and best-known animals many structures exist, which are so highly developed that no one doubts that they are of importance, yet their use has not been, or has only recently been, ascertained. *13* As Bronn gives the length of the ears and tail in the several species of mice as instances, though trifling ones, of differences in structure which can be of no special use, I may mention that, according to Dr. Schöbl, the external ears of the common mouse are supplied in an extraordinary manner with nerves, so that they no doubt serve as tactile organs; hence the length of the ears can hardly be quite unimportant. *14* We shall, also, presently see that the tail is a highly useful prehensile organ to some of the species; and its use would be much influenced by its length.

VII.382.65.0.12.15-6:f[¶] With respect to plants, to which on account of Nägeli's essay I shall confine myself in the following remarks, it will be admitted that the flowers of orchids present a multitude of curious structures, which a few years ago would have been considered as mere morphological differences without any special function; but they are now known to be of the highest importance for the fertilisation of the species through the aid of insects, and have probably been gained through natural selection. *16* No one until lately would have imagined that in dimorphic and trimorphic plants the different lengths of the stamens and pistils, and their arrangement, could have been of any service, but now we know this to be the case.

382.65.0.13:e[¶] In certain whole groups of plants the ovules stand erect, and in others they are suspended; and in some few plants within the same ovarium one ovule holds the former and a second ovule the latter position.

VII.382.65.0.13:f suspended; and within the same ovarium of some few plants, one

382.65.0.14:e These positions seem at first purely morphological and of no physiological signification; but Dr. Hooker informs me that of the ovules within the same ovarium, in some cases the upper ones alone and in other cases the lower ones alone are fertilised; and he suggests that this probably depends on the direction in which the pollen-tubes enter.

382.65.0.14:f or of no/that within the same ovarium, the upper ovules alone in some cases, and in other cases/enter the ovarium.

382.65.0.15:e If so, the position of the ovules, even when one is erect and the other suspended, would follow from the selection of any slight deviation in position which might favour their fertilisation and the production of seed.

VII.382.65.0.15:f suspended within the same ovarium, would follow from the selection of any slight deviations in position which favoured their fertilisation, and the production

382.65.0.16:e[¶] Several plants belonging to distinct orders habitually produce flowers of two kinds,—the one open and of the ordinary structure, the other closed and imperfect.

VII.382.65.0.16.1-4:f These two kinds of flowers sometimes differ wonderfully in structure, yet may be seen to graduate into each other on the same plant. *2* The ordinary and open flowers can be intercrossed; and the benefits which certainly are derived from this process are thus secured. *3* The closed and imperfect flowers are, however, manifestly of high importance, as they yield with the utmost safety a large stock of seed, with the expenditure of wonderfully little pollen. *4* The two kinds of flowers often differ much, as just stated, in structure.

382.65.0.17:e In the latter the petals are almost always reduced to the merest rudiments; the pollen-grains are reduced in diameter; five of the alternate stamens are rudimentary in Ononis columnæ; and in some species of Viola three stamens are in this state, two retaining their proper function, but being of very small size.

VII.382.65.0.17.x-y:f The petals in the imperfect flowers almost always consist of mere rudiments, and the pollen-grains are reduced in diameter. *y* In Ononis columnæ five of the alternate stamens are rudimentary; and in some species

382.65.0.18:e In six out of thirty of the closed flowers in an Indian violet (name unknown, for the plants have not as yet produced perfect flowers), the sepals were reduced from the normal number of five to three.

VII.382.65.0.18:f plants have never produced with me perfect flowers), the sepals are reduced

382.65.0.19-21:e In one section of the Malpighiaceæ the closed flowers, according to A. de Jussieu, are still further modified, for the five stamens which stand opposite to the sepals are all aborted, a sixth stamen standing opposite to a petal being alone developed; and this stamen is not present in the ordinary flowers of these species; the style is aborted; and the ovaria are reduced from three to two. *20* In all the foregoing plants the minute closed flowers are of high service, for they yield with perfect security, and with the expenditure of extremely little pollen, or other organised matter, a large supply of seed; whilst the perfect flowers permit occasional crosses with distinct individuals.

21 Therefore, these changes may have been, and no doubt have been, effected through natural selection; and I may add that nearly all the gradations between the perfect and imperfect flowers may sometimes be observed on the same plant.

VII.382.65.0.20-1:[f]

VII.382.65.0.21.1:f Now although natural selection may well have had the power to prevent some of the flowers from expanding, and to reduce the amount of pollen, when rendered by the closure of the flower superfluous, yet hardly any of the above special modifications can have been thus determined, but must have followed from the laws of growth, including the functional inactivity of parts, during the progress of the reduction of the pollen and the closure of the flowers.

382.65.0.22:e[¶] With respect to modifications which necessarily follow from other changes—through the withdrawal or excess of nutriment—through pressure and other unknown influences —there is space here only for a few brief illustrations.

VII.382.65.0.22:[f]

VII.382.65.0.22.1:f[¶] It is so necessary to appreciate the important effects of the laws of growth, that I will give some additional cases of another kind, namely of differences in the same part or organ, due to differences in relative position on the same plant.

382.65.0.23-5:e In the Spanish chesnut, and in certain fir-trees, the angles of divergence of the leaves differ, according to Schacht, in the nearly horizontal and in the upright branches. *24* In the common rue and some other plants, one flower, usually the central or terminal one, opens first, and has five sepals and petals, and five divisions to the ovarium; whilst all the other flowers on the plant are tetramerous. *25* In the British Adoxa the uppermost flower generally has two calyx-lobes with the other organs tetramerous, whilst the surrounding flowers generally have three calyx-lobes with the other organs pentamerous; and this difference appears to follow from the manner in which the flowers are closely packed together.

VII.382.65.0.25:f pentamerous.

382.65.0.26-7:e In many Compositæ and Umbelliferæ, and in some other plants, the circumferential flowers have their corollas much more developed than those of the centre; and this is probably the result of natural selection, for all the flowers are thus rendered much more conspicuous to those insects which are useful or even necessary for their fertilisation. *27* In connection with the greater development of the corolla, the reproductive organs are frequently more or less aborted.

VII.382.65.0.26 + 7:f Umbelliferæ (and in some other plants) the circumferential/centre; and this seems often connected with the abortion of the reproductive organs.

382.65.0.28:e It is a more curious fact that the achenes or seeds of the circumference and of the centre sometimes differ greatly in form, colour, and other characters.

VII.382.65.0.28:f fact, previously referred to, that the achenes or seeds of the circumference and centre

382.65.0.29-30:e In Carthamus and some other Compositæ the central achenes alone are furnished with a pappus; and in Hyoseris the same head yields achenes of three different forms. *30* In certain Umbelliferæ the exterior seeds, according to Tausch, are orthospermous, and the central one cœlospermous, and this difference has been considered by De Candolle as of the highest systematic importance in the family.

VII.382.65.0.30:f cœlospermous, and this is a character which was considered by De Candolle to be in other species of the highest systematic importance.

VII.382.65.0.30.1-2:f Prof. Braun mentions a Fumariaceous genus, in which the flowers in the lower part of the spike bear oval, ribbed, one-seeded nutlets; and in the upper part of the spike, lanceolate, two-valved, and two-seeded siliques. *2* In these several cases, with the exception of that of the well developed ray-florets, which are of service in making the flowers conspicuous to insects, natural selection cannot, as far as we can judge, have come into play, or only in a quite subordinate manner.

382.65.0.31:e If in such cases as the foregoing all the leaves, flowers, fruits, &c., on the same plant had been subjected to precisely the same external and internal conditions, all no doubt would have presented the same morphological characters; and there clearly would have been no need to call in the aid of the principle of progressive development.

VII.382.65.0.31:f All these modifications follow from the relative position and inter-action of the parts; and it can hardly be doubted that if all the flowers and leaves on the same plant had been subjected to the same external and internal condition, as are the flowers and leaves in certain positions, all would have been modified in the same manner.

382.65.0.32:e With the minute closed flowers, as well as with many degraded parasitic animals, if it be assumed that any such aid is requisite, we should have to call in an innate tendency to retrogressive development.

VII.382.65.0.32:[f]

382.65.0.33:e[¶] Many instances could be given of morphological characters varying greatly in plants of the same species growing close together, or even on the same individual plant; and some of these characters are considered as systematically important.

VII.382.65.0.33:f In numerous other cases we find modifications of structure, which are considered by botanists to be generally of a highly important nature, affecting only some of the flowers on the same plant, or occurring on distinct plants, which grow close together under the same conditions.

VII.382.65.0.33.1-2:f As these variations seem of no special use to the plants, they cannot have been influenced by natural selection. *2* Of their cause we are quite ignorant; we cannot even attribute them, as in the last class of cases, to any proximate agency, such as relative position.

382.65.0.34:e I will specify only a few cases which have first occurred to me.

VII.382.65.0.34:f I will give only a few instances.

382.65.0.35:e It is not necessary to give instances of flowers on the same plant being indifferently tetramerous, pentamerous, &c.; but as when the parts are few, numerical variations are in all cases comparatively rare, I may mention that, according to De Candolle, the flowers of Papaver bracteatum offer two sepals with four petals (and this is the common type with poppies), or three sepals with six petals.

VII.382.65.0.35:f It is so common to observe on the same plant, flowers indifferently tetramerous, pentamerous, &c., that I need not give examples; but as numerical variations are comparatively rare when the parts are few, I/offer either two sepals with four petals (which is

382.65.0.36:e The manner in which the petals are folded in the bud is in most groups a constant morphological character; but Professor Asa Gray states that with some species of Mimulus, the æstivation is almost as frequently that of the Rhinanthideæ as of the Antirrhinideæ, to which tribe the genus belongs.

VII.382.65.0.36:f groups a very constant/which latter tribe

382.65.0.37:e Aug. St. Hilaire gives the following cases: the genus Zanthoxylon belongs to a division of the Rutaceæ with a single ovary, but in some of the species flowers may be found on the same plant, and even in the same panicle, with either one or two ovaries.

VII.382.65.0.37:f some species

382.65.0.38:e In Helianthemum the capsule has been described as unilocular or 3-locular; and in H. mutabile, "Une lame, plus ou moins large, s'étend entre le pericarpe et le placenta."

VII.382.65.0.38:f plus ou moins large

382.65.0.39:e In the flowers of Saponaria officinalis, Dr. Masters also observed instances of both marginal and free central placentation.

VII.382.65.0.39:f Masters has observed

382.65.0.40:e Lastly, St. Hilaire found towards the southern extreme of the range of Gomphia oleæformis two forms which he did not at first doubt were distinct species, but he subsequently saw them growing on the same bush; and he then adds, "Voilà donc dans un même individu des loges et un style qui se rattachent tantôt à un axe verticale et tantôt à un gynobase."

VII.382.65.0.40.1:f[¶] We thus see that with plants many morphological changes may be attributed to the laws of growth and the inter-action of parts, independently of natural selection.

382.65.0.41:e[¶] In the case of these plants, will it be said that they have been detected in the act of progressing towards a higher state of development?

VII.382.65.0.41:f[No ¶] But with respect to Nägeli's doctrine of an innate tendency towards perfection or progressive development, can it be said in the case of these strongly pronounced variations, that the plants have been caught in the act

382.65.0.42:e On the contrary, I should infer from such characters varying so greatly, that they were of extremely small importance to the plants themselves, of whatever importance they may be to us in our classifications.

VII.382.65.0.42:f from the mere fact of the parts in question differing or varying greatly on the same plant, that such modifications were/may generally be to us for our

VII.382.65.0.42.1:f The acquisition of a useless part can hardly be said to raise an organism in the natural scale; and in the case of the imperfect, closed flowers above described, if any new principle has to be invoked, it must be one of retrogression rather than of progression; and so it must be with many parasitic and degraded animals.

382.65.0.43:e Although we are quite ignorant of the exciting cause of each modification, yet it seems probable from what we know of the relations of variability to changed conditions, that under certain conditions the one structure would have prevailed over the other, and thus might have been rendered almost or quite constant.

VII.382.65.0.43:f We are ignorant of the exciting cause of the above specified modifications; but if the unknown cause were to act almost uniformly for a length of time, we may infer that the result would be almost uniform; and in this case all the individuals of the species would be modified in the same manner.

382.65.0.44:e From the very fact of such differences being unimportant for the welfare of the species, any slight deviations which did occur would not be augmented or accumulated through natural selection; and they would be liable to obliteration through the occasional intercrossing of distinct individuals.

VII.382.65.0.44:f[¶] From the fact of the above characters being unimportant for the welfare of the species, any slight varations which occurred in them would not have been accumulated and augmented through natural selection.

382.65.0.45:e A structure which has been developed through long-continued selection, when it ceases to be of service to the species, will generally become variable, as we see with rudimentary organs; for it will no longer be regulated by this same power of selection; but on the other hand, when from the nature of the organism and from a change in the conditions definite modifications have been produced which are unimportant for the welfare of the species, they may be, and apparently often have been, transmitted in nearly the same state to numerous, otherwise modified descendants.

VII.382.65.0.45.x-y:f service to a species, generally becomes/selection. *y* But when, from the nature of the organism and of the conditions, modifications have been induced which/modified, descendants.

382.65.0.46:e Hair has been transmitted to almost all mammals, feathers to all birds, and scales to all true reptiles.

VII.382.65.0.46:f It cannot have been of much importance to the greater number of mammals, birds, or reptiles, whether they were clothed with hair, feathers, or scales; yet hair

382.65.0.47:e A structure, whatever it may be, which is common to many allied forms, is ranked by us as of high systematic importance, and consequently is often assumed by us to be of high vital importance to the species.

VII.382.65.0.47:f assumed to be

382.65.0.48:e Thus, as I am inclined to believe, morphological differences, which we consider as important—such as the arrangement of the leaves, the divisions of the ovarium, the position of the ovules, &c.—first appeared in many cases as fluctuating variations, which sooner or later became almost constant through the nature of the organism and of the surrounding conditions, as well as through intercrossing; for as these morphological characters do not affect the welfare of the species, any slight deviations in them would not be acted on or accumulated through natural selection.

VII.382.65.0.48:f divisions of the flower or of the ovarium/conditions, as well as through the intercrossing of distinct individuals, but not through natural selection; for/them could not have been governed or accumulated through this latter agency.

382.65.0.49:e It is a strange result which we thus arrive at, namely that characters of slight vital importance to the species, are the most important to the systematist; but, as we shall hereafter see when we treat of the genetic principle of classification, this is by no means so paradoxical as it at first appears.

VII.382.65.0.49:f paradoxical as it may at first appear.

382.65.0.50:e Finally, whatever may be thought of this view, in none of the foregoing cases do the facts, as far as I can judge, afford any evidence of the existence of an innate tendency towards perfectibility or progressive development.

VII.382.65.0.50:f[¶] Although we have no good evidence of the existence in organic beings of an innate tendency towards progressive development, yet this necessarily follows, as I have attempted to show in the fourth chapter, through the continued action of natural selection.

VII.382.65.0.50.1:f For the best definition which has ever been given of a high standard of organisation, is the degree to which the parts have been specialised or differentiated; and natural selection tends towards this end, inasmuch as the parts are thus enabled to perform their functions more efficiently. [*Space*]

VII.382.65.0.50.2-6:f[¶] A distinguished zoologist, Mr. St. George Mivart, has recently collected all the objections which have ever been advanced by myself and others against the theory of natural selection, as propounded by Mr. Wallace and myself, and has illustrated them with admirable art and force. *3* When thus marshalled, they make a formidable array; and as it forms no part of Mr. Mivart's plan to give the various facts and considerations opposed to his conclusions, no slight effort of reason and memory is left to the reader, who may wish to weigh the evidence on both sides. *4* When discussing special cases, Mr. Mivart passes over the effects of the increased use and disuse of parts, which I have always maintained to be highly important, and have treated in my 'Variation under Domestication' at greater length than, as I believe, any other writer. *5* He likewise often assumes that I attribute nothing to variation, independently of natural selection, whereas in the work just referred to I have collected a greater number of well-established cases than can be found in any other work known to me. *6* My judgment may not be trustworthy, but after reading with care Mr. Mivart's book, and comparing each section with what I have said on the same head, I never before felt so strongly convinced of the general truth of the conclusions here arrived at, subject, of course, in so intricate a subject, to much partial error.

VII.382.65.0.50.7-10:f[¶] All Mr. Mivart's objections will be, or have been, considered in the present volume. *8* The one new point which appears to have struck many readers is, "that natural selection is incompetent to account for the incipient stages of useful structures." *9* This subject is intimately connected with that of the gradation of characters, often accompanied by a change of function,—for instance, the conversion of a swim-bladder into lungs,—points which were discussed in the last chapter under two headings. *10* Nevertheless, I will here consider in some detail several of the cases advanced by Mr. Mivart, selecting those which are the most illustrative, as want of space prevents me from considering all.

VII.382.65.0.50.11-21:f[¶] The giraffe, by its lofty stature, much elongated neck, fore-legs, head and tongue, has its whole frame beautifully adapted for browsing on the higher branches of trees. *12* It can thus obtain food beyond the reach of the other Ungulata or hoofed animals inhabiting the same country; and this must be a great advantage to it during dearths. *13* The Niata cattle in S. America show us how small a difference in structure may make, during such periods, a great difference in preserving an animal's life. *14* These cattle can browse as well as others on grass, but from the projection of the lower jaw they cannot, during the often recurrent droughts, browse on the twigs of trees, reeds, &c., to which food the common cattle and horses are then driven; so that at these times the Niatas perish, if not fed by their owners. *15* Before coming to Mr. Mivart's objections, it may be well to explain once again how natural selection will act in all ordinary cases. *16* Man has modified some of his animals, without necessarily having attended to special points of structure, by simply preserving and breeding from the fleetest individuals, as with the race-horse and greyhound, or as with the game-cock, by breeding from the victorious birds. *17* So under nature with the nascent giraffe, the individuals which were the highest browsers and were able during dearths to reach even an inch or two above the others, will often have been preserved; for they will have roamed over the whole country in search of food. *18* That the individuals of the same species often differ slightly in the relative lengths of all their parts may be seen in many works of natural history, in which careful measurements are given. *19* These slight proportional differences, due to the laws of growth and variation, are not of the slightest use or importance to most species. *20* But it will have been otherwise with the nascent giraffe, considering its probable habits of life; for those individuals which had some one part or several parts of their bodies rather more elongated than usual, would generally have survived. *21* These will have intercrossed and left offspring, either inheriting the same bodily peculiarities, or with a tendency to vary again in the same manner; whilst the individuals, less favoured in the same respects, will have been the most liable to perish.

VII.382.65.0.50.22-3:f[¶] We here see that there is no need to separate single pairs, as man does, when he methodically improves a breed: natural selection will preserve and thus separate all the superior individuals, allowing them freely to intercross, and will destroy all the inferior individuals. *23* By this process long-continued, which exactly corresponds with what I have called unconscious selection by man, combined no doubt in a most important manner with the inherited effects of the increased use of parts, it seems to me almost certain that an ordinary hoofed quadruped might be converted into a giraffe.

VII.382.65.0.50.24-31:f[¶] To this conclusion Mr. Mivart brings forward two objections. *25* One is that the increased size of the body would obviously require an increased supply of food, and he considers it as "very problematical whether the disadvantges thence arising would not, in times of scarcity, more than counterbalance the advantages." *26* But as the giraffe does actually exist in large numbers in S. Africa, and as some of the largest antelopes in the world, taller than an ox, abound there, why should we doubt that, as far as size is concerned, intermediate gradations could formerly have existed there, subjected as now to severe dearths. *27* Assuredly the being able to reach, at each stage of increased size, to a supply of food, left untouched by the other hoofed quadrupeds of the country, would have been of some advantage to the nascent giraffe. *28* Nor must we overlook the fact, that increased bulk would act as a protection against almost all beasts of prey excepting the lion; and against this animal, its tall neck,—and the taller the better,—would, as Mr.

Chauncey Wright has remarked, serve as a watch-tower. 29 It is from this cause, as Sir S. Baker remarks, that no animal is more difficult to stalk than the giraffe. 30 This animal also uses its long neck as a means of offence or defence, by violently swinging its head armed with stump-like horns. 31 The preservation of each species can rarely be determined by any one advantage, but by the union of all, great and small.

VII.382.65.0.50.32-7:f[¶] Mr. Mivart then asks (and this is his second objection), if natural selection be so potent, and if high browsing be so great an advantage, why has not any other hoofed quadruped acquired a long neck and lofty stature, besides the giraffe, and, in a lesser degree, the camel, guanaco, and macrauchenia? 33 Or, again, why has not any member of the group acquired a long proboscis? 34 With respect to S. Africa, which was formerly inhabited by numerous herds of the giraffe, the answer is not difficult, and can best be given by an illustration. 35 In every meadow in England in which trees grow, we see the lower branches trimmed or planed to an exact level by the browsing of the horses or cattle; and what advantage would it be, for instance, to sheep, if kept there, to acquire slightly longer necks? 36 In every district some one kind of animal will almost certainly be able to browse higher than the others; and it is almost equally certain that this one kind alone could have its neck elongated for this purpose, through natural selection and the effects of increased use. 37 In S. Africa the competition for browsing on the higher branches of the acacias and other trees must be between giraffe and giraffe, and not with the other ungulate animals.

VII.382.65.0.50.38-43:f[¶] Why, in other quarters of the world, various animals belonging to this same order have not acquired either an elongated neck or a proboscis, cannot be distinctly answered; but it is as unreasonable to expect a distinct answer to such a question, as why some event in the history of mankind did not occur in one country, whilst it did in another. 39 We are ignorant with respect to the conditions which determine the numbers and range of each species; and we cannot even conjecture what changes of structure would be favourable to its increase in some new country. 40 We can, however, see in a general manner that various causes might have interfered with the development of a long neck or proboscis. 41 To reach the foliage at a considerable height (without climbing, for which hoofed animals are singularly ill-constructed) implies greatly increased bulk of body; and we know that some areas support singularly few large quadrupeds, for instance S. America, though it is so luxuriant; whilst S. Africa abounds with them to an unparalleled degree. 42 Why this should be so, we do not know; nor why the later tertiary periods should have been much more favourable for their existence than the present time. 43 Whatever the causes may have been, we can see that certain districts and times would have been much more favourable than others for the development of so large a quadruped as the giraffe.

VII.382.65.0.50.44-50:f[¶] In order that an animal should acquire some structure specially and largely developed, it is almost indispensable that several other parts should be modified and co-adapted. *45* Although every part of the body varies slightly, it does not follow that the necessary parts should always vary in the right direction and to the right degree. *46* With the different species of our domesticated animals we know that the parts vary in a different manner and degree; and that some species are much more variable than others. *47* Even if the fitting variations did arise, it does not follow that natural selection would be able to act on them, and produce a structure which apparently would be beneficial to the species. *48* For instance, if the number of individuals existing in a country is determined chiefly through destruction by beasts of prey,—by external or internal parasites, &c.,—as seems often to be the case, then natural selection will be able to do little, or will be greatly retarded, in modifying any particular structure for obtaining food. *49* Lastly, natural selection is a slow process, and the same favourable conditions must long endure in order that any marked effect should thus be produced. *50* Except by assigning such general and vague reasons, we cannot explain why, in many quarters of the world, hoofed quadrupeds have not acquired much elongated necks or other means for browsing on the higher branches of trees.

VII.382.65.0.50.51-63:f[¶] Objections of the same nature as the foregoing have been advanced by many writers. *52* In each case various causes, besides the general ones just indicated, have probably interfered with the acquisition through natural selection of structures, which it is thought would be beneficial to certain species. *53* One writer asks, why has not the ostrich acquired the power of flight? *54* But a moment's reflection will show what an enormous supply of food would be necessary to give to this bird of the desert force to move its huge body through the air. *55* Oceanic islands are inhabited by bats and seals, but by no terrestrial mammals; yet as some of these bats are peculiar species, they must have long inhabited their present homes. *56* Therefore Sir C. Lyell asks, and assigns certain reasons in answer, why have not seals and bats given birth on such islands to forms fitted to live on the land? *57* But seals would necessarily be first converted into terrestrial carnivorous animals of considerable size, and bats into terrestrial insectivorous animals; for the former there would be no prey; for the bats ground-insects would serve as food, but these would already be largely preyed on by the reptiles or birds, which first colonise and abound on most oceanic islands. *58* Gradations of structure, with each stage beneficial to a changing species, will be favoured only under certain peculiar conditions. *59* A strictly terrestrial animal, by occasionally hunting for food in shallow water, then in streams or lakes, might at last be converted into an animal so thoroughly aquatic as to brave the open ocean. *60* But seals would not find on oceanic islands the conditions favourable to their gradual reconversion into a terrestrial form. *61* Bats, as formerly shown, probably acquired their wings by at first gliding through the air from tree to tree, like the so-called flying-squirrels, for the sake of escaping from their enemies, or for avoiding falls; but when the power of true flight had once been acquired, it would never be reconverted back, at least for the above purposes, into the less efficient power of gliding through the air. *62* Bats might, indeed, like many birds, have had their wings greatly reduced in size, or completely lost, through disuse; but in this case it would be necessary that they should first have acquired the power of running quickly on the ground, by the aid of their hind legs alone, so as to compete with birds or other ground animals; and for such a change a bat seems singularly ill-fitted. *63* These conjectural remarks have been made merely to show that a transition of structure, with each step beneficial, is a highly complex affair; and that there is nothing strange in a transition not having occurred in any particular case.

VII.382.65.0.50.64-7:f[¶] Lastly, more than one writer has asked, why have some animals had their mental powers more highly developed than others, as such development would be advantageous to all? *65* Why have not apes acquired the intellectual powers of man? *66* Various causes could be assigned; but as they are conjectural, and their relative probability cannot be weighed, it would be useless to give them. *67* A definite answer to the latter question ought not to be expected, seeing that no one can solve the simpler problem why, of two races of savages, one has risen higher in the scale of civilisation than the other; and this apparently implies increased brain-power. [*Space*]

VII.382.65.0.50.68-73:f[¶] We will return to Mr. Mivart's other objections. *69* Insects often resemble for the sake of protection various objects, such as green or decayed leaves, dead twigs, bits of lichen, flowers, spines, excrement of birds, and living insects; but to this latter point I shall hereafter recur. *70* The resemblance is often wonderfully close, and is not confined to colour, but extends to form, and even to the manner in which the insects hold themselves. *71* The caterpillars which project motionless like dead twigs from the bushes on which they feed, offer an excellent instance of a resemblance of this kind. *72* The cases of the imitation of such objects as the excrement of birds, are rare and exceptional. *73* On this head, Mr. Mivart remarks, "As, according to Mr. Darwin's theory, there is a constant tendency to indefinite variation, and as the minute incipient variations will be in *all directions*, they must tend to neutralize each other, and at first to form such unstable modifications that it is difficult, if not impossible, to see how such indefinite oscillations of infinitesimal beginnings can ever build up a sufficiently appreciable resemblance to a leaf, bamboo, or other object, for Natural Selection to seize upon and perpetuate."

VII.382.65.0.50.74-8:f[¶] But in all the foregoing cases the insects in their original state no doubt presented some rude and accidental resemblance to an object commonly found in the stations frequented by them. 75 Nor is this at all improbable, considering the almost infinite number of surrounding objects and the diversity in form and colour of the hosts of insects which exist. 76 As some rude resemblance is necessary for the first start, we can understand how it is that the larger and higher animals do not (with the exception, as far as I know, of one fish) resemble for the sake of protection special objects, but only the surface which commonly surrounds them, and this chiefly in colour. 77 Assuming that an insect originally happened to resemble in some degree a dead twig or a decayed leaf, and that it varied slightly in many ways, then all the variations which rendered the insect at all more like any such object, and thus favoured its escape, would be preserved, whilst other variations would be neglected and ultimately lost; or, if they rendered the insect at all less like the imitated object, they would be eliminated. 78 There would indeed be force in Mr. Mivart's objection, if we were to attempt to account for the above resemblances, independently of natural selection, through mere fluctuating variability; but as the case stands there is none.

VII.382.65.0.50.79-82:f[¶] Nor can I see any force in Mr. Mivart's difficulty with respect to "the last touches of perfection in the mimicry;" as in the case given by Mr. Wallace, of a walking-stick insect (Ceroxylus laceratus), which resembles "a stick grown over by a creeping moss or jungermannia." 80 So close was this resemblance, that a native Dyak maintained that the foliaceous excrescences were really moss. 81 Insects are preyed on by birds and other enemies, whose sight is probably sharper than ours, and every grade in resemblance which aided an insect to escape notice or detection, would tend towards its preservation; and the more perfect the resemblance so much the better for the insect. 82 Considering the nature of the differences between the species in the group which includes the above Ceroxylus, there is nothing improbable in this insect having varied in the irregularities on its surface, and in these having become more or less green-coloured; for in every group the characters which differ in the several species are the most apt to vary, whilst the generic characters, or those common to all the species, are the most constant. [Space]

VII.382.65.0.50.83-8:f[¶] The Greenland whale is one of the most wonderful animals in the world, and the baleen, or whale-bone, one of its greatest peculiarities. 84 The baleen consists of a row, on each side, of the upper jaw, of about 300 plates or laminæ, which stand close together transversely to the longer axis of the mouth. 85 Within the main row there are some subsidiary rows. 86 The extremities and inner margins of all the plates are frayed into stiff bristles, which clothe the whole gigantic palate, and serve to strain or sift the water, and thus to secure the minute prey on which these great animals subsist. *87* The middle and longest lamina in the Greenland whale is ten, twelve, or even fifteen feet in length; but in the different species of Cetaceans there are gradations in length; the middle lamina being in one species, according to Scoresby, four feet, in another three, in another eighteen inches, and in the Balænoptera rostrata only about nine inches in length. *88* The quality of the whalebone also differs in the different species.

VII.382.65.0.50.89-94:f[¶] With respect to the baleen, Mr. Mivart remarks that if it "had once attained such a size and development as to be at all useful, then its preservation and augmentation within serviceable limits would be promoted by natural selection alone. *90* But how to obtain the beginning of such useful development?" *91* In answer, it may be asked, why should not the early progenitors of the whales with baleen have possessed a mouth constructed something like the lamellated beak of a duck? *92* Ducks, like whales, subsist by sifting the mud and water; and the family has sometimes been called *Criblatores,* or sifters. *93* I hope that I may not be misconstrued into saying that the progenitors of whales did actually possess mouths lamellated like the beak of a duck. *94* I wish only to show that this is not incredible, and that the immense plates of baleen in the Greenland whale might have been developed from such lamellæ by finely graduated steps, each of service to its possessor.

VII.382.65.0.50.94'-104:f[¶] The beak of a shoveller-duck (Spatula clypeata) is a more beautiful and complex structure than the mouth of a whale. *95* The upper mandible is furnished on each side (in the specimen examined by me) with a row or comb formed of 188 thin, elastic lamellæ, obliquely bevelled so as to be pointed, and placed transversely to the longer axis of the mouth. *96* They arise from the palate, and are attached by flexible membrane to the sides of the mandible. *97* Those standing towards the middle are the longest, being about one-third of an inch in length, and they project .14 of an inch beneath the edge. *98* At their bases there is a short subsidiary row of obliquely transverse lamellæ. *99* In these several respects they resemble the plates of baleen in the mouth of a whale. *100* But towards the extremity of the beak they differ much, as they project inwards, instead of straight downwards. *101* The entire head of the shoveller, though incomparably less bulky, is about one-eighteenth of the length of the head of a moderately large Balænoptera rostrata, in which species the baleen is only nine inches long; so that if we were to make the head of the shoveller as long as that of the Balænoptera, the lamellæ would be six inches in length,—that is, two-thirds of the length of the baleen in this species of whale. *102* The lower mandible of the shoveller-duck is furnished with lamellæ of equal length with those above, but finer; and in being thus furnished it differs conspicuously from the lower jaw of a whale, which is destitute of baleen. *103* On the other hand, the extremities of these lower lamellæ are frayed into fine bristly points, so that they thus curiously resemble the plates of baleen. *104* In the genus Prion, a member of the distinct family of the Petrels, the upper mandible alone is furnished with lamellæ, which are well developed and project beneath the margin; so that the beak of this bird resembles in this respect the mouth of a whale.

VII.382.65.0.50.105-10:f[¶] From the highly developed structure of the shoveller's beak we may proceed (as I have learnt from information and specimens sent to me by Mr. Salvin), without any great break, as far as fitness for sifting is concerned, through the beak of the Merganetta armata, and in some respects through that of the Aix sponsa, to the beak of the common duck. *106* In this latter species, the lamellæ are much coarser than in the shoveller, and are firmly attached to the sides of the mandible; they are only about 50 in number on each side, and do not project at all beneath the margin. *107* They are square-topped, and are edged with translucent hardish tissue, as if for crushing food. *108* The edges of the lower mandible are crossed by numerous fine ridges, which project very little. *109* Although the beak is thus very inferior as a sifter to that of the shoveller, yet this bird, as every one knows, constantly uses it for this purpose. *110* There are other species, as I hear from Mr. Salvin, in which the lamellæ are considerably less developed than in the common duck; but I do not know whether they use their beaks for sifting the water.

VII.382.65.0.50.111-8:f[¶] Turning to another group of the same family. *112* In the Egyptian goose (Chenalopex) the beak closely resembles that of the common duck; but the lamellæ are not so numerous, nor so distinct from each other, nor do they project so much inwards; yet this goose, as I am informed by Mr. E. Bartlett, "uses its bill like a duck by throwing the water out at the corners." *113* Its chief food, however, is grass, which it crops like the common goose. *114* In this latter bird, the lamellæ of the upper mandible are much coarser than in the common duck, almost confluent, about 27 in number on each side, and terminating upwards in teeth-like knobs. *115* The palate is also covered with hard rounded knobs. *116* The edges of the lower mandible are serrated with teeth much more prominent, coarser, and sharper than in the duck. *117* The common goose does not sift the water, but uses its beak exclusively for tearing or cutting herbage, for which purpose it is so well fitted, that it can crop grass closer than almost any other animal. *118* There are other species of geese, as I hear from Mr. Bartlett, in which the lamellæ are less developed than in the common goose.

VII.382.65.0.50.119-20:f[¶] We thus see that a member of the duck family, with a beak constructed like that of the common goose and adapted solely for grazing, or even a member with a beak having less well-developed lamellæ, might be converted by small changes into a species like the Egyptian goose,—this into one like the common duck,—and, lastly, into one like the shoveller, provided with a beak almost exclusively adapted for sifting the water; for this bird could hardly use any part of its beak, except the hooked tip, for seizing or tearing solid food. *120* The beak of a goose, as I may add, might also be converted by small changes into one provided with prominent, recurved teeth, like those of the Merganser (a member of the same family), serving for the widely different purpose of securing live fish.

VII.382.65.0.50.121-6:f[¶] Returning to the whales. *121′* The Hyperoodon bidens is destitute of true teeth in an efficient condition, but its palate is roughened, according to Lacepède, with small, unequal, hard points of horn. *122* There is, therefore, nothing improbable in supposing that some early Cetacean form was provided with similar points of horn on the palate, but rather more regularly placed, and which, like the knobs on the beak of the goose, aided it in seizing or tearing its food. *123* If so, it will hardly be denied that the points might have been converted through variation and natural selection into lamellæ as well-developed as those of the Egyptian goose, in which case they would have been used both for seizing objects and for sifting the water; then into lamellæ like those of the domestic duck; and so onwards, until they became as well constructed as those of the shoveller, in which case they would have served exclusively as a sifting apparatus. *124* From this stage, in which the lamellæ would be two-thirds of the length of the plates of baleen in the Balænoptera rostrata, gradations, which may be observed in still-existing Cetaceans, lead us onwards to the enormous plates of baleen in the Greenland whale. *125* Nor is there the least reason to doubt that each step in this scale might have been as serviceable to certain ancient Cetaceans, with the functions of the parts slowly changing during the progress of development, as are the gradations in the beaks of the different existing members of the duck-family. *126* We should bear in mind that each species of duck is subjected to a severe struggle for existence, and that the structure of every part of its frame must be well adapted to its conditions of life. [*Space*]

VII.382.65.0.50.127-37:f[¶] The Pleuronectidæ, or Flat-fish, are remarkable for their asymmetrical bodies. *128* They rest on one side,—in the greater number of species on the left, but in some on the right side; and occasionally reversed adult specimens occur. *129* The lower, or resting-surface, resembles at first sight the ventral surface of an ordinary fish: it is of a white colour, less developed in many ways than the upper side, with the lateral fins often of smaller size. *130* But the eyes offer the most remarkable peculiarity; for they are both placed on the upper side of the head. *131* During early youth, however, they stand opposite to each other, and the whole body is then symmetrical, with both sides equally coloured. *132* Soon the eye proper to the lower side begins to glide slowly round the head to the upper side; but does not pass right through the skull, as was formerly thought to be the case. *133* It is obvious that unless the lower eye did thus travel round, it could not be used by the fish whilst lying in its habitual position on one side. *134* The lower eye would, also, have been liable to be abraded by the sandy bottom. *135* That the Pleuronectidæ are admirably adapted by their flattened and asymmetrical structure for their habits of life, is manifest from several species, such as soles, flounders, &c., being extremely common. *136* The chief advantages thus gained seem to be protection from their enemies, and facility for feeding on the ground. *137* The different members, however, of the family present, as Schiödte remarks, "a long series of forms exhibiting a gradual transition from Hippoglossus pinguis, which does not in any considerable degree alter the shape in which it leaves the ovum, to the soles, which are entirely thrown to one side."

VII.382.65.0.50.138-46:f[¶] Mr. Mivart has taken up this case, and remarks that a sudden spontaneous transformation in the position of the eyes is hardly conceivable, in which I quite agree with him. *139* He then adds: "if the transit was gradual, then how such transit of one eye a minute fraction of the journey towards the other side of the head could benefit the individual is, indeed, far from clear. *140* It seems, even, that such an incipient transformation must rather have been injurious." *141* But he might have found an answer to this objection in the excellent observations published in 1867 by Malm. *142* The Pleuronectidæ, whilst very young and still symmetrical, with their eyes standing on opposite sides of the head, cannot long retain a vertical position, owing to the excessive depth of their bodies, the small size of their lateral fins, and to their being destitute of a swimbladder. *143* Hence soon growing tired, they fall to the bottom on one side. *144* Whilst thus at rest they often twist, as Malm observed, the lower eye upwards, to see above them; and they do this so vigorously that the eye is pressed hard against the upper part of the orbit. *145* The forehead between the eyes consequently becomes, as could be plainly seen, temporarily contracted in breadth. *146* On one occasion Malm saw a young fish raise and depress the lower eye through an angular distance of about seventy degrees.

VII.382.65.0.50.147-56:f[¶] We should remember that the skull at this early age is cartilaginous and flexible, so that it readily yields to muscular action. *148* It is also known with the higher animals, even after early youth, that the skull yields and is altered in shape, if the skin or muscles be permanently contracted through disease or some accident. *149* With long-eared rabbits, if one ear lops forwards and downwards, its weight drags forward all the bones of the skull on the same side, of which I have given a figure. *150* Malm states that the newly-hatched young of perches, salmon, and several other symmetrical fishes, have the habit of occasionally resting on one side at the bottom; and

he has observed that they often then strain their lower eyes so as to look upwards; and their skulls are thus rendered rather crooked. *151* These fishes, however, are soon able to hold themselves in a vertical position, and no permanent effect is thus produced. *152* With the Pleuronectidæ, on the other hand, the older they grow the more habitually they rest on one side, owing to the increasing flatness of their bodies, and a permanent effect is thus produced on the form of the head, and on the position of the eyes. *153* Judging from analogy, the tendency to distortion would no doubt be increased through the principle of inheritance. *154* Schiödte believes, in opposition to some other naturalists, that the Pleuronectidæ are not quite symmetrical even in the embryo; and if this be so, we could understand how it is that certain species, whilst young, habitually fall over and rest on the left side, and other species on the right side. *155* Malm adds, in confirmation of the above view, that the adult Trachypterus arcticus, which is not a member of the Pleuronectidæ, rests on its left side at the bottom, and swims diagonally through the water; and in this fish, the two sides of the head are said to be somewhat dissimilar. *156* Our great authority on Fishes, Dr. Günther, concludes his abstract of Malm's paper, by remarking that "the author gives a very simple explanation of the abnormal condition of the Pleuronectoids."
[*Note: No sentence 157*]

VII.382.65.0.50.158-67:f[¶] We thus see that the first stages of the transit of the eye from one side of the head to the other, which Mr. Mivart considers would be injurious, may be attributed to the habit, no doubt beneficial to the individual and to the species, of endeavouring to look upwards with both eyes, whilst resting on one side at the bottom. *159* We may also attribute to the inherited effects of use the fact of the mouth in several kinds of flat-fish being bent towards the lower surface, with the jaw bones stronger and more effective on this, the eyeless side of the head, than on the other, for the sake, as Dr. Traquair supposes, of feeding with ease on the ground. *160* Disuse, on the other hand, will account for the less developed condition of the whole inferior half of the body, including the lateral fins; though Yarrell thinks that the reduced size of these fins is advantageous to the fish, as "there is so much less room for their action, than with the larger fins above." *161* Perhaps the lesser number of teeth in the proportion of four to seven in the upper halves of the two jaws of the plaice, to twenty-five to thirty in the lower halves, may likewise be accounted for by disuse. *162* From the colourless state of the ventral surface of most fishes and of many other animals, we may reasonably suppose that the absence of colour in flat-fish on the side, whether it be the right or left, which is undermost, is due to the exclusion of light. *163* But it cannot be supposed that the peculiar speckled appearance of the upper side of the sole, so like the sandy bed of the sea, or the power in some species, as recently shown by Pouchet, of changing their colour in accordance with the surrounding surface, or the presence of bony tubercles on the upper side of the turbot, are due to the action of the light. *164* Here natural selection has probably come into play, as well as in adapting the general shape of the body of these fishes, and many other peculiarities, to their habits of life. *165* We should keep in mind, as I have before insisted, that the inherited effects of the increased use of parts, and perhaps of their disuse, will be strengthened by natural selection. *166* For all spontaneous variations in the right direction will thus be preserved; as will those individuals which inherit in the highest degree the effects of the increased and beneficial use of any part. *167* How much to attribute in each particular case to the effects of use, and how much to natural selection, it seems impossible to decide.

VII.382.65.0.50.168-78:f[¶] I may give another instance of a structure which apparently owes its origin exclusively to use or habit. *169* The extremity of the tail in some American monkeys has been converted into a wonderfully perfect prehensile organ, and serves as a fifth hand. *170* A reviewer who agrees with Mr. Mivart in every detail, remarks on this structure: "It is impossible to believe that in any number of ages the first slight incipient tendency to grasp could preserve the lives of the individuals possessing it, or favour their chance of having and of rearing offspring." *171* But there is no necessity for any such belief. *172* Habit, and this almost implies that some benefit great or small is thus derived, would in all probability suffice for the work. *173* Brehm saw the young of an African monkey (Cercopithecus) clinging to the under surface of their mother by their hands, and at the same time they hooked their little tails round that of their mother. *174* Professor Henslow kept in confinement some harvest mice (Mus messorius) which do not possess a structurally prehensile tail; but he frequently observed that they curled their tails round the branches of a bush placed in the cage, and thus aided themselves in climbing. *175* I have received an analogous account from Dr. Günther, who has seen a mouse thus suspend itself. *176* If the harvest mouse had been more strictly arboreal, it would perhaps have had its tail rendered structurally prehensile, as is the case with some members of the same order. *177* Why Cercopithecus, considering its habits whilst young, has not become thus provided, it would be difficult to say. *178* It is, however, possible that the long tail of this monkey may be of more service to it as a balancing organ in making its prodigious leaps, than as a prehensile organ. [*Space*]

VII.382.65.50.0.179-88:f[¶] The mammary glands are common to the whole class of mammals, and are indispensable for their existence; they must, therefore, have been developed at an extremely remote period, and we can know nothing positively about their manner of development. *180* Mr. Mivart asks: "Is it conceivable that the young of any animal was ever saved from destruction by accidentally sucking a drop of scarcely nutritious fluid from an accidentally hypertrophied cutaneous gland of its mother? *181* And even if one was so, what chance was there of the perpetuation of such a variation?" *182* But the case is not here put fairly. *183* It is admitted by most evolutionists that mammals are descended from a marsupial form; and if so, the mammary glands will have been at first developed within the marsupial sack. *184* In the case of the fish (Hippocampus) the eggs are hatched, and the young are reared for a time, within a sack of this nature; and an American naturalist, Mr. Lockwood, believes from what he has seen of the development of the young, that they are nourished by a secretion from the cutaneous glands of the sack. *185* Now with the early progenitors of mammals, almost before they deserved to be thus designated, is it not at least possible that the young might have been similarly nourished? *186* And in this case, the individuals which secreted a fluid, in some degree or manner the most nutritious, so as to partake of the nature of milk, would in the long run have reared a larger number of well-nourished offspring, than would the individuals which secreted a poorer fluid; and thus the cutaneous glands, which are the homologues of the mammary glands, would have been improved or rendered more effective. *187* It accords with the widely extended principle of specialisation, that the glands over a certain space of the sack should have become more highly developed than the remainder; and they would then have formed a breast, but at first without a nipple, as we see in the Ornithorhyncus, at the base of the mammalian series. *188* Through what agency the glands over a certain space became more highly specialised than the others, I will not pretend to decide, whether in part through compensation of growth, the effects of use, or of natural selection.

VII.382.65.50.0.189-97:f[¶] The development of the mammary glands would have been of no service, and could not have been effected through natural selection, unless the young at the same time were able to partake of the secretion. *190* There is no greater difficulty in understanding how young mammals have instinctively learnt to suck the breast, than in understanding how unhatched chickens have learnt to break the egg-shell by tapping against it with their specially adapted beaks; or how a few hours after leaving the shell they have learnt to pick up grains of food. *191* In such cases the most probable solution seems to be, that the habit was at first acquired by practice at a more advanced age, and afterwards transmitted to the offspring at an earlier age. *192* But the young kangaroo is said not to suck, only to cling to the nipple of its mother, who has the power of injecting milk into the mouth of her helpess, half-formed offspring. *193* On this head Mr. Mivart remarks: "Did no special provision exist, the young one must infallibly be choked by the intrusion of the milk into the windpipe. *194* But there *is* a special provision. *195* The larynx is so elongated that it rises up into the posterior end of the nasal passage, and is thus enabled to give free entrance to the air for the lungs, while the milk passes harmlessly on each side of this elongated larynx, and so safely attains the gullet behind it." *196* Mr. Mivart then asks how did natural selection remove in the adult kangaroo (and in most other mammals, on the assumption that they are descended from a marsupial form), "this at least perfectly innocent and harmless structure?" *197* It may be suggested in answer that the voice, which is certainly of high importance to many animals, could hardly have been used with full force as long as the larynx entered the nasal passage; and Professor Flower has suggested to me that this structure would have greatly interfered with an animal swallowing solid food.

VII.382.65.50.0.198-201:f[¶] We will now turn for a short space to the lower divisions of the animal kingdom. *199* The Echinodermata (star-fishes, sea-urchins, &c.,) are furnished with remarkable organs, called pedicellariæ, which consist, when well developed, of a tridactyle forceps—that is, of one formed of three serrated arms, neatly fitting together and placed on the summit of a flexible stem, moved by muscles. *200* These forceps can seize firmly hold of any object; and Alexander Agassiz has seen an Echinus or sea-urchin rapidly passing particles of excrement from forceps to forceps down certain lines of its body, in order that its shell should not be fouled. *201* But there is no doubt that besides removing dirt of all kinds, they subserve other functions; and one of these apparently is defence.

VII.382.65.50.0.202-6:f[¶] With respect to these organs, Mr. Mivart, as on so many previous occasions, asks: "What would be the utility of the *first rudimentary beginnings* of such structures, and how could such incipient buddings have ever preserved the life of a single Echinus?" *203* He adds, "not even the *sudden* development of the snapping action could have been beneficial without the freely moveable stalk, nor could the latter have been efficient without the snapping jaws, yet no minute merely indefinite variations could simultaneously evolve these complex co-ordinations of structure; to deny this seems to do no less than to affirm a startling paradox." *204* Paradoxical as this may appear to Mr. Mivart, tridactyle forcepses, immovably fixed at the base, but capable of a snapping action, certainly exist on some star-fishes; and this is intelligible if they serve, at least in part, as a means of defence. *205* Mr. Agassiz, to whose great kindness I am indebted for much information on the subject, informs me that there are other star-fishes, in which one of the three arms of the forceps is reduced to a support for the other two; and again, other genera in which the third arm is completely lost. *206* In Echinoneus, the shell is described by M. Perrier as bearing two kinds of pedicellariæ, one resembling those of Echinus, and the other those of Spatangus; and such cases are always interesting as affording the means of apparently sudden transitions, through the abortion of one of the two states of an organ.

VII.382.65.50.0.207-15:f[¶] With respect to the steps by which these curious organs have been evolved, Mr. Agassiz infers from his own researches and those of Müller, that both in star-fishes and sea-urchins the pedicellariæ must undoubtedly be looked at as modified spines. *208* This may be inferred from their manner of development in the individual, as well as from a long and perfect series of gradations in different species and genera, from simple granules to ordinary spines, to perfect tridactyle pedicellariæ. *209* The gradation extends even to the manner in which ordinary spines and the pedicellariæ with their supporting calcareous rods are articulated to the shell. *210* In certain genera of star-fishes, "the very combinations needed to show that the pedicellariæ are only modified branching spines" may be found. *211* Thus we have fixed spines, with three equi-distant, serrated, moveable branches, articulated to near their bases; and higher up, on the same spine, three other moveable branches. *212* Now when the latter arise from the summit of a spine they form in fact a rude tridactyle pedicellaria, and such may be seen on the same spine together with the three lower branches. *213* In this case the identity in nature between the arms of the pedicellariæ and the moveable branches of a spine, is unmistakeable. *214* It is generally admitted that the ordinary spines serve as a protection; and if so, there can be no reason to doubt that those furnished with serrated and moveable branches likewise serve for the same purpose; and they would thus serve still more effectively as soon as by meeting together they acted as a prehensile or snapping apparatus. *215* Thus every gradation, from an ordinary fixed spine to a fixed pedicellaria, would be of service.

VII.382.65.0.50.216-9:f[¶] In certain genera of star-fishes these organs, instead of being fixed or borne on an immovable support, are placed on the summit of a flexible and muscular, though short, stem; and in this case they probably subserve some additional function besides defence. *217* In the sea-urchins the steps can be followed by which a fixed spine becomes articulated to the shell, and is thus rendered moveable. *218* I wish I had space here to give a fuller abstract of Mr. Agassiz's interesting observations on the development of the pedicellariæ. *219* All possible gradations, as he adds, may likewise be found between the pedicellariæ of the star-fishes and the hooks of the Ophiurians, another group of the Echinodermata; and again between the pedicellariæ of sea-urchins and the anchors of the Holothuriæ, also belonging to the same great class. [*Space*]

VII.382.65.0.50.220-4:f[¶] Certain compound animals, or zoophytes as they have been termed, namely the Polyzoa, are provided with curious organs called avicularia. *221* These differ much in structure in the different species. *222* In their most perfect condition, they curiously resemble the head and beak of a vulture in miniature, seated on a neck and capable of movement, as is likewise the lower jaw or mandible. *223* In one species observed by me all the avicularia on the same branch often moved simultaneously backwards and forwards, with the lower jaw widely open, through an angle of about 90°, in the course of five seconds; and their movement caused the whole polyzoary to tremble. *224* When the jaws are touched with a needle they seize it so firmly that the branch can thus be shaken.

VII.382.65.0.50.225-30:f[¶] Mr. Mivart adduces this case, chiefly on account of the supposed difficulty of organs, namely the avicularia of the Polyzoa and the pedicellariæ of the Échinodermata, which he considers as "essentially similar," having been developed through natural selection in widely distinct divisions of the animal kingdom. *226* But, as far as structure is concerned, I can see no similarity between tridactyle pedicellariæ and avicularia. *227* The latter resemble somewhat more closely the chelæ or pincers of Crustaceans; and Mr. Mivart might have adduced with equal appropriateness this resemblance as a special difficulty; or even their resemblance to the head and beak of a bird. *228* The avicularia are believed by Mr. Busk, Dr. Smitt, and Dr. Nitsche—naturalists who have carefully studied this group—to be homologous with the zooids and their cells which compose the zoophyte; the moveable lip or lid of the cell corresponding with the lower and moveable mandible of the avicularium. *229* Mr. Busk, however, does not know of any gradations now existing between a zooid and an avicularium. *230* It is therefore impossible to conjecture by what serviceable gradations the one could have been converted into the other: but it by no means follows from this that such gradations have not existed.

VII.382.65.0.50.231-4:f[¶] As the chelæ of Crustaceans resemble in some degree the avicularia of Polyzoa, both serving as pincers, it may be worth while to show that with the former a long series of serviceable gradations still exists. *232* In the first and simplest stage, the terminal segment of a limb shuts down either on the square summit of the broad penultimate segment, or against one whole side; and is thus enabled to catch hold of an object; but the limb still serves as an organ of locomotion. *233* We next find one corner of the broad penultimate segment slightly prominent, sometimes furnished with irregular teeth; and against these the terminal segment shuts down. *234* By an increase in the size of this projection, with its shape, as well as that of the terminal segment, slightly modified and improved, the pincers are rendered more and more perfect, until we have at last an instrument as efficient as the chelæ of a lobster; and all these gradations can be actually traced.

VII.382.65.0.50.235-41:f[¶] Besides the avicularia, the Polyzoa possess curious organs called vibracula. *236* These generally consist of long bristles, capable of movement and easily excited. *237* In one species examined by me the vibracula were slightly curved and serrated along the outer margin; and all of them on the same polyzoary often moved simultaneously; so that, acting like long oars, they swept a branch rapidly across the object-glass of my microscope. *238* When a branch was placed on its face, the vibracula became entangled, and they made violent efforts to free themselves. *239* They are supposed to serve as a defence, and may be seen, as Mr. Busk remarks, "to sweep slowly and carefully over the surface of the polyzoary, removing what might be noxious to the delicate inhabitants of the cells when their tentacula are protruded." *240* The avicularia, like the vibracula, probably serve for defence, but they also catch and kill small living animals, which it is believed are afterwards swept by the currents within reach of the tentacula of the zooids. *241* Some species are provided with avicularia and vibracula; some with avicularia alone, and a few with vibracula alone.

VII.382.65.0.50.242-8:f[¶] It is not easy to imagine two objects more widely different in appearance than a bristle or vibraculum, and an avicularium like the head of a bird; yet they are almost certainly homologous and have been developed from the same common source, namely a zooid with its cell. *243* Hence we can understand how it is that these organs graduate in some cases, as I am informed by Mr. Busk, into each other. *244* Thus with the avicularia of several species of Lepralia, the moveable mandible is so much produced and is so like a bristle, that the presence of the upper or fixed beak alone serves to determine its avicularian nature. *245* The vibracula may have been directly developed from the lips of the cells, without having passed through the avicularian stage; but it seems more probable that they have passed through this stage, as during the early stages of the transformation, the other parts of the cell with the included zooid could hardly have disappeared at once. *246* In many cases the vibracula have a grooved support at the base, which seems to represent the fixed beak; though this support in some species is quite absent. *247* This view of the development of the vibracula, if trustworthy, is interesting; for supposing that all the species provided with avicularia had become extinct, no one with the most vivid imagination would ever have thought that the vibracula had originally existed as part of an organ, resembling a bird's head or an irregular box or hood. *248* It is interesting to see two such widely different organs developed from a common origin; and as the moveable lip of the cell serves as a protection to the zooid, there is no difficulty in believing that all the gradations, by which the lip became converted first into the lower mandible of an avicularium and then into an elongated bristle, likewise served as a protection in different ways and under different circumstances. [*Space*]

VII.382.65.0.50.249-56:f[¶] In the vegetable kingdom Mr. Mivart only alludes to two cases, namely the structure of the flowers of orchids, and the movements of climbing plants. *250* With respect to the former, he says, "the explanation of their *origin* is deemed thoroughly unsatisfactory—utterly insufficient to explain the incipient, infinitesimal beginnings of structures which are of utility only when they are considerably developed." *251* As I have fully treated this subject in another work, I will here give only a few details on one alone of the most striking peculiarities of the flowers of orchids, namely their pollinia. 252 A pollinium when highly developed consists of a mass of pollen-grains, affixed to an elastic foot-stalk or caudicle, and this to a little mass of extremely viscid matter. 253 The pollinia are by this means transported by insects from one flower to the stigma of another. 254 In some orchids there is no caudicle to the pollen-masses, and the grains are merely tied together by fine threads; but as these are not confined to orchids, they need not here be considered; yet I may mention that at the base of the orchidaceous series, in Cypripedium, we can see how the threads were probably first developed. 255 In other orchids the threads cohere at one end of the pollen-masses; and this forms the first or nascent trace of a caudicle. 256 That this is the origin of the caudicle, even when of considerable length and highly developed, we have good evidence in the aborted pollen-grains which can sometimes be detected embedded within the central and solid parts.

VII.382.65.0.50.257-64:f[¶] With respect to the second chief peculiarity, namely the little mass of viscid matter attached to the end of the caudicle, a long series of gradations can be specified, each of plain service to the plant. 258 In most flowers belonging to other orders the stigma secretes a little viscid matter. 259 Now in certain orchids similar viscid matter is secreted, but in much larger quantities by one alone of the three stigmas; and this stigma, perhaps in consequence of the copious secretion, is rendered sterile. 260 When an insect visits a flower of this kind, it rubs off some of the viscid matter and thus at the same time drags away some of the pollen-grains. 261 From this simple condition, which differs but little from that of a multitude of common flowers, there are endless gradations,—to species in which the pollen-mass terminates in a very short, free caudicle,—to others in which the caudicle becomes firmly attached to the viscid matter, with the sterile stigma itself much modified. 262 In this latter case we have a pollinium in its most highly developed and perfect condition. 263 He who will carefully examine the flowers of orchids for himself will not deny the existence of the above series of gradations—from a mass of pollen-grains merely tied together by threads, with the stigma differing but little from that of an ordinary flower, to a highly complex pollinium, admirably adapted for transportal by insects; nor will he deny that all the gradations in the several species are admirably adapted in relation to the general structure of each flower for its fertilisation by different insects. 264 In this, and in almost every other case, the enquiry may be pushed further backwards; and it may be asked how did the stigma of an ordinary flower become viscid, but as we do not know the full history of any one group of beings, it is as useless to ask, as it is hopeless to attempt answering, such questions.

VII.382.65.0.50.265-71:f[¶] We will now turn to climbing plants.

266 These can be arranged in a long series, from those which simply twine round a support, to those which I have called leaf-climbers, and to those provided with tendrils. *267* In these two latter classes the stems have generally, but not always, lost the power of twining, though they retain the power of revolving, which the tendrils likewise possess. *268* The gradations from leaf-climbers to tendril-bearers are wonderfully close, and certain plants may be indifferently placed in either class. *269* But in ascending the series from simple twiners to leaf-climbers, an important quality is added, namely sensitiveness to a touch, by which means the foot-stalks of the leaves or flowers, or these modified and converted into tendrils, are excited to bend round and clasp the touching object. *270* He who will read my memoir on these plants will, I think, admit that all the many gradations in function and structure between simple twiners and tendril-bearers are in each case beneficial in a high degree to the species. *271* For instance, it is clearly a great advantage to a twining plant to become a leaf-climber; and it is probable that every twiner which possessed leaves with long foot-stalks would have been developed into a leaf-climber, if the foot-stalks had possessed in any slight degree the requisite sensitiveness to a touch.

VII.382.65.0.50.272-82:f[¶] As twining is the simplest means of ascending a support, and forms the basis of our series, it may naturally be asked how did plants acquire this power in an incipient degree, afterwards to be improved and increased through natural selection. *273* The power of twining depends, firstly, on the stems whilst young being extremely flexible (but this is a character common to many plants which are not climbers); and, secondly, on their continually bending to all points of the compass, one after the other in succession, in the same order. *274* By this movement the stems are inclined to all sides, and are made to move round and round. *275* As soon as the lower part of a stem strikes against any object and is stopped, the upper part still goes on bending and revolving, and thus necessarily twines round and up the support. *276* The revolving movement ceases after the early growth of each shoot. *277* As in many widely separated families of plants, single species and single genera possess the power of revolving, and have thus become twiners, they must have independently acquired it, and cannot have inherited it from a common progenitor. *278* Hence I was led to predict that some slight tendency to a movement of this kind would be found to be far from uncommon with plants which did not climb; and that this had afforded the basis for natural selection to work on and improve. *279* When I made this prediction, I knew of only one imperfect case, namely of the young flower-peduncles of a Maurandia which revolved slightly and irregularly, like the stems of twining plants, but without making any use of this habit. *280* Soon afterwards Fritz Müller discovered that the young stems of an Alisma and of a Linum,—plants which do not climb and are widely separated in the natural system,—revolved plainly, though irregularly; and he states that he has reason to suspect that this occurs with some other plants. *281* These slight movements appear to be of no service to the plants in question; anyhow, they are not of the least use in the way of climbing, which is the point that concerns us. *282* Nevertheless we can see that if the stems of these plants had been flexible, and if under the conditions to which they are exposed it had profited them to ascend to a height, then the habit of slightly and irregularly revolving might have been increased and utilised through natural selection, until they had become converted into well-developed twining species.

VII.382.65.0.50.283-8:f[¶] With respect to the sensitiveness of the foot-stalks of the leaves and flowers, and of tendrils, nearly the same remarks are applicable as in the case of the revolving movements of twining plants. *284* As a vast number of species, belonging to widely distinct groups, are endowed with this kind of sensitiveness, it ought to be found in a nascent condition in many plants which have not become climbers. *285* This is the case: I observed that the young flower-peduncles of the above Maurandia curved themselves a little towards the side which was touched. *286* Morren found in several species of Oxalis that the leaves and their foot-stalks moved, especially after exposure to a hot sun, when they were gently and repeatedly touched, or when the plant was shaken. *287* I repeated these observations on some other species of Oxalis with the same result; in some of them the movement was distinct, but was best seen in the young leaves; in others it was extremely slight. *288* It is a more important fact that according to the high authority of Hofmeister, the young shoots and leaves of all plants move after being shaken; and with climbing plants it is, as we know, only during the early stages of growth that the foot-stalks and tendrils are sensitive.

VII.382.65.0.50.289-4:f[¶] It is scarcely possible that the above slight movements, due to a touch or shake, in the young and growing organs of plants, can be of any functional importance to them. *290* But plants possess, in obedience to various stimuli, powers of movement, which are of manifest importance to them; for instance, towards and more rarely from the light,—in opposition to, and more rarely in the direction of, the attraction of gravity. *291* When the nerves and muscles of an animal are excited by galvanism or by the absorption of strychnine, the consequent movements may be called an incidental result, for the nerves and muscles have not been rendered specially sensitive to these stimuli. *292* So with plants it appears that, from having the power of movement in obedience to certain stimuli, they are excited in an incidental manner by a touch, or by being shaken. *293* Hence there is no great difficulty in admitting that in the case of leaf-climbers and tendril-bearers, it is this tendency which has been taken advantage of and increased through natural selection. *294* It is, however, probable, from reasons which I have assigned in my memoir, that this will have occurred only with plants which had already acquired the power of revolving, and had thus become twiners.

VII.382.65.0.50.295-6:f[¶] I have already endeavoured to explain how plants became twiners, namely, by the increase of a tendency to slight and irregular revolving movements, which were at first of no use to them; this movement, as well as that due to a touch or shake, being the incidental result of the power of moving, gained for other and beneficial purposes. *296* Whether, during the gradual development of climbing plants, natural selection has been aided by the inherited effects of use, I will not pretend to decide; but we know that certain periodical movements, for instance the so-called sleep of plants, are governed by habit. [*Space*]

VII.382.65.0.50.297-9:f[¶] I have now considered enough, perhaps more than enough, of the cases, selected with care by a skilful naturalist, to prove that natural selection is incompetent to account for the incipient stages of useful structures; and I have shown, as I hope, that there is no great difficulty on this head. *298* A good opportunity has thus been afforded for enlarging a little on gradations of structure, often associated with changed functions,—an important subject, which was not treated at sufficient length in the former editions of this work. *299* I will now briefly recapitulate the foregoing cases.

VII.382.65.0.50.300-3:f[¶] With the giraffe, the continued preservation of the individuals of some extinct high-reaching ruminant, which had the longest necks, legs, &c., and could browse a little above the average height, and the continued destruction of those which could not browse so high, would have sufficed for the production of this remarkable quadruped; but the prolonged use of all the parts together with inheritance will have aided in an important manner in their co-ordination. *301* With the many insects which imitate various objects, there is no improbability in the belief that an accidental resemblance to some common object was in each case the foundation for the work of natural selection, since perfected through the occasional preservation of slight variations which made the resemblance at all closer; and this will have been carried on as long as the insect continued to vary, and as long as a more and more perfect resemblance led to its escape from sharp-sighted enemies. *302* In certain species of whales there is a tendency to the formation of irregular little points of horn on the palate; and it seems to be quite within the scope of natural selection to preserve all favourable variations, until the points were converted first into lamellated knobs or teeth, like those on the beak of a goose,— then into short lamellæ, like those of the domestic ducks,—and then into lamellæ, as perfect as those of the shoveller-duck,— and finally into the gigantic plates of baleen, as in the mouth of the Greenland whale. *303* In the family of the ducks, the lamellæ are first used as teeth, then partly as teeth and partly as a sifting apparatus, and at last almost exclusively for this latter purpose.

VII.382.65.0.50.304-10:f[¶] With such structures as the above lamellæ of horn or whalebone, habit or use can have done little or nothing, as far as we can judge, towards their development. *305* On the other hand, the transportal of the lower eye of a flat-fish to the upper side of the head, and the formation of a prehensile tail, may be attributed almost wholly to continued use, together with inheritance. *306* With respect to the mammæ of the higher animals, the most probable conjecture is that primordially the cutaneous glands over the whole surface of a marsupial sack secreted a nutritious fluid; and that these glands were improved in function through natural selection, and concentrated into a confined area, in which case they would have formed a mamma. *307* There is no more difficulty in understanding how the branched spines of some ancient Echinoderm, which served as a defence, became developed through natural selection into tridactyle pedicellariæ, than in understanding the development of the pincers of crustaceans, through slight, serviceable modifications in the ultimate and penultimate segments of a limb, which was at first used solely for locomotion. *308* In the avicularia and vibracula of the Polyzoa we have organs widely different in appearance developed from the same source; and with the vibracula we can understand how the successive gradations might have been of service. *309* With the pollinia of orchids, the threads which originally served to tie together the pollen-grains, can be traced cohering into caudicles; and the steps can likewise be followed by which viscid matter, such as that secreted by the stigmas of ordinary flowers, and still subserving nearly but not quite the same purpose, became attached to the free ends of the caudicles;—all these gradations being of manifest benefit to the plants in question. *310* With respect to climbing plants, I need not repeat what has been so lately said.

VII.382.65.0.50.311-22:f[¶] It has often been asked, if natural selection be so potent, why has not this or that structure been gained by certain species, to which it would apparently have been advantageous? *312* But it is unreasonable to expect a precise answer to such questions, considering our ignorance of the past history of each species, and of the conditions which at the present day determine its numbers and range. *313* In most cases only general reasons, but in some few cases special reasons, can be assigned. *314* Thus to adapt a species to new habits of life, many co-ordinated modifications are almost indispensable, and it may often have happened that the requisite parts did not vary in the right manner or to the right degree. *315* Many species must have been prevented from increasing in numbers through destructive agencies, which stood in no relation to certain structures, which we imagine would have been gained through natural selection from appearing to us advantageous to the species. *316* In this case, as the struggle for life did not depend on such structures, they could not have been acquired through natural selection. *317* In many cases complex and long-enduring conditions, often of a peculiar nature, are necessary for the development of a structure; and the requisite conditions may seldom have concurred. *318* The belief that any given structure, which we think, often erroneously, would have been beneficial to a species, would have been gained under all circumstances through natural selection, is opposed to what we can understand of its manner of action. *319* Mr. Mivart does not deny that natural selection has effected something; but he considers it as "demonstrably insufficient" to account for the phenomena which I explain by its agency. *320* His chief arguments have now been considered, and the others will hereafter be considered. *321* They seem to me to partake little of the character of demonstration, and to have little weight in comparison with those in favour of the power of natural selection, aided by the other agencies often specified. *322* I am bound to add, that some of the facts and arguments here used by me, have been advanced for the same purpose in an able article lately published in the 'Medico-Chirurgical Review.'

VII.382.65.0.50.323-6:f[¶] At the present day almost all naturalists admit evolution under some form. *324* Mr. Mivart believes that species change through "an internal force or tendency," about which it is not pretended that anything is known. *325* That species have a capacity for change will be admitted by all evolutionists; but there is no need, as it seems to me, to invoke any internal force beyond the tendency to ordinary variability, which through the aid of selection by man has given rise to many well-adapted domestic races, and which through the aid of natural selection would equally well give rise by graduated steps to natural races or species. *326* The final result will generally have been, as already explained, an advance, but in some few cases a retrogression, in organisation.

VII.382.65.0.50.327-30:f[¶] Mr. Mivart is further inclined to believe, and some naturalists agree with him, that new species manifest themselves "with suddenness and by modifications appearing at once." *328* For instance, he supposes that the differences between the extinct three-toed Hipparion and the horse arose suddenly. *329* He thinks it difficult to believe that the wing of a bird "was developed in any other way than by a comparatively sudden modification of a marked and important kind;" and apparently he would extend the same view to the wings of bats and pterodactyles. *330* This conclusion, which implies great breaks or discontinuity in the series, appears to me improbable in the highest degree.

VII.382.65.0.50.331-5:f[¶] Every one who believes in slow and gradual evolution, will of course admit that specific changes may have been as abrupt and as great as any single variation which we meet with under nature, or even under domestication. *332* But as species are more variable when domesticated or cultivated than under their natural conditions, it is not probable that such great and abrupt variations have often occurred under nature, as are known occasionally to arise under domestication. *333* Of these latter variations several may be attributed to reversion; and the characters which thus reappear were, it is probable, in many cases at first gained in a gradual manner. *334* A still greater number must be called monstrosities, such as six-fingered men, porcupine men, Ancon sheep, Niata cattle, &c.; and as they are widely different in character from natural species, they throw very little light on our subject. *335* Excluding such cases of abrupt variations, the few which remain would at best constitute, if found in a state of nature, doubtful species, closely related to their parental types.

VII.382.65.0.50.336-40:f[¶] My reasons for doubting whether natural species have changed as abruptly as have occasionally domestic races, and for entirely disbelieving that they have changed in the wonderful manner indicated by Mr. Mivart, are as follows. *337* According to our experience, abrupt and strongly marked variations occur in our domesticated productions, singly and at rather long intervals of time. *338* If such occurred under nature, they would be liable, as formerly explained, to be lost by accidental causes of destruction and by subsequent inter-crossing; and so it is known to be under domestication, unless abrupt variations of this kind are specially preserved and separated by the care of man. *339* Hence in order that a new species should suddenly appear in the manner supposed by Mr. Mivart, it is almost necessary to believe, in opposition to all analogy, that several wonderfully changed individuals appeared simultaneously within the same district. *340* This difficulty, as in the case of unconscious selection by man, is avoided on the theory of gradual evolution, through the preservation of a large number of individuals, which varied more or less in any favourable direction, and of the destruction of a large number which varied in an opposite manner.

VII.382.65.0.50.341-8:f[¶] That many species have been evolved in an extremely gradual manner, there can hardly be a doubt. *342* The species and even the genera of many large natural families are so closely allied together, that it is difficult to distinguish not a few of them. *343* On every continent in proceeding from north to south, from lowland to upland, &c., we meet with a host of closely related or representative species; as we likewise do on certain distinct continents, which we have reason to believe were formerly connected. *344* But in making these and the following remarks, I am compelled to allude to subjects hereafter to be discussed. *345* Look at the many outlying islands round a continent, and see how many of their inhabitants can be raised only to the rank of doubtful species. *346* So it is if we look to past times, and compare the species which have just passed away with those still living within the same areas; or if we compare the fossil species embedded in the sub-stages of the same geological formation. *347* It is indeed manifest that multitudes of species are related in the closest manner to other species that still exist, or have lately existed; and it will hardly be maintained that such species have been developed in an abrupt or sudden manner. *348* Nor should it be forgotten, when we look to the special parts of allied species, instead of to distinct species, that numerous and wonderfully fine gradations can be traced, connecting together widely different structures.

VII.382.65.0.50.349-53:f[¶] Many large groups of facts are intelligible only on the principle that species have been evolved by very small steps. *350* For instance, the fact that the species included in the larger genera are more closely related to each other, and present a greater number of varieties than do the species in the smaller genera. *351* The former are also grouped in little clusters, like varieties round species; and they present other analogies with varieties, as was shown in our second chapter. *352* On this same principle we can understand how it is that specific characters are more variable than generic characters; and how the parts which are developed in an extraordinary degree or manner are more variable than other parts of the same species. *353* Many analogous facts, all pointing in the same direction, could be added.

VII.382.65.0.50.354-9:f[¶] Although very many species have almost certainly been produced by steps not greater than those separating fine varieties; yet it may be maintained that some have been developed in a different and abrupt manner. *355* Such an admission, however, ought not to be made without strong evidence being assigned. *356* The vague and in some respects false analogies, as they have been shown to be by Mr. Chauncey Wright, which have been advanced in favour of this view, such as the sudden crystallisation of inorganic substances, or the falling of a facetted spheroid from one facet to another, hardly deserve consideration. *357* One class of facts, however, namely, the sudden appearance of new and distinct forms of life in our geological formations supports at first sight the belief in abrupt development. *358* But the value of this evidence depends entirely on the perfection of the geological record, in relation to periods remote in the history of the world. *359* If the record is as fragmentary as many geologists strenuously assert, there is nothing strange in new forms appearing as if suddenly developed.

VII.382.65.0.50.360-6:f[¶] Unless we admit transformations as prodigious as those advocated by Mr. Mivart, such as the sudden development of the wings of birds or bats, or the sudden conversion of a Hipparion into a horse, hardly any light is thrown by the belief in abrupt modifications on the deficiency of connecting links in our geological formations. *361* But against the belief in such abrupt changes, embryology enters a strong protest. *362* It is notorious that the wings of birds and bats, and the legs of horses or other quadrupeds, are undistinguishable at an early embryonic period, and that they become differentiated by insensibly fine steps. *363* Embryological resemblances of all kinds can be accounted for, as we shall hereafter see, by the progenitors of our existing species having varied after early youth, and having transmitted their newly acquired characters to their offspring, at a corresponding age. *364* The embryo is thus left almost unaffected, and serves as a record of the past condition of the species. *365* Hence it is that existing species during the early stages of their development so often resemble ancient and extinct forms belonging to the same class. *366* On this view of the meaning of embryological resemblances, and indeed on any view, it is incredible that an animal should have undergone such momentous and abrupt transformations, as those above indicated; and yet should not bear even a trace in its embryonic condition of any sudden modification; every detail in its structure being developed by insensibly fine steps.

VII.382.65.0.50.367-71:f[¶] He who believes that some ancient form was transformed suddenly through an internal force or tendency into, for instance, one furnished with wings, will be almost compelled to assume, in opposition to all analogy, that many individuals varied simultaneously. *368* It cannot be denied that such abrupt and great changes of structure are widely different from those which most species apparently have undergone. *369* He will further be compelled to believe that many structures beautifully adapted to all the other parts of the same creature and to the surrounding conditions, have been suddenly produced; and of such complex and wonderful co-adaptations, he will not be able to assign a shadow of an explanation. *370* He will be forced to admit that these great and sudden transformations have left no trace of their action on the embryo. *371* To admit all this is, as it seems to me, to enter into the realms of miracle, and to leave those of Science.

[*Note: End of VII:f*]

382.65.0.50.372:f [*Center*] Convergence of Character. [*Space*]

382.66:c[¶] A distinguished botanist, Mr. H. C. Watson, believes that I have overrated the importance of the principle of divergence of character (in which, however, he apparently believes), and that convergence of character, as it may be called, has likewise played a part.

382.66:e I need allude only to two other objections: a distinguished/ importance of divergence

382.66:f Mr. H. C. Watson thinks that I/ convergence, as

382.67:c This is an intricate subject which need not be here discussed.

382.67:e be fully discussed.

382.67:[f]

382.68:c I will only say that if two species of two closely allied genera produced a number of new and divergent species, I can believe

267

that these new forms might sometimes approach each other so closely that they would for convenience sake be classed in the same new genus, and thus two genera would converge into one; but from the strength of the principle of inheritance, it seems hardly credible that the two groups of new species would not at least form two sections of the supposed new single genus.

382.68:d only remark that if/convenience'

382.68:e of two allied genera, both produced a number of new and divergent species, I can believe that they might/inheritance, and from the two parent-species already differing and consequently tending to vary in a somewhat different manner, it seems hardly credible that the two new groups would not at least form distinct sections in the genus.

382.68:f If two species, belonging to two distinct though allied genera, had both produced a large number of new and divergent forms, it is conceivable that these might approach each other so closely that they would have all to be classed under the same genus; and thus the descendants of two distinct genera would converge into one.

382.68.0.0.1-4:f But it would in most cases be extremely rash to attribute to convergence a close and general similarity of structure in the modified descendants of widely distinct forms. *2* The shape of a crystal is determined solely by the molecular forces, and it is not surprising that dissimilar substances should sometimes assume the same form; but with organic beings we should bear in mind that the form of each depends on an infinitude of complex relations, namely on the variations which have arisen, these being due to causes far too intricate to be followed out, —on the nature of the variations which have been preserved or selected, and this depends on the surrounding physical conditions, and in a still higher degree on the surrounding organisms with which each being has come into competition,—and lastly, on inheritance (in itself a fluctuating element) from innumerable progenitors, all of which have had their forms determined through equally complex relations. *3* It is incredible that the descendants of two organisms, which had originally differed in a marked manner, should ever afterwards converge so closely as to lead to a near approach to identity throughout their whole organisation. *4* If this had occurred, we should meet with the same form, independently of genetic connection, recurring in widely separated geological formations; and the balance of evidence is opposed to any such an admission.

382.69:c[¶] Mr. Watson has also objected that the continued action of natural selection with divergence of character will tend to make an indefinite number of specific forms.

382.69:f selection, together with divergence of character, would tend

382.70:c As far as mere inorganic conditions are concerned, it seems probable that a sufficient number of species would soon become adapted to all considerable diversities of heat, moisture, &c.;

but I fully admit that the mutual relations of organic beings are more important; and as the number of species in any country goes on increasing, the organic conditions of life will become more and more complex.

382.70:d life become

382.70:f life must become

382.71:c Consequently there seems at first sight to be no limit to the amount of profitable diversification of structure, and therefore no limit to the number of species which might be produced.

382.71:d sight no limit to the amount

382.72-3:c We do not know that even the most prolific area is fully stocked with specific forms: at the Cape of Good Hope and in Australia, which support such an astonishing number of species, many European plants have become naturalised. *73* But geology shows us, at least within the whole immense tertiary period, that the number of species of shells, and, probably, of mammals, has not greatly or at all increased.

382.73:d us, that from an early part of the long tertiary period the number of species of shells, and that from the middle part of this same period the number of mammals

382.73:f of the tertiary

382.74-5:c What then checks an indefinite increase in the number of species? *75* The amount of life (I do not mean the number of specific forms) supported on any area must have a limit, depending so largely as it does on physical conditions: therefore, if an area be inhabited by very many species, each or nearly each species will be represented by few individuals; and such species will be liable to extermination from accidental fluctuations in the nature of the seasons or in the number of their enemies.

382.75:d conditions; therefore

382.75:f on an area must

382.76:c The process of extermination in these cases will be rapid, whereas the production of new species will always be slow.

382.76:e species must always

382.76:f cases would be rapid

382.77-8:c Imagine the extreme case of as many species as individuals in England, and the first severe winter or very dry summer would exterminate thousands on thousands of species. *78* Rare species, and each species will become rare if the number of species become in any country indefinitely increased, will, on the principle often explained, present within a given period few favourable variations; consequently, the process of giving birth to new specific forms will thus be retarded.

382.78:e number of species in any country becomes indefinitely

382.79:c When any species becomes very rare, close interbreeding will help in exterminating it; at least authors have thought that this comes into play in accounting for the deterioration of

Aurochs in Lithuania, of Red Deer in Scotland, and of Bears in Norway, &c.

382.79:d help to exterminate

382.79:e it; authors/of the Aurochs

382.80:c As far as animals are concerned, some species are closely adapted to prey on some one other being; but if this other being had been rare, it would not have been any advantage to the animal to have been produced in close relation to its prey: therefore, it would not have been produced by natural selection.

382.80:d other organism; but if this other organism had been rare

382.80:[e]

382.81-2:c Lastly, and this I am inclined to think is the most important element, a dominant species, which has already beaten many competitors in its owr home, will tend to spread and supplant many others. *82* Alph. de Candolle has shown that those species which spread widely tend generally to spread *very* widely; and, consequently, they will tend to exterminate several species in several areas, and thus check the inordinate increase of specific forms throughout the world.

382.82:d tend to supplant and exterminate

382.82:f widely, tend generally to spread *very* widely; consequently

382.83:c Dr. Hooker has recently shown that in the S. E. corner of Australia, where, apparently, there are many invaders from different quarters of the world, the endemic Australian species have been greatly reduced in number.

382.83:f of the globe, the endemic

382.84:c How much weight to attribute to these several considerations I do not pretend to assign; but conjointly they must limit in each country the tendency to an indefinite augmentation of specific forms.

382.84:f considerations I will not pretend to say; but

382.83 *Summary of Chapter.—*

383:d [*Center*] *Summary of Chapter.* [*Space*]

384 If during the long course of ages and under varying conditions of life, organic beings vary at all in the several parts of their organisation, and I think this cannot be disputed; if there be, owing to the high geometrical powers of increase of each species, at some age, season, or year, a severe struggle for life, and this certainly cannot be disputed; then, considering the infinite complexity of the relations of all organic beings to each other and to their conditions of existence, causing an infinite diversity in structure, constitution, and habits, to be advantageous to them, I think it would be a most extraordinary fact if no variation ever had occurred useful to each being's own welfare, in the same way as so many variations have occurred useful to man.

384:b geometrical ratio of increase of each species, a severe struggle for life at some age, season, or year, and this/same manner as so many

384:c them, it would be a most extraordinary fact if no variation had ever occurred

384:d[¶]

384:e If under changing conditions of life organic beings present individual differences in all parts of their structure, and this/ conditions of life, causing/variations ever

384:f differences in almost every part/owing to their geometrical rate of increase, a severe/variations had ever

385 But if variations useful to any organic being do occur, assuredly individuals thus characterised will have the best chance of being preserved in the struggle for life; and from the strong principle of inheritance they will tend to produce offspring similarly characterised.

385:c inheritance, they

385:d do ever occur

385:f being ever do occur/inheritance, these will

386 This principle of preservation, I have called, for the sake of brevity, Natural Selection.

386:b Selection; and it leads to the improvement of each creature in relation to its organic and inorganic conditions of life.

386.x:c preservation I/Selection.

386.x:e preservation, or the survival of the fittest, I

386.y:c It leads/life; and consequently, in most cases, to what must be regarded as an advance in organisation.

386.1:c Nevertheless low and simple forms would long endure if well fitted for their simple conditions of life.

386.1:e forms will long

386.1:f Nevertheless, low

387 Natural selection, on the principle of qualities being inherited at corresponding ages, can modify the egg, seed, or young, as easily as the adult.

387:b[¶]

388 Amongst many animals, sexual selection will give its aid to ordinary selection, by assuring to the most vigorous and best adapted males the greatest number of offspring.

388:f will have given

389 Sexual selection will also give characters useful to the males alone, in their struggles with other males.

389:e useful only to the males, in their struggles with other males; and these characters will be transmitted to one sex or to both sexes, according to the form of inheritance which prevails.

389:f useful to the males alone, in their struggles or rivalry with

390 Whether natural selection has really thus acted in nature, in

modifying and adapting the various forms of life to their several conditions and stations, must be judged of by the general tenour and balance of evidence given in the following chapters.

390:c tenor

390:e acted in adapting/balance of the evidence

390:f judged by the general tenor and balance of evidence

391 But we already see how it entails extinction; and how largely extinction has acted in the world's history, geology plainly declares.

391:f we have already seen

392 Natural selection, also, leads to divergence of character; for more living beings can be supported on the same area the more they diverge in structure, habits, and constitution, of which we see proof by looking at the inhabitants of any small spot or at naturalised productions.

392:b or to naturalised

392:c for the more organic beings diverge in structure, habits, and constitution, by so much a greater number can be supported on the same area,—of which we see proof by looking to the inhabitants

392:e much can a greater number be

392:f much the more can a large number/spot, and to the productions naturalised in foreign lands.

393 Therefore during the modification of the descendants of any one species, and during the incessant struggle of all species to increase in numbers, the more diversified these descendants become, the better will be their chance of succeeding in the battle of life.

393:b battle for life.

393:c Therefore, during the modification

393:f diversified the descendants/chance of success in

394 Thus the small differences distinguishing varieties of the same species, will steadily tend to increase till they come to equal the greater differences between species of the same genus, or even of distinct genera.

394:b species, steadily

394:e increase, till

394:f they equal

395 We have seen that it is the common, the widely-diffused, and widely-ranging species, belonging to the larger genera, which vary most; and these will tend to transmit to their modified offspring that superiority which now makes them dominant in their own countries.

395:b these tend

395:c genera within each class, which vary

396 Natural selection, as has just been remarked, leads to divergence of character and to much extinction of the less improved and intermediate forms of life.

397 On these principles, I believe, the nature of the affinities of all organic beings may be explained.

397:c principles, the nature of the affinities, and the generally well-defined distinctions of the innumerable organic beings in each class throughout the world, may

397:f distinctions between the innumerable

398 It is a truly wonderful fact—the wonder of which we are apt to overlook from familiarity—that all animals and all plants throughout all time and space should be related to each other in group subordinate to group, in the manner which we everywhere behold—namely, varieties of the same species most closely related together, species of the same genus less closely and unequally related together, forming sections and sub-genera, species of distinct genera much less closely related, and genera related in different degrees, forming sub-families, families, orders, sub-classes, and classes.

398:e in natural groups subordinate to groups, in the manner

398:f in groups subordinate/closely related, species

399 The several subordinate groups in any class cannot be ranked in a single file, but seem rather to be clustered round points, and these round other points, and so on in almost endless cycles.

399:f seem clustered

400 On the view that each species has been independently created, I can see no explanation of this great fact in the classification of all organic beings; but, to the best of my judgment, it is explained through inheritance and the complex action of natural selection, entailing extinction and divergence of character, as we have seen illustrated in the diagram.

400:e If each species has been independently created, no explanation can be given of this great fact in the classification of all organic being; but it

400:f If species had been independently created, no explanation would have been possible of this kind of classification; but

401 The affinities of all the beings of the same class have sometimes been represented by a great tree.

402 I believe this simile largely speaks the truth.

403 The green and budding twigs may represent existing species; and those produced during each former year may represent the long succession of extinct species.

403:e during former years

404 At each period of growth all the growing twigs have tried to branch out on all sides, and to overtop and kill the surrounding twigs and branches, in the same manner as species and groups

of species have tried to overmaster other species in the great battle for life.

404:e groups of species have at all times overmastered

405 The limbs divided into great branches, and these into lesser and lesser branches, were themselves once, when the tree was small, budding twigs; and this connexion of the former and present buds by ramifying branches may well represent the classification of all extinct and living species in groups subordinate to groups.

405:e connection

405:f was young, budding

406 Of the many twigs which flourished when the tree was a mere bush, only two or three, now grown into great branches, yet survive and bear all the other branches; so with the species which lived during long-past geological periods, very few now have living and modified descendants.

406:f bear the other/few have left living

407 From the first growth of the tree, many a limb and branch has decayed and dropped off; and these lost branches of various sizes may represent those whole orders, families, and genera which have now no living representatives, and which are known to us only from having been found in a fossil state.

407:d only from being found

407:f these fallen branches of various sizes/only in a fossil state.

408 As we here and there see a thin straggling branch springing from a fork low down in a tree, and which by some chance has been favoured and is still alive on its summit, so we occasionally see an animal like the Ornithorhynchus or Lepidosiren, which in some small degree connects by its affinities two large branches of life, and which has apparently been saved from fatal competition by having inhabited a protected station.

409 As buds give rise by growth to fresh buds, and these, if vigorous, branch out and overtop on all sides many a feebler branch, so by generation I believe it has been with the great Tree of Life, which fills with its dead and broken branches the crust of the earth, and covers the surface with its ever branching and beautiful ramifications.

409:d ever-branching

CHAPTER V.

Laws of Variation.

3 Effects of external conditions—Use and disuse, combined with natural selection; organs of flight and of vision—Acclimatisation —Correlation of growth—Compensation and economy of growth —False correlations—Multiple, rudimentary, and lowly organised structures variable—Parts developed in an unusual manner are highly variable: specific characters more variable than generic: secondary sexual characters variable—Species of the same genus vary in an analogous manner—Reversions to long lost characters—Summary.

3:b long-lost

3:e Effects of changed conditions/Acclimatisation—Correlated variation—Compensation

4 I HAVE hitherto sometimes spoken as if the variations—so common and multiform in organic beings under domestication, and in a lesser degree in those in a state of nature—had been due to chance.

4:e multiform with organic/degree with those

4:f those under nature—were due

5 This, of course, is a wholly incorrect expression, but it serves to acknowledge plainly our ignorance of the cause of each particular variation.

6 Some authors believe it to be as much the function of the reproductive system to produce individual differences, or very slight deviations of structure, as to make the child like its parents.

6:e or slight

7 But the much greater variability, as well as the greater frequency of monstrosities, under domestication or cultivation, than under nature, leads me to believe that deviations of structure are in some way due to the nature of the conditions of life, to which the parents and their more remote ancestors have been exposed during several generations.

7:c leads to the belief that

7:e But the fact of variations and monstrosities occurring much more frequently under domestication than under nature, and the greater variability of species having wide ranges than of those having restricted ranges, lead to the conclusion that variability is directly related to the conditions of life to which each species has been exposed during several successive generations.

7:f those with restricted ranges, lead to the conclusion that variability is generally related

8 I have remarked in the first chapter—but a long catalogue of facts which cannot be here given would be necessary to show the truth of the remark—that the reproductive system is eminently susceptible to changes in the conditions of life; and to this system being functionally disturbed in the parents, I chiefly attribute the varying or plastic condition of the offspring.

8:[e]

9 The male and female sexual elements seem to be affected before that union takes place which is to form a new being.

9:[e]

10 In the case of "sporting" plants, the bud, which in its earliest condition does not apparently differ essentially from an ovule, is alone affected.

10:d case of bud-variations, or "sporting" plants as they are often called, the bud

10:[e]

11 But why, because the reproductive system is disturbed, this or that part should vary more or less, we are profoundly ignorant.

11:[e]

12 Nevertheless, we can here and there dimly catch a faint ray of light, and we may feel sure that there must be some cause for each deviation of structure, however slight.

12:[e]

12.1-5:e In the first chapter I attempted to show that changed conditions act in two ways, directly on the whole organisation or on certain parts alone, and indirectly through the reproductive system. *2* In all cases there are two factors, the nature of the organism, which is much the most important of the two, and the nature of the conditions. *3* The direct action of changed conditions leads to definite or indefinite results. *4* In the latter case the organisation seems to become plastic, and we have much fluctuating variability. *5* In the former case the nature of the organism is such that it yields readily, when subjected to certain conditions, and all, or nearly all the individuals become modified in the same way.

13 How much direct effect difference of climate, food, &c., produces on any being is extremely doubtful.

13:e It is very difficult to decide how far changed conditions, such as of climate, food, &c., have acted in a definite manner.

14 My impression is, that the effect is extremely small in the case of animals, but perhaps rather more in that of plants.

14:c effect is small in the case of animals, but more

14:e There is some reason to believe that in the course of time the effects have been greater than can be proved to be the case by any clear evidence.

14:f is reason/proved by clear

15 We may, at least, safely conclude that such influences cannot have produced the many striking and complex co-adaptations of structure between one organic being and another, which we see everywhere throughout nature.

15:e But we may safely conclude that the innumerable complex co-adaptations of structure which we see throughout nature between various organic beings, cannot be attributed simply to such action.

15:f structure, which

16 Some little influence may be attributed to climate, food, &c.: thus, E. Forbes speaks confidently that shells at their southern limit, and when living in shallow water, are more brightly coloured than those of the same species further north or from greater depths.

16:c water, vary and become more

16:e In the following cases the conditions seem to have produced some slight definite effect: E. Forbes asserts that shells/water, are more/species from further north or from a greater depth; but these statements have lately been disputed.

16:f depth; but this certainly does not always hold good.

17 Gould believes that birds of the same species are more brightly coloured under a clear atmosphere, than when living on islands or near the coast.

18 So with insects, Wollaston is convinced that residence near the sea affects their colours.

18:d insects: Wollaston

17 + 8:e Mr. Gould/coast; and Wollaston/affects the colours of insects.

17 + 8:f living near the coast or on islands; and

19 Moquin-Tandon gives a list of plants which when growing near the sea-shore have their leaves in some degree fleshy, though not elsewhere fleshy.

19:f which, when growing near the sea-shore, have

20 Several other such cases could be given.

20:e Other similar facts could be given.

20:[f]

20.1:f These slightly varying organisms are interesting in as far as they present characters analogous to those possessed by the species which are confined to similar conditions.

21 The fact of varieties of one species, when they range into the zone of habitation of other species, often acquiring in a very slight degree some of the characters of such species, accords with

our view that species of all kinds are only well-marked and permanent varieties.

21:e into the habitations of other species, often acquiring in a slight degree some of their characters, accords with the view that species are

21:[f]

22 Thus the species of shells which are confined to tropical and shallow seas are generally brighter-coloured than those confined to cold and deeper seas.

22:[f]

23 The birds which are confined to continents are, according to Mr. Gould, brighter-coloured than those of islands.

23:[f]

24 The insect-species confined to sea-coasts, as every collector knows, are often brassy or lurid.

24:[f]

25 Plants which live exclusively on the sea-side are very apt to have fleshy leaves.

25:[f]

26 He who believes in the creation of each species, will have to say that this shell, for instance, was created with bright colours for a warm sea; but that this other shell became bright-coloured by variation when it ranged into warmer or shallower waters.

26:e this insect, for instance, was created of a brassy colour, because it was intended to live near the sea, but that this other insect became brassy through variation as soon as it reached the sea-coast.

26:[f]

27 When a variation is of the slightest use to a being, we cannot tell how much of it to attribute to the accumulative action of natural selection, and how much to the conditions of life.

27:e to any being, we cannot tell how much to attribute/much to the definite action of the conditions

28 Thus, it is well known to furriers that animals of the same species have thicker and better fur the more severe the climate is under which they have lived; but who can tell how much of this difference may be due to the warmest-clad individuals having been favoured and preserved during many generations, and how much to the direct action of the severe climate? for it would appear that climate has some direct action on the hair of our domestic quadrupeds.

28:e fur the further north they live/and how much to the action of the severe

29 Instances could be given of the same variety being produced under conditions of life as different as can well be conceived;

and, on the other hand, of different varieties being produced from the same species under the same conditions.

29:c under apparently the same

29:e given of similar varieties being produced from the same species under external conditions of life/hand, or dissimilar varieties being produced under apparently the same external conditions.

30 Such facts show how indirectly the conditions of life must act.

30:b life act.

30:[e]

31 Again, innumerable instances are known to every naturalist of species keeping true, or not varying at all, although living under the most opposite climates.

31:e naturalist, of

32 Such considerations as these incline me to lay very little weight on the direct action of the conditions of life.

32:e me not to lay much weight on the direct and definite action of the conditions of life; but I fully admit that strong arguments of a general nature may be advanced on the other side.

32:f me to lay less weight on the direct action of the surrounding conditions, than on a tendency to vary, due to causes of which we are quite ignorant.

33 Indirectly, as already remarked, they seem to play an important part in affecting the reproductive system, and in thus inducing variability; and natural selection will then accumulate all profitable variations, however slight, until they become plainly developed and appreciable by us.

33:[e]

33.1:d[¶] In a far-fetched sense, however, the conditions of life may be said, not only to cause variability, but likewise to include natural selection; for it depends on the nature of the conditions whether this or that variety shall be preserved.

33.1:e In one sense the conditions/selection; for the conditions determine whether this or that variety shall survive.

33.1:f variability, either directly or indirectly, but

33.2:d But we see in selection by man, that these two elements of change are essentially distinct; the conditions under domestication causing the variability, and the will of man, acting either consciously or unconsciously, accumulating the variations in certain definite directions.

33.2:e But when man is the selecting agent, we clearly see that the two elements of change are distinct; the conditions cause the variability; the will of man, acting either consciously or unconsciously, accumulates the variations in certain directions, and this answers to the survival of the fittest under nature.

33.2:f distinct; variability is in some manner excited, but it is the will of man which accumulates the variations in certain directions; and it is this latter agency which answers

34 *Effects of Use and Disuse.—*

34:d [*Center*] *Effects of Use and Disuse, as controlled by Natural Selection.* [*Space*]

34:f Effects of the increased Use and Disuse of Parts, as

35 From the facts alluded to in the first chapter, I think there can be little doubt that use in our domestic animals strengthens and enlarges certain parts, and disuse diminishes them; and that such modifications are inherited.

35:d[¶] be no doubt

35:f animals has strengthened and enlarged certain parts, and disuse diminished

36 Under free nature, we can have no standard of comparison, by which to judge of the effects of long-continued use or disuse, for we know not the parent-forms; but many animals have structures which can be explained by the effects of disuse.

36:e we have no

36:f animals possess structures which can be best explained

37 As Professor Owen has remarked, there is no greater anomaly in nature than a bird that cannot fly; yet there are several in this state.

38 The logger-headed duck of South America can only flap along the surface of the water, and has its wings in nearly the same condition as the domestic Aylesbury duck.

38:f Aylesbury duck: it is a remarkable fact that the young birds, according to Mr. Cunningham, can fly, while the adults have lost this power.

39 As the larger ground-feeding birds seldom take flight except to escape danger, I believe that the nearly wingless condition of several birds, which now inhabit or have lately inhabited several oceanic islands, tenanted by no beast of prey, has been caused by disuse.

39:f danger, it is probable that the nearly wingless condition of several birds, now inhabiting or which lately

40 The ostrich indeed inhabits continents and is exposed to danger from which it cannot escape by flight, but by kicking it can defend itself from enemies, as well as any of the smaller quadrupeds.

40:f continents, and/but it can defend itself by kicking its enemies, as efficiently as many quadrupeds.

41 We may imagine that the early progenitor of the ostrich had habits like those of a bustard, and that as natural selection increased in successive generations the size and weight of its body,

its legs were used more, and its wings less, until they became incapable of flight.

41:c may believe that the progenitor of the ostrich genus had

41:d bustard, and that, as

41:f those of the bustard, and that, as the size and weight of its body were increased during successive generations, its legs

42 Kirby has remarked (and I have observed the same fact) that the anterior tarsi, or feet, of many male dung-feeding beetles are very often broken off; he examined seventeen specimens in his own collection, and not one had even a relic left.

42:f are often

43 In the Onites apelles the tarsi are so habitually lost, that the insect has been described as not having them.

44 In some other genera they are present, but in a rudimentary condition.

45 In the Ateuchus or sacred beetle of the Egyptians, they are totally deficient.

46 There is not sufficient evidence to induce us to believe that mutilations are ever inherited; and I should prefer explaining the entire absence of the anterior tarsi in Ateuchus, and their rudimentary condition in some other genera, by the long-continued effects of disuse in their progenitors; for as the tarsi are almost always lost in many dung-feeding beetles, they must be lost early in life, and therefore cannot be much used by these insects.

46:b induce me to

46.x-y:c The evidence that accidental mutilations can be inherited is at present very scanty; but the remarkable case observed by Brown-Séquard of epilepsy produced by injuring the spinal chord of guinea-pigs, being inherited, should make us cautious. *y* So that it will perhaps be safest to look at the entire/genera, as due to the long-continued/cannot be of much importance or be much used by

46.x:d present not quite decisive; but/cautious in denying such power.

46.x-y:e Brown-Séquard of inherited epilepsy in guinea-pigs, caused by an operation performed on the spinal chord, should make us cautious in denying such power. *y* Hence it will perhaps/due to the effects of long-continued disuse; for as many dung-feeding beetles are generally found with their tarsi lost, this must happen early in life; therefore the tarsi cannot be

46.x-y:f not decisive; but the remarkable cases observed by Brown-Séquard in guinea-pigs, of the inherited effects of operations, should make us cautious in denying this tendency. Hence/ genera, not as cases of inherited mutilations, but as due

47 In some cases we might easily put down to disuse modifications of structure which are wholly, or mainly, due to natural selection.

48 Mr. Wollaston has discovered the remarkable fact that 200 beetles, out of the 550 species inhabiting Madeira, are so far deficient in wings that they cannot fly; and that of the twenty-nine endemic genera, no less than twenty-three genera have all their species in this condition!

48:c beetles out of the 550 species inhabiting

48:d beetles, out of the 550 species (but more are now known) which inhabit Maderia, are/fly; and that, of

48:f known) inhabiting/twenty-three have

49 Several facts, namely, that beetles in many parts of the world are very frequently blown to sea and perish; that the beetles in Madeira, as observed by Mr. Wollaston, lie much concealed, until the wind lulls and the sun shines; that the proportion of wingless beetles is larger on the exposed Dezertas than in Madeira itself; and especially the extraordinary fact, so strongly insisted on by Mr. Wollaston, of the almost entire absence of certain large groups of beetles, elsewhere excessively numerous, and which groups have habits of life almost necessitating frequent flight;—these several considerations have made me believe that the wingless condition of so many Madeira beetles is mainly due to the action of natural selection, but combined probably with disuse.

49:b are frequently blown/Desertas

49:d facts,—namely

49:f insisted on by Mr. Wollaston, that certain large groups of beetles, elsewhere excessively numerous, which absolutely require the use of their wings, are here almost entirely absent;—these/selection, combined

50 For during thousands of successive generations each individual beetle which flew least, either from its wings having been ever so little less perfectly developed or from indolent habit, will have had the best chance of surviving from not being blown out to sea; and, on the other hand, those beetles which most readily took to flight will oftenest have been blown to sea and thus have been destroyed.

50:b flight would oftenest

50:c during many successive

50:f oftenest have been blown to sea, and thus destroyed.

51 The insects in Madeira which are not ground-feeders, and which, as the flower-feeding coleoptera and lepidoptera, must habitually use their wings to gain their subsistence, have, as Mr. Wollaston suspects, their wings not at all reduced, but even enlarged.

51:f as certain flower-feeding

52 This is quite compatible with the action of natural selection.

53 For when a new insect first arrived on the island, the tendency of natural selection to enlarge or to reduce the wings, would

282

depend on whether a greater number of individuals were saved by successively battling with the winds, or by giving up the attempt and rarely or never flying.

54 As with mariners shipwrecked near a coast, it would have been better for the good swimmers if they had been able to swim still further, whereas it would have been better for the bad swimmers if they had not been able to swim at all and had stuck to the wreck.

55 The eyes of moles and of some burrowing rodents are rudimentary in size, and in some cases are quite covered up by skin and fur.

56 This state of the eyes is probably due to gradual reduction from disuse, but aided perhaps by natural selection.

57 In South America, a burrowing rodent, the tuco-tuco, or Ctenomys, is even more subterranean in its habits than the mole; and I was assured by a Spaniard, who had often caught them, that they were frequently blind; one which I kept alive was certainly in this condition, the cause, as appeared on dissection, having been inflammation of the nictitating membrane.

57.x-y:f blind. *y* One

58 As frequent inflammation of the eyes must be injurious to any animal, and as eyes are certainly not indispensable to animals with subterranean habits, a reduction in their size with the adhesion of the eyelids and growth of fur over them, might in such case be an advantage; and if so, natural selection would constantly aid the effects of disuse.

58:d size, with

58:e not necessary to animals having subterranean

59 It is well known that several animals, belonging to the most different classes, which inhabit the caves of Styria and of Kentucky, are blind.

59:e caves of Carniola and

60 In some of the crabs the foot-stalk for the eye remains, though the eye is gone; the stand for the telescope is there, though the telescope with its glasses has been lost.

60:f gone;—the stand

61 As it is difficult to imagine that eyes, though useless, could be in any way injurious to animals living in darkness, I attribute their loss wholly to disuse.

61:f darkness, their loss may be attributed to

62 In one of the blind animals, namely, the cave-rat, the eyes are of immense size; and Professor Silliman thought that it regained, after living some days in the light, some slight power of vision.

62:c cave-rat (Neotoma), two of which were captured by Professor Silliman at above half a mile distance from the mouth of the cave, and therefore not in the profoundest depths, the eyes were lustrous and of large size; but these animals, as I am informed by Professor Silliman, having been exposed for about a month to a graduated light, acquired a dim perception of objects when brought towards their eyes, and blinked.

62:e size; and these animals, as I am informed by Professor Silliman, after having/objects.

63 In the same manner as in Madeira the wings of some of the insects have been enlarged, and the wings of others have been reduced by natural selection aided by use and disuse, so in the case of the cave-rat natural selection seems to have struggled with the loss of light and to have increased the size of the eyes; whereas with all the other inhabitants of the caves, disuse by itself seems to have done its work.

63:[c]

64　It is difficult to imagine conditions of life more similar than deep limestone caverns under a nearly similar climate; so that on the common view of the blind animals having been separately created for the American and European caverns, close similarity in their organisation and affinities might have been expected; but, as Schiödte and others have remarked, this is not the case, and the cave-insects of the two continents are not more closely allied than might have been anticipated from the general resemblance of the other inhabitants of North America and Europe.

64.x:c European caverns, very close/expected.

64.x:d that, on

64.x:f that, in accordance with the old view

64.y:c This is certainly not the case if we look at the two whole faunas; and with respect to the insects alone, Schiödte has remarked, "We are accordingly prevented from considering the entire phenomenon in any other light than something purely local, and the similarity which is exhibited in a few forms between the Mammoth cave (in Kentucky) and the caves in Carniola, otherwise than as a very plain expression of that analogy which subsists generally between the fauna of Europe and of North America."

65 On my view we must suppose that American animals, having ordinary powers of vision, slowly migrated by successive generations from the outer world into the deeper and deeper recesses of the Kentucky caves, as did European animals into the caves of Europe.

65:c having in most cases ordinary

66 We have some evidence of this gradation of habit; for, as Schiödte remarks, "animals not far remote from ordinary forms, prepare the transition from light to darkness. Next follow those that are constructed for twilight; and, last of all, those destined for total darkness."

66:c remarks, "We accordingly look upon the subterranean faunas as small ramifications which have penetrated into the earth from the geographically limited faunas of the adjacent tracts, and which, as they extended themselves into darkness, have been accommodated to surrounding circumstances. Animals/ darkness, and whose formation is quite peculiar."

66.1:c These remarks of Schiödte's of course apply not to the same, but to distinct species.

66.1:d Schiödte's, it should be understood, apply

67 By the time that an animal had reached, after numberless generations, the deepest recesses, disuse will on this view have more or less perfectly obliterated its eyes, and natural selection will often have effected other changes, such as an increase in the length of the antennæ or palpi, as a compensation for blindness.

68 Notwithstanding such modifications, we might expect still to see in the cave-animals of America, affinities to the other inhabitants of that continent, and in those of Europe, to the inhabitants of the European continent.

68:d Europe to

69 And this is the case with some of the American cave-animals, as I hear from Professor Dana; and some of the European cave-insects are very closely allied to those of the surrounding country.

70 It would be most difficult to give any rational explanation of the affinities of the blind cave-animals to the other inhabitants of the two continents on the ordinary view of their independent creation.

70:f be difficult

71 That several of the inhabitants of the caves of the Old and New Worlds should be closely related, we might expect from the well-known relationship of most of their other productions.

71.1:c As a blind species of Bathyscia is found in abundance on shady rocks out of the caves, the loss of vision in the cave-species has probably had no relation to its dark habitation; and it is very natural that an insect already deprived of vision should readily become adapted to dark caverns.

71.1:d rocks far from caves, the loss of vision in the cave-species of this one genus has probably

71.1:f habitation; for it is natural

71.2:c Another blind genus (Anophthalmus) offers this remarkable peculiarity: the several distinct species, as Mr. Murray has remarked, inhabit several distinct European caves and likewise those of Kentucky, and the genus is found nowhere except in caves; but it is possible that the progenitor or progenitors of these several species may formerly have ranged widely over both continents, and have since (like the elephant on both continents) become extinct, excepting in their present secluded habitations.

71.2:d several species, whilst furnished with eyes, formerly may have ranged widely over both continents, and since (like the elephant on both continents) have become/secluded abodes.

71.2:e continents, and then have become

71.2:f peculiarity, that the species, as Mr. Murray observes, have not as yet been found anywhere except in caves; yet those which inhabit the several caves of Europe and America are distinct; but it is possible that the progenitors of these several species, whilst they were furnished with eyes, may formerly have ranged over both continents

72 Far from feeling any surprise that some of the cave-animals should be very anomalous, as Agassiz has remarked in regard to the blind fish, the Amblyopsis, and as is the case with the blind Proteus with reference to the reptiles of Europe, I am only surprised that more wrecks of ancient life have not been preserved, owing to the less severe competition to which the inhabitants of these dark abodes will probably have been exposed.

72:c feeling surprise/will have

72:d abodes must have

72:f which the scanty inhabitants

73   *Acclimatisation.—*

73:d *[Center] Acclimatisation. [Space]*

74 Habit is hereditary with plants, as in the period of flowering, in the amount of rain requisite for seeds to germinate, in the time of sleep, &c., and this leads me to say a few words on acclimatisation.

74:c *time*

74:d[¶]

74:f time of sleep, in the amount of rain requisite for seeds to germinate, &c.

75 As it is extremely common for species of the same genus to inhabit very hot and very cold countries, and as I believe that all the species of the same genus have descended from a single parent, if this view be correct, acclimatisation must be readily effected during long-continued descent.

75:d to inhabit hot and cold

75:e countries, if it be true that all the species of the same genus are descended from a single parent-form, acclimatisation must be readily effected during a long course of descent.

75:f common for distinct species belonging to the same genus to

76 It is notorious that each species is adapted to the climate of its own home: species from an arctic or even from a temperate region cannot endure a tropical climate, or conversely.

77 So again, many succulent plants cannot endure a damp climate.

78 But the degree of adaptation of species to the climates under which they live is often overrated.

79 We may infer this from our frequent inability to predict whether or not an imported plant will endure our climate, and from the number of plants and animals brought from warmer countries which here enjoy good health.

79:e brought from different countries which are here perfectly healthy.

80 We have reason to believe that species in a state of nature are limited in their ranges by the competition of other organic beings quite as much as, or more than, by adaptation to particular climates.

80:c are closely limited

81 But whether or not the adaptation be generally very close, we have evidence, in the case of some few plants, of their becoming, to a certain extent, naturally habituated to different temperatures, or becoming acclimatised: thus the pines and rhododendrons, raised from seed collected by Dr. Hooker from trees growing at different heights on the Himalaya, were found in this country to possess different constitutional powers of resisting cold.

81:d trees of the same species growing

81:e temperatures; that is, they become acclimatised/from the same

81:f adaptation is in most cases very close, we have evidence with some/found to possess in this country different

82 Mr. Thwaites informs me that he has observed similar facts in Ceylon, and analogous observations have been made by Mr. H. C. Watson on European species of plants brought from the Azores to England.

82:d Ceylon; analogous/England; and I could give other cases.

83 In regard to animals, several authentic cases could be given of species within historical times having largely extended their range from warmer to cooler latitudes, and conversely; but we do not positively know that these animals were strictly adapted to their native climate, but in all ordinary cases we assume such to be the case; nor do we know that they have subsequently become acclimatised to their new homes.

83:c climate, yet in

83:d authentic instances could be adduced of species/become specially acclimatised to their new homes, so as to be better fitted for them than they were at first.

83:e climate, though in

83:f species having largely extended, within historical times, their range

84 As I believe that our domestic animals were originally chosen

by uncivilised man because they were useful and bred readily under confinement, and not because they were subsequently found capable of far-extended transportation, I think the common and extraordinary capacity in our domestic animals of not only withstanding the most different climates but of being perfectly fertile (a far severer test) under them, may be used as an argument that a large proportion of other animals, now in a state of nature, could easily be brought to bear widely different climates.

84:c As we may infer that our domestic animals were originally/ transportation, the common/climates, but/other animals now in a state of nature could

84:e useful and because they bred

85 We must not, however, push the foregoing argument too far, on account of the probable origin of some of our domestic animals from several wild stocks: the blood, for instance, of a tropical and arctic wolf or wild dog may perhaps be mingled in our domestic breeds.

85:e wolf may

86 The rat and mouse cannot be considered as domestic animals, but they have been transported by man to many parts of the world, and now have a far wider range than any other rodent, living free under the cold climate of Faroe in the north and of the Falklands in the south, and on many islands in the torrid zones.

86:e rodent; for they live under

86:f on many an island

87 Hence I am inclined to look at adaptation to any special climate as a quality readily grafted on an innate wide flexibility of constitution, which is common to most animals.

87:e Hence adaptation to any special climate may be looked at as/constitution, common

88 On this view, the capacity of enduring the most different climates by man himself and by his domestic animals, and such facts as that former species of the elephant and rhinoceros were capable of enduring a glacial climate, whereas the living species are now all tropical or sub-tropical in their habits, ought not to be looked at as anomalies, but merely as examples of a very common flexibility of constitution, brought, under peculiar circumstances, into play.

88:c circumstances, into action.

88:e animals, and the fact of the extinct elephant and rhinoceros having formerly endured/but as examples

89 How much of the acclimatisation of species to any peculiar climate is due to mere habit, and how much to the natural selection of varieties having different innate constitutions, and how much to both means combined, is a very obscure question.

89:c combined, is an obscure

90 That habit or custom has some influence I must believe, both from analogy, and from the incessant advice given in agricultural works, even in the ancient Encyclopædias of China, to be very cautious in transposing animals from one district to another; for it is not likely that man should have succeeded in selecting so many breeds and sub-breeds with constitutions specially fitted for their own districts: the result must, I think, be due to habit.

90:c analogy and from

90:e cautious in transporting animals from one district to another; for as it/own districts, the result

90.x-y:f influence, I/another. *y* And as

91 On the other hand, I can see no reason to doubt that natural selection will continually tend to preserve those individuals which are born with constitutions best adapted to their native countries.

91:e hand natural selection would inevitably tend to preserve those individuals which were born with constitutions best adapted to any country which they inhabited.

91:f hand, natural

92 In treatises on many kinds of cultivated plants, certain varieties are said to withstand certain climates better than others: this is very strikingly shown in works on fruit trees published in the United States, in which certain varieties are habitually recommended for the northern, and others for the southern States; and as most of these varieties are of recent origin, they cannot owe their constitutional differences to habit.

92:c northern and others

92:d fruit-trees

92:f is strikingly

93 The case of the Jerusalem artichoke, which is never propagated by seed, and of which consequently new varieties have not been produced, has even been advanced—for it is now as tender as ever it was—as proving that acclimatisation cannot be effected!

93:d propagated in England by

93:f advanced, as proving that acclimatisation cannot be effected, for it is now as tender as ever it was!

94 The case, also of the kidney-bean has been often cited for a similar purpose, and with much greater weight; but until some one will sow, during a score of generations, his kidney-beans so early that a very large proportion are destroyed by frost, and then collect seed from the few survivors, with care to prevent accidental crosses, and then again get seed from these seedlings, with the same precautions, the experiment cannot be said to have been even tried.

94:e said to have been tried.

95 Nor let it be supposed that no differences in the constitution of seedling kidney-beans ever appear, for an account has been published how much more hardy some seedlings appeared to be than others.

95:d some seedlings were than others; and of this fact I have myself observed a striking instance.

95:e that differences in the constitution of seedling kidney-beans never appear/observed striking instances.

95:f some seedlings are than

96   On the whole, I think we may conclude that habit, use, and disuse, have, in some cases, played a considerable part in the modification of the constitution, and of the structure of various organs; but that the effects of use and disuse have often been largely combined with, and sometimes overmastered by, the natural selection of innate differences.

96:b by the natural

96:d by, the natural

96:e habit, or use and disuse, have, in some

96:f whole, we/constitution and structure; but that the effects have

97   *Correlation of Growth.—*

97:d [*Center*] *Correlation of Growth.* [*Space*]

97:e *Correlated Variation.*

98 I mean by this expression that the whole organisation is so tied together during its growth and development, that when slight variations in any one part occur, and are accumulated through natural selection, other parts become modified.

98:d [¶]

99 This is a very important subject, most imperfectly understood.

99:d understood, and no doubt totally different classes of facts may be here easily confounded together: we shall presently see that simple inheritance often gives the false appearance of correlation.

99.x-y:f together. *y* We

100 The most obvious case is, that modifications accumulated solely for the good of the young or larva, will, it may safely be concluded, affect the structure of the adult; in the same manner as any malconformation affecting the early embryo, seriously affects the whole organisation of the adult.

100:d obvious instance of real correlation is, that variations of structure arising in the young or in the larvæ naturally tend to affect the structure of the mature animal; in the same manner as any malconformation in the early embryo is known seriously to affect

100:f One of the most obvious real cases is, that/animal.

101 The several parts of the body which are homologous, and

which, at an early embryonic period, are alike, seem liable to vary in an allied manner: we see this in the right and left sides of the body varying in the same manner; in the front and hind legs, and even in the jaws and limbs, varying together, for the lower jaw is believed to be homologous with the limbs.

101:d period, are identical in structure, and which are necessarily exposed to similar conditions, seem eminently liable

101:e believed by some anatomists to

102 These tendencies, I do not doubt, may be mastered more or less completely by natural selection: thus a family of stags once existed with an antler only on one side; and if this had been of any great use to the breed it might probably have been rendered permanent by natural selection.

102:c breed, it

102:e permanent by selection.

103 Homologous parts, as has been remarked by some authors, tend to cohere; this is often seen in monstrous plants; and nothing is more common than the union of homologous parts in normal structures, as the union of the petals of the corolla into a tube.

103:e plants; and

103:f structures, as in the union of the petals into

104 Hard parts seem to affect the form of adjoining soft parts; it is believed by some authors that the diversity in the shape of the pelvis in birds causes the remarkable diversity in the shape of their kidneys.

104:d forms

104:f authors that with birds the diversity in the shape of the pelvis causes

105 Others believe that the shape of the pelvis in the human mother influences by pressure the shape of the head of the child.

106 In snakes, according to Schlegel, the shape of the body and the manner of swallowing determine the position of several of the most important viscera.

106:e position and form of several

106:f Schlegel, the form of the body

107 The nature of the bond of correlation is very frequently quite obscure.

107:e is frequently

107:f bond is

108 M. Is. Geoffroy St. Hilaire has forcibly remarked, that certain malconformations very frequently, and that others rarely co-exist, without our being able to assign any reason.

108:f malconformations frequently, and that others rarely, co-exist

109 What can be more singular than the relation between blue eyes

and deafness in cats, and the tortoise-shell colour with the female sex; the feathered feet and skin between the outer toes in pigeons, and the presence of more or less down on the young birds when first hatched, with the future colour of their plumage; or, again, the relation between the hair and teeth in the naked Turkish dog, though here probably homology comes into play?

109:d singular than in cats the relation between complete whiteness with blue eyes and deafness, or between the tortoise-shell colour and the female sex; or in pigeons between their feathered feet and skin betwixt the outer toes, and between the presence of more or less down on the young bird when first hatched, with the future colour of its plumage/here no doubt homology

109:e toes, or between the presence

109:f than the relation in cats between complete whiteness and blue eyes with deafness/young pigeon when

110 With respect to this latter case of correlation, I think it can hardly be accidental, that if we pick out the two orders of mammalia which are most abnormal in their dermal covering, viz. Cetacea (whales) and Edentata (armadilloes, scaly anteaters, &c.), that these are likewise the most abnormal in their teeth.

110:d orders of mammals which

110:e that, if

110:f accidental, that the two/&c.), are likewise on the whole the most abnormal in their teeth; but there are so many exceptions to this rule, as Mr. Mivart has remarked, that it has little value.

111 I know of no case better adapted to show the importance of the laws of correlation in modifying important structures, independently of utility and, therefore, of natural selection, than that of the difference between the outer and inner flowers in some Compositous and Umbelliferous plants.

111:d correlation in leading to modifications of important structures, independently of utility and therefore

111:e laws of variation and correlation, independently of utility and therefore of natural selection, than that before referred to, of the difference

111:f laws of correlation and variation, independently of utility and therefore of natural selection, than that of the **difference**

112 Every one knows the difference in the ray and central florets of, for instance, the daisy, and this difference is often accompanied with the abortion of parts of the flower.

112:e accompanied with the partial or complete abortion of the reproductive organs.

112:f one is familiar with the difference between the ray

113 But, in some Compositous plants, the seeds also differ in shape

and sculpture; and even the ovary itself, with its accessory parts, differs, as has been described by Cassini.

113:e some of these plants, the seeds also differ in shape and sculpture.

114 These differences have been attributed by some authors to pressure, and the shape of the seeds in the ray-florets in some Compositæ countenances this idea; but, in the case of the corolla of the Umbelliferæ, it is by no means, as Dr. Hooker informs me, in species with the densest heads that the inner and outer flowers most frequently differ.

114:e authors to the pressure of the involucra on the florets, or to their mutual pressure, and the shape/but with the Umbelliferæ, it is by no means, as Dr. Hooker informs me, the species with the densest heads which most frequently differ in their inner and outer flowers.

114:f have sometimes been attributed to the pressure

115 It might have been thought that the development of the ray-petals by drawing nourishment from certain other parts of the flower had caused their abortion; but in some Compositæ there is a difference in the seeds of the outer and inner florets without any difference in the corolla.

115:e nourishment from the reproductive organs had caused their abortion; but this can hardly be the sole cause, for in some Compositæ the seeds of the outer and inner florets differ without

115:f organs causes/differ, without

116 Possibly, these several differences may be connected with some difference in the flow of nutriment towards the central and external flowers: we know, at least, that in irregular flowers, those nearest to the axis are oftenest subject to peloria, and become regular.

116:c Possibly these/are said to be oftenest

116:e connected with the different flow/that with normally irregular flowers, those nearest to the axis are most subject to peloria, that is they become symmetrical.

116:f that with irregular/is to become abnormally symmetrical.

117 I may add, as an instance of this, and of a striking case of correlation, that I have recently observed in some garden pelargoniums, that the central flower of the truss often loses the patches of darker colour in the two upper petals; and that when this occurs, the adherent nectary is quite aborted; when the colour is absent from only one of the two upper petals, the nectary is only much shortened.

117:d petals; and that, when

117.x-y:e in many pelargoniums, that in the central flower of the truss, the two upper petals often lose their patches of darker colour; and when this occurs, the adherent nectary is quite aborted. *y* When the colour/nectary is not quite aborted but is much shortened.

117.x-y:f this fact, and as a striking case of correlation, that in many pelargoniums, the two upper petals in the central flower of the truss often/aborted; the central flower thus becoming peloric or regular. *y* When

118 With respect to the difference in the corolla of the central and exterior flowers of a head or umbel, I do not feel at all sure that C. C. Sprengel's idea that the ray-florets serve to attract insects, whose agency is highly advantageous in the fertilisation of plants of these two orders, is so far-fetched, as it may at first appear: and if it be advantageous, natural selection may have come into play.

118:d far-fetched as

118:e to the development of the corolla in the central and exterior flowers, Sprengel's idea that the ray-florets serve to attract insects, whose agency is highly advantageous or necessary for the fertilisation of these plants, is highly probable; and if so, natural

118:f corolla, Sprengel's

119 But in regard to the differences both in the internal and external structure of the seeds, which are not always correlated with any differences in the flowers, it seems impossible that they can be in any way advantageous to the plant: yet in the Umbelliferæ these differences are of such apparent importance— the seeds being in some cases, according to Tausch, orthospermous in the exterior flowers and cœlospermous in the central flowers,—that the elder De Candolle founded his main divisions of the order on analogous differences.

119:e But with respect to the seeds, it seems impossible that their differences in shape, which are not always correlated with any difference in the corolla, can be in any way beneficial: yet/ importance—the seeds being sometimes orthospermous/divisions in the order on such characters.

120 Hence we see that modifications of structure, viewed by systematists as of high value, may be wholly due to unknown laws of correlated growth, and without being, as far as we can see, of the slightest service to the species.

120:e Hence, as before remarked, we/due to the laws of variation and correlation, without being, as far as we can judge, of

120:f Hence modifications

121 We may often falsely attribute to correlation of growth, structures which are common to whole groups of species, and which in truth are simply due to inheritance; for an ancient progenitor may have acquired through natural selection some one modification in structure, and, after thousands of generations, some other and independent modification; and these two modifications, having been transmitted to a whole group of

descendants with diverse habits, would naturally be thought to be correlated in some necessary manner.

121:c growth structures

121:e attribute to correlated variation structures

121:f thought to be in some necessary manner correlated.

122 So, again, I do not doubt that some apparent correlations, occurring throughout whole orders, are entirely due to the manner alone in which natural selection can act.

122:c So, again, some correlations, occurring throughout whole orders, are apparently due

122:e Some correlations are apparently due to the manner in which natural selection acts.

122:f Some other correlations/selection can alone act.

123 For instance, Alph. De Candolle has remarked that winged seeds are never found in fruits which do not open: I should explain the rule by the fact that seeds could not gradually become winged through natural selection, except in fruits which opened; so that the individual plants producing seeds which were a little better fitted to be wafted further, might get an advantage over those producing seed less fitted for dispersal; and this process could not possibly go on in fruit which did not open.

123:d open. [*Space*]

123:e explain this rule by the impossibility of seeds gradually becoming winged through natural selection, unless the capsules first opened themselves; for in this case alone could the seeds, which were a little better adapted to be wafted by the wind, gain an advantage over those less well fitted for wide dispersal.

123:f capsules were open; for/over others less

123.1:d [*Center*] *Compensation and Economy of Growth.* [*Space*]

124 The elder Geoffroy and Goethe propounded, at about the same period, their law of compensation or balancement of growth; or, as Goethe expressed it, "in order to spend on one side, nature is forced to economise on the other side."

124:f same time their

125 I think this holds true to a certain extent with our domestic productions: if nourishment flows to one part or organ in excess, it rarely flows, at least in excess, to another part; thus it is difficult to get a cow to give much milk and to fatten readily.

126 The same varieties of the cabbage do not yield abundant and nutritious foliage and a copious supply of oil-bearing seeds.

127 When the seeds in our fruits become atrophied, the fruit itself gains largely in size and quality.

128 In our poultry, a large tuft of feathers on the head is generally accompanied by a diminished comb, and a large beard by diminished wattles.

129 With species in a state of nature it can hardly be maintained that the law is of universal application; but many good observers, more especially botanists, believe in its truth.

130 I will not, however, here give any instances, for I see hardly any way of distinguishing between the effects, on the one hand, of a part being largely developed through natural selection and another and adjoining part being reduced by this same process or by disuse, and, on the other hand, the actual withdrawal of nutriment from one part owing to the excess of growth in another and adjoining part.

131 I suspect, also, that some of the cases of compensation which have been advanced, and likewise some other facts, may be merged under a more general principle, namely, that natural selection is continually trying to economise in every part of the organisation.

131:f economise every

132 If under changed conditions of life a structure before useful becomes less useful, any diminution, however slight, in its development, will be seized on by natural selection, for it will profit the individual not to have its nutriment wasted in building up an useless structure.

132:f structure, before useful, becomes less useful, its diminution will be favoured, for

133 I can thus only understand a fact with which I was much struck when examining cirripedes, and of which many other instances could be given: namely, that when a cirripede is parasitic within another and is thus protected, it loses more or less completely its own shell or carapace.

133:f many analogous instances

134 This is the case with the male Ibla, and in a truly extraordinary manner with the Proteolepas: for the carapace in all other cirripedes consists of the three highly-important anterior segments of the head enormously developed, and furnished with great nerves and muscles; but in the parasitic and protected Proteolepas, the whole anterior part of the head is reduced to the merest rudiment attached to the bases of the prehensile antennæ.

135 Now the saving of a large and complex structure, when rendered superfluous by the parasitic habits of the Proteolepas, though effected by slow steps, would be a decided advantage to each successive individual of the species; for in the struggle for life to which every animal is exposed, each individual Proteolepas would have a better chance of supporting itself, by less nutriment being wasted in developing a structure now become useless.

135:f superfluous, would be/each would/wasted.

136 Thus, as I believe, natural selection will always succeed in

the long run in reducing and saving every part of the organisation, as soon as it is rendered superfluous, without by any means causing some other part to be largely developed in a corresponding degree.

136:d rendered by changed habits of life superfluous

136:f will tend in the long run to reduce any part of the organisation, as soon as it becomes, through changed habits, superfluous

137 And, conversely, that natural selection may perfectly well succeed in largely developing any organ, without requiring as a necessary compensation the reduction of some adjoining part.

137:d part. [*Space*]

137:f developing an organ without

137.1:d [*Center*] *Multiple, Rudimentary, and Lowly-organised Structures are Variable.*

138 It seems to be a rule, as remarked by Is. Geoffroy St. Hilaire, both in varieties and in species, that when any part or organ is repeated many times in the structure of the same individual (as the vertebræ in snakes, and the stamens in polyandrous flowers) the number is variable; whereas the number of the same part or organ, when it occurs in lesser numbers, is constant.

138:f both with varieties and species, that when any part or organ is repeated many times in the same/whereas the same

139 The same author and some botanists have further remarked that multiple parts are also very liable to variation in structure.

139:f author as well as some/parts are extremely liable to vary in structure.

140 Inasmuch as this "vegetative repetition," to use Prof. Owen's expression, seems to be a sign of low organisation; the foregoing remark seems connected with the very general opinion of naturalists, that beings low in the scale of nature are more variable than those which are higher.

140:b organisation, the foregoing

140:f As "vegetative repetition," to use Prof. Owen's expression, is a sign of low organisation, the foregoing statements accord with the common opinion of naturalists, that beings which stand low

141 I presume that lowness in this case means that the several parts of the organisation have been but little specialised for particular functions; and as long as the same part has to perform diversified work, we can perhaps see why it should remain variable, that is, why natural selection should have preserved or rejected each little deviation of form less carefully than when the part has to serve for one special purpose alone.

141:e should not have preserved or rejected each little deviation of form so carefully as when the part had to

297

141:f lowness here means/when the part has to serve for some one special purpose.

142 In the same way that a knife which has to cut all sorts of things may be of almost any shape: whilst a tool for some particular object had better be of some particular shape.

142:e particular purpose had

142:f purpose must be of

143 Natural selection, it should never be forgotten, can act on each part of each being, solely through and for its advantage.

143:f act solely through and for the advantage of each being.

144 Rudimentary parts, it has been stated by some authors, and I believe with truth, are apt to be highly variable.

144:f parts, as it is generally admitted, are

145 We shall have to recur to the general subject of rudimentary and aborted organs; and I will here only add that their variability seems to be owing to their uselessness, and therefore to natural selection having no power to check deviations in their structure.

145:f recur to this subject; and I will here only add that their variability seems to result from their uselessness, and consequently from natural selection having had no power

146 Thus rudimentary parts are left to the free play of the various laws of growth, to the effects of long-continued disuse, and to the tendency to reversion.

146:[f]

147 *A part developed in any species in an extraordinary degree or manner, in comparison with the same part in allied species, tends to be highly variable.—*

147:d [Center] *A Part/Species/same Part/Species/variable.* [Space]

148 Several years ago I was much struck with a remark, nearly to the above effect, published by Mr. Waterhouse.

148:d[¶]

149 I infer also from an observation made by Professor Owen, with respect to the length of the arms of the ourang-outang, that he has come to a nearly similar conclusion.

148 + 9:e struck by a remark, to the above effect, made by Mr. Waterhouse; Professor Owen, also, seems to have come

148/149:f Waterhouse. Professor

150 It is hopeless to attempt to convince any one of the truth of this proposition without giving the long array of facts which I have collected, and which cannot possibly be here introduced.

150:e truth of the above proposition

151 I can only state my conviction that it is a rule of high generality.

152 I am aware of several causes of error, but I hope that I have made due allowance for them.

153 It should be understood that the rule by no means applies to any part, however unusually developed, unless it be unusually developed in comparison with the same part in closely allied species.

153:f unless it be unusually developed in one species or in a few species in comparison with the same part in many closely

154 Thus, the bat's wing is a most abnormal structure in the class mammalia; but the rule would not here apply, because there is a whole group of bats having wings; it would apply only if some one species of bat had its wings developed in some remarkable manner in comparison with the other species of the same genus.

154:e Thus, the wing of the bat is a most/not apply here, because the whole group of bats possesses wings; it would apply only if some one species had wings developed in a remarkable

154:f class of mammals; but

155 The rule applies very strongly in the case of secondary sexual characters, when displayed in any unusual manner.

156 The term, secondary sexual characters, used by Hunter, applies to characters which are attached to one sex, but are not directly connected with the act of reproduction.

156:f Hunter, relates to characters

157 The rule applies to males and females; but as females more rarely offer remarkable secondary sexual characters, it applies more rarely to them.

157:f but more rarely to the females, as they seldom offer remarkable secondary sexual characters.

158. The rule being so plainly applicable in the case of secondary sexual characters, may be due to the great variability of these characters, whether or not displayed in any unusual manner— of which fact I think there can be little doubt.

159 But that our rule is not confined to secondary sexual characters is clearly shown in the case of hermaphrodite cirripedes; and I may here add, that I particularly attended to Mr. Waterhouse's remark, whilst investigating this Order, and I am fully convinced that the rule almost invariably holds good with cirripedes.

159:e cirripedes; I particularly/good.

159:f almost always holds

160 I shall, in my future work, give a list of the more remarkable cases; I will here only briefly give one, as it illustrates the rule in its largest application.

160:e in a future work, give a list of all the more remarkable cases; I will here only give

160:f here give only one

161 The opercular valves of sessile cirripedes (rock barnacles) are, in every sense of the word, very important structures, and they differ extremely little even in different genera; but in the several species of one genus, Pyrgoma, these valves present a marvellous amount of diversification: the homologous valves in the different species being sometimes wholly unlike in shape; and the amount of variation in the individuals of several of the species is so great, that it is no exaggeration to state that the varieties differ more from each other in the characters of these important valves than do other species of distinct genera.

161:c diversification; the homologous

161:e individuals of the same species is so/characters derived from these important valves than do other species belonging to distinct

161:f even in distinct genera; but/varieties of the same species differ more from each other in the characters derived from these important organs, than do the species belonging to other distinct

162 As birds within the same country vary in a remarkably small degree, I have particularly attended to them, and the rule seems to me certainly to hold good in this class.

162:f As with birds the individuals of the same species, inhabiting the same country, vary extremely little, I have particularly attended to them; and the rule certainly seems to

163 I cannot make out that it applies to plants, and this would seriously have shaken my belief in its truth, had not the great variability in plants made it particularly difficult to compare their relative degrees of variability.

164 When we see any part or organ developed in a remarkable degree or manner in any species, the fair presumption is that it is of high importance to that species; nevertheless the part in this case is eminently liable to variation.

164:e nevertheless it is in this case eminently

164:f manner in a species, the fair

165 Why should this be so?

166 On the view that each species has been independently created, with all its parts as we now see them, I can see no explanation.

167 But on the view that groups of species have descended from other species, and have been modified through natural selection, I think we can obtain some light.

167:f species are descended

168 In our domestic animals, if any part, or the whole animal, be neglected and no selection be applied, that part (for instance, the comb in the Dorking fowl) or the whole breed will cease to have a nearly uniform character.

168:e First let me remark that if any part in our domestic animals, of the whole/have a uniform

169 The breed will then be said to have degenerated.

168.x-y + 169:f First let me make some preliminary remarks. *y* If, in our domestic animals, any part or the whole animal be neglected, and no/character; and the breed may be said to be degenerating.

170 In rudimentary organs, and in those which have been but little specialised for any particular purpose, and perhaps in polymorphic groups, we see a nearly parallel natural case; for in such cases natural selection either has not or cannot come into full play, and thus the organisation is left in a fluctuating condition.

170:e cannot **have come**

171 But what here more especially concerns us is, that in our domestic animals those points, which at the present time are undergoing rapid change by continued selection, are also eminently liable to variation.

171:e here concerns us is, that those points in our domestic animals, which

171:f here more particularly concerns

172 Look at the breeds of the pigeon; see what a prodigious amount of difference there is in the beak of the different tumblers, in the beak and wattle of the different carriers, in the carriage and tail of our fantails, &c., these being the points now mainly attended to by English fanciers.

172:e beaks of tumblers, in the beaks and wattle of carriers, in the carriage and tail of fantails

172:f at the individuals of the same breed of the pigeon, and see what a prodigious

173 Even in the sub-breeds, as in the short-faced tumbler, it is notoriously difficult to breed them nearly to perfection, and frequently individuals are born which depart widely from the standard.

173:c tumblers

173:e Even in sub-breeds, as in that of the short-faced tumbler, it is notoriously difficult to breed nearly perfect birds, some frequently departing widely

173:f in the same sub-breed/birds, many departing

174 There may be truly said to be a constant struggle going on between, on the one hand, the tendency to reversion to a less modified state, as well as an innate tendency to further variability of all kinds, and, on the other hand, the power of steady selection to keep the breed true.

174:e may truly be said/less perfect state, as well as an innate tendency to further variability, and

174:f innate tendency to new variations, and

175 In the long run selection gains the day, and we do not expect to fail so far as to breed a bird as coarse as a common tumbler from a good short-faced strain.

175:f so completely as to breed a bird as coarse as a common tumbler pigeon from

176 But as long as selection is rapidly going on, there may always be expected to be much variability in the structure undergoing modification.

176:e on, much variability in the parts undergoing modification may always be expected.

177 It further deserves notice that these variable characters, produced by man's selection, sometimes become attached, from causes quite unknown to us, more to one sex than to the other, generally to the male sex, as with the wattle of carriers and the enlarged crop of pouters.

177:e notice that characters, modified through selection by man, are sometimes transmitted, from

177:[f]

178   Now let us turn to nature.

179 When a part has been developed in an extraordinary manner in any one species, compared with the other species of the same genus, we may conclude that this part has undergone an extraordinary amount of modification, since the period when the species branched off from the common progenitor of the genus.

179:b modification since

179:f period when the several species

180 This period will seldom be remote in any extreme degree, as species very rarely endure for more than one geological period.

180:e species rarely

181 An extraordinary amount of modification implies an unusually large and long-continued amount of variability, which has continually been accumulated by natural selection for the benefit of the species.

182 But as the variability of the extraordinarily-developed part or organ has been so great and long-continued within a period not excessively remote, we might, as a general rule, expect still to find more variability in such parts than in other parts of the organisation, which have remained for a much longer period nearly constant.

182:b organisation which

182:c extraordinarily developed

182:e rule, still expect to

183 And this, I am convinced, is the case.

184 That the struggle between natural selection on the one hand,

and the tendency to reversion and variability on the other hand, will in the course of time cease; and that the most abnormally developed organs may be made constant, I can see no reason to doubt.

185 Hence when an organ, however abnormal it may be, has been transmitted in approximately the same condition to many modified descendants, as in the case of the wing of the bat, it must have existed, according to my theory, for an immense period in nearly the same state; and thus it comes to be no more variable than any other structure.

185:d Hence, when

185:f according to our theory/thus it has come not to be more

186 It is only in those cases in which the modification has been comparatively recent and extraordinarily great that we ought to find the *generative variability,* as it may be called, still present in a high degree.

187 For in this case the variability will seldom as yet have been fixed by the continued selection of the individuals varying in the required manner and degree, and by the continued rejection of those tending to revert to a former and less modified condition.

187:d less-modified condition. [*Space*]

187.1:d [*Center*] Specific Characters more Variable than Generic Characters. [*Space*]

188 The principle included in these remarks may be extended.

188:e in the above remarks

188:f principle discussed under the last heading may be applied to our present subject.

189 It is notorious that specific characters are more variable than generic.

190 To explain by a simple example what is meant.

191 If some species in a large genus of plants had blue flowers and some had red, the colour would be only a specific character, and no one would be surprised at one of the blue species varying into red, or conversely; but if all the species had blue flowers, the colour would become a generic character, and its variation would be a more unusual circumstance.

191:e If in a large genus of plants some species had blue

190 + 1:f meant: if in

192 I have chosen this example because an explanation is not in this case applicable, which most naturalists would advance, namely, that specific characters are more variable than generic, because they are taken from parts of less physiological importance than those commonly used for classing genera.

192:f because the explanation which most naturalists would advance is not here applicable, namely

193 I believe this explanation is partly, yet only indirectly, true; I shall, however, have to return to this subject in our chapter on Classification.

193:d in the chapter

193:f to this point in

194 It would be almost superfluous to adduce evidence in support of the above statement, that specific characters are more variable than generic; but I have repeatedly noticed in works on natural history, that when an author has remarked with surprise that some *important* organ or part, which is generally very constant throughout large groups of species, has *differed* considerably in closely-allied species, that it has, also, been *variable* in the individuals of some of the species.

194:f that ordinary specific characters are more variable than generic; but with respect to important characters, I have repeatedly noticed in works on natural history, that when an author remarks with surprise that some important organ or part, which is generally very constant throughout a large group of species, *differs* considerably in closely-allied species, it is often *variable* in the individuals of the same species.

195 And this fact shows that a character, which is generally of generic value, when it sinks in value and becomes only of specific value, often becomes variable, though its physiological importance may remain the same.

196 Something of the same kind applies to monstrosities: at least Is. Geoffroy St. Hilaire seems to entertain no doubt, that the more an organ normally differs in the different species of the same group, the more subject it is to individual anomalies.

196:f Hilaire apparently entertains/is to anomalies in the individuals.

197 On the ordinary view of each species having been independently created, why should that part of the structure, which differs from the same part in other independently-created species of the same genus, be more variable than those parts which are closely alike in the several species?

198 I do not see that any explanation can be given.

199 But on the view of species being only strongly marked and fixed varieties, we might surely expect to find them still often continuing to vary in those parts of their structure which have varied within a moderately recent period, and which have thus come to differ.

199:e view that species are only strongly marked and fixed varieties, we might expect/structure which had varied/and which had thus

199:f expect often to find them still continuing to vary in those parts of their structure which have varied

200 Or to state the case in another manner:—the points in which

all the species of a genus resemble each other, and in which they differ from the species of some other genus, are called generic characters; and these characters in common I attribute to inheritance from a common progenitor, for it can rarely have happened that natural selection will have modified several species, fitted to more or less widely-different habits, in exactly the same manner: and as these so-called generic characters have been inherited from a remote period, since that period when the species first branched off from their common progenitor, and subsequently have not varied or come to differ in any degree, or only in a slight degree, it is not probable that they should vary at the present day.

200:e differ from allied genera, are/been inherited from before the period when the different species

200:f these characters may be attributed/several distinct species/ when the several species

201 On the other hand, the points in which species differ from other species of the same genus, are called specific characters; and as these specific characters have varied and come to differ within the period of the branching off of the species from a common progenitor, it is probable that they should still often be in some degree variable,—at least more variable than those parts of the organisation which have for a very long period remained constant.

201:d constant. [*Space*]

201:e genus are/to differ since the period when the species branched off from

201:f constant. [*No space*]

201.1:d [*Center*] Secondary Sexual Characters Variable. [*Space*]

201.1:f[¶] *Secondary Sexual Characters Variable.—*[*No space*]

202　In connexion with the present subject, I will make only two other remarks.

202:[*f*]

203 I think it will be admitted, without my entering on details, that secondary sexual characters are very variable; I think it also will be admitted that species of the same group differ from each other more widely in their secondary sexual characters, than in other parts of their organisation; compare, for instance, the amount of difference between the males of gallinaceous birds, in which secondary sexual characters are strongly displayed, with the amount of difference between their females; and the truth of this proposition will be granted.

203:e with the amount of difference between the females

203.x-y:f I think it will be admitted by naturalists, without/are highly variable. *y* It will also be admitted/females.

204 The cause of the original variability of secondary sexual characters is not manifest; but we can see why these characters

should not have been rendered as constant and uniform as other parts of the organisation; for secondary sexual characters have been accumulated by sexual selection, which is less rigid in its action than ordinary selection, as it does not entail death, but only gives fewer offspring to the less favoured males.

204:f variability of these characters is not manifest; but we can see why they should not have been rendered as constant and uniform as others, for they are accumulated

205 Whatever the cause may be of the variability of secondary sexual characters, as they are highly variable, sexual selection will have had a wide scope for action, and may thus readily have succeeded in giving to the species of the same group a greater amount of difference in their sexual characters, than in other parts of their structure.

205:f thus have/difference in these than in other respects.

206 It is a remarkable fact, that the secondary sexual differences between the two sexes of the same species are generally displayed in the very same parts of the organisation in which the different species of the same genus differ from each other.

206:f secondary differences/which the species

207 Of this fact I will give in illustration two instances, the first which happen to stand on my list; and as the differences in these cases are of a very unusual nature, the relation can hardly be accidental.

207:e give two instances in illustration, the first

207:f give in illustration the two first instances which

208 The same number of joints in the tarsi is a character generally common to very large groups of beetles, but in the Engidæ, as Westwood has remarked, the number varies greatly; and the number likewise differs in the two sexes of the same species: again in fossorial hymenoptera, the manner of neuration of the wings is a character of the highest importance, because common to large groups; but in certain genera the neuration differs in the different species, and likewise in the two sexes of the same species.

208.x-y:f character common to very/species. *y* Again in the fossorial hymenoptera, the neuration

208.1:c Mr. Lubbock has recently remarked, that several minute crustaceans offer excellent illustrations of this law.

208.1:d Sir J. Lubbock

208.2:c "In Pontella, for instance, the sexual characters are afforded mainly by the anterior antennæ and by the fifth pair of legs: the specific differences also are principally given by these organs."

209 This relation has a clear meaning on my view of the subject: I look at all the species of the same genus as having as certainly

descended from the same progenitor, as have the two sexes of any one of the species.

209:f view: I/from a common progenitor/one species.

210 Consequently, whatever part of the structure of the common progenitor, or of its early descendants, became variable; variations of this part would, it is highly probable, be taken advantage of by natural and sexual selection, in order to fit the several species to their several places in the economy of nature, and likewise to fit the two sexes of the same species to each other, or to fit the males and females to different habits of life, or the males to struggle with other males for the possession of the females.

210:e variable, variations

210:f males to struggle with other males for the possession of the females. [*Space*]

211 Finally, then, I conclude that the greater variability of specific characters, or those which distinguish species from species, than of generic characters, or those which the species possess in common;—that the frequent extreme variability of any part which is developed in a species in an extraordinary manner in comparison with the same part in its congeners; and the not great degree of variability in a part, however extraordinarily it may be developed, if it be common to a whole group of species;—that the great variability of secondary sexual characters, and the great amount of difference in these same characters between closely allied species;—that secondary sexual and ordinary specific differences are generally displayed in the same parts of the organisation,—are all principles closely connected together.

211:b congeners; and the slight degree

211:f generic characters, or those which are possessed by all the species;—that the frequent/group of species;—that the great variability of secondary sexual characters, and their great difference in closely allied species;—that secondary

212 All being mainly due to the species of the same group having descended from a common progenitor, from whom they have inherited much in common,—to parts which have recently and largely varied being more likely still to go on varying than parts which have long been inherited and have not varied,—to natural selection having more or less completely, according to the lapse of time, overmastered the tendency to reversion and to further variability,—to sexual selection being less rigid than ordinary selection,—and to variations in the same parts having been accumulated by natural and sexual selection, and thus adapted for secondary sexual, and for ordinary specific purposes.

212:b accumulated by natural and sexual selection, and having been thus adapted

307

212:d purposes. [*Space*]

212:e ordinary purposes.

212:f group being the descendants of a common progenitor/pur-
poses. [*No space*]

213   *Distinct species present analogous variations; and a variety
of one species often assumes some of the characters of an allied
species, or reverts to some of the characters of an early progenitor.—*

213:d [*Center*] Distinct Species/Variations/Variety/Species/Char-
acters/Characters/Progenitor. [*Space*]

213:f[¶] Distinct/Variations, so that a Variety of one Species
often assumes a Character proper to an allied/Progenitor.—[*No
space*]

214 These propositions will be most readily understood by looking
to our domestic races.

214:d[¶]

214:f[*No* ¶]

215 The most distinct breeds of pigeons, in countries most widely
apart, present sub-varieties with reversed feathers on the head
and feathers on the feet,—characters not possessed by the aboriginal rock-pigeon; these then are analogous variations in two
or more distinct races.

215:f breeds of the pigeon, in countries widely/head, and with
feathers

216 The frequent presence of fourteen or even sixteen tail-feathers
in the pouter, may be considered as a variation representing
the normal structure of another race, the fantail.

216:d pouter may

217 I presume that no one will doubt that all such analogous variations are due to the several races of the pigeon having inherited from a common parent the same constitution and
tendency to variation, when acted on by similar unknown influences.

218 In the vegetable kingdom we have a case of analogous variation, in the enlarged stems, or roots as commonly called, of the
Swedish turnip and Ruta baga, plants which several botanists
rank as varieties produced by cultivation from a common parent: if this be not so, the case will then be one of analogous
variation in two so-called distinct species; and to these a third
may be added, namely, the common turnip.

218:f stems, or as commonly called roots, of the Swedish

219 According to the ordinary view of each species having been independently created, we should have to attribute this similarity
in the enlarged stems of these three plants, not to the *vera
causa* of community of descent, and a consequent tendency to

vary in a like manner, but to three separate yet closely related acts of creation.

219.1-2:d Many similar cases of analogous variation have been observed by Naudin in the great gourd-family, and by various authors in our cereals. 2 Similar cases occurring with insects under their natural conditions have lately been discussed with much ability by Mr. Walsh, who has grouped them under his law of Equable Variability.

219.2:f under natural

220 With pigeons, however, we have another case, namely, the occasional appearance in all the breeds, of slaty-blue birds with two black bars on the wings, a white rump, a bar at the end of the tail, with the outer feathers externally edged near their bases with white.

220:c white croup, a bar

220:e white loins, a bar

221 As all these marks are characteristic of the parent rock-pigeon, I presume that no one will doubt that this is a case of reversion, and not of a new yet analogous variation appearing in the several breeds.

222 We may I think confidently come to this conclusion, because, as we have seen, these coloured marks are eminently liable to appear in the crossed offspring of two distinct and differently coloured breeds; and in this case there is nothing in the external conditions of life to cause the reappearance of the slaty-blue, with the several marks, beyond the influence of the mere act of crossing on the laws of inheritance.

222:c may, I think, confidently

223 No doubt it is a very surprising fact that characters should reappear after having been lost for many, perhaps for hundreds of generations.

223:e many, probably for

224 But when a breed has been crossed only once by some other breed, the offspring occasionally show a tendency to revert in character to the foreign breed for many generations—some say, for a dozen or even a score of generations.

224:f show for many generations a tendency to revert in character to the foreign breed—some

225 After twelve generations, the proportion of blood, to use a common expression, of any one ancestor, is only 1 in 2048; and yet, as we see, it is generally believed that a tendency to reversion is retained by this very small proportion of foreign blood.

225:f expression, from one/this remnant of

226 In a breed which has not been crossed, but in which *both* parents have lost some character which their progenitor possessed, the tendency, whether strong or weak, to reproduce the lost

character might be, as was formerly remarked, for all that we can see to the contrary, transmitted for almost any number of generations.

226:f might, as/contrary, be transmitted

227 When a character which has been lost in a breed, reappears after a great number of generations, the most probable hypothesis is, not that the offspring suddenly takes after an ancestor some hundred generations distant, but that in each successive generation there has been a tendency to reproduce the character in question, which at last, under unknown favourable conditions, gains an ascendancy.

227:e ancestor removed by some hundred generations, but that in each successive generation the character in question has been lying latent, and at last, under unknown favourable conditions, is developed.

227:f not that one individual suddenly

228 For instance, it is probable that in each generation of the barb-pigeon, which produces most rarely a blue and black-barred bird, there has been a tendency in each generation in the plumage to assume this colour.

228:e With the barb-pigeon, for instance, which very rarely produces a blue bird, it is probable that a latent tendency exists in each generation to produce blue plumage.

228:f that there is a latent tendency in

228.1:e The possibility of characters long lying latent can be understood according to the hypothesis of pangenesis, which I have given in another work.

228.1:[f].

229 This view is hypothetical, but could be supported by some facts; and I can see no more abstract improbability in a tendency to produce any character being inherited for an endless number of generations, than in quite useless or rudimentary organs being, as we all know them to be, thus inherited.

229:e The abstract improbability of a latent tendency being transmitted through a vast number of generations, is not greater than that of quite useless or rudimentary organs being thus transmitted.

229:f of such a tendency/being similarly transmitted.

230 Indeed, we may sometimes observe a mere tendency to produce a rudiment inherited: for instance, in the common snapdragon (Antirrhinum) a rudiment of a fifth stamen so often appears, that this plant must have an inherited tendency to produce it.

230:d observe that a mere tendency to produce a rudiment is inherited.

230:e A mere tendency to produce a rudiment is indeed sometimes inherited.

230:f sometimes thus inherited.

231 As all the species of the same genus are supposed, on my theory, to have descended from a common parent, it might be expected that they would occasionally vary in an analogous manner; so that a variety of one species would resemble in some of its characters another species; this other species being on my view only a well-marked and permanent variety.

231:d common progenitor, it/so that the varieties of two or more species would resemble each other, or that a variety of one species would resemble in some of its characters another and distinct species,—this other

231:e on our theory, to be descended/ each other, or that a variety of some one species would resemble in certain characters another and distinct species,—this other species being according to our view

231:f supposed to/each other, or that a variety of one species would/being, according to our view, only

232 But characters thus gained would probably be of an unimportant nature, for the presence of all important characters will be governed by natural selection, in accordance with the diverse habits of the species, and will not be left to the mutual action of the conditions of life and of a similar inherited constitution.

232:e with the different habits/action of the nature of the organism and of the conditions of life.

232:f characters exclusively due to analogous variation would probably be of an unimportant nature, for the preservation of all functionally important characters will have been determined through natural/species.

233 It might further be expected that the species of the same genus would occasionally exhibit reversions to lost ancestral characters.

233:e to long lost

233:f lost characters.

234 As, however, we never know the exact character of the common ancestor of a group, we could not distinguish these two cases: if, for instance, we did not know that the rock-pigeon was not feather-footed or turn-crowned, we could not have told, whether these characters in our domestic breeds were reversions or only analogous variations; but we might have inferred that the blueness was a case of reversion, from the number of the markings, which are correlated with the blue tint, and which it does not appear probable would all appear together from simple variation.

234:d case of reversion from the number

234:e ancestor of a natural group/inferred that the blue colour was a case/correlated with this tint

234.x-y:f we do not know the common ancestor of any natural group, we cannot distinguish between reversionary and analogous characters. *y* If/that the parent rock-pigeon/whether such characters/tint, and which would not probably have all appeared together

235 More especially we might have inferred this, from the blue colour and marks so often appearing when distinct breeds of diverse colours are crossed.

235:e colour and the several marks/distinct colours

235:f when differently coloured breeds are

236 Hence, though under nature it must generally be left doubtful, what cases are reversions to an anciently existing character, and what are new but analogous variations, yet we ought, on my theory, sometimes to find the varying offspring of a species assuming characters (either from reversion or from analogous variation) which already occur in some other members of the same group.

236:d are reversions to a formerly existing

236:e Hence, although under/on our theory/already are present in other

236:f to formerly existing characters, and/characters which are already present

237 And this undoubtedly is the case in nature.

237:e case.

238 A considerable part of the difficulty in recognising a variable species in our systematic works, is due to its varieties mocking, as it were, some of the other species of the same genus.

238:d recognising in our systematic works a variable species, is

238:e were, other

238:f The difficulty in distinguishing variable species is largely due to the varieties

239 A considerable catalogue, also, could be given of forms intermediate between two other forms, which themselves must be doubtfully ranked as either varieties or species; and this shows, unless all these forms be considered as independently created species, that the one in varying has assumed some of the characters of the other, so as to produce the intermediate form.

239:e themselves can only doubtfully be ranked as species/intermediate forms.

239:f these closely allied forms be considered as independently created species, that they have in varying assumed some of the characters of the others.

240 But the best evidence is afforded by parts or organs of an important and uniform nature occasionally varying so as to ac-

240:e important and generally uniform
quire, in some degree, the character of the same part or organ in an allied species.

240:f evidence of analogous variations is afforded by parts or organs which are generally constant in character, but which occasionally vary so as to resemble, in some degree, the same part or

241 I have collected a long list of such cases; but here, as before, I lie under a great disadvantage in not being able to give them.

241:e under the great disadvantage of not

242 I can only repeat that such cases certainly do occur, and seem to me very remarkable.

242:f certainly occur

243   I will, however, give one curious and complex case, not indeed as affecting any important character, but from occurring in several species of the same genus, partly under domestication and partly under nature.

244 It is a case apparently of reversion.

244:d case almost certainly of

245 The ass not rarely has very distinct transverse bars on its legs, like those of the legs of the zebra: it has been asserted that these are plainest in the foal, and from inquiries which I have made, I believe this to be true.

245:d ass sometimes has/and, from

246 It has also been asserted that the stripe on each shoulder is sometimes double.

247 The shoulder-stripe is certainly very variable in length and outline.

246 + 7:d The stripe on the shoulder is sometimes double and is very

246 + 7:f double, and is

248 A white ass, but *not* an albino, has been described without either spinal or shoulder stripe; and these stripes are sometimes very obscure, or actually quite lost, in dark-coloured asses.

248:f stripe: and

249 The koulan of Pallas is said to have been seen with a double shoulder-stripe.

250 The hemionus has no shoulder-stripe; but traces of it, as stated by Mr. Blyth and others, occasionally appear: and I have been informed by Colonel Poole that the foals of this species are generally striped on the legs, and faintly on the shoulder.

250:d appear; and I

250:e [shoulder-stripe.] Mr. Blyth has seen a specimen of the hemionus with a distinct shoulder-stripe, though it properly has none; and I

251 The quagga, though so plainly barred like a zebra over the body, is without bars on the legs; but Dr. Gray has figured one specimen with very distinct zebra-like bars on the hocks.

252 With respect to the horse, I have collected cases in England of the spinal stripe in horses of the most distinct breeds, and of *all* colours; transverse bars on the legs are not rare in duns, mouse-duns, and in one instance in a chestnut: a faint shoulder-stripe may sometimes be seen in duns, and I have seen a trace in a bay horse.

252:e colours: transverse

253 My son made a careful examination and sketch for me of a dun Belgian cart-horse with a double stripe on each shoulder and with leg-stripes; and a man, whom I can implicitly trust, has examined for me a small dun Welch pony with *three* short parallel stripes on each shoulder.

253:c leg-stripes; I have myself seen a dun Devonshire pony, and a small dun Welch pony has been carefully described to me, both with *three* parallel

253:f Welsh

254 In the north-west part of India the Kattywar breed of horses is so generally striped, that, as I hear from Colonel Poole, who examined the breed for the Indian Government, a horse without stripes is not considered as purely-bred.

254:f examined this breed

255 The spine is always striped; the legs are generally barred; and the shoulder-stripe, which is sometimes double and sometimes treble, is common; the side of the face, moreover, is sometimes striped.

256 The stripes are plainest in the foal; and sometimes quite disappear in old horses.

256:c are often plainest

257 Colonel Poole has seen both gray and bay Kattywar horses striped when first foaled.

258 I have, also, reason to suspect, from information given me by Mr. W. W. Edwards, that with the English race-horse the spinal stripe is much commoner in the foal than in the full-grown animal.

258:d have also reason

258.1:d I have myself recently bred a foal from a bay mare (offspring of a Turcoman horse and a Flemish mare) by a bay English race-horse; this foal when a week old was marked on its hinder quarters and on its forehead with numerous, very narrow, dark zebra-like bars, and its legs were feebly striped: all the stripes soon disappeared completely.

258.1:e dark, zebra-like

259 Without here entering on further details, I may state that I have collected cases of leg and shoulder stripes in horses of very different breeds, in various countries from Britain to Eastern China; and from Norway in the north to the Malay Archipelago in the south.

259:f breeds in various

260 In all parts of the world these stripes occur far oftenest in duns and mouse-duns; by the term dun a large range of colour is included, from one between brown and black to a close approach to cream-colour.

261 I am aware that Colonel Hamilton Smith, who has written on this subject, believes that the several breeds of the horse have descended from several aboriginal species—one of which, the dun, was striped; and that the above-described appearances are all due to ancient crosses with the dun stock.

261:d horse are descended

262 But I am not at all satisfied with this theory, and should be loth to apply it to breeds so distinct as the heavy Belgian cart-horse, Welch ponies, cobs, the lanky Kattywar race, &c., inhabiting the most distant parts of the world.

262:d this view, and

262:e But this view may be safely rejected; for it is highly improbable that the heavy/world, should all have been crossed with one supposed aboriginal stock.

262:f Welsh ponies, Norwegian cobs, the lanky

263 Now let us turn to the effects of crossing the several species of the horse-genus.

264 Rollin asserts, that the common mule from the ass and horse is particularly apt to have bars on its legs.

264:b legs: according to Mr. Gosse, in certain parts of the United States about nine out of ten mules have striped legs.

265 I once saw a mule with its legs so much striped that any one at first would have thought that it must have been the product of a zebra; and Mr. W. C. Martin, in his excellent treatise on the horse, has given a figure of a similar mule.

265:b one would at first have

265:e one might have thought that it was a hybrid-zebra; and

266 In four coloured drawings, which I have seen, of hybrids between the ass and zebra, the legs were much more plainly barred than the rest of the body; and in one of them there was a double shoulder-stripe.

267 In Lord Moreton's famous hybrid from a chestnut mare and male quagga, the hybrid, and even the pure offspring subsequently produced from the mare by a black Arabian sire, were much more plainly barred across the legs than is even the pure quagga.

267:b Morton's

268 Lastly, and this is another most remarkable case, a hybrid has been figured by Dr. Gray (and he informs me that he knows of a second case) from the ass and the hemionus; and this hybrid, though the ass seldom has stripes on its legs and the

hemionus has none and has not even a shoulder-stripe, nevertheless had all four legs barred, and had three short shoulder-stripes, like those on the dun Welch pony, and even had some zebra-like stripes on the sides of its face.

268:b on his legs and the hemionus

268:c dun Devonshire and Welch ponies, and

268:d though the ass only occasionally has stripes on his legs and the hemionus

268:f Welsh

269 With respect to this last fact, I was so convinced that not even a stripe of colour appears from what would commonly be called an accident, that I was led solely from the occurrence of the face-stripes on this hybrid from the ass and hemionus, to ask Colonel Poole whether such face-stripes ever occur in the eminently striped Kattywar breed of horses, and was, as we have seen, answered in the affirmative.

269:b hemionus to

269:d what is commonly called chance, that/occurred

270 What now are we to say to these several facts?

271 We see several very distinct species of the horse-genus becoming, by simple variation, striped on the legs like a zebra, or striped on the shoulders like an ass.

271:f several distinct

272 In the horse we see this tendency strong whenever a dun tint appears—a tint which approaches to that of the general colouring of the other species of the genus.

273 The appearance of the stripes is not accompanied by any change of form or by any other new character.

274 We see this tendency to become striped most strongly displayed in hybrids from between several of the most distinct species.

275 Now observe the case of the several breeds of pigeons: they are descended from a pigeon (including two or three sub-species or geographical races) of a bluish colour, with certain bars and other marks; and when any breed assumes by simple variation a bluish tint, these bars and other marks invariably reappear; but without any other change of form or character.

276 When the oldest and truest breeds of various colours are crossed, we see a strong tendency for the blue tint and bars and marks to reappear in the mongrels.

277 I have stated that the most probable hypothesis to account for the reappearance of very ancient characters, is—that there is a *tendency* in the young of each successive generation to produce the long-lost character, and that this tendency, from unknown causes, sometimes prevails.

278 And we have just seen that in several species of the horse-genus

the stripes are either plainer or appear more commonly in the young than in the old.

279 Call the breeds of pigeons, some of which have bred true for centuries, species; and how exactly parallel is the case with that of the species of the horse-genus!

280 For myself, I venture confidently to look back thousands on thousands of generations, and I see an animal striped like a zebra, but perhaps otherwise very differently constructed, the common parent of our domestic horse, whether or not it be descended from one or more wild stocks, of the ass, the hemionus, quagga, and zebra.

280:d horse (whether/stocks) of

281 He who believes that each equine species was independently created, will, I presume, assert that each species has been created with a tendency to vary, both under nature and under domestication, in this particular manner, so as often to become striped like other species of the genus; and that each has been created with a strong tendency, when crossed with species inhabiting distant quarters of the world, to produce hybrids resembling in their stripes, not their own parents, but other species of the genus.

281:f like the other

282 To admit this view is, as it seems to me, to reject a real for an unreal, or at least for an unknown, cause.

283 It makes the works of God a mere mockery and deception; I would almost as soon believe with the old and ignorant cosmogonists, that fossil shells had never lived, but had been created in stone so as to mock the shells now living on the sea-shore.

284 *Summary.—*
284:d [Center] Summary. [Space]
284:f[¶] Summary.—[No space]

285 Our ignorance of the laws of variation is profound.
285:d[¶]
285:f[No ¶]

286 Not in one case out of a hundred can we pretend to assign any reason why this or that part differs, more or less, from the same part in the parents.

286:e part has varied.

287 But whenever we have the means of instituting a comparison, the same laws appear to have acted in producing the lesser differences between varieties of the same species, and the greater differences between species of the same genus.

288 The external conditions of life, as climate and food, &c., seem to have induced some slight modifications.

288:e Changed conditions generally induce mere fluctuating variability, but sometimes they cause direct and definite effects; and these may become strongly marked in the course of time, though we have not sufficient evidence on this head.

289 Habit in producing constitutional differences, and use in strengthening, and disuse in weakening and diminishing organs, seem to have been more potent in their effects.

289:b strengthening and disuse

289:e constitutional peculiarities and use/organs, appear in many cases to have been potent

290 Homologous parts tend to vary in the same way, and homologous parts tend to cohere.

290:f same manner, and

291 Modifications in hard parts and in external parts sometimes affect softer and internal parts.

292 When one part is largely developed, perhaps it tends to draw nourishment from the adjoining parts; and every part of the structure which can be saved without detriment to the individual, will be saved.

292:e detriment will

293 Changes of structure at an early age will generally affect parts subsequently developed; and there are very many other correlations of growth, the nature of which we are utterly unable to understand.

293:d will often affect

293:e age may affect parts subsequently developed; and many cases of correlated variation, the nature of which we are unable to understand, undoubtedly occur.

294 Multiple parts are variable in number and structure, perhaps arising from such parts not having been closely specialised to any particular function, so that their modifications have not been closely checked by natural selection.

294:e specialised for any

295 It is probably from this same cause that organic beings low in the scale of nature are more variable than those which have their whole organisation more specialised, and are higher in the scale.

295:e It follows probably from this same cause, that organic beings low in the scale are more variable than those standing higher in the scale, and which have their whole organisation more specialised.

296 Rudimentary organs, from being useless, will be disregarded by natural selection, and hence probably are variable.

296:e useless, are not regulated by natural selection, and hence are

297 Specific characters—that is, the characters which have come to differ since the several species of the same genus branched off

from a common parent—are more variable than generic characters, or those which have long been inherited, and have not differed within this same period.

298 In these remarks we have referred to special parts or organs being still variable, because they have recently varied and thus come to differ; but we have also seen in the second Chapter that the same principle applies to the whole individual; for in a district where many species of any genus are found—that is, where there has been much former variation and differentiation, or where the manufactory of new specific forms has been actively at work—there, on an average, we now find most varieties or incipient species.

298:c work—in that district and amongst these species, we now find, on an average, most varieties.

298:e chapter

298:f species of a genus

299 Secondary sexual characters are highly variable, and such characters differ much in the species of the same group.

300 Variability in the same parts of the organisation has generally been taken advantage of in giving secondary sexual differences to the sexes of the same species, and specific differences to the several species of the same genus.

300:f to the two sexes

301 Any part or organ developed to an extraordinary size or in an extraordinary manner, in comparison with the same part or organ in the allied species, must have gone through an extraordinary amount of modification since the genus arose; and thus we can understand why it should often still be variable in a much higher degree than other parts; for variation is a long-continued and slow process, and natural selection will in such cases not as yet have had time to overcome the tendency to further variability and to reversion to a less modified state.

302 But when a species with any extraordinarily-developed organ has become the parent of many modified descendants—which on my view must be a very slow process, requiring a long lapse of time—in this case, natural selection may readily have succeeded in giving a fixed character to the organ, in however extraordinary a manner it may be developed.

302:d manner it may have been developed.

302:e on our view/selection has succeeded

303 Species inheriting nearly the same constitution from a common parent and exposed to similar influences will naturally tend to present analogous variations, and these same species may occasionally revert to some of the characters of their ancient progenitors.

303:e influences, naturally tend to present analogous variations, or these

303:f parent, and

304 Although new and important modifications may not arise from reversion and analogous variation, such modifications will add to the beautiful and harmonious diversity of nature.

305 Whatever the cause may be of each slight difference in the offspring from their parents—and a cause for each must exist— it is the steady accumulation, through natural selection, of such differences, when beneficial to the individual, that gives rise to all the more important modifications of structure, by which the innumerable beings on the face of this earth are enabled to struggle with each other, and the best adapted to survive.

305:e difference between the offspring and their/selection, of beneficial differences that has given rise to all those modifications of structure which are the most important for the welfare of each species.

305:f exist—we have reason to believe that it is the steady accumulation of beneficial differences which has given rise to all the more important modifications of structure in relation to the habits of each species.

CHAPTER VI.

DIFFICULTIES ON THEORY.

2:e   ### DIFFICULTIES OF THE THEORY.

3 Difficulties on the theory of descent with modification—Transitions—Absence or rarity of transitional varieties—Transitions in habits of life—Diversified habits in the same species—Species with habits widely different from those of their allies—Organs of extreme perfection—Means of transition—Cases of difficulty—Natura non facit saltum—Organs of small importance—Organs not in all cases absolutely perfect—The law of Unity of Type and of the Conditions of Existence embraced by the theory of Natural Selection.

3:e Difficulties of the theory

3:f modification—Absence/perfection—Modes of transition

4 LONG before having arrived at this part of my work, a crowd of difficulties will have occurred to the reader.

4:f Long before the reader has arrived/occurred to him.

5 Some of them are so grave that to this day I can never reflect on them without being staggered; but, to the best of my judgment, the greater number are only apparent, and those that are real are not, I think, fatal to my theory.

5:c so serious that to this

5:d can hardly reflect

5:f being in some degree staggered/judgment, the number are only/fatal to the theory.

6   These difficulties and objections may be classed under the following heads:—Firstly, why, if species have descended from other species by insensibly fine gradations, do we not everywhere see innumerable transitional forms?

6:e First

6:f by fine

7 Why is not all nature in confusion instead of the species being, as we see them, well defined?

7:c confusion, instead

8   Secondly, is it possible that an animal having, for instance, the structure and habits of a bat, could have been formed by the modification of some animal with wholly different habits?

8:e animal with widely-different structure and habits?

8:f some other animal with widely-different habits and structure?

9 Can we believe that natural selection could produce, on the one hand, organs of trifling importance, such as the tail of a giraffe, which serves as a fly-flapper, and, on the other hand, organs of such wonderful structure, as the eye, of which we hardly as yet fully understand the inimitable perfection?

9:f one hand, an organ of trifling/other hand, an organ so wonderful as the eye?

10 Thirdly, can instincts be acquired and modified through natural selection?

11 What shall we say to so marvellous an instinct as that which leads the bee to make cells, which have practically anticipated the discoveries of profound mathematicians?

11:b which has practically

11:f to the instinct which leads the bee to make cells, and which

12 Fourthly, how can we account for species, when crossed, being sterile and producing sterile offspring, whereas, when varieties are crossed, their fertility is unimpaired?

13 The two first heads shall be here discussed—Instinct and Hybridism in separate chapters.

13:f heads will here be discussed; some miscellaneous objections in the following chapter; Instinct and Hybridism in the two succeeding chapters.

14 *On the absence or rarity of transitional varieties.—*

14:d [Center] On the Absence or Rarity of Transitional Varieties. [Space]

14:d[¶] On/Varieties.—

15 As natural selection acts solely by the preservation of profitable modifications, each new form will tend in a fully-stocked country to take the place of, and finally to exterminate, its own less improved parent or other less-favoured forms with which it comes into competition.

15:d[¶] improved parent-form and other

15:f[No ¶]

16 Thus extinction and natural selection will, as we have seen, go hand in hand.

16:d selection go

17 Hence, if we look at each species as descended from some other unknown form, both the parent and all the transitional varieties will generally have been exterminated by the very process of formation and perfection of the new form.

17:d process of the formation

17:f some unknown

18 But, as by this theory innumerable transitional forms must have existed, why do we not find them embedded in countless numbers in the crust of the earth?

19 It will be much more convenient to discuss this question in the chapter on the Imperfection of the geological record; and I will here only state that I believe the answer mainly lies in the record being incomparably less perfect than is generally supposed; the imperfection of the record being chiefly due to organic beings not inhabiting profound depths of the sea, and to their remains being embedded and preserved to a future age only in masses of sediment sufficiently thick and extensive to withstand an enormous amount of future degradation; and such fossiliferous masses can be accumulated only where much sediment is deposited on the shallow bed of the sea, whilst it slowly subsides.

19:c Geological Record

19:e be more/supposed.

20 These contingencies will concur only rarely, and after enormously long intervals.

20:[e]

21 Whilst the bed of the sea is stationary or is rising, or when very little sediment is being deposited, there will be blanks in our geological history.

21:[e]

22 The crust of the earth is a vast museum; but the natural collections have been made only at intervals of time immensely remote.

22:e been imperfectly made, and only at long intervals of time.

23 But it may be urged that when several closely-allied species inhabit the same territory we surely ought to find at the present time many transitional forms.

23:c territory, we

24 Let us take a simple case: in travelling from north to south over a continent, we generally meet at successive intervals with closely allied or representative species, evidently filling nearly the same place in the natural economy of the land.

25 These representative species often meet and interlock; and as the one becomes rarer and rarer, the other becomes more and more frequent, till the one replaces the other.

26 But if we compare these species where they intermingle, they are generally as absolutely distinct from each other in every detail of structure as are specimens taken from the metropolis inhabited by each.

27 By my theory these allied species have descended from a common parent; and during the process of modification, each has become adapted to the conditions of life of its own region, and has supplanted and exterminated its original parent and all the transitional varieties between its past and present states.

27:d original parent-form and all

27:e species are descended

28 Hence we ought not to expect at the present time to meet with numerous transitional varieties in each region, though they must have existed there, and may be embedded there in a fossil condition.

29 But in the intermediate region, having intermediate conditions of life, why do we not now find closely-linking intermediate varieties?

30 This difficulty for a long time quite confounded me.

31 But I think it can be in large part explained.

32   In the first place we should be extremely cautious in inferring, because an area is now continuous, that it has been continuous during a long period.

33 Geology would lead us to believe that almost every continent has been broken up into islands even during the later tertiary periods; and in such islands distinct species might have been separately formed without the possibility of intermediate varieties existing in the intermediate zones.

33:e that most continents have been broken

34 By changes in the form of the land and of climate, marine areas now continuous must often have existed within recent times in a far less continuous and uniform condition than at present.

35 But I will pass over this way of escaping from the difficulty; for I believe that many perfectly defined species have been formed on strictly continuous areas; though I do not doubt that the formerly broken condition of areas now continuous has played an important part in the formation of new species, more especially with freely-crossing and wandering animals.

35:f now continuous, has

36   In looking at species as they are now distributed over a wide area, we generally find them tolerably numerous over a large territory, then becoming somewhat abruptly rarer and rarer on the confines, and finally disappearing.

37 Hence the neutral territory between two representative species is generally narrow in comparison with the territory proper to each.

38 We see the same fact in ascending mountains, and sometimes it is quite remarkable how abruptly, as Alph. De Candolle has observed, a common alpine species disappears.

39 The same fact has been noticed by Forbes in sounding the depths of the sea with the dredge

39:b by E. Forbes

40 To those who look at climate and the physical conditions of life as the all-important elements of distribution, these facts

ought to cause surprise, as climate and height or depth graduate away insensibly.

41 But when we bear in mind that almost every species, even in its metropolis, would increase immensely in numbers, were it not for other competing species; that nearly all either prey on or serve as prey for others; in short, that each organic being is either directly or indirectly related in the most important manner to other organic beings, we must see that the range of the inhabitants of any country by no means exclusively depends on insensibly changing physical conditions, but in large part on the presence of other species, on which it depends, or by which it is destroyed, or with which it comes into competition; and as these species are already defined objects (however they may have become so), not blending one into another by insensible gradations, the range of any one species, depending as it does on the range of others, will tend to be sharply defined.

41:e presence of other species, on which it lives, or by which it is destroyed/objects, not

41:f to other organic beings,—we see/conditions, but in a large

42 Moreover, each species on the confines of its range, where it exists in lessened numbers, will, during fluctuations in the number of its enemies or of its prey, or in the seasons, be extremely liable to utter extermination; and thus its geographical range will come to be still more sharply defined.

42:e numbers, will during/geographical range will come to be still more sharply defined.

42:f numbers, will, during/prey, or in the nature of the seasons

43   If I am right in believing that allied or representative species, when inhabiting a continuous area, are generally so distributed that each has a wide range, with a comparatively narrow neutral territory between them, in which they become rather suddenly rarer and rarer; then, as varieties do not essentially differ from species, the same rule will probably apply to both; and if we in imagination adapt a varying species to a very large area, we shall have to adapt two varieties to two large areas, and a third variety to a narrow intermediate zone.

43:e and if we take a varying species inhabiting a very

43:f As allied or representative species, when inhabiting a continuous area, are generally distributed in such a manner that

44 The intermediate variety, consequently, will exist in lesser numbers from inhabiting a narrow and lesser area; and practically, as far as I can make out, this rule holds good with varieties in a state of nature.

45 I have met with striking instances of the rule in the case of varieties intermediate between well-marked varieties in the genus Balanus.

46 And it would appear from information given me by Mr. Wat-

son, Dr. Asa Gray, and Mr. Wollaston, that generally when varieties intermediate between two other forms occur, they are much rarer numerically than the forms which they connect.

46:d generally, when

47 Now, if we may trust these facts and inferences, and therefore conclude that varieties linking two other varieties together have generally existed in lesser numbers than the forms which they connect, then, I think, we can understand why intermediate varieties should not endure for very long periods;—why as a general rule they should be exterminated and disappear, sooner than the forms which they originally linked together.

47:d periods;—why, as

47:f inferences, and conclude that varieties linking two other varieties together generally have existed in lesser numbers than the forms which they connect, then we/rule, they

48　For any form existing in lesser numbers would, as already remarked, run a greater chance of being exterminated than one existing in large numbers; and in this particular case the intermediate form would be eminently liable to the inroads of closely allied forms existing on both sides of it.

48:e closely-allied

49 But a far more important consideration, as I believe, is that, during the process of further modification, by which two varieties are supposed on my theory to be converted and perfected into two distinct species, the two which exist in larger numbers from inhabiting larger areas, will have a great advantage over the intermediate variety, which exists in smaller numbers in a narrow and intermediate zone.

49:d numbers, from

49:e supposed to

49:f But it is a far more important consideration, that during

50 For forms existing in larger numbers will always have a better chance, within any given period, of presenting further favourable variations for natural selection to seize on, than will the rarer forms which exist in lesser numbers.

50:f will **have**

51 Hence, the more common forms, in the race for life, will tend to beat and supplant the less common forms, for these will be more slowly modified and improved.

52 It is the same principle which, as I believe, accounts for the common species in each country, as shown in the second chapter, presenting on an average a greater number of well-marked varieties than do the rarer species.

53 I may illustrate what I mean by supposing three varieties of sheep to be kept, one adapted to an extensive mountainous region; a second to a comparatively narrow, hilly tract; and a

third to wide plains at the base; and that the inhabitants are all trying with equal steadiness and skill to improve their stocks by selection; the chances in this case will be strongly in favour of the great holders on the mountains or on the plains improving their breeds more quickly than the small holders on the intermediate narrow, hilly tract; and consequently the improved mountain or plain breed will soon take the place of the less improved hill breed; and thus the two breeds, which originally existed in greater numbers, will come into close contact with each other, without the interposition of the supplanted, intermediate hill-variety.

53:f third to the wide plains at/or on the plains, improving their

54   To sum up, I believe that species come to be tolerably well-defined objects, and do not at any one period present an inextricable chaos of varying and intermediate links: firstly, because new varieties are very slowly formed, for variation is a very slow process, and natural selection can do nothing until favourable variations chance to occur, and until a place in the natural polity of the country can be better filled by some modification of some one or more of its inhabitants.

54:e first/is a slow/favourable individual differences or variations occur

55 And such new places will depend on slow changes of climate, or on the occasional immigration of new inhabitants, and, probably, in a still more important degree, on some of the old inhabitants becoming slowly modified, with the new forms thus produced and the old ones acting and reacting on each other.

56 So that, in any one region and at any one time, we ought only to see a few species presenting slight modifications of structure in some degree permanent; and this assuredly we do see.

56:f ought to see only a few

57   Secondly, areas now continuous must often have existed within the recent period in isolated portions, in which many forms, more especially amongst the classes which unite for each birth and wander much, may have separately been rendered sufficiently distinct to rank as representative species.

57:e period as isolated

58 In this case, intermediate varieties between the several representative species and their common parent, must formerly have existed in each broken portion of the land, but these links will have been supplanted and exterminated during the process of natural selection, so that they will no longer exist in a living state.

58:d existed within each isolated portion of the land, but these links during the process of natural selection will have been supplanted and exterminated, so

58:e must have existed formerly within

58:f must formerly have existed within/longer be found in

59   Thirdly, when two or more varieties have been formed in different portions of a strictly continuous area, intermediate varieties will, it is probable, at first have been formed in the intermediate zones, but they will generally have had a short duration.

60 For these intermediate varieties will, from reasons already assigned (namely from what we know of the actual distribution of closely allied or representative species, and likewise of acknowledged varieties), exist in the intermediate zones in lesser numbers than the varieties which they tend to connect.

61   From this cause alone the intermediate varieties will be liable to accidental extermination; and during the process of further modification through natural selection, they will almost certainly be beaten and supplanted by the forms which they connect; for these from existing in greater numbers will, in the aggregate, present more variation, and thus be further improved through natural selection and gain further advantages.

61:f more varieties, and thus

62   Lastly, looking not to any one time, but to all time, if my theory be true, numberless intermediate varieties, linking most closely all the species of the same group together, must assuredly have existed; but the very process of natural selection constantly tends, as has been so often remarked, to exterminate the parent-forms and the intermediate links.

62:c linking closely

62:d closely together all the species of the same group, must

63 Consequently evidence of their former existence could be found only amongst fossil remains, which are preserved, as we shall in a future chapter attempt to show, in an extremely imperfect and intermittent record.

63:f shall attempt to show in a future chapter, in [*No space*]

64   *On the origin and transitions of organic beings with peculiar habits and structure.—*

64:d [*Center*] *On the Origin and Transitions of Organic Beings with peculiar Habits and Structure.* [*Space*]

64:f[¶] *Structure.—*

65 It has been asked by the opponents of such views as I hold, how, for instance, a land carnivorous animal could have been converted into one with aquatic habits; for how could the animal in its transitional state have subsisted?

65:d[¶]

65:f[*No* ¶] instance, could a land carnivorous animal have been

66 It would be easy to show that within the same group carnivorous animals exist having every intermediate grade between truly aquatic and strictly terrestrial habits; and as each exists by a

struggle for life, it is clear that each is well adapted in its habits to its place in nature.

66:d exist, having

66:f that there now exist carnivorous animals presenting close intermediate grades from strictly terrestrial to aquatic habits; and as each exists by a struggle for life, it is clear that each must be well adapted to

67 Look at the Mustela vison of North America, which has webbed feet and which resembles an otter in its fur, short legs, and form of tail; during summer this animal dives for and preys on fish, but during the long winter it leaves the frozen waters, and preys like other pole-cats on mice and land animals.

67:d feet, and which/waters, and preys, like other pole-cats, on

67.x-y:f tail. *y* During the summer

68 If a different case had been taken, and it had been asked how an insectivorous quadruped could possibly have been converted into a flying bat, the question would have been far more difficult, and I could have given no answer.

68:d difficult to answer.

69 Yet I think such difficulties have very little weight.

69:c have little

70 Here, as on other occasions, I lie under a heavy disadvantage, for out of the many striking cases which I have collected, I can give only one or two instances of transitional habits and structures in closely allied species of the same genus; and of diversified habits, either constant or occasional, in the same species.

70:d for, out

70:f in allied species; and

71 And it seems to me that nothing less than a long list of such cases is sufficient to lessen the difficulty in any particular case like that of the bat.

72 Look at the family of squirrels; here we have the finest gradation from animals with their tails only slightly flattened, and from others, as Sir J. Richardson has remarked, with the posterior part of their bodies rather wide and with the skin on their flanks rather full, to the so-called flying squirrels; and flying squirrels have their limbs and even the base of the tail united by a broad expanse of skin, which serves as a parachute and allows them to glide through the air to an astonishing distance from tree to tree.

73 We cannot doubt that each structure is of use to each kind of squirrel in its own country, by enabling it to escape birds or beasts of prey, or to collect food more quickly, or, as there is reason to believe, by lessening the danger from occasional falls.

73:f believe, to lessen

74 But it does not follow from this fact that the structure of each squirrel is the best that it is possible to conceive under all natural conditions.

74:f all possible conditions.

75 Let the climate and vegetation change, let other competing rodents or new beasts of prey immigrate, or old ones become modified, and all analogy would lead us to believe that some at least of the squirrels would decrease in numbers or become exterminated, unless they also became modified and improved in structure in a corresponding manner.

76 Therefore, I can see no difficulty, more especially under changing conditions of life, in the continued preservation of individuals with fuller and fuller flank-membranes, each modification being useful, each being propagated, until by the accumulated effects of this process of natural selection, a perfect so-called flying squirrel was produced.

76:d until, by

77　Now look at the Galeopithecus or flying lemur, which formerly was falsely ranked amongst bats.

77:e was ranked

77:f or so-called flying lemur, which formerly was ranked amongst bats, but is now believed to belong to the Insectivora.

78 It has an extremely wide flank-membrane, stretching from the corners of the jaw to the tail, and including the limbs and the elongated fingers: the flank-membrane is, also, furnished with an extensor muscle.

78.x-y:f An extremely wide flank-membrane stretches from the corners of the jaw to the tail, and includes the limbs with the elongated fingers. *y* This flank-membrane is furnished

79 Although no graduated links of structure, fitted for gliding through the air, now connect the Galeopithecus with the other Lemuridæ, yet I can see no difficulty in supposing that such links formerly existed, and that each had been formed by the same steps as in the case of the less perfectly gliding squirrels; and that each grade of structure had been useful to its possessor.

79:b grade of structure was useful

79:e yet there is no

79:f other Insectivora, yet there is no difficulty in supposing that such links formerly existed, and that each was developed in the same manner as with the less perfectly gliding squirrels; each grade of structure having been

80 Nor can I see any insuperable difficulty in further believing it possible that the membrane-connected fingers and fore-arm of the Galeopithecus might be greatly lengthened by natural selection; and this, as far as the organs of flight are concerned, would convert it into a bat.

80:c forearm

80:f believing that the membrane-connected fingers and fore-arm of the Galeopithecus might have been greatly/would have converted the animal into

81 In bats which have the wing-membrane extended from the top of the shoulder to the tail, including the hind-legs, we perhaps see traces of an apparatus originally constructed for gliding through the air rather than for flight.

81:e perhaps yet see actual traces of an apparatus originally fitted for

81:f In certain bats in which the wing-membrane extends from the top of the shoulder to the tail and includes the hind-legs, we perhaps see traces

82 If about a dozen genera of birds had become extinct or were unknown, who would have ventured to have surmised that birds might have existed which used their wings solely as flappers, like the logger-headed duck (Micropterus of Eyton); as fins in the water and front legs on the land, like the penguin; as sails, like the ostrich; and functionally for no purpose, like the Apteryx?

82:c to surmise

82:f birds were to become/water and as front-legs

83 Yet the structure of each of these birds is good for it, under the conditions of life to which it is exposed, for each has to live by a struggle; but it is not necessarily the best possible under all possible conditions.

84 It must not be inferred from these remarks that any of the grades of wing-structure here alluded to, which perhaps may all have resulted from disuse, indicate the natural steps by which birds have acquired their perfect power of flight; but they serve, at least, to show what diversified means of transition are possible.

84:f all be the result of disuse, indicate the steps by which birds actually acquired their perfect power of flight; but they serve to show what diversified means of transition are at least possible.

85 Seeing that a few members of such water-breathing classes as the Crustacea and Mollusca are adapted to live on the land, and seeing that we have flying birds and mammals, flying insects of the most diversified types, and formerly had flying reptiles, it is conceivable that flying-fish, which now glide far through the air, slightly rising and turning by the aid of their fluttering fins, might have been modified into perfectly winged animals.

85:b land; and seeing

86 If this had been effected, who would have ever imagined that in an early transitional state they had been inhabitants of the open ocean, and had used their incipient organs of flight exclusively, as far as we know, to escape being devoured by other fish?

87 When we see any structure highly perfected for any partic-

ular habit, as the wings of a bird for flight, we should bear in mind that animals displaying early transitional grades of the structure will seldom continue to exist to the present day, for they will have been supplanted by the very process of perfection through natural selection.

87:e seldom exist at the present day, for they will have been supplanted by their successors, which were gradually rendered more perfect through

87:f seldom have survived to the present

88 Furthermore, we may conclude that transitional grades between structures fitted for very different habits of life will rarely have been developed at an early period in great numbers and under many subordinate forms.

88:e transitional states between

89 Thus, to return to our imaginary illustration of the flying-fish, it does not seem probable that fishes capable of true flight would have been developed under many subordinate forms, for taking prey of many kinds in many ways, on the land and in the water, until their organs of flight had come to a high stage of perfection, so as to have given them a decided advantage over other animals in the battle for life.

90 Hence the chance of discovering species with transitional grades of structure in a fossil condition will always be less, from their having existed in lesser numbers, than in the case of species with fully developed structures.

91 I will now give two or three instances of diversified and of changed habits in the individuals of the same species.

91:f instances both of diversified

92 When either case occurs, it would be easy for natural selection to fit the animal, by some modification of its structure, for its changed habits, or exclusively for one of its several different habits.

92:e In either case it would be easy for natural selection to adapt the structure of the animal to its changed habits, or exclusively to one of its several habits.

93 But it is difficult to tell, and immaterial for us, whether habits generally change first and structure afterwards; or whether slight modifications of structure lead to changed habits; both probably often change almost simultaneously.

93:e It is, however, difficult to decide, and immaterial/often occurring almost

94 Of cases of changed habits it will suffice merely to allude to that of the many British insects which now feed on exotic plants, or exclusively on artificial substances.

95 Of diversified habits innumerable instances could be given: I have often watched a tyrant flycatcher (Saurophagus sulphura-

tus) in South America, hovering over one spot and then proceeding to another, like a kestrel, and at other times standing stationary on the margin of water, and then dashing like a kingfisher at a fish.

95:d dashing into it like

96 In our own country the larger titmouse (Parus major) may be seen climbing branches, almost like a creeper; it often, like a shrike, kills small birds by blows on the head; and I have many times seen and heard it hammering the seeds of the yew on a branch, and thus breaking them like a nuthatch.

96:e it sometimes, like

97 In North America the black bear was seen by Hearne swimming for hours with widely open mouth, thus catching, like a whale, insects in the water.

97:b catching, almost like

98 Even in so extreme a case as this, if the supply of insects were constant, and if better adapted competitors did not already exist in the country, I can see no difficulty in a race of bears being rendered, by natural selection, more and more aquatic in their structure and habits, with larger and larger mouths, till a creature was produced as monstrous as a whale.

98:[b]

99   As we sometimes see individuals of a species following habits widely different from those both of their own species and of the other species of the same genus, we might expect, on my theory, that such individuals would occasionally have given rise to new species, having anomalous habits, and with their structure either slightly or considerably modified from that of their proper type.

99:b those of their

99:e expect that such individuals would occasionally give rise

99:f following habits different from those proper to their species and to the other/from that of their type.

100 And such instances do occur in nature.

100:f instances occur

101 Can a more striking instance of adaptation be given than that of a woodpecker for climbing trees and for seizing insects in the chinks of the bark?

101:e and seizing

102 Yet in North America there are woodpeckers which feed largely on fruit, and others with elongated wings which chase insects on the wing; and on the plains of La Plata, where not a tree grows, there is a woodpecker, which in every essential part of its organisation, even in its colouring, in the harsh tone of its voice, and undulatory flight, told me plainly of its close blood-

relationship to our common species; yet it is a woodpecker which never climbs a tree!

102.x-y:c insects on the wing. *y* On the plains of La Plata, where not a tree grows, there is a woodpecker *(Colaptes campestris)* which has two toes before and two behind, a long pointed tongue, stiff pointed tail feathers, but not so stiff as in the typical woodpeckers (yet I have seen it use its tail when alighting vertically on a post), and a straight strong beak.

102.y:e tongue, pointed tail-feathers, sufficiently stiff to support the bird in a vertical position on a post, but not so stiff as in the typical woodpeckers, and a straight

102.y:f where hardly a tree/(Colaptes campestris)

102.y.1:c The beak, however, is not so straight or strong as in the typical woodpeckers; but it is strong enough to bore into wood; and I may mention, as another illustration of the varied habits of the tribe, that a Mexican Colaptes has been described by De Saussure as boring holes into hard wood in order to lay up a store of acorns for its future consumption!

102.y.1.x:e into wood. [and I]

102.y.1.y:[e] = *102.y.3.1:e*

102.y.2:c Hence the Colaptes of La Plata in all the essential parts of its organization is a woodpecker, and until recently was classed in the same typical genus.

102.y.2:e Hence this Colaptes in all the essential parts of its structure is a woodpecker.

102.y.3:c Even such trifling characters as its colouring, the harsh tone of its voice, and undulatory flight, all told me plainly of its close blood-relationship to our common species; yet it is a woodpecker, as I can assert not only from my own observation, but from that of the accurate Azara, which never climbs a tree!

102.y.3:e Even in such trifling characters as the colouring, the harsh tone of the voice, and undulatory flight, all plainly declared its close blood-relationship to our common woodpecker; yet, as I can assert, not only from my own observation, but from that of the accurate Azara, it never climbs a tree!

102.y.3:f flight, its close blood-relationship to our common woodpecker is plainly declared; yet, as I can assert, not only from my own observations, but from those of the accurate Azara, in certain large districts it does not climb trees, and it makes its nest in holes in banks!

102.y.3.0.1:f In certain other districts, however, this same woodpecker, as Mr. Hudson states, frequents trees, and bores holes in the trunk for its nest.

102.y.3.1:e = *102.y.1.y:[e]* I may mention as another illustration of the varied habits of the tribe, that a Mexican Colaptes has been described by De Saussure as boring holes into hard wood in order to lay up a store of acorns, but for what use is not yet known.

334

102.y.3.1:f this genus, that/acorns.

103   Petrels are the most aërial and oceanic of birds, yet in the quiet Sounds of Tierra del Fuego, the Puffinuria berardi, in its general habits, in its astonishing power of diving, its manner of swimming, and of flying when unwillingly it takes flight, would be mistaken by any one for an auk or grebe; nevertheless, it is essentially a petrel, but with many parts of its organisation profoundly modified.

103:c birds, but in the quiet/modified in relation to its new habits of life; whereas the woodpecker of La Plata has its structure only slightly modified.

103:e sounds/diving, its manner of swimming, and of flying when made to take flight, would be mistaken by any one for an auk or a grebe; nevertheless it is/La Plata has had its

103:f diving, in its manner of swimming and of flying

104   On the other hand, the acutest observer by examining the dead body of the water-ouzel would never have suspected its sub-aquatic habits; yet this anomalous member of the strictly terrestrial thrush family wholly subsists by diving,—grasping the stones with its feet and using its wings under water.

104:c In the case of the water-ouzel, the acutest observer by examining its dead body would/member of the terrestrial/feet, and

104:e this bird, which is allied to the thrush family, wholly subsists by diving—using its wings under water, and grasping stones with its feet.

104:f family, subsists

104.1:d All the members of the great order of Hymenoptera are terrestrial, excepting the genus Proctotrupes, which Sir John Lubbock has recently discovered to be aquatic in its habits; it often enters the water and dives about by the use not of its legs but of its wings, and remains as long as four hours beneath the surface; yet not the least modification in its structure can be detected in accordance with such abnormal habits.

104.1:e order of Hymenopterous insects are

104.1:f has discovered/yet it exhibits no modification in structure in accordance with its abnormal habits.

105   He who believes that each being has been created as we now see it, must occasionally have felt surprise when he has met with an animal having habits and structure not at all in agreement.

105:f not in

106   What can be plainer than that the webbed feet of ducks and geese are formed for swimming? yet there are upland geese with webbed feet which rarely or never go near the water; and no one except Audubon has seen the frigate-bird, which has all its four toes webbed, alight on the surface of the sea.

106.x-y:e swimming? *y* Yet/surface of the ocean.

335

106.x-y:f rarely go

107 On the other hand, grebes and coots are eminently aquatic, although their toes are only bordered by membrane.

107:b hand grebes

107:c hand, grebes

108 What seems plainer than that the long toes of grallatores are formed for walking over swamps and floating plants, yet the water-hen is nearly as aquatic as the coot; and the landrail nearly as terrestrial as the quail or partridge.

108:c plants?—yet

108:d toes, not furnished with membrane, of the Grallatores are formed for walking over swamps and floating plants?—the water-hen and landrail are members of this order, yet the first is nearly as aquatic as the coot, and the second nearly

109 In such cases, and many others could be given, habits have changed without a corresponding change of structure.

110 The webbed feet of the upland goose may be said to have become rudimentary in function, though not in structure.

110:f become almost rudimentary

111 In the frigate-bird, the deeply-scooped membrane between the toes shows that structure has begun to change.

111:e deeply scooped

112 He who believes in separate and innumerable acts of creation will say, that in these cases it has pleased the Creator to cause a being of one type to take the place of one of another type; but this seems to me only restating the fact in dignified language.

112:e creation may say/place of one belonging to another

112:f place of one belonging to another/re-stating

113 He who believes in the struggle for existence and in the principle of natural selection, will acknowledge that every organic being is constantly endeavouring to increase in numbers; and that if any one being vary ever so little, either in habits or structure, and thus gain an advantage over some other inhabitant of the country, it will seize on the place of that inhabitant, however different it may be from its own place.

113:d one varies/gains

113:f one being varies/inhabitant of the same country/different that may

114 Hence it will cause him no surprise that there should be geese and frigate-birds with webbed feet, either living on the dry land or most rarely alighting on the water; that there should be long-toed corncrakes living in meadows instead of in swamps; that there should be woodpeckers where not a tree grows; that there should be diving thrushes, and petrels with the habits of auks.

114:b feet, living on the dry

114:d thrushes and diving Hymenoptera, and petrels

114:e corncrakes, living in meadows

114:f land and rarely/where hardly a tree

115   *Organs of extreme perfection and complication.—*

115:d [*Center*] *Organs of extreme Perfection and Complication.*
[*Space*]

116 To suppose that the eye, with all its inimitable contrivances
for adjusting the focus to different distances, for admitting
different amounts of light, and for the correction of spherical
and chromatic aberration, could have been formed by natural
selection, seems, I freely confess, absurd in the highest possible
degree.

116:d[¶] highest degree.

116:e eye with

116.1:c When it was first said that the sun stood still and the world
turned round, the common sense of mankind declared the
doctrine false; but the old saying of *Vox populi, vox Dei,* as
every philosopher knows, can never be trusted in science.

116.1:d knows, cannot be

117 Yet reason tells me, that if numerous gradations from a perfect
and complex eye to one very imperfect and simple, each grade
being useful to its possessor, can be shown to exist; if further,
the eye does vary ever so slightly, and the variations be inherited, which is certainly the case; and if any variation or
modification in the organ be ever useful to an animal under
changing conditions of life, then the difficulty of believing that
a perfect and complex eye could be formed by natural selection,
though insuperable by our imagination, can hardly be considered real.

117:c Yet Reason

117:d one imperfect/slightly and the variations be inherited/could
have been formed

117:e from an imperfect and simple eye to one perfect and complex,
each grade being useful to its possessor, can be shown to exist,
as is certainly the case; if further, the eye ever slightly varies,
and the variations be inherited, as is likewise certainly the case;
and if such variations should ever be useful to any animal under
changing conditions of life, then the difficulty of believing that
a perfect and complex eye could be formed by natural selection,
though insuperable by our imagination, cannot be considered
real.

117:f from a simple and imperfect eye to one complex and perfect
can be shown to exist, each grade being useful to its possessor,
as is certainly the case; if further, the eye ever varies and the
variations be inherited, as is likewise certainly the case; and if

337

such variations should be useful/imagination, should not be considered as subversive of the theory.

118 How a nerve comes to be sensitive to light, hardly concerns us more than how life itself first originated; but I may remark that several facts make me suspect that any sensitive nerve may be rendered sensitive to light, and likewise to those coarser vibrations of the air which produce sound.

118:c suspect that nerves sensitive to touch may

118:d remark that, as some of the lowest organisms, in which nerves cannot be detected, are known to be sensitive to light, it does not seem impossible that certain elements in their tissues or sarcode should have become aggregated and developed into nerves endowed with special sensibility to its action.

118:e elements in the sarcode, of which they are mainly composed, should become aggregated and developed into nerves endowed with this special sensibility.

118:f itself originated/are capable of perceiving light, it does not seem impossible that certain sensitive elements in their sarcode should become aggregated and developed into nerves, endowed

119 In looking for the gradations by which an organ in any species has been perfected, we ought to look exclusively to its lineal ancestors; but this is scarcely ever possible, and we are forced in each case to look to species of the same group, that is to the collateral descendants from the same original parent-form, in order to see what gradations are possible, and for the chance of some gradations having been transmitted from the earlier stages of descent, in an unaltered or little altered condition.

119:d In searching for the gradations through which any organ in any species has been perfected, we ought to look exclusively to its lineal progenitors; but this is scarcely ever possible, and we are forced in each case to look to other species and genera of the same group

119:e which an organ/forced to look/from the same parent-form/ transmitted in an unaltered

120 Amongst existing Vertebrata, we find but a small amount of gradation in the structure of the eye, and from fossil species we can learn nothing on this head.

120:c eye (though in the fish Amphioxus, the eye is in an extremely simple condition without a lens), and from

120:[d]

121 In this great class we should probably have to descend far beneath the lowest known fossiliferous stratum to discover the earlier stages, by which the eye has been perfected.

121:[d]

121.1:d But the state of the same organ even in the other main

338

divisions of the organic world may incidentally throw light on the steps by which it has been perfected.

121.1:e of the organ even in distinct classes may incidentally throw light on the steps by which it has been perfected in any one species.

121.1:f of the same organ in/perfected.

121.2:d[¶] The simplest organ which can be called an eye consists of an optic nerve, surrounded by pigment-cells, covered by translucent skin, but without any lens or other refractive body.

121.2:f pigment-cells and covered

121.3:d We may, however, according to M. Jourdain, descend even a step lower and find aggregates of pigment-cells, apparently serving as an organ of vision, but which rest merely on sarcodic tissue not furnished with any nerve.

121.3:e but without any nerve, and resting merely on sarcodic tissue.

121.3:f as organs of vision, without any nerves

121.4:d Eyes of the above simple nature are not capable of distinct vision, but serve merely to distinguish light from darkness.

121.4:e vision, and serve only to

121.5:d In certain star-fishes, small depressions in the layer of pigment which surrounds the nerve are filled, as described by the author just quoted, with transparent gelatinous matter, and this projects outwardly with a convex surface, like the cornea in the higher animals.

121.5:e matter, projecting with

121.6:d He suggests that this structure serves not to form an image, but only to concentrate the luminous rays and render their perception more perfect.

121.6:e this serves/more easy.

121.7:d In this concentration of the rays we gain the first and by far the most important step towards the formation of a true or picture-forming eye; for we have only to place the naked extremity of the optic nerve, which in some of the lower animals lies deeply buried in the body and in some near the surface, at the right distance from the concentrating apparatus, and an image must be formed on it.

121.7:e true, picture-forming/body, and in/image will be

1 2 2 In the Articulata we can commence a series with an optic nerve merely coated with pigment, and without any other mechanism; and from this low stage, numerous gradations of structure, branching off in two fundamentally different lines, can be shown to exist, until we reach a moderately high stage of perfection.

122.x:c In the great kingdom of the Articulata, we can start from an optic nerve, simply coated with pigment, which sometimes forms a sort of pupil, but is destitute of a lens or any other optical mechanism.

122.x:d In the great class of the Articulata, if we look for gradations, we may start from an optic nerve simply/or other optical contrivance.

122.x:e Articulata, we may start from an optic nerve simply coated with pigment, the latter sometimes forming a sort of pupil, but destitute

122.y:c From this rudimentary eye, which can distinguish light from darkness, but nothing else, there is an advance towards perfection along two lines of structure, which Müller thought were fundamentally different; namely,—firstly, stemmata, or the so-called "simple eyes," which have a lens and cornea; and secondly, "compound eyes," which seem to act mainly by excluding all the rays from each point of the object viewed, except the pencil that comes in a line perpendicular to the convex retina.

122.y.1:c In compound eyes, besides endless differences in the form, proportion, number, and position of the transparent cones coated by pigment, and which act by exclusion, we have additions of a more or less perfect concentrating apparatus: thus in the eye of the Meloe the facets of the cornea are "slightly convex both externally and internally—that is, lens-shaped."

122.y.1:[d]

123 In certain crustaceans, for instance, there is a double cornea, the inner one divided into facets, within each of which there is a lens-shaped swelling.

123:c In many crustaceans there are two corneæ—the external smooth, and the internal divided into facets—within the substance of which, as Milne Edwards says, "renflemens lenticulaires paraissent s'être développés;" and sometimes these lenses can be detached in a layer distinct from the cornea.

122.y + 123:d From this point we have to make a rather wider stride than in the case of the above-mentioned star-fish, and we come to certain Crustaceans in which the eyes are covered by a double cornea,—the external membrane smooth and the internal one divided into facets,—within the substance of which, as Milne Edwards states, "renflemens lenticulaires paraissent s'être développés;" and these lenses can sometimes be

122.y + 123:[e]

124 In other crustaceans the transparent cones which are coated by pigment, and which properly act only by excluding lateral pencils of light, are convex at their upper ends and must act by convergence; and at their lower ends there seems to be an imperfect vitreous substance.

124:c The transparent cones coated with pigment, which were supposed by Müller to act solely by excluding divergent pencils of light, usually adhere to the cornea, but not rarely they are separate from it, and have their free ends convex; and in this case they must act as converging lenses.

124:d With insects it is now known that the numerous cones surrounded by pigment, which form the great compound eyes, are filled with transparent refractive matter, and these cones produce images; but in addition, in certain beetles the facets of the cornea are slightly convex both externally and internally,— that is, are lens-shaped.

124:e numerous facets on the cornea of the great compound eyes form true lenses, and that the cones include curiously modified nervous filaments.

124:f of their great

124.1:c Altogether so diversified is the structure of the compound eyes, that Müller makes three main classes, with no less than seven subdivisions of structure; he makes a fourth main class, namely, "aggregates" of stemmata, and he adds that "this is the transition-form between the mosaic-like compound eyes unprovided with a concentrating apparatus, and organs of vision with such an apparatus."

124.1:d structure of the eye in the Articulata that Müller makes three main classes, with seven subdivisions, of compound eyes, and he adds a fourth main class of aggregated simple-eyes.

124.1:e But these organs in the Articulata are so much diversified that Müller formerly made three main classes of compound eyes with seven subdivisions, besides a fourth main class of aggregated simple eyes.

124.1:f classes with

125 With these facts, here far too briefly and imperfectly given, which show that there is much graduated diversity in the eyes of living crustaceans, and bearing in mind how small the number of living animals is in proportion to those which have become extinct, I can see no very great difficulty (not more than in the case of many other structures) in believing that natural selection has converted the simple apparatus of an optic nerve merely coated with pigment and invested by transparent membrane, into an optical instrument as perfect as is possessed by any member of the great Articulate class.

125:c here too briefly and imperfectly given, which show how much graduated diversity there is in the eyes of our existing crustaceans, and bearing

125:d[¶] When we reflect on these facts, here given too briefly and imperfectly, with respect to the wide, diversified, and graduated range of structure in the eyes of the existing Articulata; and when we bear in mind how small the number of all living forms must be in comparison with those which have become extinct, the difficulty ceases to be very great (not more so than in the case of many other structures) in believing that natural selection may have converted the simple apparatus of an optic nerve, coated/Class.

125:e briefly, with respect to the wide, diversified, and graduated range of structure in the eyes of the lower animals; and when we bear in mind how small the number of all the forms now living must/great in believing

125:f given much too/all living forms must/member of the Articulate Class.

126   He who will go thus far, if he find on finishing this treatise that large bodies of facts, otherwise inexplicable, can be explained by the theory of descent, ought not to hesitate to go further, and to admit that a structure even as perfect as the eye of an eagle might be formed by natural selection, although in this case he does not know any of the transitional grades.

126:d this volume that large bodies of facts, otherwise inexplicable, can be explained by the theory of descent with modification, ought/might have been formed/know the transitional steps.

126:e far, ought not to hesitate to go one step further, if he finds on finishing/modification; he ought to admit that a structure even as perfect as an eagle's eye might be formed/transitional states.

126:f theory of modification through natural selection; he/might thus be formed, although

126.0.1:e It has been objected that in order to modify the eye and still preserve it as a perfect instrument, many changes would have to be effected simultaneously, which, it is assumed, could not be done through natural selection; but as I have attempted to show in my work on the variation of domestic animals, it is not necessary to suppose that all the modifications were simultaneous, if they were extremely slight and gradual.

126.0.1:f suppose that the modifications were all simultaneous

126.0.1.1-2:f Different kinds of modification would, also, serve for the same general purpose: as Mr. Wallace has remarked, "if a lens has too short or too long a focus, it may be amended either by an alteration of curvature, or an alteration of density; if the curvature be irregular, and the rays do not converge to a point, then any increased regularity of curvature will be an improvement. 2 So the contraction of the iris and the muscular movements of the eye are neither of them essential to vision, but only improvements which might have been added and perfected at any stage of the construction of the instrument."

126.1:d Even in the Vertebrata, so manifestly the most highly organized division of the animal kingdom, we can start, as in the former cases, from an eye, such as exists in the fish called the lancelet, which is so simple that it consists only of a little fold-like sack of skin, lined with pigment and furnished with a nerve, but destitute of any other apparatus, being merely covered by transparent membrane.

126.1:e Even in the most highly organised division of the animal kingdom, namely the Vertebrata, we can start from an eye so simple, that it consists, as in the lancelet, of a little sack of transparent skin, furnished with a nerve and lined with pigment, but destitute of any other apparatus.

126.1:f Within the highest division

126.2:d In the class both of fishes and reptiles, as Owen has remarked, "the range of gradations of dioptric structures is very great."

126.2:e In both fishes

126.2:f In fishes

126.3:d It is a significant fact that even in man, according to the high authority of Virchow, the beautiful crystalline lens is originally formed merely by an accumulation of cells of the epidermis, lying in a sack-like fold of the skin; and the vitreous body is formed from embryonic sub-cutaneous tissue.

126.3:e lens is formed in the embryo by an accumulation of epidermic cells, lying

127 His reason ought to conquer his imagination; though I have felt the difficulty far too keenly to be surprised at any degree of hesitation in extending the principle of natural selection to such startling lengths.

127:d It is indeed indispensable that the naturalist who reflects on the origin and manner of formation of the eye, with all its marvellously perfect attributes, should make his reason/selection to so startling a length.

127:e indispensable, in order to arrive at a just conclusion regarding the formation of the eye, with all its marvellously perfect characters, that the reason should conquer the imagination; but I have felt this difficulty

127:f To arrive, however, at a just conclusion regarding the formation of the eye, with all its marvellous yet not absolutely perfect characters, it is indispensable that the reason should conquer the imagination; but I have felt the difficulty far too keenly to be surprised at others hesitating to extend

128 It is scarcely possible to avoid comparing the eye to a telescope.

128:e eye with a telescope.

129 We know that this instrument has been perfected by the long-continued efforts of the highest human intellects; and we naturally infer that the eye has been formed by a somewhat analogous process.

130 But may not this inference be presumptuous?

131 Have we any right to assume that the Creator works by intellectual powers like those of man?

132 If we must compare the eye to an optical instrument, we ought in imagination to take a thick layer of transparent tissue, with a nerve sensitive to light beneath, and then suppose every part of this layer to be continually changing slowly in density, so

as to separate into layers of different densities and thicknesses, placed at different distances from each other, and with the surfaces of each layer slowly changing in form.

132:c tissue, with spaces filled with fluid, and with a nerve

133 Further we must suppose that there is a power always intently watching each slight accidental alteration in the transparent layers; and carefully selecting each alteration which, under varied circumstances, may in any way, or in any degree, tend to produce a distincter image.

133:c power (natural selection) always

133:e power, represented by natural selection or the survival of the fittest, always intently watching each slight alteration in the transparent layers; and carefully preserving each which, under varied circumstances, in any way or in any degree, tends

134 We must suppose each new state of the instrument to be multiplied by the million; and each to be preserved till a better be produced, and then the old ones to be destroyed.

134:e million; each to be preserved until a better one is produced, and then the old ones to be all destroyed.

135 In living bodies, variation will cause the slight alterations, generation will multiply them almost infinitely, and natural selection will pick out with unerring skill each improvement.

136 Let this process go on for millions on millions of years; and during each year on millions of individuals of many kinds; and may we not believe that a living optical instrument might thus be formed as superior to one of glass, as the works of the Creator are to those of man?

136.1:d [*Center*] *Modes of Transition.* [*Space*]

137 If it could be demonstrated that any complex organ existed, which could not possibly have been formed by numerous, successive, slight modifications, my theory would absolutely break down.

138 But I can find out no such case.

139 No doubt many organs exist of which we do not know the transitional grades, more especially if we look to much-isolated species, round which, according to my theory, there has been much extinction.

139:e according to the theory

140 Or again, if we look to an organ common to all the members of a large class, for in this latter case the organ must have been first formed at an extremely remote period, since which all the many members of the class have been developed; and in order to discover the early transitional grades through which the organ has passed, we should have to look to very ancient ancestral forms, long since become extinct.

140:d been originally formed

140:f Or again, if we take an organ common to all the members of a class/at a remote

141 We should be extremely cautious in concluding that an organ could not have been formed by transitional gradations of some kind.

142 Numerous cases could be given amongst the lower animals of the same organ performing at the same time wholly distinct functions; thus the alimentary canal respires, digests, and excretes in the larva of the dragon-fly and in the fish Cobites.

142:f thus in the larva of the dragon-fly and in the fish Cobites the alimentary canal respires, digests, and excretes.

143 In the Hydra, the animal may be turned inside out, and the exterior surface will then digest and the stomach respire.

144 In such cases natural selection might easily specialise, if any advantage were thus gained, a part or organ, which had performed two functions, for one function alone, and thus wholly change its nature by insensible steps.

144:c might specialise

144:d gained, the whole or part of an organ/thus greatly change

144:e had previously performed two functions, for one function alone, and thus by insensible steps greatly change its nature.

144.1:c Certain plants, as some Leguminosæ, Violaceæ, &c., bear two kinds of flowers; one having the normal structure of the order, the other kind being degraded, though sometimes more fertile than the perfect kind: if the plant ceased to bear its perfect flowers, and this did occur during several years with an imported specimen of Aspicarpa in France, a great and sudden transition would apparently be effected in the nature of the plant.

144.1:d Many cases are known of plants which regularly produce at different parts of their inflorescence, as on the summit of a spike and lower down, or at the centre and circumference of an umbel, corymb, &c., or during different periods of the year, differently constructed flowers; and if the plant were to cease producing both kinds and bore one alone, a great change would suddenly be effected in its specific character.

144.1:e Many plants are known which regularly produce at the same time differently constructed flowers; and if such plants were to produce one kind alone, a great change would in some cases be effected in the character of the species.

144.1:f would be effected with comparative suddenness in

144.1.1:d It is a distinct question how the same plant has come to produce two kinds of flowers; but it can be shown in some cases to be probable, and in other cases to be almost certain, that this has been effected by finely graduated steps.

144.1.1:e It can also be shown that the production of the two sorts of flowers by the same plant has

345

144.1.1:f It is, however, probable that the two sorts of flowers borne by the same plant were originally differentiated by finely graduated steps, which may still be followed in some few cases.

145 Two distinct organs sometimes perform simultaneously the same function in the same individual; to give one instance, there are fish with gills or branchiæ that breathe the air dissolved in the water, at the same time that they breathe free air in their swimbladders, this latter organ having a ductus pneumaticus for its supply, and being divided by highly vascular partitions.

145:d Again, two distinct organs in the same individual sometimes perform simultaneously the same function, and this is a highly important means of transition: to give one instance,—there/ supply and

145:e individual may simultaneously perform the same/organ being divided by highly vascular partitions, and having a ductus pneumaticus for the supply of air.

145:f[¶] Again, two distinct organs, or the same organ under two very different forms, may simultaneously perform in the same individual the same function, and this is an extremely important

145.1:d To give another instance from the vegetable kingdom: plants climb by three distinct means, by spirally twining, by clasping a support with their sensitive tendrils, and by the emission of aërial rootlets; these three means are usually found in distinct genera or families, but some few plants exhibit two of the means, or even all three, combined in the same individual.

145.1:f in distinct groups, but some few species exhibit

146 In these cases, one of the two organs might with ease be modified and perfected so as to perform all the work by itself, being aided during the process of modification by the other organ; and then this other organ might be modified for some other and quite distinct purpose, or be quite obliterated.

146:b cases one

146:c might be modified

146:d In all such cases one of the two organs or means of performing the same function might be modified and perfected so as to perform all the work, being aided/or be wholly obliterated.

146:e organs for performing

146:f organs might readily be modified

147 The illustration of the swimbladder in fishes is a good one, because it shows us clearly the highly important fact that an organ originally constructed for one purpose, namely flotation, may be converted into one for a wholly different purpose, namely respiration.

147:d into one for a widely different

147:f namely, flotation/namely, respiration.

148 The swimbladder has, also, been worked in as an accessory to the auditory organs of certain fish, or, for I do not know which view is now generally held, a part of the auditory apparatus has been worked in as a complement to the swimbladder.

148:e fish.

148:f fishes.

149 All physiologists admit that the swimbladder is homologous, or "ideally similar," in position and structure with the lungs of the higher vertebrate animals: hence there seems to me to be no great difficulty in believing that natural selection has actually converted a swimbladder into a lung, or organ used exclusively for respiration.

149:b similar" in position

149:c be no extreme difficulty

149:e there is no reason to doubt that the swimbladder has actually been converted into lungs, or an organ

150 I can, indeed, hardly doubt that all vertebrate animals having true lungs have descended by ordinary generation from an ancient prototype, of which we know nothing, furnished with a floating apparatus or swimbladder.

150:c On this view it may be inferred that

150:e According to this view it may be inferred that all vertebrate animals with true lungs have descended by ordinary generation from an ancient and unknown prototype, which was furnished

150:f lungs are descended

151 We can thus, as I infer from Professor Owen's interesting description of these parts, understand the strange fact that every particle of food and drink which we swallow has to pass over the orifice of the trachea, with some risk of falling into the lungs, notwithstanding the beautiful contrivance by which the glottis is closed.

151:e from Owen's

152 In the higher Vertebrata the branchiæ have wholly disappeared —the slits on the sides of the neck and the loop-like course of the arteries still marking in the embryo their former position.

152:d disappeared—in the embryo the slits/still marking their

152:e disappeared—but in the embryo/mark

153 But it is conceivable that the now utterly lost branchiæ might have been gradually worked in by natural selection for some quite distinct purpose: in the same manner as, on the view entertained by some naturalists that the branchiæ and dorsal scales of Annelids are homologous with the wings and wing-covers of insects, it is probable that organs which at a very

347

ancient period served for respiration have been actually converted into organs of flight.

153:d some distinct purpose: for instance, the branchiæ and dorsal scales of Annelids are believed to be homologous with the wings and wing-covers of insects, and it is not improbable that with our existing insects, organs, which at an ancient period served for respiration, have actually been converted

153:f instance, Landois has shown that the wings of insects are developed from the tracheæ; it is therefore highly probable that in this great class organs which once served for respiration have been actually converted into organs for flight.

154 In considering transitions of organs, it is so important to bear in mind the probability of conversion from one function to another, that I will give one more instance.

154:e give another instance.

155 Pedunculated cirripedes have two minute folds of skin, called by me the ovigerous frena, which serve, through the means of a sticky secretion, to retain the eggs until they are hatched within the sack.

156 These cirripedes have no branchiæ, the whole surface of the body and sack, including the small frena, serving for respiration.

156:d and of the sack

156:e sack, together with the small

157 The Balanidæ or sessile cirripedes, on the other hand, have no ovigerous frena, the eggs lying loose at the bottom of the sack, in the well-enclosed shell; but they have large folded branchiæ.

157:c they have, in the same relative position, large, much folded membranes, which freely communicate with the circulatory lacunæ of the sack and body, and which have been considered to be branchiæ by Prof. Owen and all other naturalists who have treated on the subject.

157:d position with the frena, large, much-folded

157:e sack, within the well-enclosed

157:f considered by all naturalists to act as branchiæ.

158 Now I think no one will dispute that the ovigerous frena in the one family are strictly homologous with the branchiæ of the other family; indeed, they graduate into each other.

159 Therefore I do not doubt that little folds of skin, which originally served as ovigerous frena, but which, likewise, very slightly aided the act of respiration, have been gradually converted by natural selection into branchiæ, simply through an increase in their size and the obliteration of their adhesive glands.

159:c that the two little

159:d Therefore it need not be doubted that

159:e slightly aid in the act

160 If all pedunculated cirripedes had become extinct, and they have already suffered far more extinction than have sessile cirripedes, who would ever have imagined that the branchiæ in this latter family had originally existed as organs for preventing the ova from being washed out of the sack?

160:d sack? [*Space*]

160:f they have suffered [*No space*]

160.0.1-8:f[¶] There is another possible mode of transition, namely, through the acceleration or retardation of the period of reproduction. *2* This has lately been insisted on by Prof. Cope and others in the United States. *3* It is now known that some animals are capable of reproduction at a very early age, before they have acquired their perfect characters; and if this power became thoroughly well developed in a species, it seems probable that the adult stage of development would sooner or later be lost; and in this case, especially if the larva differed much from the mature form, the character of the species would be greatly changed and degraded. *4* Again, not a few animals, after arriving at maturity, go on changing in character during nearly their whole lives. *5* With mammals, for instance, the form of the skull is often much altered with age, of which Dr. Murie has given some striking instances with seals; every one knows how the horns of stags become more and more branched, and the plumes of some birds become more finely developed, as they grow older. *6* Prof. Cope states that the teeth of certain lizards change much in shape with advancing years; with crustaceans not only many trivial, but some important parts assume a new character, as recorded by Fritz Müller, after maturity. *7* In all such cases,—and many could be given,—if the age for reproduction were retarded, the character of the species, at least in its adult state, would be modified; nor is it improbable that the previous and earlier stages of development would in some cases be hurried through and finally lost. *8* Whether species have often or ever been modified through this comparatively sudden mode of transition, I can form no opinion; but if this has occurred, it is probable that the differences between the young and mature, and between the mature and the old, were primordially acquired by graduated steps. [*Space*]

160.1:d [*Center*] Cases of special Difficulty on the Theory of Natural Selection. [*Space*]

160.1:e Special Difficulties of the Theory

161 Although we must be extremely cautious in concluding that any organ could not possibly have been produced by successive transitional gradations, yet, undoubtedly, grave cases of difficulty occur, some of which will be discussed in my future work.

161:d not have/yet undoubtedly serious cases

161:f successive, small, transitional/occur.

349

162   One of the gravest is that of neuter insects, which are often very differently constructed from either the males or fertile females; but this case will be treated of in the next chapter.

162:d One of the most serious is that of neuter insects, which are often differently

163   The electric organs of fishes offer another case of special difficulty; it is impossible to conceive by what steps these wondrous organs have been produced; but, as Owen and others have remarked, their intimate structure closely resembles that of common muscle; and as it has lately been shown that Rays have an organ closely analogous to the electric apparatus, and yet do not, as Matteucci asserts, discharge any electricity, we must own that we are far too ignorant to argue that no transition of any kind is possible.

163.x-y:d difficulty; for it/produced. *y* As Owen has remarked, there is much analogy between them and ordinary muscles, in their manner of action, in the influence on them of the nervous power and other stimulants such as strychnine, and as some believe in their intimate structure.

163.y:e power and of stimulants

163.y:[f]

163.z:d We do not even in all cases know of what use these organs are; though in the Gymnotus and Torpedo they no doubt serve as powerful means of defence and perhaps for securing prey; yet in the Ray an analogous organ in the tail, even when greatly irritated, manifests, as lately observed by Matteucci, but little electricity; so little that it can hardly be of much use for these ends.

163.z:e even know/be of use for such purposes.

163.z.x-y:f But this is not surprising, for we do not even know of what use they are. *y* In the Gymnotus and Torpedo they no doubt serve as powerful means of defence, and perhaps for securing prey; yet in the Ray, as observed by Matteucci, an analogous organ in the tail manifests but little electricity, even when the animal is greatly irritated; so little, that it can hardly be of any use for the above purposes.

163.1:d Moreover, in the Ray, besides the organ just referred to, there is, as Dr. R. M'Donnell has shown, another organ near the head, not known to be electrical, but which apparently is the real homologue of the electric battery in the torpedo.

163.1:e which appears to be the real

163.1:f Torpedo.

163.1.1-2:f It is generally admitted that there exists between these organs and ordinary muscle a close analogy, in intimate structure, in the distribution of the nerves, and in the manner in which they are acted on by various reagents. 2 It should, also, be especially observed that muscular contraction is accompanied by an electrical discharge; and, as Dr. Radcliffe insists, "in the electrical apparatus of the torpedo during rest, there would seem to be a charge in every respect like that which is met with in muscle and nerve during rest, and the discharge of the torpedo, instead of being peculiar, may be only another form of the discharge which attends upon the action of muscle and motor nerve."

163.2:d And lastly, as we know nothing about the lineal progenitors of any of these fishes, it must be admitted that we are too ignorant to be enabled to affirm that no transitions are possible, through which the electric organs might have been developed.

163.2:e Lastly/progenitors of these fishes

163.2:f Beyond this we cannot at present go in the way of explanation; but as we know so little about the uses of these organs, and as we know nothing about the habits and structure of the progenitors of the existing electric fishes, it would be extremely bold to maintain that no serviceable transitions are possible by which these organs might have been gradually developed.

164   The electric organs offer another and even more serious difficulty; for they occur in only about a dozen fishes, of which several are widely remote in their affinities.

164:d These same organs at first appear to offer another and far more serious difficulty; for they occur in about a dozen kinds of fish, of

164:f These organs appear at first to

165   Generally when the same organ appears in several members of the same class, especially if in members having very different habits of life, we may attribute its presence to inheritance from a common ancestor; and its absence in some of the members to its loss through disuse or natural selection.

165:d organ is found in several/some of the members to loss

165:f When/may generally attribute

166   But if the electric organs had been inherited from one ancient progenitor thus provided, we might have expected that all electric fishes would have been specially related to each other.

166:d So that, if the electric organs had been inherited from some one ancient progenitor, we/other; but this is far from the case.

167   Nor does geology at all lead to the belief that formerly most fishes had electric organs, which most of their modified descendants have lost.

167:d fishes formerly possessed electric organs, which their modified descendants have now lost.

167.1:d But when we look closer to the subject, we find in the several fishes provided with electric organs that these are situated in different parts of the body,—that they differ in construction, as in the arrangement of the plates, and, according to Pacini, in the process or means by which the electricity is excited,—and lastly, in the requisite nervous power (and this is perhaps the most important of all the differences) being supplied through different nerves from widely different sources.

167.1:e look at the subject more closely, we/excited—and lastly, in the requisite nervous power being supplied through different nerves from widely different sources, and this is perhaps the most important of all the differences.

167.1:f organs, that these/lastly, in being supplied with nerves proceeding from different

167.2:d Hence in the several remotely allied fishes furnished with electric organs, these cannot be considered as homologous, but only as analogous in function.

167.2:f several fishes

167.3-4:d Consequently there is no reason to suppose that they have been inherited from a common progenitor; for had this been the case they would have closely resembled each other in all respects. *4* Thus the greater difficulty disappears, leaving only the lesser yet still great difficulty; namely, by what graduated steps these organs have arisen and been developed in each separate fish.

167.4:e separate group of fishes.

167.4:f Thus the difficulty of an organ, apparently the same, arising in several remotely allied species, disappears/organs have been

168 The presence of luminous organs in a few insects, belonging to different families and orders, offers a parallel case of difficulty.

168:d[¶] The luminous organs which occur only in a few insects, belonging to widely different families and orders, but which are situated in different parts of their bodies, offer a difficulty almost exactly parallel with that of the electric organs.

168:f occur in a few insects, belonging to widely different families, and which are situated in different parts of the body, offer, under our present state of ignorance, a difficulty

169 Other cases could be given; for instance in plants, the very curious contrivance of a mass of pollen-grains, borne on a footstalk with a sticky gland at the end, is the same in Orchis and Asclepias,—genera almost as remote as possible amongst flowering plants.

169:d with an adhesive gland, is apparently the same/remote as is possible

169:f Other similar cases/flowering plants; but here again the parts are not homologous.

170 In all these cases of two very distinct species furnished with apparently the same anomalous organ, it should be observed that, although the general appearance and function of the organ may be the same, yet some fundamental difference can generally be detected.

170:d two species, far removed from each other in the scale of organisation, being furnished with a similar anomalous organ, it should be observed that although the general appearance and function of the organ may be identically the same, yet some fundamental difference between them can always, or almost always, be detected.

170:e all such cases/furnished with similar anomalous organs, it

170:f In all cases of beings, far removed from each other in the scale of organisation, which are furnished with similar and peculiar organs, it will be found that although the general appearance and function of the organs may be the same, yet fundamental differences between them can always be detected.

170.1-10:f For instance, the eyes of cephalopods or cuttle-fish and of vertebrate animals appear wonderfully alike; and in such widely sundered groups no part of this resemblance can be due to inheritance from a common progenitor. *2* Mr. Mivart has advanced this case as one of special difficulty, but I am unable to see the force of his argument. *3* An organ for vision must be formed of transparent tissue, and must include some sort of lens for throwing an image at the back of a darkened chamber. *4* Beyond this superficial resemblance, there is hardly any real similarity between the eyes of cuttle-fish and vertebrates, as may be seen by consulting Hensen's admirable memoir on these organs in the Cephalopoda. *5* It is impossible for me here to enter on details, but I may specify a few of the points of difference. *6* The crystalline lens in the higher cuttle-fish consists of two parts, placed one behind the other like two lenses, both having a very different structure and disposition to what occurs in the vertebrata. *7* The retina is wholly different, with an actual inversion of the elemental parts, and with a large nervous ganglion included within the membranes of the eye. *8* The relations of the muscles are as different as it is possible to conceive, and so in other points. *9* Hence it is not a little difficult to decide how far even the same terms ought to be employed in describing the eyes of the Cephalopoda and Vertebrata. *10* It is, of course, open to any one to deny that the eye in either case could have been developed through the natural selection of successive, slight variations; but if this be admitted in the one case, it is clearly possible in the other; and fundamental differences of structure in the visual organs of two groups might have been anticipated, in accordance with this view of their manner of formation.

171 I am inclined to believe that in nearly the same way as two men have sometimes independently hit on the very same invention, so natural selection, working for the good of each being and taking advantage of analogous variations, has sometimes modified in very nearly the same manner two parts in two organic beings, which owe but little of their structure in common to inheritance from the same ancestor.

171:b which beings owe

171:d that, in the same manner as two men have sometimes independently hit on the same invention/modified in nearly the same way two organs in two organic beings, which owe but little of their structure in common to inheritance from a common progenitor.

171:e organs in two distinct organic

171:f As two men have sometimes independently hit on the same invention, so in the several foregoing cases it appears that natural selection, working for the good of each being, and taking advantage of all favourable variations, has produced similar organs, as far as function is concerned, in distinct organic beings, which owe none of their structure

171.1:d[¶] Fritz Müller, in a remarkable work recently published, has discussed a case nearly parallel with that of electric fishes, luminous insects, &c.; he undertook the laborious examination of this case in order to test the views advanced by me in this volume.

171.1:e has investigated a nearly parallel case, in order to test the views advanced in this volume.

171.1:f Fritz Müller, in order to test the conclusions arrived at in this volume, has followed out with much care a nearly similar line of argument.

171.2:d Several families of crustaceans include a few members which are fitted to live out of the water and possess an air-breathing apparatus.

171.2:e few species which possess an air-breathing apparatus, and are fitted to live out of the water.

171.2:f species, possessing an air-breathing apparatus and fitted

171.3:d In two of these families, which were more especially examined by Müller, and which are nearly related to each other, the species agree most closely in all important characters: namely in the structure of their sense-organs, in their heart and system of circulation, in the position of every tuft of hair with which their stomachs, equally complicated in both cases, are lined, and lastly in the water-breathing branchiæ, even to the microscopical hooks by which they are cleansed.

171.3:e characters; namely in their sense-organs, circulating system, in the position of the tufts of hair with which their complex stomachs are lined, and lastly in the whole structure of the water-breathing

171.3:f hair within their complex stomachs, and lastly

171.4:d Hence it might have been expected from mere analogy that the equally important air-breathing apparatus would have been the same in the few species in both families which are thus furnished; and this might have been the more confidently expected by those who believe in the creation of each separate species; for why should this one apparatus, given for the same special purpose to a few species which are so closely similar or rather identical in all other important points, have been made to differ?

171.4:e expected that/families which live on the land; and/believe in distinct creations; for/to these species, have been made to differ, whilst all the other important organs are closely similar or rather identical.

171.4:f expected that in the few species belonging to both families which live on the land, the equally-important air-breathing apparatus would have been the same; for why should this one apparatus, given for the same purpose, have been made to differ, whilst all the other important organs were closely similar or rather identical.

171.5:d[¶] Fritz Müller then argued to himself that this close similarity in so many points of structure must, in accordance with the views advanced by me, be accounted for by inheritance from a common progenitor.

171.5:e Fritz Müller argues that

171.6:d But as the vast majority of the species in the above two families, as well as the main body of crustaceans of all orders, are aquatic in their habits, it is improbable in the highest degree, that their common progenitor should have been adapted for breathing air.

171.6:e families, as well as most crustaceans

171.6:f most other crustaceans, are

171.7:d Müller was thus led carefully to examine and describe the apparatus in the few air-breathing species; and in each he found it to differ in several important points, as in the position of the orifices, in the manner in which they are opened and closed, and in some accessory details.

171.7:e in the air-breathing

171.7:f species; and he found it to differ in each in several

171.8:d Now, on the belief that species belonging to distinct families, already differing in some characters, and which whenever they varied would probably have varied in different manners, have been slowly adapted through natural selection to live more and more out of water and to breathe the air, it is quite intelligible, and might even have been confidently expected, that the structural contrivances thus acquired would in each case have materially differed, although serving for the same purpose.

171.8.x:e Now such differences are intelligible, and might even have been anticipated, on the supposition that species belonging to distinct families had slowly become adapted to live more and more out of water, and to breathe the air.

171.8.x:f been expected, on

171.8.y:e For these species, from belonging to distinct families, would differ to a certain extent, and in accordance with the principle that the nature of each variation depends on two factors, viz. the nature of the organism and that of the conditions, the variability of these crustaceans assuredly would not have been exactly the same.

171.8.y:f would have differed/organism and that of the surrounding conditions, their variability assuredly

171.8.z:e Consequently natural selection would have had different materials or variations to work on, in order to arrive at the same functional result; and the structures thus acquired would almost necessarily have differed.

171.9:d On the hypothesis of separate acts of creation the whole case must remain unintelligible, and we can only say, so it is.

171.9:e case remains unintelligible.

171.10:d This line of argument seems to have had great weight in leading this distinguished naturalist fully to accept the views maintained by me in this volume. [*Space*]

171.10:e The above line of argument, as advanced by Fritz Müller, seems/naturalist to

171.10:f This line of argument seems to have had great weight in leading Fritz Müller to [*No space*]

171.10.1-3:f[¶] Another distinguished zoologist, the late Professor Claparède, has argued in the same manner, and has arrived at the same result. *2* He shows that there are parasitic mites (Acaridæ), belonging to distinct sub-families and families, which are furnished with hair-claspers. *3* These organs must have been independently developed, as they could not have been inherited from a common progenitor; and in the several groups they are formed by the modification of the fore-legs,—of the hind-legs,— of the maxillæ or lips,—and of appendages on the under side of the hind part of the body. [*Space*]

171.11:d[¶] In the several cases just discussed, we have seen that in beings more or less remotely allied, the same end is gained and the same function performed by organs in appearance, though not in truth, closely similar.

171.11:f In the foregoing cases, we see the same end gained and the same function performed, in beings not at all or only remotely allied, by organs in appearance, though not in development, closely similar.

171.12:d But the common rule throughout nature is that the same end is gained, even sometimes in the case of beings closely related to each other, by the most diversified means.

171.12:f On the other hand, it is a common rule throughout nature that the same end should be gained, even sometimes in the case of closely-related beings, by

171.13:d How differently constructed is the feathered wing of a bird and the membrane-covered wing of a bat with all its fingers developed; and still more so the four wings of a butterfly, the two wings of a fly, and the two of a beetle with their elytra.

171.13:e bat with all the digits largely developed/beetle, together with the elytra.

171.13:f bat; and still more so the four wings of a butterfly, the two wings of a fly, and the two wings with the elytra of a beetle.

171.14:d Bivalve shells have only to open and shut, but on what a number of patterns is the hinge constructed, from the long row of neatly interlocking teeth in a Nucula to the simple ligament of a Mussel.

171.14:e shells are made to open/Mussel!

171.14:f constructed,—from

171.15:d Seeds are disseminated by their minuteness or by their capsule being converted into a light balloon-like envelope; or by being embedded in pulp or flesh, formed of the most diverse parts, and rendered nutritious as well as conspicuously coloured, so as to attract and be devoured by birds; or by having hooks and grapnels of many kinds and serrated awns, so as to adhere to the fur of quadrupeds; or by being furnished with wings and plumes, as diversified in shape as elegant in structure, so as to be wafted by every breeze.

171.15:e minuteness,—by their/envelope,—by being embedded/nutritious, as well/birds,—by having/quadrupeds,—and by being furnished with wings and plumes, as different in shape

171.15:f shape as they are elegant

171.16:d I will give one other instance; for the subject is worthy of reflection by those who are not able to credit that organic beings have been formed in many ways for the sake of mere variety, like toys in a shop.

171.16.x-y:e for this subject of the same end being gained by the most diversified means well deserves attention. *y* Some authors maintain that organic beings have been formed in many ways for the sake of mere variety, almost like toys in a shop, but such a view of nature is incredible.

171.17-18:d With plants having separated sexes, and with those in which, though hermaphrodites, the pollen does not spontaneously fall on the stigma, some aid is necessary for their fertilisation. *18* With several kinds this is effected by the light and incoherent pollen-grains being blown by the wind through mere chance on to the stigma; and this is the simplest plan which can well be conceived.

171.18:e effected by the pollen-grains, which are light and incoherent, being

171.19:d An almost equally simple, though very different, plan occurs in many cases, in which a symmetrical flower secretes a few drops of nectar, and is consequently visited by insects; and these carry the pollen from the anthers to the stigma.

171.19:e occurs in many plants in which

171.20:d[¶] From this simple stage we may pass through an inexhaustible number of contrivances, all for the same purpose and effected in essentially the same manner, but entailing changes in every part of the flower; with the nectar stored in variously shaped receptacles, with the stamens and pistils modified in many ways, sometimes forming trap-like contrivances, and sometimes capable of neatly adapted movements through irritability or elasticity.

171.20.x-y:e flower. *y* The nectar may be stored

171.21:d From such structures we may advance till we come to such an acme of perfect adaptation, as has lately been described by Dr. Crüger in the case of Coryanthes.

171.21:e to such a case of extraordinary adaptation as that lately described by Dr. Crüger in the Coryanthes.

171.22:d This orchid has its labellum or lower lip hollowed out into a great bucket, into which drops of almost pure water, not nectar, continually fall from two secreting horns which stand above it; and when the bucket is half full, the water overflows by a spout on one side.

171.22:e has part of its/water continually

171.23:d The basal part of the labellum curves over the bucket, and is itself hollowed out into a sort of chamber with two lateral entrances, within which and outside there are some curious fleshy ridges.

171.23:e labellum stands over/within this chamber there are curious fleshy

171.24:d The most ingenious man, if he had not witnessed what takes place, could never have imagined what purpose all these parts served.

171.24:e serve.

171.25:d But Dr. Crüger saw crowds of large humble-bees visiting the gigantic flowers of this orchid in the early morning, and they came, not to suck nectar, but to gnaw off the ridges above the bucket; in doing this they frequently pushed each other into the bucket, and thus their wings were wetted, so that they could not fly out, but had to crawl out through the passage formed by the spout or overflow.

171.25:e orchid, not in order to suck nectar, but to gnaw off the ridges within the chamber above the bucket; in doing this they frequently pushed each other into the bucket, and their wings being thus wetted they could not fly away, but

171.25:f away, but were compelled to

171.26:d Dr. Crüger has seen a "continual procession" of bees thus crawling out of their involuntary bath.

171.26:e Dr. Crüger saw a "continual

171.27-8:d The passage is narrow, and is roofed over by the column, so that a bee, in forcing its way out, first rubs its back against the viscid stigma and then against the viscid glands of the pollen-masses. *28* The pollen-masses are thus glued to the back of

358

the bee which first happens to crawl through the passage of a lately expanded flower, and are thus carried away.

171.28:e of that bee which first happens to crawl out through

171.28:f of the bee

171.29:d Dr. Crüger sent me a flower in spirits of wine, with a bee which he had killed before it had quite crawled out of the passage with a pollen-mass fastened to its back.

171.29:e out with a pollen-mass still fastened

171.30-1:d When the bee, thus provided, flies to another flower, or to the same flower a second time, and is pushed by its comrades into the bucket and then crawls out by the passage, the pollen-mass necessarily comes first into contact with the viscid stigma, and adheres to it, and the flower is fertilised. *31* Now at last we see the full use of the water-secreting horns, of the bucket with its spout, and of the shape of every part of the flower!

171.31:e of every part of the flower, of the water-secreting horns, of the bucket half full of water, which prevents the bees from flying away and forces them to crawl out through the spout, and rub against the properly placed viscid pollen-masses and viscid stigma.

171.31:f away, and forces/masses and the viscid

171.32:d The construction of the flower of another closely allied orchid, namely Catasetum, is widely different, though serving the same end; and is equally curious.

171.32:e[¶]

171.33:d Bees visit this flower, as in the case of the Coryanthes, in order to gnaw the labellum; in doing this they inevitably touch a long, tapering, sensitive projection, or, as I have called it, antenna.

171.33:e visit these flowers, like those of the Coryanthes/it, the antenna.

171.34:d The antenna being touched causes a certain membrane to rupture through its own irritability, and this sets free a spring by which the pollen-mass is shot forth, like an arrow, in the right direction, and adheres by its viscid extremity to the back of the bee.

171.34:e This antenna, when touched, transmits a sensation or vibration to a certain membrane which is instantly ruptured; this sets

171.35:d The pollen-mass is thus carried to another flower, where it is brought into contact with the stigma, which is viscid enough to break certain elastic threads, and to retain the pollen-mass, which then performs its office of fertilisation.

171.35:e The pollen-mass of a male plant is thus carried to the flower of a female plant, where/threads, and retaining the pollen, fertilisation is effected.

171.35:f plant (for the sexes are separate in this orchid) is thus carried to the flower of the female

171.36:d[¶] How, it may be asked, in the foregoing and in innumerable other and similar cases, can we understand the cause of such a wide scale of complexity and of such multifarious means for gaining the same end, both in the case of forms widely remote from each other in affinity, and with forms so closely allied as are the two orchids last described?

171.36:e other instances, can we understand the graduated scale of complexity and the multifarious means for gaining the same end.

171.37:d It was shown, when we discussed the air-breathing apparatus of certain crustaceans, that the process of adaptation for any purpose may start from two or more forms already differing from each other to a considerable degree, and that in almost all cases the nature of the variability, on which natural selection has to work, will be different; consequently, the final structure gained through natural selection, though serving for the same purpose, will be different.

171.37:e The answer no doubt is, as already remarked, that when two forms vary, which already differ from each other even in a slight degree, the variability will not be of the same exact nature, and consequently the results obtained through natural selection for the same general purpose will not be the same.

171.37:f other in some slight

171.38:d We must also bear in mind that every well-developed organism has already passed through a long course of modification; and that each modified structure tends to be inherited, so that it will not readily be lost, but may be modified again and again.

171.38:e We should also bear in mind that every highly developed organism has passed/readily be wholly lost

171.38:f through many changes; and that each modified structure tends to be inherited, so that each modification will not readily be quite lost, but may be again and again further altered.

171.39:d Hence the structure of each part of each species, for whatever purpose used, will be the sum of the many inherited changes, through which that species has passed during its successive adaptations to changed habits and conditions of life.

171.39:e used, is the sum

171.39:f purpose it may serve, is the sum of many inherited changes, through which the species

172   Although in many cases it is most difficult to conjecture by what transitions an organ could have arrived at its present state; yet, considering that the proportion of living and known forms to the extinct and unknown is very small, I have been astonished how rarely an organ can be named, towards which no transitional grade is known to lead.

172:b transitions organs could have arrived at their present

172:d Finally then, although in many cases it is most difficult to conjecture by what transitions organs have arrived at their present state; yet, considering that the proportion of living and known forms is very small compared with the extinct and unknown forms, I

172:e difficult even to conjecture by what transitions many organs have arrived at their present state; yet, considering how small the proportion of living and known forms is to the extinct and unknown, I

172:f transitions organs have arrived

173 The truth of this remark is indeed shown by that old canon in natural history of "Natura non facit saltum."

173:b old but somewhat exaggerated canon

173:c It certainly is not true, that new organs often appear suddenly in any class, as if created for some special purpose; as indeed is shown by that old, but somewhat

173:d is true, that new organs very rarely or never suddenly appear in any

173:e organs appearing as if specially created for some purpose, rarely or never appear suddenly in any class; as

173:f if created for some special purpose, rarely or never appear in any being;—as

174 We meet with this admission in the writings of almost every experienced naturalist; or, as Milne Edwards has well expressed it, nature is prodigal in variety, but niggard in innovation.

174:b Nature

174:c naturalist; as Milne

174:e naturalist; or as Milne

175 Why, on the theory of Creation, should this be so?

175:d Creation, should there be so much variety and so little novelty?

175:f little real novelty?

176 Why should all the parts and organs of many independent beings, each supposed to have been separately created for its proper place in nature, be so invariably linked together by graduated steps?

176:b so commonly linked

177 Why should not Nature have taken a leap from structure to structure?

177:d Nature take a sudden leap

178 On the theory of natural selection, we can clearly understand why she should not; for natural selection can act only by taking advantage of slight successive variations; she can never take a leap, but must advance by the shortest and slowest steps.

178:c advance by short and slow steps.

178:d for natural selection acts/take a sudden leap, but must advance by short and sure though slow

178:f take a great and sudden leap, but must advance by short and sure, though

179   *Organs of little apparent importance.—*

179:d [*Center*] *Organs of little apparent Importance, as affected by Natural Selection.* [*Space*]

180 As natural selection acts by life and death,—by the preservation of individuals with any favourable variation, and by the destruction of those with any unfavourable deviation of structure,—I have sometimes felt much difficulty in understanding the origin of simple parts, of which the importance does not seem sufficient to cause the preservation of successively varying individuals.

180:d[¶]

181 I have sometimes felt as much difficulty, though of a very different kind, on this head, as in the case of an organ as perfect and complex as the eye.

180 + 1:e death,—by the survival of the fittest, and by the destruction of the less well fitted individuals,—I have sometimes felt great difficulty in understanding the origin or formation of parts of little importance; almost as great, though of a very different kind, as in the case of the most perfect and complex organs.

180 + 1:f well-fitted

182   In the first place, we are much too ignorant in regard to the whole economy of any one organic being, to say what slight modifications would be of importance or not.

183 In a former chapter I have given instances of most trifling characters, such as the down on fruit and the colour of the flesh, which, from determining the attacks of insects or from being correlated with constitutional differences, might assuredly be acted on by natural selection.

183:c colour of its flesh, the colour of the skin and hair of quadrupeds, which, from being correlated with constitutional differences or from determining the attacks of insects, might

183:e of very trifling

184 The tail of the giraffe looks like an artificially constructed fly-flapper; and it seems at first incredible that this could have been adapted for its present purpose by successive slight modifications, each better and better, for so trifling an object as driving away flies; yet we should pause before being too positive even in this case, for we know that the distribution and existence of cattle and other animals in South America absolutely depends on their power of resisting the attacks of insects: so that individuals which could by any means defend themselves

from these small enemies, would be able to range into new pastures and thus gain a great advantage.

184:c depend

184:e better and better fitted, for so trifling an object as to drive away

185 It is not that the larger quadrupeds are actually destroyed (except in some rare cases) by the flies, but they are incessantly harassed and their strength reduced, so that they are more subject to disease, or not so well enabled in a coming dearth to search for food, or to escape from beasts of prey.

185:b by flies

186 Organs now of trifling importance have probably in some cases been of high importance to an early progenitor, and, after having been slowly perfected at a former period, have been transmitted in nearly the same state, although now become of very slight use; and any actually injurious deviations in their structure will always have been checked by natural selection.

186:e transmitted to existing species in nearly the same state, although now of very slight use; but any actually injurious deviations in their structure will of course have

186:f structure would of

187 Seeing how important an organ of locomotion the tail is in most aquatic animals, its general presence and use for many purposes in so many land animals, which in their lungs or modified swimbladders betray their aquatic origin, may perhaps be thus accounted for.

188 A well-developed tail having been formed in an aquatic animal, it might subsequently come to be worked in for all sorts of purposes, as a fly-flapper, an organ of prehension, or as an aid in turning, as with the dog, though the aid must be slight, for the hare, with hardly any tail, can double quickly enough.

188:e though the aid in this latter respect must

188:f purposes,—as a fly-flapper, an organ of prehension, or as an aid in turning, as in the case of the dog/double still more quickly.

189 In the second place, we may sometimes attribute importance to characters which are really of very little importance, and which have originated from quite secondary causes, independently of natural selection.

189:e sometimes wrongly attribute importance to characters which have originated

189:f may easily err in attributing importance to characters, and in believing that they have been developed through natural selection.

190 We should remember that climate, food, &c., probably have some little direct influence on the organisation; that characters reappear from the law of reversion; that correlation of growth

will have had a most important influence in modifying various structures; and finally, that sexual selection will often have largely modified the external characters of animals having a will, to give one male an advantage in fighting with another or in charming the females.

190:e some, perhaps a considerable, direct/correlation is an important element of change; and finally, that sexual selection has often largely modified the external characters of the higher animals, so as to give one male an advantage in fighting with other males, or in charming the female; and characters gained through sexual selection may be transmitted to both sexes.

190:f We must by no means overlook the effects of the definite action of changed conditions of life,—of so-called spontaneous variations, which seem to depend in a quite subordinate degree on the nature of the conditions,—of the tendency to reversion to long-lost characters,—of the complex laws of growth, such as of correlation, compensation, of the pressure of one part on another, &c.,—and finally of sexual selection, by which characters of use to one sex are often gained and then transmitted more or less perfectly to the other sex, though of no use to this sex.

191 Moreover when a modification of structure has primarily arisen from the above or other unknown causes, it may at first have been of no advantage to the species, but may subsequently have been taken advantage of by the descendants of the species under new conditions of life and with newly acquired habits.

191:e Moreover a modification, caused in any of the above specified ways, may at first have been of no direct advantage to a species, but may subsequently have been taken advantage of by its descendants under new conditions of life and newly

191:f But structures thus indirectly gained, although at first of no advantage to a species, may subsequently have been taken advantage of by its modified descendants, under

192   To give a few instances to illustrate these latter remarks.

192:[e]

193 If green woodpeckers alone had existed, and we did not know that there were many black and pied kinds, I dare say that we should have thought that the green colour was a beautiful adaptation to hide this tree-frequenting bird from its enemies; and consequently that it was a character of importance and might have been acquired through natural selection; as it is, I have no doubt that the colour is due to some quite distinct cause, probably to sexual selection.

193:c as it is, we cannot doubt

193:e[¶] If, for instance, green woodpeckers/importance and had been acquired through natural selection; as it is, the colour is probably in chief part due to sexual selection.

364

193:f If green woodpeckers/adaptation to conceal this tree-frequenting/importance, and

194 A trailing bamboo in the Malay Archipelago climbs the loftiest trees by the aid of exquisitely constructed hooks clustered around the ends of the branches, and this contrivance, no doubt, is of the highest service to the plant; but as we see nearly similar hooks on many trees which are not climbers, the hooks on the bamboo may have arisen from unknown laws of growth, and have been subsequently taken advantage of by the plant undergoing further modification and becoming a climber.

194:c trailing palm in the Malay/climbers, the hooks on the palm may

194:e climbers, and which there is reason to believe from the distribution of the thorn-bearing species in Africa and South America, serves as a defence against browsing quadrupeds, so the hooks on the palm may first have been developed for this object, and subsequently been taken advantage of by the plant as it underwent further modification and became a climber.

194:f climbers, and which, as there/America, serve as a defence against browsing quadrupeds, so the spikes on the palm may at first have been developed for this object, and subsequently have been improved and taken advantage of by the plant, as

195 The naked skin on the head of a vulture is generally looked at as a direct adaptation for wallowing in putridity; and so it may be, or it may possibly be due to the direct action of putrid matter; but we should be very cautious in drawing any such inference, when we see that the skin on the head of the clean-feeding male turkey is likewise naked.

195:c generally considered as

195:e Turkey

196 The sutures in the skulls of young mammals have been advanced as a beautiful adaptation for aiding parturition, and no doubt they facilitate, or may be indispensable for this act; but as sutures occur in the skulls of young birds and reptiles, which have only to escape from a broken egg, we may infer that this structure has arisen from the laws of growth, and has been taken advantage of in the parturition of the higher animals.

197 We are profoundly ignorant of the causes producing slight and unimportant variations; and we are immediately made conscious of this by reflecting on the differences in the breeds of our domesticated animals in different countries,—more especially in the less civilized countries where there has been but little artificial selection.

197:e cause of each slight variation or individual difference; and/ little methodical selection.

197:f differences between the breeds

197.1:c = *202:[c]* climates

197.1:e and are exposed

197.2:c = *203:[c]* observer states

197.2:f With cattle

198 Careful observers are convinced that a damp climate affects the growth of the hair, and that with the hair the horns are correlated.

198:c Other observers

198:f Some observers

199 Mountain breeds always differ from lowland breeds; and a mountainous country would probably affect the hind limbs from exercising them more, and possibly even the form of the pelvis; and then by the law of homologous variation, the front limbs and even the head would probably be affected.

199:c front limbs and the head

200 The shape, also, of the pelvis might affect by pressure the shape of the head of the young in the womb.

200:e shape of certain parts of the young

201 The laborious breathing necessary in high regions would, we have some reason to believe, increase the size of the chest; and again correlation would come into play.

201:f regions tends, as we have good reason to believe, to increase

202 Animals kept by savages in different countries often have to struggle for their own subsistence, and would be exposed to a certain extent to natural selection, and individuals with slightly different constitutions would succeed best under different climates; and there is reason to believe that constitution and colour are correlated.

202:[c] = *197.1:c*

203 A good observer, also, states that in cattle susceptibility to the attacks of flies is correlated with colour, as is the liability to be poisoned by certain plants; so that colour would be thus subjected to the action of natural selection.

203:[c] = *197.2:c*

203.1:c The effects on the whole organisation of lessened exercise with abundant food is probably still more important; and this, as H. von Nathusius has lately shown in his excellent Treatise, is apparently one chief cause of the great modification which the breeds of swine have undergone.

203.1:e exercise together with

203.1:f The effects of lessened exercise together with abundant food on the whole organisation is

204 But we are far too ignorant to speculate on the relative importance of the several known and unknown laws of variation; and I have here alluded to them only to show that, if we are unable to account for the characteristic differences of our domestic breeds, which nevertheless we generally admit to have

arisen through ordinary generation, we ought not to lay too much stress on our ignorance of the precise cause of the slight analogous differences between species.

204:e unknown causes of variation; and I have made these remarks only/nevertheless are generally admitted to have arisen through ordinary generation from one or a few parent-stocks, we

204:f our several domestic/between true species.

205 I might have adduced for this same purpose the differences between the races of man, which are so strongly marked; I may add that some little light can apparently be thrown on the origin of these differences, chiefly through sexual selection of a particular kind, but without here entering on copious details my reasoning would appear frivolous.

205:c frivolous. [*Space*]

205:e on these differences, through sexual selection of a particular kind, but without entering on full details

205:[f]

205.1:d [*Center*] *Utilitarian Doctrine how far true: Beauty how acquired.* [*Space*]

206 The foregoing remarks lead me to say a few words on the protest lately made by some naturalists, against the utilitarian doctrine that every detail of structure has been produced for the good of its possessor.

207 They believe that very many structures have been created for beauty in the eyes of man, or for mere variety.

207:d man, or, as already mentioned and discussed, for the sake of mere

207:e already discussed

207:f created for the sake of beauty, to delight man or the Creator (but this latter point is beyond the scope of scientific discussion), or for the sake of mere variety, a view already discussed.

208 This doctrine, if true, would be absolutely fatal to my theory.

208:d Such doctrines

209 Yet I fully admit that many structures are of no direct use to their possessors.

209:e are now of no direct use to their possessors, and may never have been of any use to their progenitors.

209:f [theory.] I fully/but this does not prove that they were formed solely for beauty or variety.

210 Physical conditions probably have had some little effect on structure, quite independently of any good thus gained.

210:d little direct effect

211 Correlation of growth has no doubt played a most important part, and a useful modification of one part will often have entailed on other parts diversified changes of no direct use.

211:d growth no doubt has largely come into action, and a useful modification of one part has often entailed on other parts changes of structure of

212 So again characters which formerly were useful, or which formerly had arisen from correlation of growth, or from other unknown cause, may reappear from the law of reversion, though now of no direct use.

210 + 11 + 12:e No doubt, as recently remarked, the definite action of changed conditions, correlated variation, and reversion have all produced their effects.

210 + 11 + 12:f No doubt the definite action of changed conditions, and the various causes of modifications, lately specified, have all produced an effect, probably a great effect, independently of any advantage thus gained.

213 The effects of sexual selection, when displayed in beauty to charm the females, can be called useful only in rather a forced sense.

213:[d]

214 But by far the most important consideration is that the chief part of the organisation of every being is simply due to inheritance; and consequently, though each being assuredly is well fitted for its place in nature, many structures now have no direct relation to the habits of life of each species.

214:e But the most/organisation of every living creature is simply due to inheritance; and consequently, though each assuredly/ relation to existing habits of life.

214:f But a still more important consideration is that the chief part of the organisation of every living creature is due to inheritance; and consequently, though each being assuredly is well fitted for its place in nature, many structures have now no very close and direct relation to present habits of life.

215 Thus, we can hardly believe that the webbed feet of the upland goose or of the frigate-bird are of special use to these birds; we cannot believe that the same bones in the arm of the monkey, in the fore leg of the horse, in the wing of the bat, and in the flipper of the seal, are of special use to these animals.

215:b fore-leg

215:c birds; we cannot believe that the similar bones

216 We may safely attribute these structures to inheritance.

217 But to the progenitor of the upland goose and of the frigate-bird, webbed feet no doubt were as useful as they now are to the most aquatic of existing birds.

217:e aquatic of living birds.

217:f But webbed feet no doubt were as useful to the progenitor of the upland goose and of the frigate-bird, as

218 So we may believe that the progenitor of the seal had not a

flipper, but a foot with five toes fitted for walking or grasping; and we may further venture to believe that the several bones in the limbs of the monkey, horse, and bat, which have been inherited from a common progenitor, were formerly of more special use to that progenitor, or its progenitors, than they now are to these animals having such widely diversified habits.

218:e seal did not possess a flipper/inherited from some ancient progenitor, were formerly of more special use than they now are to these animals with their widely diversified habits, and might consequently have been modified through natural selection.

218:f bat, were originally developed, on the principle of utility, probably through the reduction of more numerous bones in the fin of some ancient fish-like progenitor of the whole class.

219 Therefore we may infer that these several bones might have been acquired through natural selection, subjected formerly, as now, to the several laws of inheritance, reversion, correlation of growth, &c.

220 Hence every detail of structure in every living creature (making some little allowance for the direct action of physical conditions) may be viewed, either as having been of special use to some ancestral form, or as being now of special use to the descendants of this form—either directly, or indirectly through the complex laws of growth.

219 + 20:e Making due allowance for the definite action of changed conditions, correlation, reversion, &c., we may conclude that every detail of structure in every living creature is either now or was formerly of use,—directly or

219 + 20:f It is scarcely possible to decide how much allowance ought to be made for such causes of change, as the definite action of external conditions, so-called spontaneous variations, and the complex laws of growth; but with these important exceptions, we may conclude that the structure of every living creature either now is, or was formerly, of some direct or indirect use to its possessor.

220.1:d[¶] With respect to the view that organic beings have been created beautiful for the delight of man,—a view which it has lately been pronounced may safely be accepted as true, and as subversive of my whole theory,—I may first remark that the idea of the beauty of any particular object obviously depends on the mind of man, irrespective of any real quality in the admired object; and that the idea is not an innate and unalterable element in the mind.

220.1:e to the belief that organic/has been/any object

220.1:f man,—a belief which it has been pronounced is subversive of my whole theory,—I may first remark that the sense of beauty obviously depends on the nature of the mind, irrespective of any real quality in the admired object; and that the idea of what is beautiful, is not innate or unalterable.

220.2:*d* We see this in men of different races admiring an entirely different standard of beauty in their women; neither the Negro nor the Chinese admires the Caucasian beau-ideal.

220.2:*f* this, for instance, in the men/women.

220.3:*d* The idea also of beauty in natural scenery has arisen only within modern times.

220.3:*e* of picturesque beauty in scenery

220.3:[*f*]

220.4:*d* On the view of beautiful objects having been created for man's gratification, it ought to be shown that there was less beauty on the face of the earth before man appeared than since he came on the stage.

220.4:*f* If beautiful objects had been created solely for man's gratification, it ought to be shown that before man appeared, there was less beauty on the face of the earth than

220.5-8:*d* Were the beautiful volute and cone shells of the Eocene epoch, and the gracefully sculptured ammonites of the Secondary period, created that man might ages afterwards admire them in his cabinet? *6* Few objects are more beautiful than the minute siliceous cases of the diatomaceæ: were these created that they might be examined and admired under the higher powers of the miscroscope? *7* The beauty in this latter case, and in many others, is apparently wholly due to symmetry of growth. *8* Flowers rank amongst the most beautiful productions of nature; and they have become through natural selection beautiful, or rather conspicuous in contrast with the greenness of the leaves, that they might be easily observed and visited by insects, so that their fertilisation might be favoured.

220.8:*e* contrast with the green leaves, that they might easily be observed

220.8:*f* nature; but they have been rendered conspicuous in contrast with the green leaves, and in consequence at the same time beautiful, so that they may be easily observed by insects.

220.9-10:*d* I have come to this conclusion from finding it an invariable rule that when a flower is fertilised by the wind it never has a gaily-coloured corolla. *10* Again, several plants habitually produce two kinds of flowers; one kind open and coloured so as to attract insects; the other closed and not coloured, destitute of nectar, and never visited by insects.

220.10:*f* Several/closed, not

220.11:*d* We may safely conclude that, if insects had never existed on the face of the earth, the vegetation would not have been decked with beautiful flowers, but would have produced only such poor flowers as are now borne by our firs, oaks, nut and ash trees, by the grasses, by spinach, docks, and nettles.

220.11:*e* Hence we may conclude/as we now see on our firs, oaks, nut and ash trees, on grasses, spinach

220.11:*f* had not been developed on the face of the earth, our plants would not/we see/nettles, which are all fertilised through the agency of the wind.

220.12:d A similar line of argument holds good with the many kinds of beautiful fruits; that a ripe strawberry or cherry is as pleasing to the eye as to the palate, that the gaily-coloured fruit of the spindle-wood tree and the scarlet berries of the holly are beautiful objects, will be admitted by every one.

220.12:f with fruits/palate,—that/objects,—will

220.13:d But this beauty serves merely as a guide to birds and beasts, that the fruit may be devoured and the seeds thus disseminated: I infer that this is the case from having as yet found in every instance that seeds, which are embedded within a fruit of any kind, that is within a fleshy or pulpy envelope, if it be coloured of any brilliant tint, or merely rendered conspicuous by being coloured white or black, are always disseminated by being first devoured.

220.13:e devoured and the manured seeds thus

220.13:f beasts, in order that the fruit may/found no exception to the rule that seeds are always thus disseminated when embedded within a fruit of any kind (that is within a fleshy or pulpy envelope), if it be coloured of any brilliant tint, or rendered conspicuous by being white or black.

220.14:d[¶] On the other hand, I willingly admit that a great number of male animals, as all our most gorgeous birds, certainly some fishes, perhaps some mammals, and a host of magnificently coloured butterflies and some other insects, have been rendered beautiful for beauty's sake; but this has been effected not for the delight of man, but through sexual selection, that is from the more beautiful males having been continually preferred by their less ornamented females.

220.14:e birds, some fishes, some mammals

220.14:f fishes, reptiles, and mammals, and a host of magnificently coloured butterflies, have been rendered beautiful for beauty's sake; but this has been effected through sexual selection, that is, by the more beautiful males having been continually preferred by the females, and not for the delight of man.

220.15-6:d So it is with the music of birds. *16* We may infer from all this that a similar taste for beautiful colours and for musical sounds runs through a large part of the animal kingdom.

220.16:f that a nearly similar

220.17:d When the female is as beautifully coloured as the male, which is not rarely the case with birds and butterflies, the cause simply lies in the colours acquired through sexual selection having been inherited by both sexes, instead of by the males alone.

220.17:e been transmitted to both sexes, instead of to the males

220.17:f cause apparently lies

220.18:d We can sometimes plainly see the proximate cause of the transmission of ornaments to the males alone; for a pea-hen with the long tail of the male bird would be badly fitted to sit on her eggs, and a coal-black female capercailzie would be far more conspicuous on her nest and more exposed to danger than in her present modest attire.

220.18:e In some instances, however, the acquirement of conspicuous colours by the female may have been checked through natural selection, on account of the danger to which she would thus have been exposed during incubation.

220.18:[f]

220.18.1-3:f How the sense of beauty in its simplest form—that is, the reception of a peculiar kind of pleasure from certain colours, forms, and sounds—was first developed in the mind of man and of the lower animals, is a very obscure subject. *2* The same sort of difficulty is presented, if we enquire how it is that certain flavours and odours give pleasure, and others displeasure. *3* Habit in all these cases appears to have come to a certain extent into play; but there must be some fundamental cause in the constitution of the nervous system in each species. [*Space*]

221   Natural selection cannot possibly produce any modification in any one species exclusively for the good of another species; though throughout nature one species incessantly takes advantage of, and profits by, the structure of another.

221:e structure of others.

221:f modification in a species exclusively/structures of others.

222   But natural selection can and does often produce structures for the direct injury of other species, as we see in the fang of the adder, and in the ovipositor of the ichneumon, by which its eggs are deposited in the living bodies of other insects.

222:e other animals, as

223   If it could be proved that any part of the structure of any one species had been formed for the exclusive good of another species, it would annihilate my theory, for such could not have been produced through natural selection.

224   Although many statements may be found in works on natural history to this effect, I cannot find even one which seems to me of any weight.

225   It is admitted that the rattlesnake has a poison-fang for its own defence and for the destruction of its prey; but some authors suppose that at the same time this snake is furnished with a rattle for its own injury, namely, to warn its prey to escape.

225:f defence, and/time it is/prey.

226   I would almost as soon believe that the cat curls the end of its tail when preparing to spring, in order to warn the doomed mouse.

226.1:f It is a much more probable view that the rattlesnake uses its rattle, the cobra expands its frill, and the puff-adder swells whilst hissing so loudly and harshly, in order to alarm the many birds and beasts which are known to attack even the most venomous species.

227 But I have not space here to enter on this and other such cases.

227:f Snakes act on the same principle which makes the hen ruffle her feathers and expand her wings when a dog approaches her chickens; but I have not space here to enlarge on the many ways by which animals endeavour to frighten away their enemies.

228 Natural selection will never produce in a being anything injurious to itself, for natural selection acts solely by and for the good of each.

228:f being any structure more injurious than beneficial to that being, for

229 No organ will be formed, as Paley has remarked, for the purpose of causing pain or for doing an injury to its possessor.

230 If a fair balance be struck between the good and evil caused by each part, each will be found on the whole advantageous.

231 After the lapse of time, under changing conditions of life, if any part comes to be injurious, it will be modified; or if it be not so, the being will become extinct, as myriads have become extinct.

231:f extinct as myriads

232 Natural selection tends only to make each organic being as perfect as, or slightly more perfect than, the other inhabitants of the same country with which it has to struggle for existence.

232:f it comes into competition.

233 And we see that this is the degree of perfection attained under nature.

233:f is the standard of

234 The endemic productions of New Zealand, for instance, are perfect one compared with another; but they are now rapidly yielding before the advancing legions of plants and animals introduced from Europe.

235 Natural selection will not produce absolute perfection, nor do we always meet, as far as we can judge, with this high standard under nature.

236 The correction for the aberration of light is said, on high authority, not to be perfect even in that most perfect organ, the eye.

236:d said by Müller not/organ, the human eye.

236.1:f Helmholtz, whose judgment no one will dispute, after describing in the strongest terms the wonderful powers of the human eye, adds these remarkable words: "That which we have discovered in the way of inexactness and imperfection in the optical machine and in the image on the retina, is as nothing in comparison with the incongruities which we have just come across in the domain of the sensations. One might say that nature has taken delight in accumulating contradictions in order to remove all foundation from the theory of a pre-existing harmony between the external and internal worlds."

237 If our reason leads us to admire with enthusiasm a multitude of inimitable contrivances in nature, this same reason tells us, though we may easily err on both sides, that some other contrivances are less perfect.

238 Can we consider the sting of the wasp or of the bee as perfect, which, when used against many attacking animals, cannot be withdrawn, owing to the backward serratures, and so inevitably causes the death of the insect by tearing out its viscera?

238:d sting of the bee

238:f many kinds of enemies, cannot/and thus inevitably

239   If we look at the sting of the bee, as having originally existed in a remote progenitor as a boring and serrated instrument, like that in so many members of the same great order, and which has been modified but not perfected for its present purpose, with the poison originally adapted to cause galls subsequently intensified, we can perhaps understand how it is that the use of the sting should so often cause the insect's own death: for if on the whole the power of stinging be useful to the community, it will fulfil all the requirements of natural selection, though it may cause the death of some few members.

239:c adapted for some purpose, such as to produce galls, subsequently

239:d useful to the social community

239:e some other purpose

239:f having existed/order, and that it has since been modified/ other object, such as to produce galls, since intensified

240 If we admire the truly wonderful power of scent by which the males of many insects find their females, can we admire the production for this single purpose of thousands of drones, which are utterly useless to the community for any other end, and which are ultimately slaughtered by their industrious and sterile sisters?

240:e other purpose, and

241 It may be difficult, but we ought to admire the savage instinctive hatred of the queen-bee, which urges her instantly to destroy the young queens her daughters as soon as born, or to perish herself in the combat; for undoubtedly this is for the good of the community; and maternal love or maternal hatred,

though the latter fortunately is most rare, is all the same to the inexorable principle of natural selection.

241:d her to destroy

241:e young queens, her daughters, as soon as they are born

242 If we admire the several ingenious contrivances, by which the flowers of the orchis and of many other plants are fertilised through insect agency, can we consider as equally perfect the elaboration by our fir-trees of dense clouds of pollen, in order that a few granules may be wafted by a chance breeze on to the ovules?

242:d which orchids and many other/elaboration of dense clouds of pollen by our fir-trees, so that

243 *Summary of Chapter.—*

243:d [Center] *Summary: the Law of Unity of Type and of the Conditions of Existence embraced by the Theory of Natural Selection.* [Space]

244 We have in this chapter discussed some of the difficulties and objections which may be urged against my theory.

244:d[¶]

244:f against the theory.

245 Many of them are very grave; but I think that in the discussion light has been thrown on several facts, which on the theory of independent acts of creation are utterly obscure.

245:b very serious; but

245:e are serious; but/which on the belief of independent

246 We have seen that species at any one period are not indefinitely variable, and are not linked together by a multitude of inter-mediate gradations, partly because the process of natural selec-tion will always be very slow, and will act, at any one time, only on a very few forms; and partly because the very process of natural selection almost implies the continual supplanting and extinction of preceding and intermedate gradations.

246:d only on a few

246:e forms; and partly because the very process of natural selec-tion implies

246:f because the process of natural selection is always very slow, and at any one time acts only

247 Closely allied species, now living on a continuous area, must often have been formed when the area was not continuous, and when the conditions of life did not insensibly graduate away from one part to another.

248 When two varieties are formed in two districts of a continuous area, an intermediate variety will often be formed, fitted for an intermediate zone; but from reasons assigned, the intermedi-ate variety will usually exist in lesser numbers than the two

forms which it connects; consequently the two latter, during the course of further modification, from existing in greater numbers, will have a great advantage over the less numerous intermediate variety, and will thus generally succeed in supplanting and exterminating it.

249   We have seen in this chapter how cautious we should be in concluding that the most different habits of life could not graduate into each other; that a bat, for instance, could not have been formed by natural selection from an animal which at first could only glide through the air.

249:f first only glided

250   We have seen that a species may under new conditions of life change its habits, or have diversified habits, with some habits very unlike those of its nearest congeners.

250:e some very

250:f species under new conditions of life may change its habits; or it may have

251 Hence we can understand, bearing in mind that each organic being is trying to live wherever it can live, how it has arisen that there are upland geese with webbed feet, ground woodpeckers, diving thrushes, and petrels with the habits of auks.

252   Although the belief that an organ so perfect as the eye could have been formed by natural selection, is more than enough to stagger any one; yet in the case of any organ, if we know of a long series of gradations in complexity, each good for its possessor, then, under changing conditions of life, there is no logical impossibility in the acquirement of any conceivable degree of perfection through natural selection.

252:f is enough

253 In the cases in which we know of no intermediate or transitional states, we should be very cautious in concluding that none could have existed, for the homologies of many organs and their intermediate states show that wonderful metamorphoses in function are at least possible.

253:d none have/show what wonderful

253:f be extremely cautious in concluding that none can have existed, for the metamorphoses of many organs show what wonderful changes in

254 For instance, a swim-bladder has apparently been converted into an air-breathing lung.

255 The same organ having performed simultaneously very different functions, and then having been specialised for one function; and two very distinct organs having performed at the same time the same function, the one having been perfected whilst aided by the other, must often have largely facilitated transitions.

255:d been in part or in whole specialised for one function; and two distinct organs

255.1:d[¶] We have seen in two beings widely remote from each other in the natural scale, that an organ serving in both for the same purpose and appearing closely similar may have been separately and independently formed; but when such organs are closely examined, essential differences in their structure can almost always be detected; and this naturally follows from the principle of natural selection.

255.1:f seen that in two beings widely remote from each other in the natural scale, organs serving for the same purpose and in external appearance closely similar

255.2:d On the other hand, the common rule throughout nature is infinite diversity of structure for gaining the same end; and this again naturally follows on the same great principle.

255.2:f follows from the same

256   We are far too ignorant, in almost every case, to be enabled to assert that any part or organ is so unimportant for the welfare of a species, that modifications in its structure could not have been slowly accumulated by means of natural selection.

256:d In almost every case we are far too ignorant to be enabled

256:f In many cases/that a part

257 But we may confidently believe that many modifications, wholly due to the laws of growth, and at first in no way advantageous to a species, have been subsequently taken advantage of by the still further modified descendants of this species.

257.x-y:f In many other cases, modifications are probably the direct result of the laws of variation or of growth, independently of any good having been thus gained. *y* But even such structures have often, as we may feel assured, been subsequently taken advantage of, and still further modified, for the good of species under new conditions of life.

258 We may, also, believe that a part formerly of high importance has often been retained (as the tail of an aquatic animal by its terrestrial descendants), though it has become of such small importance that it could not, in its present state, have been acquired by natural selection,—a power which acts solely by the preservation of profitable variations in the struggle for life.

258:e solely through the survival of the best-fitted individuals in the struggle

258:f has frequently been retained/acquired by means of natural selection.

259   Natural selection will produce nothing in one species for the exclusive good or injury of another; though it may well produce parts, organs, and excretions highly useful or even indispensable, or highly injurious to another species, but in all cases at the same time useful to the owner.

259:*f* selection can produce nothing/indispensable, or again highly

260 Natural selection in each well-stocked country, must act chiefly through the competition of the inhabitants one with another, and consequently will produce perfection, or strength in the battle for life, only according to the standard of that country.

260:*c* country must

260:*f* In each well-stocked country natural selection acts through the competition of the inhabitants, and consequently leads to success in the battle for life, only in accordance with the standard of that particular country.

261 Hence the inhabitants of one country, generally the smaller one, will often yield, as we see they do yield, to the inhabitants of another and generally larger country.

261:*d* smaller one, often yield to the inhabitants

261:*e* smaller one, will often

261:*f* smaller one, often/generally the larger

262 For in the larger country there will have existed more individuals, and more diversified forms, and the competition will have been severer, and thus the standard of perfection will have been rendered higher.

262:*f* individuals and more

263 Natural selection will not necessarily produce absolute perfection; nor, as far as we can judge by our limited faculties, can absolute perfection be everywhere found.

263:*f* necessarily lead to absolute perfection; nor/everywhere predicated.

264 On the theory of natural selection we can clearly understand the full meaning of that old canon in natural history, "Natura non facit saltum."

265 This canon, if we look only to the present inhabitants of the world, is not strictly correct, but if we include all those of past times, it must by my theory be strictly true.

265:*d* correct; but if we include all those of past times, whether known or not yet known, it

265:*f* inhabitants alone of the world/known or unknown, it must on this theory

266 It is generally acknowledged that all organic beings have been formed on two great laws—Unity of Type, and the Conditions of Existence.

267 By unity of type is meant that fundamental agreement in structure, which we see in organic beings of the same class, and which is quite independent of their habits of life.

267:*c* structure which we

268 On my theory, unity of type is explained by unity of descent.

269 The expression of conditions of existence, so often insisted on

378

by the illustrious Cuvier, is fully embraced by the principle of natural selection.

270 For natural selection acts by either now adapting the varying parts of each being to its organic and inorganic conditions of life; or by having adapted them during long-past periods of time: the adaptations being aided in some cases by use and disuse, being slightly affected by the direct action of the external conditions of life, and being in all cases subjected to the several laws of growth.

270:e disuse, being affected

270:f during past periods of time: the adaptations being aided in many cases by the increased use or disuse of parts, being affected by the direct action of the external conditions of life, and subjected in all cases to the several laws of growth and variation.

271 Hence, in fact, the law of the Conditions of Existence is the higher law; as it includes, through the inheritance of former adaptations, that of Unity of Type.

271:f former variations and adaptations

CHAPTER VII.

1:f VIII.

2   INSTINCT.

3 Instincts comparable with habits, but different in their origin—
   Instincts graduated—Aphides and ants—Instincts variable—Domestic instincts, their origins—Natural instincts of the cuckoo,
   ostrich, and parasitic bees—Slave-making ants—Hive-bee, its
   cell-making instinct—Difficulties on the theory of the Natural
   Selection of instincts—Neuter or sterile insects—Summary.

3:c cell-making instincts—Changes of instinct and structure not
   necessarily simultaneous—Difficulties

3:e Difficulties of the theory

3:f cuckoo, molothrus, ostrich

4 THE subject of instinct might have been worked into the previous
   chapters; but I have thought that it would be more convenient
   to treat the subject separately, especially as so wonderful an instinct as that of the hive-bee making its cells will probably have
   occurred to many readers, as a difficulty sufficient to overthrow
   my whole theory.

4:e INSTINCTS might/I thought/as an instinct so wonderful as that of
   the construction of the comb by the hive-bee will/overthrow
   the whole

4:f MANY instincts are so wonderful that their development will
   probably appear to the reader a difficulty sufficient to overthrow my whole theory.

5 I must premise, that I have nothing to do with the origin of the
   primary mental powers, any more than I have with that of life
   itself.

5:f I may here premise

6 We are concerned only with the diversities of instinct and of the
   other mental qualities of animals within the same class.

6:f mental faculties in animals of the same

7   I will not attempt any definition of instinct.

8 It would be easy to show that several distinct mental actions are
   commonly embraced by this term; but every one understands
   what is meant, when it is said that instinct impels the cuckoo to
   migrate and to lay her eggs in other birds' nests.

9 An action, which we ourselves should require experience to enable us to perform, when performed by an animal, more especially by a very young one, without any experience, and when

performed by many individuals in the same way, without their knowing for what purpose it is performed, is usually said to be instinctive.

9:f without experience

10 But I could show that none of these characters of instinct are universal.

10:f characters are

11 A little dose, as Pierre Huber expresses it, of judgment or reason, often comes into play, even in animals very low in the scale of nature.

11:e animals low

11:f dose of judgment or reason, as Pierre Huber expresses it, often comes into play, even with animals

12 Frederick Cuvier and several of the older metaphysicians have compared instinct with habit.

13 This comparison gives, I think, a remarkably accurate notion of the frame of mind under which an instinctive action is performed, but not of its origin.

13:e think, an accurate/not necessarily of

14 How unconsciously many habitual actions are performed, indeed not rarely in direct opposition to our conscious will! yet they may be modified by the will or reason.

15 Habits easily become associated with other habits, and with certain periods of time and states of the body.

15:f other habits, with certain periods of time, and states

16 When once acquired, they often remain constant throughout life.

17 Several other points of resemblance between instincts and habits could be pointed out.

18 As in repeating a well-known song, so in instincts, one action follows another by a sort of rhythm; if a person be interrupted in a song, or in repeating anything by rote, he is generally forced to go back to recover the habitual train of thought: so P. Huber found it was with a caterpillar, which makes a very complicated hammock; for if he took a caterpillar which had completed its hammock up to, say, the sixth stage of construction, and put it into a hammock completed up only to the third stage, the caterpillar simply re-performed the fourth, fifth, and sixth **stages of construction.**

19 If, however, a caterpillar were taken out of a hammock made up, for instance, to the third stage, and were put into one finished up to the sixth stage, so that much of its work was already done for it, far from feeling the benefit of this, it was much embarrassed, and, in order to complete its hammock, seemed forced to start from the third stage, where it had left off, and thus tried to complete the already finished work.

19:f from deriving any benefit from this

20   If we suppose any habitual action to become inherited—and I think it can be shown that this does sometimes happen—then the resemblance between what originally was a habit and an instinct becomes so close as not to be distinguished.

21   If Mozart, instead of playing the pianoforte at three years old with wonderfully little practice, had played a tune with no practice at all, he might truly be said to have done so instinctively.

22   But it would be the most serious error to suppose that the greater number of instincts have been acquired by habit in one generation, and then transmitted by inheritance to succeeding generations.

22:e be a serious

23 It can be clearly shown that the most wonderful instincts with which we are acquainted, namely, those of the hive-bee and of many ants, could not possibly have been thus acquired.

23:d been acquired by habit.

24   It will be universally admitted that instincts are as important as corporeal structure for the welfare of each species, under its present conditions of life.

24:f structures

25 Under changed conditions of life, it is at least possible that slight modifications of instinct might be profitable to a species; and if it can be shown that instincts do vary ever so little, then I can see no difficulty in natural selection preserving and continually accumulating variations of instinct to any extent that may be profitable.

25:c extent that was profitable.

26 It is thus, as I believe, that all the most complex and wonderful instincts have originated.

27 As modifications of corporeal structure arise from, and are increased by, use or habit, and are diminished or lost by disuse, so I do not doubt it has been with instincts.

28 But I believe that the effects of habit are of quite subordinate importance to the effects of the natural selection of what may be called accidental variations of instincts;—that is of variations produced by the same unknown causes which produce slight deviations of bodily structure.

28:e called spontaneous variations of instincts
28:f are in many cases of subordinate

29   No complex instinct can possibly be produced through natural selection, except by the slow and gradual accumulation of numerous, slight, yet profitable, variations.

30 Hence, as in the case of corporeal structures, we ought to find

in nature, not the actual transitional gradations by which each complex instinct has been acquired—for these could be found only in the lineal ancestors of each species—but we ought to find in the collateral lines of descent some evidence of such gradations; or we ought at least to be able to show that gradations of some kind are possible; and this we certainly can do.

31 I have been surprised to find, making allowance for the instincts of animals having been but little observed except in Europe and North America, and for no instinct being known amongst extinct species, how very generally gradations, leading to the most complex instincts, can be discovered.

32 The canon of "Natura non facit saltum" applies with almost equal force to instincts as to bodily organs.

32:[b]

33 Changes of instinct may sometimes be facilitated by the same species having different instincts at different periods of life, or at different seasons of the year, or when placed under different circumstances, &c.; in which case either one or the other instinct might be preserved by natural selection.

34 And such instances of diversity of instinct in the same species can be shown to occur in nature.

35 Again as in the case of corporeal structure, and comformably with my theory, the instinct of each species is good for itself, but has never, as far as we can judge, been produced for the exclusive good of others.

35:d Again, as

35:e conformably to my

36 One of the strongest instances of an animal apparently performing an action for the sole good of another, with which I am acquainted, is that of aphides voluntarily yielding their sweet excretion to ants: that they do so voluntarily, the following facts show.

36:c yielding, as was first observed by Huber, their

37 I removed all the ants from a group of about a dozen aphides on a dock-plant, and prevented their attendance during several hours.

38 After this interval, I felt sure that the aphides would want to excrete.

39 I watched them for some time through a lens, but not one excreted; I then tickled and stroked them with a hair in the same manner, as well as I could, as the ants do with their antennæ; but not one excreted.

40 Afterwards I allowed an ant to visit them, and it immediately seemed, by its eager way of running about, to be well aware what a rich flock it had discovered; it then began to play with its antennæ on the abdomen first of one aphis and then of

another; and each aphis, as soon as it felt the antennæ, immediately lifted up its abdomen and excreted a limpid drop of sweet juice, which was eagerly devoured by the ant.

40:c each, as soon

41 Even the quite young aphides behaved in this manner, showing that the action was instinctive, and not the result of experience.

41.1:c It is certain, from the observations of Huber, that the aphides show no dislike to the ants: if the latter be not present, they are at last compelled to eject their excretion.

41.1:e present they

42 But as the excretion is extremely viscid, it is probably a convenience to the aphides to have it removed; and therefore probably the aphides do not instinctively excrete for the sole good of the ants.

42:c probably they do

42:e is no doubt a convenience to the aphides to have it removed; therefore probably they do not excrete solely for the good

43 Although I do not believe that any animal in the world performs an action for the exclusive good of another of a distinct species, yet each species tries to take advantage of the instincts of others, as each takes advantage of the weaker bodily structure of others.

43:c Although there is no evidence that any animal performs

43:e another species, yet each tries/structure of other species.

44 So again, in some few cases, certain instincts cannot be considered as absolutely perfect; but as details on this and other such points are not indispensable, they may be here passed over.

44:e So again certain

45   As some degree of variation in instincts under a state of nature, and the inheritance of such variations, are indispensable for the action of natural selection, as many instances as possible ought to have been here given; but want of space prevents me.

45:b to be here

45:f be given

46 I can only assert, that instincts certainly do vary—for instance, the migratory instinct, both in extent and direction, and in its total loss.

47 So it is with the nests of birds, which vary partly in dependence on the situations chosen, and on the nature and temperature of the country inhabited, but often from causes wholly unknown to us: Audubon has given several remarkable cases of differences in nests of the same species in the northern and southern United States.

47:b differences in the nests

47.1:c Why, it has been asked, if instinct be variable, has it not given to the bee "the ability to use some other material when wax was deficient?"

47.1:d not granted to the bee

47.2:c But what other material could bees use?

47.2:f other natural material

47.3:c They will work with and use, as I have seen, wax hardened with vermilion and softened with lard.

47.3:f work, as I have seen, with wax hardened with vermilion or softened

47.4-5:c Andrew Knight observed that his bees, instead of laboriously collecting propolis, used a cement of wax and turpentine, with which he had covered decorticated trees. 5 It has lately been shown that bees, instead of searching flowers for their pollen, will gladly use a very different substance, namely, oatmeal.

47.5:f searching for pollen/namely oatmeal.

48 Fear of any particular enemy is certainly an instinctive quality, as may be seen in nestling birds, though it is strengthened by experience, and by the sight of fear of the same enemy in other animals.

49 But fear of man is slowly acquired, as I have elsewhere shown, by various animals inhabiting desert islands; and we may see an instance of this, even in England, in the greater wildness of all our large birds than of our small birds; for the large birds have been most persecuted by man.

49:e The fear/by the various animals which inhabit desert islands; and we see an instance of this even in England, in the greater wildness of all our large birds in comparison with our

50 We may safely attribute the greater wildness of our large birds to this cause; for in uninhabited islands large birds are not more fearful than small; and the magpie, so wary in England, is tame in Norway, as is the hooded crow in Egypt.

51   That the general disposition of individuals of the same species, born in a state of nature, is extremely diversified, can be shown by a multitude of facts.

51:e That the mental qualities of animals of the same kind, born in a state of nature, vary much, could be shown by many facts.

52 Several cases also, could be given, of occasional and strange habits in certain species, which might, if advantageous to the species, give rise, through natural selection, to quite new instincts.

52:d given of

52:e cases could also be given of occasional and strange habits in wild animals, which, if advantageous to the species, might give rise, through natural selection, to new

52:f be adduced of

53 But I am well aware that these general statements, without facts given in detail, can produce but a feeble effect on the reader's mind.

53:e without the facts in detail, will produce

54 I can only repeat my assurance, that I do not speak without good evidence.

54:d evidence. [*Space*]

54.1:d[*Center*] Inherited Changes of Habit or Instinct in Domesticated Animals. [*Space*]

55   The possibility, or even probability, of inherited variations of instinct in a state of nature will be strengthened by briefly considering a few cases under domestication.

56 We shall thus also be enabled to see the respective parts which habit and the selection of so-called accidental variations have played in modifying the mental qualities of our domestic animals.

56:e see the part which habit and the selection of so-called accidental or spontaneous variations

56.1-3:e It is notorious how much domestic animals vary in their mental qualities. *2* With cats, for instance, one naturally takes to catching rats, and another mice, and these tendencies are known to be inherited. *3* One cat, according to Mr. St. John, always brought home game-birds, another hares or rabbits, and another hunted on marshy ground and almost nightly caught woodcocks or snipes.

57 A number of curious and authentic instances could be given of the inheritance of all shades of disposition and tastes, and likewise of the oddest tricks, associated with certain frames of mind or periods of time.

57:e given of various shades of disposition and of taste, and likewise/time, being inherited.

58 But let us look to the familiar case of the several breeds of dogs: it cannot be doubted that young pointers (I have myself seen a striking instance) will sometimes point and even back other dogs the very first time that they are taken out; retrieving is certainly in some degree inherited by retrievers; and a tendency to run round, instead of at, a flock of sheep, by shepherd-dogs.

58:e case of the breeds of the dog: it

59 I cannot see that these actions, performed without experience by the young, and in nearly the same manner by each individual, performed with eager delight by each breed, and without the end being known,—for the young pointer can no more know that he points to aid his master, than the white butterfly knows why she lays her eggs on the leaf of the cabbage,—I cannot see that these actions differ essentially from true instincts.

59:c known—for/cabbage—I

60 If we were to see one kind of wolf, when young and without any training, as soon as it scented its prey, stand motionless like a statue, and then slowly crawl forward with a peculiar gait; and another kind of wolf rushing round, instead of at, a herd of deer, and driving them to a distant point, we should assuredly call these actions instinctive.

60:c to behold one

61 Domestic instincts, as they may be called, are certainly far less fixed or invariable than natural instincts; but they have been acted on by far less rigorous selection, and have been transmitted for an incomparably shorter period, under less fixed conditions of life.

62 How strongly these domestic instincts, habits, and dispositions are inherited, and how curiously they become mingled, is well shown when different breeds of dogs are crossed.

63 Thus it is known that a cross with a bull-dog has affected for many generations the courage and obstinacy of greyhounds; and a cross with a greyhound has given to a whole family of shepherd-dogs a tendency to hunt hares.

64 These domestic instincts, when thus tested by crossing, resemble natural instincts, which in a like manner become curiously blended together, and for a long period exhibit traces of the instincts of either parent: for example, Le Roy describes a dog, whose great-grandfather was a wolf, and this dog showed a trace of its wild parentage only in one way, by not coming in a straight line to his master when called.

64:e master, when

65 Domestic instincts are sometimes spoken of as actions which have become inherited solely from long-continued and compulsory habit, but this, I think, is not true.

65:c this is

65:f habit; but

66 No one would ever have thought of teaching, or probably could have taught, the tumbler-pigeon to tumble,—an action which, as I have witnessed, is performed by young birds, that have never seen a pigeon tumble.

67 We may believe that some one pigeon showed a slight tendency to this strange habit, and that the long-continued selection of the best individuals in successive generations made tumblers what they now are; and near Glasgow, there are house-tumblers, as I hear from Mr. Brent, which cannot fly eighteen inches high without going head over heels.

68 It may be doubted whether any one would have thought of training a dog to point, had not some one dog naturally shown a tendency in this line; and this is known occasionally to happen, as I once saw in a pure terrier.

68:b terrier: the act of pointing is probably, as many have thought, only the exaggerated pause of an animal preparing to spring on its prey.

68:c saw, in

69 When the first tendency was once displayed, methodical selection and the inherited effects of compulsory training in each successive generation would soon complete the work; and unconscious selection is still at work, as each man tries to procure, without intending to improve the breed, dogs which will stand and hunt best.

69:b tendency to point was once

69:c still in progress, as

69:f which stand

70 On the other hand, habit alone in some cases has sufficed; no animal is more difficult to tame than the young of the wild rabbit; scarcely any animal is tamer than the young of the tame rabbit; but I do not suppose that domestic rabbits have ever been selected for tameness; and I presume that we must attribute the whole of the inherited change from extreme wildness to extreme tameness, simply to habit and long-continued close confinement.

70:c sufficed; hardly any animal is more difficult, in most cases, to tame than the young of the wild

70:d but I can hardly suppose that domestic rabbits have often been selected for tameness alone; so that we must attribute the inherited change from extreme wildness to extreme tameness, chiefly to habit

70:e attribute at least the greater part of the inherited change from extreme wildness to extreme tameness, to

70:f difficult to tame the young of the wild

71 Natural instincts are lost under domestication: a remarkable instance of this is seen in those breeds of fowls which very rarely or never become "broody," that is, never wish to sit on their eggs.

72 Familiarity alone prevents our seeing how universally and largely the minds of our domestic animals have been modified by domestication.

72:e how largely and how permanently the minds of our domestic animals have been modified.

73 It is scarcely possible to doubt that the love of man has become instinctive in the dog.

74 All wolves, foxes, jackals, and species of the cat genus, when kept tame, are most eager to attack poultry, sheep, and pigs; and this tendency has been found incurable in dogs which have been brought home as puppies from countries, such as Tierra del Fuego and Australia, where the savages do not keep these domestic animals.

75 How rarely, on the other hand, do our civilised dogs, even when quite young, require to be taught not to attack poultry, sheep, and pigs!

76 No doubt they occasionally do make an attack, and are then beaten; and if not cured, they are destroyed; so that habit, with some degree of selection, has probably concurred in civilising by inheritance our dogs.

76:e selection, have probably

76:f habit and some degree of selection have

77 On the other hand, young chickens have lost, wholly by habit, that fear of the dog and cat which no doubt was originally instinctive in them, in the same way as it is so plainly instinctive in young pheasants, though reared under a hen.

77.x-y:d them; for I am informed by Captain Hutton that the young chickens of the parent-stock, the Gallus bankiva, when reared in India under a hen, are at first excessively wild. *y* So it is with young pheasants reared in England under a hen.

78 It is not that chickens have lost all fear, but fear only of dogs and cats, for if the hen gives the danger-chuckle, they will run (more especially young turkeys) from under her, and conceal themselves in the surrounding grass or thickets; and this is evidently done for the instinctive purpose of allowing, as we see in wild ground-birds, their mother to fly away.

78:d for, if

78:e for if

79 But this instinct retained by our chickens has become useless under domestication, for the mother-hen has almost lost by disuse the power of flight.

80 Hence, we may conclude, that domestic instincts have been acquired and natural instincts have been lost partly by habit, and partly by man selecting and accumulating during successive generations, peculiar mental habits and actions, which at first appeared from what we must in our ignorance call an accident.

80:c that under domestication instincts have been acquired

80:d acquired, and natural instincts have been lost, partly by habit/ accumulating, during

81 In some cases compulsory habit alone has sufficed to produce such inherited mental changes; in other cases compulsory habit has done nothing, and all has been the result of selection, pursued both methodically and unconsciously; but in most cases, probably, habit and selection have acted together.

81:d together. [*Space*]

81:e produce inherited/unconsciously: but in most cases habit and selection have probably acted

81:f probably concurred.

81.1:d [*Center*] *Special Instincts.* [*Space*]

82 We shall, perhaps, best understand how instincts in a state of nature have become modified by selection, by considering a few cases.

83 I will select only three, out of the several which I shall have to discuss in my future work,—namely, the instinct which leads the cuckoo to lay her eggs in other birds' nests; the slave-making instinct of certain ants; and the comb-making power of the hive-bee: these two latter instincts have generally, and most justly, been ranked by naturalists as the most wonderful of all known instincts.

83:e of those which I/ants; and the cell-making power of the hive-bee: these two latter instincts have generally and justly been

83.x-y:f three,—namely, the instinct which/hive-bee. *y* These

83.1:d[¶] *Instincts of the Cuckoo.—*

84 It is now commonly admitted that the more immediate and final cause of the cuckoo's instinct is, that she lays her eggs, not daily, but at intervals of two or three days; so that, if she were to make her own nest and sit on her own eggs, those first laid would have to be left for some time unincubated, or there would be eggs and young birds of different ages in the same nest.

84:d[No ¶] of the most remarkable of the instincts of the cuckoo is, that she lays

84:e is supposed by some naturalists that the more immediate cause of the instinct of the cuckoo is, that she lays

85 If this were the case, the process of laying and hatching might be inconveniently long, more especially as she has to migrate at a very early period; and the first hatched young would probably have to be fed by the male alone.

85:d she migrates

86 But the American cuckoo is in this predicament; for she makes her own nest and has eggs and young successively hatched, all at the same time.

86:d nest, and has

87 It has been asserted that the American cuckoo occasionally lays her eggs in other birds' nests; but I hear on the high authority of Dr. Brewer, that this is a mistake.

87:e been both asserted and denied that the American/nests; but I have lately heard from Dr. Merrell, of Iowa, that he once found in Illinois a young cuckoo together with a young jay in the nest of a Blue jay (*Garrulus cristatus*); and as both were nearly fully feathered, there could be no mistake in their identification.

87:f Garrulus cristatus

88 Nevertheless, I could give several instances of various birds which have been known occasionally to lay their eggs in other birds' nests.

88:e I could also give

89 Now let us suppose that the ancient progenitor of our European cuckoo had the habits of the American cuckoo; but that occasionally she laid an egg in another bird's nest.

89:e American cuckoo, and that she occasionally laid

90 If the old bird profited by this occasional habit, or if the young were made more vigorous by advantage having been taken of the mistaken maternal instinct of another bird, than by their own mother's care, encumbered as she can hardly fail to be by having eggs and young of different ages at the same time; then the old birds or the fostered young would gain an advantage.

90:d habit through being enabled to migrate earlier or through any other cause; or if the young were/mistaken instinct of another species, than by their own mother's care, encumbered as she could hardly/time, and by having to migrate at an early period; then

90:e than when reared by their own mother, encumbered/time; then

91 And analogy would lead me to believe, that the young thus reared would be apt to follow by inheritance the occasional and aberrant habit of their mother, and in their turn would be apt to lay their eggs in other birds' nests, and thus be successful in rearing their young.

91:d nests, and thus be more successful

91:e lead us to believe

92 By a continued process of this nature, I believe that the strange instinct of our cuckoo could be, and has been, generated.

92:e cuckoo has been generated.

93 I may add that, according to Dr. Gray and to some other observers, the European cuckoo has not utterly lost all maternal love and care for her own offspring.

93:e It has, also, recently been ascertained that the cuckoo occasionally lays her eggs on the bare ground, sits on them and feeds her young; this rare and strange event evidently is a case of reversion to the long-lost aboriginal instinct of nidification.

93.x-y:f ascertained on sufficient evidence, by Adolf Müller, that/ them, and feeds her young. *y* This rare event is probably a case

93.1:d[¶] It has been objected by some authors that I have not noticed other related instincts and points of structure in the cuckoo, which are fasely spoken of as necessarily co-ordinated.

93.1:e objected that I have not noticed other related instincts and adaptations in

93.1:f adaptations of structure in the cuckoo, which are spoken

93.2:d But in all cases, speculation on any instinct or character known in only a single species, is useless, for we have no facts to guide us.

93.2:e on an instinct known

93.2:f known to us only/have hitherto had no

93.3:d Until quite recently the instincts of the European and of the non-parasitic American cuckoo alone were known; now, owing to Mr. E. Ramsay's observations, we know something about three Australian species, which lay their eggs in other birds' nests.

93.3:f Until recently/Mr. Ramsay's observations, we have learnt something

93.4:d The chief points referred to are three: firstly, that the cuckoo, with rare exceptions, lays only one egg in a nest, so that the large and voracious young cuckoo receives ample food.

93.4:e points are three: first, that the cuckoo, with/young bird may receive

93.4:f points to be referred to are three: first, that the common cuckoo/bird receives

93.5:d Secondly, that the egg is so remarkably small, that it does not exceed in size that of the skylark,—a bird not more than one-fourth of the size of the cuckoo; that this is a real case of adaptation we may infer from the fact of the non-parasitic American cuckoo laying eggs of full size proportionally with her body.

93.5.x:e Secondly, that the eggs are of remarkably small size, not exceeding those of the skylark,—a bird about one-fourth as large as the cuckoo.

93.5.x:f eggs are remarkably small, not

93.5.y:e That the small size of the egg is a real/laying full-sized eggs.

93.6:d Thirdly and lastly, that the young cuckoo, soon after birth, has the instinct, the strength, and a properly shaped back for ejecting its foster-brothers, which then perish from cold and hunger.

93.6:e Thirdly, that the young

93.7:d This, it has been boldly maintained, is beneficently designed, in order that the young cuckoo may get sufficient food, and that its foster-brothers may perish, before, as it is supposed, they have acquired much feeling!

93.7:e is a beneficent arrangement, in/perish before they have acquired much feeling.

93.7:f This has been boldly called a beneficent/they had acquired much feeling!

93.8:d[¶] Turning now to the Australian species; though these birds generally lay only one egg in a nest, it is not rare to find two and even three eggs of the same species of cuckoo in the same nest.

93.8:e three eggs in

93.9-10:d In the Bronze cuckoo the eggs vary greatly in size, from eight to ten lines in length. *10* Now if it had been of any advantage to this species to have laid eggs even smaller than those now laid by her, so as to have deceived certain foster-parents, or, as is more probable, to have been hatched within some shorter period (for it is asserted that there is a relation between the size of eggs and the period of incubation), then there is no difficulty in believing that a race or species might have been formed which would have laid smaller and smaller eggs; for these would have been more safely hatched and reared.

93.10:e those now laid, so/within a shorter period (for it is asserted that there is a relation between size and the period

93.10:f of an advantage/between the size of eggs and the period of their incubation

93.11:d Mr. Ramsay remarks that two of the Australian cuckoos, when they lay their eggs in an open or not domed nest, manifest a decided preference for nests containing eggs similar to their own.

93.11:e open nest/similar in colour to

93.12:d The European species certainly manifests some tendency towards a similar instinct, but not rarely departs from it, as is shown by her laying her dull and pale-coloured eggs in the nest of the Hedge-warbler with its bright greenish-blue eggs: had she invariably displayed the above instinct, it would assuredly have been added to those which it is assumed must all have been acquired together.

93.12.x:e greenish-blue eggs.

93.12.x:f species apparently manifests

93.12.y:e Had our cuckoo invariably

93.13:d The eggs of the Australian Bronze cuckoo vary, according to Mr. Ramsay, in an extraordinary manner in colour; so that in this respect, as well as in size, natural selection assuredly might have secured and fixed any advantageous variation.

93.13:e Ramsay, to an extraordinary degree in colour

93.13:f selection might

93.14:d[¶] With respect to the last point insisted on—namely, of the young European cuckoo ejecting its foster-brothers—it must first be remarked that Mr. Gould, who has paid particular attention to this subject, is convinced that the belief is an error; he asserts that the young foster-birds are generally ejected during the first three days, when the young cuckoo is quite powerless; he maintains that the young cuckoo exerts, by its hunger-cries, or by some other means, such a fascination over its foster-parents, that it alone receives food, so that the others are starved to death, and are then thrown out, like the egg-shells or the excrement, by the old birds.

93.14:e With reference to the young European

93.14:f In the case of the European cuckoo, the offspring of the foster-parents are commonly ejected from the nest within three days after the cuckoo is hatched; and as the latter at this age is in a most helpless condition, Mr. Gould was formerly inclined to believe that the act of ejection was performed by the foster-parents themselves.

93.14.0.1-2:f But he has now received a trustworthy account of a young cuckoo which was actually seen, whilst still blind and not able even to hold up its own head, in the act of ejecting its foster-brothers. *2* One of these was replaced in the nest by the observer, and was again thrown out.

93.15:d He admits, however, that the young cuckoo when grown older and stronger may have the power, and perhaps the instinct, of ejecting its foster-brothers, if they happen to escape starvation during the first few days after birth.

93.15:e power and

93.15:[f]

93.16:d Mr. Ramsay has arrived at a similar conclusion with respect to the Australian species, which he especially observed: he states that the young cuckoo is at first a little helpless fat creature, but, "as it grows rapidly, it soon fills up the greater part of the nest, and its unfortunate companions, either smothered by its weight, or starved to death through its greediness, are thrown out by their parents."

93.16:e at the same conclusion regarding the Australian species: he

93.16:[f]

93.16.1:e Nevertheless there is so much evidence, both ancient and recent, that the young European cuckoo does eject its foster-brothers, that this can hardly be doubted.

93.16.1:[f]

93.17:d Now, if it had been of great importance to the young cuckoo to have received as much food as possible during the first few days after birth, I can see no especial difficulty, if it possessed sufficient strength, in its gradually acquiring, during successive generations, the habit (first, perhaps, through mere unintentional restlessness) and the structure best fitted for ejecting its foster-brothers; for those young cuckoos which had such habits and structure would have been the best fed and most securely reared.

93.17:e Now, if it were of great importance to the young cuckoo to receive as much food as possible soon after birth, I can see no special difficulty in its gradually acquiring during successive generations, the habit (perhaps through mere unintentional restlessness) the strength and the structure/would be the best

93.17:f With respect to the means by which this strange and odious instinct was acquired, if it were of great importance for the young cuckoo, as is probably the case, to receive as much food as possible soon after birth, I can see no special difficulty in its having gradually acquired, during successive generations, the blind desire, the strength, and structure necessary for the work

394

of ejection; for those young cuckoos which had such habits and structure best developed would be the most securely reared.

93.17.0.1:f The first step towards the acquisition of the proper instinct might have been mere unintentional restlessness on the part of the young bird, when somewhat advanced in age and strength; the habit having been afterwards improved, and transmitted to an earlier age.

93.18:d I can see no more difficulty in this, than in young birds acquiring the instinct and the temporary hard tips to their beaks for breaking through their own shells;—or than in the young snake having in its upper jaw, as Owen has remarked, a transitory sharp tooth for cutting through the tough egg-shell.

93.18:e or than in young snakes having in their upper jaws, as

93.18:f this, than in the unhatched young of other birds acquiring the instinct to break through their own shells;—or than in the young snake acquiring in

93.19:d For if each part is liable to variation at any age, and the variations tend to be inherited at a corresponding age,—propositions which cannot, as we shall hereafter see, rightfully be disputed,—then the instincts and structure of the young can be slowly modified as well as those of the adult, and both cases must stand or fall together with the whole theory of natural selection.

93.19:e to individual variations at any/cannot be disputed,—then the instincts and structure of the young could be slowly modified as well as those of the adult; and both cases must stand or fall with

93.19:f at all ages, and the variations tend to be inherited at a corresponding or earlier age,—propositions/modified as surely as/ fall together with

93.19.1-13:f[¶] Some species of Molothrus, a widely distinct genus of American birds, allied to our starlings, have parasitic habits like those of the cuckoo; and the species present an interesting gradation in the perfection of their instincts. *2* The sexes of Molothrus badius are stated by an excellent observer, Mr. Hudson, sometimes to live promiscuously together in flocks, and sometimes to pair. *3* They either build a nest of their own, or seize on one belonging to some other bird, occasionally throwing out the nestlings of the stranger. *4* They either lay their eggs in the nest thus appropriated, or oddly enough build one for themselves on the top of it. *5* They usually sit on their own eggs and rear their own young; but Mr. Hudson says it is probable that they are occasionally parasitic, for he has seen the young of this species following old birds of a distinct kind and clamouring to be fed by them. *6* The parasitic habits of another species of Molothrus, the M. bonariensis, are much more highly developed than those of the last, but are still far from perfect. *7* This bird, as far as it is known, invariably lays its eggs in the nests of strangers; but it is remarkable that several together sometimes commence to build an irregular untidy nest of their own, placed in singularly ill-adapted situations, as on the leaves of a large thistle. *8* They never, however, as far as Mr. Hudson has ascertained, complete a nest for themselves. *9* They often lay so many eggs—from fifteen to twenty—in the same foster-nest, that few or none can possibly be hatched. *10* They have, moreover, the extraordinary habit of pecking holes in the eggs, whether of their own species or of their foster-parents, which they find in the appropriated nests. *11* They drop also many eggs on the bare ground, which are thus wasted. *12* A third species, the M. pecoris of North America, has acquired instincts as perfect as those of the cuckoo, for it never lays more than one egg in a foster-nest, so that the young bird is securely reared. *13* Mr. Hudson is a strong disbeliever in evolution, but he appears to have been so much struck by the imperfect instincts of the Molothrus bonariensis that he quotes my words, and asks, "Must we consider these habits, not as especially endowed or created instincts, but as small consequences of one general law, namely, transition?"

94 The occasional habit of birds laying their eggs in other birds' nests, either of the same or of a distinct species, is not very uncommon with the Gallinaceæ; and this perhaps explains the origin of a singular instinct in the allied group of ostriches.

94:e instinct in the nearest allied group, that of ostriches.

94.x-y:f Various birds, as has already been remarked, occasionally lay their eggs in the nests of other birds. *y* This habit is not very uncommon with the Gallinaceæ, and throws some light on the singular instinct of the ostrich.

95 For several hen ostriches, at least in the case of the American species, unite and lay first a few eggs in one nest and then in another; and these are hatched by the males.

95:d ostriches unite

95:f In this family several hen-birds unite

96 This instinct may probably be accounted for by the fact of the hens laying a large number of eggs; but, as in the case of the cuckoo, at intervals of two or three days.

96:d eggs, but

96:f but, as with the cuckoo

97 This instinct, however, of the American ostrich has not as yet been perfected; for a surprising number of eggs lie strewed over the plains, so that in one day's hunting I picked up no less than twenty lost and wasted eggs.

97:e The instinct

97:f ostrich, as in the case of the Molothrus bonariensis, has

98 Many bees are parasitic, and always lay their eggs in the nests of bees of other kinds.

98:e and regularly lay their eggs in the nests of other kinds of bees.

99 This case is more remarkable than that of the cuckoo; for these bees have not only their instincts but their structure modified in accordance with their parasitic habits; for they do not possess the pollen-collecting apparatus which would be necessary if they had to store food for their own young.

99:e only had their instincts/apparatus which would have been indispensable if they had stored

99:f stored up food

100 Some species, likewise, of Sphegidæ (wasp-like insects) are parasitic on other species; and M. Fabre has lately shown good reason for believing that although the Tachytes nigra generally makes its own burrow and stores it with paralysed prey for its own larvæ to feed on, yet that when this insect finds a burrow already made and stored by another sphex, it takes advantage of the prize, and becomes for the occasion parasitic.

100:d believing that, although/yet that, when

100:e parasitic in the same manner on other/larvæ, yet

100:f species of Sphegidæ (wasp-like insects) are likewise parasitic; and M. Fabre

101 In this case, as with the supposed case of the cuckoo, I can see no difficulty in natural selection making an occasional habit permanent, if of advantage to the species, and if the insect whose nest and stored food are thus feloniously appropriated, be not thus exterminated.

101:d exterminated. [*No space*]

101:e appropriated be

101:f with that of the Molothrus or cuckoo, I/are feloniously appropriated, be

102 *Slave-making instinct.—*

103 This remarkable instinct was first discovered in the Formica (Polyerges) rufescens by Pierre Huber, a better observer even than his celebrated father.

104 This ant is absolutely dependent on its slaves; without their aid, the species would certainly become extinct in a single year.

105 The males and fertile females do no work.

106 The workers or sterile females, though most energetic and courageous in capturing slaves, do no other work.

105 + 6:c work of any kind, and the workers

107 They are incapable of making their own nests, or of feeding their own larvæ.

108 When the old nest is found inconvenient, and they have to migrate, it is the slaves which determine the migration, and actually carry their masters in their jaws.

109 So utterly helpless are the masters, that when Huber shut up

thirty of them without a slave, but with plenty of food which they like best, and with their larvæ and pupæ to stimulate them to work, they did nothing; they could not even feed themselves, and many perished of hunger.

109:e their own larvæ

110 Huber then introduced a single slave (F. fusca), and she instantly set to work, fed and saved the survivors; made some cells and tended the larvæ, and put all to rights.

111 What can be more extraordinary than these well-ascertained facts?

112 If we had not known of any other slave-making ant, it would have been hopeless to have speculated how so wonderful an instinct could have been perfected.

112:e to speculate

113 Formica sanguinea was likewise first discovered by P. Huber to be a slave-making ant.

113:b Another species, Formica sanguinea, was

114 This species is found in the southern parts of England, and its habits have been attended to by Mr. F. Smith, of the British Museum, to whom I am much indebted for information on this and other subjects.

115 Although fully trusting to the statements of Huber and Mr. Smith, I tried to approach the subject in a sceptical frame of mind, as any one may well be excused for doubting the truth of so extraordinary and odious an instinct as that of making slaves.

115:f doubting the existence of so extraordinary an instinct

116 Hence I will give the observations which I have myself made, in some little detail.

116:c which I made, in

116:f Hence, I will

117 I opened fourteen nests of F. sanguinea, and found a few slaves in all.

118 Males and fertile females of the slave-species are found only in their own proper communities, and have never been observed in the nests of F. sanguinea.

118:b slave-species (F. fusca) are

119 The slaves are black and not above half the size of their red masters, so that the contrast in their appearance is very great.

119:e is great.

120 When the nest is slightly disturbed, the slaves occasionally come out, and like their masters are much agitated and defend the nest: when the nest is much disturbed and the larvæ and pupæ are exposed, the slaves work energetically with their masters in carrying them away to a place of safety.

120:d disturbed, and

121 Hence, it is clear, that the slaves feel quite at home.

122 During the months of June and July, on three successive years, I have watched for many hours several nests in Surrey and Sussex, and never saw a slave either leave or enter a nest.

122:f I watched

123 As, during these months, the slaves are very few in number, I thought that they might behave differently when more numerous; but Mr. Smith informs me that he has watched the nests at various hours during May, June and August, both in Surrey and Hampshire, and has never seen the slaves, though present in large numbers in August, either leave or enter the nest.

123:c June, and August, both

124 Hence he considers them as strictly household slaves.

125 The masters, on the other hand, may be constantly seen bringing in materials for the nest, and food of all kinds.

126 During the present year, however, in the month of July, I came across a community with an unusually large stock of slaves, and I observed a few slaves mingled with their masters leaving the nest, and marching along the same road to a tall Scotch-fir-tree, twenty-five yards distant, which they ascended together, probably in search of aphides or cocci.

126:c During the year 1860, however

127 According to Huber, who had ample opportunities for observation, in Switzerland the slaves habitually work with their masters in making the nest, and they alone open and close the doors in the morning and evening; and, as Huber expressly states, their principal office is to search for aphides.

127:d observation, the slaves in Switzerland habitually

128 This difference in the usual habits of the masters and slaves in the two countries, probably depends merely on the slaves being captured in greater numbers in Switzerland than in England.

129 One day I fortunately chanced to witness a migration from one nest to another, and it was a most interesting spectacle to behold the masters carefully carrying, as Huber has described, their slaves in their jaws.

129:b fortunately witnessed a migration of F. sanguinea from/carrying (instead of being carried by, as in the case of F. rufescens) their

129:c carrying their slaves in their jaws instead of being carried by them, as in the case of F. rufescens.

130 Another day my attention was struck by about a score of the slave-makers haunting the same spot, and evidently not in search of food; they approached and were vigorously repulsed by an independent community of the slave species (F. fusca);

sometimes as many of three of these ants clinging to the legs of the slave-making F. sanguinea.

130:b slave-species

131 The latter ruthlessly killed their small opponents, and carried their dead bodies as food to their nest, twenty-nine yards distant; but they were prevented from getting any pupæ to rear as slaves.

132 I then dug up a small parcel of the pupæ of F. fusca from another nest, and put them down on a bare spot near the place of combat; they were eagerly seized, and carried off by the tyrants, who perhaps fancied that, after all, they had been victorious in their late combat.

132:d seized and

133 At the same time I laid on the same place a small parcel of the pupæ of another species, F. flava, with a few of these little yellow ants still clinging to the fragments of the nest.

133:e fragments of their nest.

134 This species is sometimes, though rarely, made into slaves, as has been described by Mr. Smith.

135 Although so small a species, it is very courageous, and I have seen it ferociously attack other ants.

136 In one instance I found to my surprise an independent community of F. flava under a stone beneath a nest of the slave-making F. sanguinea; and when I had accidentally disturbed both nests, the little ants attacked their big neighbours with surprising courage.

137 Now I was curious to ascertain whether F. sanguinea could distinguish the pupæ of F. fusca, which they habitually make into slaves, from those of the little and furious F. flava, which they rarely capture, and it was evident that they did at once distinguish them: for we have seen that they eagerly and instantly seized the pupæ of F. fusca, whereas they were much terrified when they came across the pupæ, or even the earth from the nest of F. flava, and quickly ran away; but in about a quarter of an hour, shortly after all the little yellow ants had crawled away, they took heart and carried off the pupæ.

137:d nest, of F. flava

137:e them; for

138 One evening I visited another community of F. sanguinea, and found a number of these ants entering their nest, carrying the dead bodies of F. fusca (showing that it was not a migration) and numerous pupæ.

138:b ants returning home and entering their nests, carrying

139 I traced the returning file burthened with booty, for about forty yards, to a very thick clump of heath, whence I saw the last

individual of F. sanguinea emerge, carrying a pupa; but I was not able to find the desolated nest in the thick heath.

139:b traced a long file of ants burthened

139:e yards back, to a very

140 The nest, however, must have been close at hand, for two or three individuals of F. fusca were rushing about in the greatest agitation, and one was perched motionless with its own pupa in its mouth on the top of a spray of heath over its ravaged home.

140:b heath, an image of despair, over

141 Such are the facts, though they did not need confirmation by me, in regard to the wonderful instinct of making slaves.

142 Let it be observed what a contrast the instinctive habits of F. sanguinea present with those of the F. rufescens.

142:b those of the continental F. rufescens.

143 The latter does not build its own nest, does not determine its own migrations, does not collect food for itself or its young, and cannot even feed itself: it is absolutely dependent on its numerous slaves.

144 Formica sanguinea, on the other hand, possesses much fewer slaves, and in the early part of the summer extremely few.

145 The masters determine when and where a new nest shall be formed, and when they migrate, the masters carry the slaves.

144 + 5:b extremely few: the masters

146 Both in Switzerland and England the slaves seem to have the exclusive care of the larvæ, and the masters alone go on slave-making expeditions.

147 In Switzerland the slaves and masters work together, making and bringing materials for the nest: both, but chiefly the slaves, tend, and milk as it may be called, their aphides; and thus both collect food for the community.

147:d nest; both, but

148 In England the masters alone usually leave the nest to collect building materials and food for themselves, their slaves and larvæ.

149 So that the masters in this country receive much less service from their slaves than they do in Switzerland.

150 By what steps the instinct of F. sanguinea originated I will not pretend to conjecture.

151 But as ants, which are not slave-makers, will, as I have seen, carry off pupæ of other species, if scattered near their nests, it is possible that pupæ originally stored as food might become developed; and the ants thus unintentionally reared would then follow their proper instincts, and do what work they could.

151:b that such pupæ originally stored as food might become developed; and the foreign ants

151:f slave-makers will

152 If their presence proved useful to the species which had seized them—if it were more advantageous to this species to capture workers than to procreate them—the habit of collecting pupæ originally for food might by natural selection be strengthened and rendered permanent for the very different purpose of raising slaves.

152:f pupæ, originally for food, might

153 When the instinct was once acquired, if carried out to a much less extent even than in our British F. sanguinea, which, as we have seen, is less aided by its slaves than the same species in Switzerland, I can see no difficulty in natural selection increasing and modifying the instinct—always supposing each modification to be of use to the species—until an ant was formed as abjectly dependent on its slaves as is the Formica rufescens.

153:c Switzerland, natural selection might increase and modify

153:d rufescens. [*No space*]

154   *Cell-making instinct of the Hive-Bee.—*

155 I will not here enter on minute details on this subject, but will merely give an outline of the conclusions at which I have arrived.

156 He must be a dull man who can examine the exquisite structure of a comb, so beautifully adapted to its end, without enthusiastic admiration.

157 We hear from mathematicians that bees have practically solved a recondite problem, and have made their cells of the proper shape to hold the greatest possible amount of honey, with the least possible consumption of precious wax in their construction.

158 It has been remarked that a skilful workman, with fitting tools and measures, would find it very difficult to make cells of wax of the true form, though this is perfectly effected by a crowd of bees working in a dark hive.

158:f is effected

159 Grant whatever instincts you please, and it seems at first quite inconceivable how they can make all the necessary angles and planes, or even perceive when they are correctly made.

159:e Granting whatever instincts you please, it

160 But the difficulty is not nearly so great as it at first appears: all this beautiful work can be shown, I think, to follow from a few very simple instincts.

160:c few simple

161   I was led to investigate this subject by Mr. Waterhouse, who

has shown that the form of the cell stands in close relation to the presence of adjoining cells; and the following view may, perhaps, be considered only as a modification of his theory.

162 Let us look to the great principle of gradation, and see whether Nature does not reveal to us her method of work.

163 At one end of a short series we have humble-bees, which use their old cocoons to hold honey, sometimes adding to them short tubes of wax, and likewise making separate and very irregular cells of wax.

164 At the other end of the series we have the cells of the hive-bee, placed in a double layer: each cell, as is well known, is an hexagonal prism, with the basal edges of its six sides bevelled so as to join on to a pyramid, formed of three rhombs.

164:b to fit on

164:f to join an inverted pyramid, of three rhombs.

165 These rhombs have certain angles, and the three which form the pyramidal base of a single cell on one side of the comb, enter into the composition of the bases of three adjoining cells on the opposite side.

165:f comb enter

166 In the series between the extreme perfection of the cells of the hive-bee and the simplicity of those of the humble-bee, we have the cells of the Mexican Melipona domestica, carefully described and figured by Pierre Huber.

166:f humble-bee we

167 The Melipona itself is intermediate in structure between the hive and humble bee, but more nearly related to the latter: it forms a nearly regular waxen comb of cylindrical cells, in which the young are hatched, and, in addition, some large cells of wax for holding honey.

167:e latter; it

168 These latter cells are nearly spherical and of nearly equal sizes, and are aggregated into an irregular mass.

169 But the important point to notice, is that these cells are always made at that degree of nearness to each other, that they would have intersected or broken into each other, if the spheres had been completed; but this is never permitted, the bees building perfectly flat walls of wax between the spheres which thus tend to intersect.

169:f notice is, that these/nearness to each other that they would have intersected or broken into each other if

170 Hence each cell consists of an outer spherical portion and of two, three, or more perfectly flat surfaces, according as the cell adjoins two, three, or more other cells.

170:d portion, and

170:f more flat

171 When one cell comes into contact with three other cells, which, from the spheres being nearly of the same size, is very frequently and necessarily the case, the three flat surfaces are united into a pyramid; and this pyramid, as Huber has remarked, is manifestly a gross imitation of the three-sided pyramidal basis of the cell of the hive-bee.

171:b bases

171:d cell rests on three other

171:f base

172 As in the cells of the hive-bee, so here, the three plane surfaces in any one cell necessarily enter into the construction of three adjoining cells.

173 It is obvious that the Melipona saves wax by this manner of building; for the flat walls between the adjoining cells are not double, but are of the same thickness as the outer spherical portions, and yet each flat portion forms a part of two cells.

173:e wax, and what is more important, labour, by

174 Reflecting on this case, it occurred to me that if the Melipona had made its spheres at some given distance from each other, and had made them of equal sizes and had arranged them symmetrically in a double layer, the resulting structure would probably have been as perfect as the comb of the hive-bee.

175 Accordingly I wrote to Professor Miller, of Cambridge, and this geometer has kindly read over the following statement, drawn up from his information, and tells me that it is strictly correct:—

176 If a number of equal spheres be described with their centres placed in two parallel layers; with the centre of each sphere at the distance of radius $\times \sqrt{2}$, or radius $\times 1 \cdot 41421$ (or at some lesser distance), from the centres of the six surrounding spheres in the same layer; and at the same distance from the centres of the adjoining spheres in the other and parallel layer; then, if planes of intersection between the several spheres in both layers be formed, there will result a double layer of hexagonal prisms united together by pyramidal bases formed of three rhombs; and the rhombs and the sides of the hexagonal prisms will have every angle identically the same with the best measurements which have been made of the cells of the hive-bee.

176.1:d But I hear from Prof. Wyman, who has made numerous careful measurements, that the accuracy of the workmanship of the bee has been greatly exaggerated; so much so, that, as he adds, whatever the typical form of the cell may be, it is rarely, if ever, realised.

176.1:f much so, that whatever

177 Hence we may safely conclude that if we could slightly modify the instincts already possessed by the Melipona, and in

themselves not very wonderful, this bee would make a structure as wonderfully perfect as that of the hive-bee.

177:d that, if

178 We must suppose the Melipona to make her cells truly spherical, and of equal sizes; and this would not be very surprising, seeing that she already does so to a certain extent, and seeing what perfectly cylindrical burrows in wood many insects can make, apparently by turning round on a fixed point.

178:d to have the power of forming her

178:f burrows many insects make in wood, apparently

179 We must suppose the Melipona to arrange her cells in level layers, as she already does her cylindrical cells; and we must further suppose, and this is the greatest difficulty, that she can somehow judge accurately at what distance to stand from her fellow-labourers when several are making their spheres; but she is already so far enabled to judge of distance, that she always describes her spheres so as to intersect largely; and then she unites the points of intersection by perfectly flat surfaces.

179:f intersect to a certain extent; and

180 We have further to suppose, but this is no difficulty, that after hexagonal prisms have been formed by the intersection of adjoining spheres in the same layer, she can prolong the hexagon to any length requisite to hold the stock of honey; in the same way as the rude humble-bee adds cylinders of wax to the circular mouths of her old cocoons.

180:[f]

181 By such modifications of instincts in themselves not very wonderful,—hardly more wonderful than those which guide a bird to make its nest,—I believe that the hive-bee has acquired, through natural selection, her inimitable architectural powers.

181:f instincts which in themselves are not

182 But this theory can be tested by experiment.

183 Following the example of Mr. Tegetmeier, I separated two combs, and put between them a long, thick, square strip of wax: the bees instantly began to excavate minute circular pits in it; and as they deepened these little pits, they made them wider and wider until they were converted into shallow basins, appearing to the eye perfectly true or parts of a sphere, and of about the diameter of a cell.

183:c thick, rectangular strip

184 It was most interesting to me to observe that wherever several bees had begun to excavate these basins near together, they had begun their work at such a distance from each other, that by the time the basins had acquired the above stated width (*i. e.* about the width of an ordinary cell), and were in depth about one sixth of the diameter of the sphere of which they

formed a part, the rims of the basins intersected or broke into each other.

184:d that, wherever

184:f interesting to observe

185 As soon as this occurred, the bees ceased to excavate, and began to build up flat walls of wax on the lines of intersection between the basins, so that each hexagonal prism was built upon the festooned edge of a smooth basin, instead of on the straight edges of a three-sided pyramid as in the case of ordinary cells.

185:b upon the scalloped edge of a smooth

186 I then put into the hive, instead of a thick, square piece of wax, a thin and narrow, knife-edged ridge, coloured with vermilion.

186:c thick, rectangular piece

187 The bees instantly began on both sides to excavate little basins near to each other, in the same way as before; but the ridge of wax was so thin, that the bottoms of the basins, if they had been excavated to the same depth as in the former experiment, would have broken into each other from the opposite sides.

188 The bees, however, did not suffer this to happen, and they stopped their excavations in due time; so that the basins, as soon as they had been a little deepened, came to have flat bottoms; and these flat bottoms, formed by thin little plates of the vermilion wax having been left ungnawed, were situated, as far as the eye could judge, exactly along the planes of imaginary intersection between the basins on the opposite sides of the ridge of wax.

188:e have bottoms with flat sides; and these flat sides, formed by little thin plates of the vermilion wax left ungnawed

188:f flat bases; and these flat bases, formed

189 In parts, only little bits, in other parts, large portions of a rhombic plate had been left between the opposed basins, but the work from the unnatural state of things, had not been neatly performed.

189:e In some parts, only small portions, in other

189:f plate were thus left

190 The bees must have worked at very nearly the same rate on the opposite sides of the ridge of vermilion wax, as they circularly gnawed away and deepened the basins on both sides, in order to have succeeded in thus leaving flat plates between the basins, by stopping work along the intermediate planes or planes of intersection.

190:e rate in circularly gnawing away and deepening the basins on both sides of the ridge of vermilion wax, in order to have thus succeeded in leaving flat plates between the basins, by stopping work at the planes of intersection.

191   Considering how flexible thin wax is, I do not see that there is any difficulty in the bees, whilst at work on the two sides of a strip of wax, perceiving when they have gnawed the wax away to the proper thinness, and then stopping their work.

192 In ordinary combs it has appeared to me that the bees do not always succeed in working at exactly the same rate from the opposite sides; for I have noticed half-completed rhombs at the base of a just-commenced cell, which were slightly concave on one side, where I suppose that the bees had excavated too quickly, and convex on the opposed side, where the bees had worked less quickly.

192:e opposed side where

193 In one well-marked instance, I put the comb back into the hive, and allowed the bees to go on working for a short time, and again examined the cell, and I found that the rhombic plate had been completed, and had become *perfectly flat:* it was absolutely impossible, from the extreme thinness of the little rhombic plate, that they could have effected this by gnawing away the convex side; and I suspect that the bees in such cases stand in the opposed cells and push and bend the ductile and warm wax (which as I have tried is easily done) into its proper intermediate plane, and thus flatten it.

193:e well marked/little plate/stand on opposite sides and push

194   From the experiment of the ridge of vermilion wax, we can clearly see that if the bees were to build for themselves a thin wall of wax, they could make their cells of the proper shape, by standing at the proper distance from each other, by excavating at the same rate, and by endeavouring to make equal spherical hollows, but never allowing the spheres to break into each other.

194:c wax we

194:d that, if

194:e can see

195 Now bees, as may be clearly seen by examining the edge of a growing comb, do make a rough, circumferential wall or rim all round the comb; and they gnaw into this from the opposite sides, always working circularly as they deepen each cell.

195:e gnaw this away from

196 They do not make the whole three-sided pyramidal base of any one cell at the same time, but only the one rhombic plate which stands on the extreme growing margin, or the two plates, as the case may be; and they never complete the upper edges of the rhombic plates, until the hexagonal walls are commenced.

196:c only that one

197 Some of these statements differ from those made by the justly celebrated elder Huber, but I am convinced of their accuracy;

and if I had space, I could show that they are conformable with my theory.

198   Huber's statement that the very first cell is excavated out of a little parallel-sided wall of wax, is not, as far as I have seen, strictly correct; the first commencement having always been a little hood of wax; but I will not here enter on these details.

198:d statement, that

198:f on details.

199 We see how important a part excavation plays in the construction of the cells; but it would be a great error to suppose that the bees cannot build up a rough wall of wax in the proper position—that is, along the plane of intersection between two adjoining spheres.

200 I have several specimens showing clearly that they can do this.

201 Even in the rude circumferential rim or wall of wax round a growing comb, flexures may sometimes be observed, corresponding in position to the planes of the rhombic basal plates of future cells.

202 But the rough wall of wax has in every case to be finished off, by being largely gnawed away on both sides.

203 The manner in which the bees build is curious; they always make the first rough wall from ten to twenty times thicker than the excessively thin finished wall of the cell, which will ultimately be left.

.04 We shall understand how they work, by supposing masons first to pile up a broad ridge of cement, and then to begin cutting it away equally on both sides near the ground, till a smooth, very thin wall is left in the middle; the masons always piling up the cut-away cement, and adding fresh cement, on the summit of the ridge.

204:e fresh cement on

205 We shall thus have a thin wall steadily growing upward; but always crowned by a gigantic coping.

205:e upward but

206 From all the cells, both those just commenced and those completed, being thus crowned by a strong coping of wax, the bees can cluster and crawl over the comb without injuring the delicate hexagonal walls, which are only about one four-hundredth of an inch in thickness; the plates of the pyramidal basis being about twice as thick.

206.x-y:c hexagonal walls. *y* These walls, as Professor Miller has kindly ascertained for me, vary greatly in thickness; being, on an average of twelve measurements made near the border of the comb, $\frac{1}{353}$ of an inch in thickness; whereas the basal rhomboidal plates are thicker nearly in the proportion of three to two, having a mean thickness, from twenty-one measurements, of $\frac{1}{229}$ of an inch.

206.y:f thicker, nearly

207 By this singular manner of building, strength is continually given to the comb, with the utmost ultimate economy of wax.

207:c By the above singular

208 It seems at first to add to the difficulty of understanding how the cells are made, that a multitude of bees all work together; one bee after working a short time at one cell going to another, so that, as Huber has stated, a score of individuals work even at the commencement of the first cell.

209 I was able practically to show this fact, by covering the edges of the hexagonal walls of a single cell, or the extreme margin of the circumferential rim of a growing comb, with an extremely thin layer of melted vermilion wax; and I invariably found that the colour was most delicately diffused by the bees—as delicately as a painter could have done with his brush—by atoms of the coloured wax having been taken from the spot on which it had been placed, and worked into the growing edges of the cells all round.

209:f done it with

210 The work of construction seems to be a sort of balance struck between many bees, all instinctively standing at the same relative distance from each other, all trying to sweep equal spheres, and then building up, or leaving ungnawed, the planes of intersection between these spheres.

211 It was really curious to note in cases of difficulty, as when two pieces of comb met at an angle, how often the bees would entirely pull down and rebuild in different ways the same cell, sometimes recurring to a shape which they had at first rejected.

211:b would pull

212 When bees have a place on which they can stand in their proper positions for working,—for instance, on a slip of wood, placed directly under the middle of a comb growing downwards so that the comb has to be built over one face of the slip—in this case the bees can lay the foundations of one wall of a new hexagon, in its strictly proper place, projecting beyond the other completed cells.

212:f downwards, so

213 It suffices that the bees should be enabled to stand at their proper relative distances from each other and from the walls of the last completed cells, and then, by striking imaginary spheres, they can build up a wall intermediate between two adjoining spheres; but, as far as I have seen, they never gnaw away and finish off the angles of a cell till a large part both of that cell and of the adjoining cells has been built.

214 This capacity in bees of laying down under certain circum-

stances a rough wall in its proper place between two just-commenced cells, is important, as it bears on a fact, which seems at first quite subversive of the foregoing theory; namely, that the cells on the extreme margin of wasp-combs are sometimes strictly hexagonal; but I have not space here to enter on this subject.

214:f first subversive

215 Nor does there seem to me any great difficulty in a single insect (as in the case of a queen-wasp) making hexagonal cells, if she work alternately on the inside and outside of two or three cells commenced at the same time, always standing at the proper relative distance from the parts of the cells just begun, sweeping spheres or cylinders, and building up intermediate planes.

215:d she were to work

216 It is even conceivable that an insect might, by fixing on a point at which to commence a cell, and then moving outside, first to one point, and then to five other points, at the proper relative distances from the central point and from each other, strike the planes of intersection, and so make an isolated hexagon: but I am not aware that any such case has been observed; nor would any good be derived from a single hexagon being built, as in its construction more materials would be required than for a cylinder.

216:[d]

217 As natural selection acts only by the accumulation of slight modifications of structure or instinct, each profitable to the individual under its conditions of life, it may reasonably be asked, how a long and graduated succession of modified architectural instincts, all tending towards the present perfect plan of construction, could have profited the progenitors of the hive-bee?

218 I think the answer is not difficult: it is known that bees are often hard pressed to get sufficient nectar; and I am informed by Mr. Tegetmeier that it has been experimentally found that no less than from twelve to fifteen pounds of dry sugar are consumed by a hive of bees for the secretion of each pound of wax; so that a prodigious quantity of fluid nectar must be collected and consumed by the bees in a hive for the secretion of the wax necessary for the construction of their combs.

218.x:d difficult: cells constructed like those of the bee or the wasp gain in strength, and save much in labour and space, and especially in the materials of which they are constructed.

218.x:e space, and in the materials

218.y:d With respect to the formation of wax, it is known that bees are often hard pressed to get sufficient nectar, and I am informed by Mr. Tegetmeier that it has been experimentally proved that from

219 Moreover, many bees have to remain idle for many days during the process of secretion.

220 A large store of honey is indispensable to support a large stock of bees during the winter; and the security of the hive is known mainly to depend on a large number of bees being supported.

221 Hence the saving of wax by largely saving honey must be a most important element of success in any family of bees.

221:e honey and the time consumed in collecting the honey must be an important element of success to any

222 Of course the success of any species of bees may be dependent on the number of its parasites or other enemies, or on quite distinct causes, and so be altogether independent of the quantity of honey which the bees could collect.

222:e success of the species may be dependent on the number of its enemies, or parasites, or on

222:f which the bees can collect.

223 But let us suppose that this latter circumstance determined, as it probably often does determine, the numbers of a humble-bee which could exist in a country; and let us further suppose that the community lived throughout the winter, and consequently required a store of honey: there can in this case be no doubt that it would be an advantage to our humble-bee, if a slight modification of her instinct led her to make her waxen cells near together, so as to intersect a little; for a wall in common even to two adjoining cells, would save some little wax.

223:c further suppose (differently to what really is the case) that the community

223:d often has determined, the numbers of a bee allied to our humble-bees, which existed in any country/modification in her instincts/some little wax and labour.

223:e often has determined, whether a bee allied to our humble-bees could exist in large numbers in any country/lived through the winter/adjoining cells would save some little labour and wax.

224 Hence it would continually be more and more advantageous to our humble-bee, if she were to make her cells more and more regular, nearer together, and aggregated into a mass, like the cells of the Melipona; for in this case a large part of the bounding surface of each cell would serve to bound other cells, and much wax would be saved.

224:d wax and labour would

224:e humble-bees, if they were to make their cells more and more/ bound the adjoining cells, and much labour and wax would

225 Again, from the same cause, it would be advantageous to the Melipona, if she were to make her cells closer together, and more regular in every way than at present; for then, as we have seen, the spherical surfaces would wholly disappear, and would

411

all be replaced by plane surfaces; and the Melipona would make a comb as perfect as that of the hive-bee.

225:e disappear and be

226 Beyond this stage of perfection in architecture, natural selection could not lead; for the comb of the hive-bee, as far as we can see, is absolutely perfect in economising wax.

226:d economising labour and wax.

227 Thus, as I believe, the most wonderful of all known instincts, that of the hive-bee, can be explained by natural selection having taken advantage of numerous, successive, slight modifications of simpler instincts; natural selection having by slow degrees, more and more perfectly, led the bees to sweep equal spheres at a given distance from each other in a double layer, and to build up and excavate the wax along the planes of intersection.

228 The bees, of course, no more knowing that they swept their spheres at one particular distance from each other, than they know what are the several angles of the hexagonal prisms and of the basal rhombic plates.

229 The motive power of the process of natural selection having been economy of wax; that individual swarm which wasted least honey in the secretion of wax, having succeeded best, and having transmitted by inheritance its newly acquired economical instinct to new swarms, which in their turn will have had the best chance of succeeding in the struggle for existence.

229:c wax, together with cells of due strength, and of the proper size and shape for the larvæ; that individual swarm which made the best cells, and wasted/inheritance their newly acquired/ existence. [*No space*]

227 + 8 + 9:d intersection; the bees/plates; the motive power of the process of natural selection having been the construction of cells of due strength and of the proper size and shape for the larvæ, this being effected with the greatest possible economy of wax and labour; that individual swarm which thus made the best cells with least labour, and least waste of honey

227 + 8 + 9:e having, by slow degrees more and more perfectly led/economy of labour and wax; that

227 + 8 + 9:f transmitted their newly-acquired

229.0.1:d [*Center*] *Objections to the Theory of Natural Selection as applied to Instincts: Neuter and Sterile Insects.* [*Space*]

229.1:c[¶] It has been objected to the foregoing view on the origin of instinct that "the variations of structure and of instinct must have been simultaneous and accurately adjusted to each other, as a modification in the one without an immediate corresponding change in the other would have been fatal."

229.1:d instincts that

229.1:e view of the origin

229.2:c The force of this objection seems entirely to rest on the assumption that the changes in both instinct and structure are abrupt.

229.2:e objection rests entirely on

229.2:f changes in the instincts

229.3:c To take as an illustration the case of the larger titmouse (Parus major) alluded to in the last chapter: this bird often holds the seeds of the yew between its feet on a branch, and hammers away till it gets into the kernel.

229.3:f in a previous chapter/hammers with its beak till it gets at the kernel.

229.4:c Now what special difficulty would there be in natural selection preserving each slight variation of beak, better and better adapted to break open seeds, until a beak was formed, as well constructed for this purpose as that of the nuthatch, at the same time that hereditary habit, or compulsion from the want of other food, or the preservation of chance variations of taste, made the bird more and more of a seed-eater?

229.4:e preserving all the slight individual variations in the shape of the beak, which were better and better adapted to break open the seeds/time that habit, or compulsion, or spontaneous variations of taste, led the bird to become more and more of a seed-eater?

229.5:c In this case the beak is supposed to be slowly modified by natural selection, subsequently to, but in accordance with, slowly changing habit; but let the feet of the titmouse vary and grow larger from correlation with the beak, or from any other unknown cause, and is it very improbable that such larger feet might lead the bird to climb more and more until it acquired even the remarkable climbing instinct and capacity of the nuthatch?

229.5:d such larger feet would lead

229.5:e changing habits or taste/cause, and it is not improbable/acquired the remarkable climbing instinct and power of the nuthatch.

229.6:c In this case a gradual change of structure is supposed to lead to changed instinctive habits of life.

229.6:d habits.

229.7-11:c To take one more case: few instincts are more remarkable than that which leads the swift of the Eastern Islands to make its nest wholly of inspissated saliva. *8* Some birds build their nests of mud, believed to be moistened with saliva; and one of the swifts of North America makes its nest (as I have seen) of sticks agglutinated with saliva, and even with flakes of this substance. *9* Is it then very improbable that the natural selection of individual swifts, which secreted more and more saliva, should at last produce a species with instincts leading it to neglect other materials, and to make its nest exclusively of inspissated saliva? *10* And so in other cases. *11* It must be admitted that in many instances we cannot conjecture whether instinct or structure has first slightly changed; nor can we conjecture by what gradations many instincts have been developed when they relate to organs (such as the mammary glands) on the first origin of which we know nothing.

229.11:e must, however, be/whether it was instinct or structure which first varied.

230 No doubt many instincts of very difficult explanation could be opposed to the theory of natural selection,—cases, in which we cannot see how an instinct could possibly have originated; cases, in which no intermediate gradations are known to exist; cases of instinct of apparently such trifling importance, that they could hardly have been acted on by natural selection; cases of instincts almost identically the same in animals so remote in the scale of nature, that we cannot account for their similarity by inheritance from a common parent, and must therefore believe that they have been acquired by independent acts of natural selection.

230:d believe that they have been independently acquired by natural selection.

230:e selection—cases, in which we cannot see/exist; cases of instinct of such/common progenitor, and consequently must believe that they were independently acquired through natural

230:f how an instinct could have originated

231 I will not here enter on these several cases, but will confine myself to one special difficulty, which at first appeared to me insuperable, and actually fatal to my whole theory.

231:e fatal to the whole

232 I allude to the neuters or sterile females in insect-communities: for these neuters often differ widely in instinct and in structure from both the males and fertile females, and yet, from being sterile, they cannot propagate their kind.

232:g insect-communities; for

233 The subject well deserves to be discussed at great length, but I will here take only a single case, that of working or sterile ants.

234 How the workers have been rendered sterile is a difficulty; but not much greater than that of any other striking modification of structure; for it can be shown that some insects and other articulate animals in a state of nature occasionally become sterile; and if such insects had been social, and it had been profitable to the community that a number should have been annually born capable of work, but incapable of procreation, I can see no very great difficulty in this being effected by natural selection.

234:e no especial difficulty in this having been effected through natural

235 But I must pass over this preliminary difficulty.

236 The great difficulty lies in the working ants differing widely from both the males and the fertile females in structure, as in the shape of the thorax and in being destitute of wings and sometimes of eyes, and in instinct.

236:d thorax, and in being

237 As far as instinct alone is concerned, the prodigious difference in this respect between the workers and the perfect females, would have been far better exemplified by the hive-bee.

237:e concerned, the wonderful difference/been better

238 If a working ant or other neuter insect had been an animal in the ordinary state, I should have unhesitatingly assumed that all its characters had been slowly acquired through natural selection; namely, by an individual having been born with some slight profitable modification of structure, this being inherited by its offspring, which again varied and were again selected, and so onwards.

238:e been an ordinary animal, I/by individuals having been born with slight profitable modifications, which were inherited by the offspring; and that these again varied and again were selected

239 But with the working ant we have an insect differing greatly from its parents, yet absolutely sterile; so that it could never have transmitted successively acquired modifications of structure or instinct to its progeny.

240 It may well be asked how is it possible to reconcile this case with the theory of natural selection?

241   First, let it be remembered that we have innumerable instances, both in our domestic productions and in those in a state of nature, of all sorts of differences of structure which have become correlated to certain ages, and to either sex.

241:e differences of inherited structure which are correlated with certain ages, and with either sex.

242 We have differences correlated not only to one sex, but to that short period alone when the reproductive system is active, as in the nuptial plumage of many birds, and in the hooked jaws of the male salmon.

242:e only with one sex, but with that short period when

243 We have even slight differences in the horns of different breeds of cattle in relation to an artifically imperfect state of the male sex; for oxen of certain breeds have longer horns than in other breeds, in comparison with the horns of the bulls or cows of these same breeds.

243:e longer horns than the oxen of other breeds, relatively to the length of the horns in both the bulls and cows

244 Hence I can see no real difficulty in any character having become correlated with the sterile condition of certain members of insect-communities: the difficulty lies in understanding how such correlated modifications of structure could have been slowly accumulated by natural selection.

244:e no great difficulty in any character becoming

245 This difficulty, though appearing insuperable, is lessened, or, as I believe, disappears, when it is remembered that selection may be applied to the family, as well as to the individual, and may thus gain the desired end.

246 Thus, a well-flavoured vegetable is cooked, and the individual is destroyed; but the horticulturist sows seeds of the same stock, and confidently expects to get nearly the same variety; breeders of cattle wish the flesh and fat to be well marbled together; the animal has been slaughtered, but the breeder goes with confidence to the same family.

246:c same family, and confidently

246:d Thus, breeders of/but the breeder has gone with confidence to the same stock and has succeeded.

246:f Breeders/together: an animal thus characterised has been

247 I have such faith in the powers of selection, that I do not doubt that a breed of cattle, always yielding oxen with extraordinarily long horns, could be slowly formed by carefully watching which individual bulls and cows, when matched, produced oxen with the longest horns; and yet no one ox could ever have propagated its kind.

247:d Such faith may be placed in the power of selection, that probably a breed

247:f that a breed/could, it is probable, be formed/one ox would ever

247.1:d Here is a better and real illustration: according to M. Verlot, some varieties of the double annual stock of various colours, from having been long carefully selected to the right degree, always produce by seed a large proportion of plants bearing double and quite sterile flowers; so that, if the variety had not yielded others, it would at once have become extinct; but it likewise always yields some single and fertile plants, which differ only in their power of producing two forms, from ordinary single varieties.

247.1:e not likewise yielded others, it would at once have become extinct; but it always yields some single and fertile plants, which differ from ordinary single varieties only in their power of producing the two forms.

247.1:f annual Stock from having been long and carefully selected to the right degree, always produce a large proportion of seed-

416

.lings bearing double and quite sterile flowers; but they likewise yield some single and fertile plants.

247.2:*d* Thus these single and fertile plants may be compared with the males and females of an ant-community, and the sterile double-flowered plants, which are regularly produced in large numbers, with the many sterile neuters of the same community.

247.2:*e* Thus the fertile plants producing single flowers may

247.2:*f* These latter, by which alone the variety can be propagated, may be compared with the fertile male and female ants, and the double sterile plants with the neuters of the same community.

248 Thus I believe it has been with social insects: a slight modification of structure, or instinct, correlated with the sterile condition of certain members of the community, has been advantageous to the community: consequently the fertile males and females of the same community flourished, and transmitted to their fertile offspring a tendency to produce sterile members having the same modification.

248:*d* or of instinct

248.*x-y:f* As with the varieties of the stock, so with social insects, selection has been applied to the family, and not to the individual, for the sake of gaining a serviceable end. *y* Hence we may conclude that slight modifications of structure or of instinct, correlated with the sterile condition of certain members of the community, have proved advantageous: consequently the fertile males and females have flourished, and transmitted/members with the same modifications.

249 And I believe that this process has been repeated, until that prodigious amount of difference between the fertile and sterile females of the same species has been produced, which we see in many social insects.

249:*d* in so many

249:*f* This process must have been repeated many times, until/in many

250 But we have not as yet touched on the climax of the difficulty; namely, the fact that the neuters of several ants differ, not only from the fertile females and males, but from each other, sometimes to an almost incredible degree, and are thus divided into two or even three castes.

250:*g* on the acme of the difficulty

251 The castes, moreover, do not generally graduate into each other, but are perfectly well defined; being as distinct from each other, as are any two species of the same genus, or rather as any two genera of the same family.

251:*e* from each other as are

251:*f* not commonly graduate

252 Thus in Eciton, there are working and soldier neuters, with

jaws and instincts extraordinarily different: in Cryptocerus, the workers of one caste alone carry a wonderful sort of shield on their heads, the use of which is quite unknown: in the Mexican Myrmecocystus, the workers of one caste never leave the nest; they are fed by the workers of another caste, and they have an enormously developed abdomen which secretes a sort of honey, supplying the place of that excreted by the aphides, or the domestic cattle as they may be called, which our European ants guard or imprison.

253 It will indeed be thought that I have an overweening confidence in the principle of natural selection, when I do not admit that such wonderful and well-established facts at once annihilate my theory.

253:e annihilate the theory.

254 In the simpler case of neuter insects all of one caste or of the same kind, which have been rendered by natural selection, as I believe to be quite possible, different from the fertile males and females,—in this case, we may safely conclude from the analogy of ordinary variations, that each successive, slight, profitable modification did not probably at first appear in all the individual neuters in the same nest, but in a few alone; and that by the long-continued selection of the fertile parents which produced most neuters with the profitable modification, all the neuters ultimately came to have the desired character.

254:d caste, which/females—in this/not at

254:e caste, which, as I believe, have been rendered different from the fertile males and females through natural selection, we may conclude from the analogy of ordinary variations, that the successive, slight, profitable modifications did not first arise in all the neuters in the same nest, but in some few alone; and that by the survival of the communities with females which produce most neuters having the advantageous modification, all the neuters ultimately come to be thus characterized.

254:f produced

255 On this view we ought occasionally to find neuter-insects of the same species, in the same nest, presenting gradations of structure; and this we do find, even often, considering how few neuter-insects out of Europe have been carefully examined.

255:c neuter insects

255:d even frequently, considering

255:e According to this view we ought occasionally to find in the same nest neuter insects, presenting

255:f even not rarely, considering

256 Mr. F. Smith has shown how surprisingly the neuters of several British ants differ from each other in size and sometimes in colour; and that the extreme forms can sometimes be perfectly linked together by individuals taken out of the same nest: I have myself compared perfect gradations of this kind.

256:e shown that the neuters of several British ants differ surprisingly from/can be linked

257 It often happens that the larger or the smaller sized workers are the most numerous; or that both large and small are numerous, with those of an intermediate size scanty in numbers.

257:e It sometimes happens/both large and small are numerous, whilst those of an intermediate size are scanty

258 Formica flava has larger and smaller workers, with some of intermediate size; and, in this species, as Mr. F. Smith has observed, the larger workers have simple eyes (ocelli), which though small can be plainly distinguished, whereas the smaller workers have their ocelli rudimentary.

258:e some few of

259 Having carefully dissected several specimens of these workers, I can affirm that the eyes are far more rudimentary in the smaller workers than can be accounted for merely by their proportionately lesser size; and I fully believe, though I dare not assert so positively, that the workers of intermediate size have their ocelli in an exactly intermediate condition.

260 So that we here have two bodies of sterile workers in the same nest, differing not only in size, but in their organs of vision, yet connected by some few members in an intermediate condition.

260:e that here we have

261 I may digress by adding, that if the smaller workers had been the most useful to the community, and those males and females had been continually selected, which produced more and more of the smaller workers, until all the workers had come to be in this condition; we should then have had a species of ant with neuters very nearly in the same condition with those of Myrmica.

261:e all the workers were in this/neuters nearly

261:f neuters in nearly the same condition as those

262 For the workers of Myrmica have not even rudiments of ocelli, though the male and female ants of this genus have well-developed ocelli.

263 I may give one other case: so confidently did I expect to find gradations in important points of structure between the different castes of neuters in the same species, that I gladly availed myself of Mr. F. Smith's offer of numerous specimens from the same nest of the driver ant (Anomma) of West Africa.

263:f expect occasionally to find gradations of important structures

264 The reader will perhaps best appreciate the amount of difference in these workers, by my giving not the actual measurements, but a strictly accurate illustration: the difference was the same as if we were to see a set of workmen building a house of whom many were five feet four inches high, and many sixteen

feet high; but we must suppose that the larger workmen had heads four instead of three times as big as those of the smaller men, and jaws nearly five times as big.

264:d house, of whom

264:f must in addition suppose

265 The jaws, moreover, of the working ants of the several sizes differed wonderfully in shape, and in the form and number of the teeth.

266 But the important fact for us is, that though the workers can be grouped into castes of different sizes, yet they graduate insensibly into each other, as does the widely-different structure of their jaws.

266:d that, though

267 I speak confidently on this latter point, as Mr. Lubbock made drawings for me with the camera lucida of the jaws which I had dissected from the workers of the several sizes.

267:d as Sir J. Lubbock/me, with the camera lucida, of

267.1:d Mr. Bates, in his most interesting 'Naturalist on the Amazons,' has described some analogous cases.

267.1:f his interesting/described analogous

268   With these facts before me, I believe that natural selection, by acting on the fertile parents, could form a species which should regularly produce neuters, either all of large size with one form of jaw, or all of small size with jaws having a widely different structure; or lastly, and this is our climax of difficulty, one set of workers of one size and structure, and simultaneously another set of workers of a different size and structure;—a graduated series having been first formed, as in the case of the driver ant, and then the extreme forms, from being the most useful to the community, having been produced in greater and greater numbers through the natural selection of the parents which generated them; until none with an intermediate structure were produced.

268:d extreme forms having been produced through the natural selection of the parents which generated them, in greater and greater numbers, until

268:e fertile ants or parents, could/workers of one size and structure, and at the same time another/having first been formed, as/produced in greater and greater numbers, through the survival of the parents which generated them, until

268:f neuters, all of large size with one form of jaw, or all of small size with widely different jaws; or lastly, and this is the climax of difficulty, one set of workers of one size and structure, and simultaneously another

268.1:d[¶] An analogous explanation has been given by Mr. Wallace, of the equally complex case, of certain Malayan Butterflies regularly appearing at the same time and place under two or even three distinct female forms; and by Fritz Müller, of certain Brazilian crustaceans likewise appearing under two widely distinct male forms.

268.1:e appearing under

268.2:d But the subject need not here be discussed.

268.2:e But this subject

269   Thus, as I believe, the wonderful fact of two distinctly defined castes of sterile workers existing in the same nest, both widely different from each other and from their parents, has originated.

269:d I have now explained how, as I believe, the wonderful

270   We can see how useful their production may have been to a social community of insects, on the same principle that the division of labour is useful to civilised man.

270:d of ants, on

271   As ants work by inherited instincts and by inherited tools or weapons, and not by acquired knowledge and manufactured instruments, a perfect division of labour could be effected with them only by the workers being sterile; for had they been fertile, they would have intercrossed, and their instincts and structure would have become blended.

271:b instincts and by inherited organs or tools, and not

271:d Ants, however, work by inherited instincts and by inherited organs or tools, whilst man works by acquired knowledge and manufactured instruments.

272   And nature has, as I believe, effected this admirable division of labour in the communities of ants, by the means of natural selection.

272:[d]

273   But I am bound to confess, that, with all my faith in this principle, I should never have anticipated that natural selection could have been efficient in so high a degree, had not the case of these neuter insects convinced me of the fact.

273:d But I must confess, that, with all my faith in natural selection, I should never have anticipated that this principle could

273:f insects led me to this conclusion.

274   I have, therefore, discussed this case, at some little but wholly insufficient length, in order to show the power of natural selection, and likewise because this is by far the most serious special difficulty, which my theory has encountered.

274:d difficulty which

275   The case, also, is very interesting, as it proves that with animals, as with plants, any amount of modification in structure can be effected by the accumulation of numerous, slight, and as we must call them accidental, variations, which are in any manner profitable, without exercise or habit having come into play.

275:*d* and spontaneous variations

275:*e* modification may be/slight, spontaneous/having been brought into

276 For no amount of exercise, or habit, or volition, in the utterly sterile members of a community could possibly have affected the structure or instincts of the fertile members, which alone leave descendants.

276:*b* possibly affect

276:*e* For peculiar habits confined to the workers or sterile females, however long they might be followed, could not possibly affect the males and fertile female, which

276:*f* females

277 I am surprised that no one has advanced this demonstrative case of neuter insects, against the well-known doctrine of Lamarck.

277:*c* doctrine of inherited habit advanced by Lamarck.

277:*e* has hitherto advanced this/habit, as advanced

278   *Summary.—*

278:*d* [*Center*] *Summary.* [*Space*]

279 I have endeavoured briefly in this chapter to show that the mental qualities of our domestic animals vary, and that the variations are inherited.

279:*d*[¶] endeavoured in this chapter briefly to

280 Still more briefly I have attempted to show that instincts vary slightly in a state of nature.

281 No one will dispute that instincts are of the highest importance to each animal.

282 Therefore I can see no difficulty, under changing conditions of life, in natural selection accumulating slight modifications of instinct to any extent, in any useful direction.

282:*e* Therefore there is no real difficulty/accumulating to any extent slight modifications of instinct which are in any way useful.

283 In some cases habit or use and disuse have probably come into play.

283:*f* In many cases

284 I do not pretend that the facts given in this chapter strengthen in any great degree my theory; but none of the cases of difficulty, to the best of my judgment, annihilate it.

285 On the other hand, the fact that instincts are not always absolutely perfect and are liable to mistakes;—that no instinct has been produced for the exclusive good of other animals, but that each animal takes advantage of the instincts of others;—that the canon in natural history, of "natura non facit saltum" is applicable to instincts as well as to corporeal structure, and

is plainly explicable on the foregoing views, but is otherwise inexplicable,—all tend to corroborate the theory of natural selection.

285:b Natura non facit saltum," is

285:e no instinct can be shown to have been produced for the good of other animals, though animals take

286   This theory is, also, strengthened by some few other facts in regard to instincts; as by that common case of closely allied, but certainly distinct, species, when inhabiting distant parts of the world and living under considerably different conditions of life, yet often retaining nearly the same instincts.

286:d but distinct

286:f is also strengthened

287 For instance, we can understand on the principle of inheritance, how it is that the thrush of South America lines its nest with mud, in the same peculiar manner as does our British thrush: how it is that the male wrens (Troglodytes) of North America, build "cock-nests," to roost in, like the males of our distinct Kitty-wrens,—a habit wholly unlike that of any other known bird.

287:c thrush: how it is that the Hornbills of Africa and India have the same extraordinary instinct of plastering up and imprisoning the females in a hole in a tree, with only a small hole left in the plaster through which the males feed them and their young when hatched: how it is that the male wrens

287:d thrush of tropical South

287:f British thrush; how it is that the Hornbills/hatched; how it is that the male wrens

288 Finally, it may not be a logical deduction, but to my imagination it is far more satisfactory to look at such instincts as the young cuckoo ejecting its foster-brothers,—ants making slaves— the larvæ of ichneumonidæ feeding within the live bodies of caterpillars,—not as specially endowed or created instincts, but as small consequences of one general law, leading to the advancement of all organic beings, namely, multiply, vary, let the strongest live and the weakest die.

288:d beings,—namely

288:f law leading

CHAPTER VIII.

3 Distinction between the sterility of first crosses and of hybrids—
   Sterility various in degree, not universal, affected by close interbreeding, removed by domestication—Laws governing the
   sterility of hybrids—Sterility not a special endowment, but incidental on other differences—Causes of the sterility of first crosses
   and of hybrids—Parallelism between the effects of changed conditions of life and crossing—Fertility of varieties when crossed
   and of their mongrel offspring not universal—Hybrids and
   mongrels compared independently of their fertility—Summary.

3:d differences, not accumulated by natural selection—Causes/life
   and of crossing—Dimorphism and trimorphism—Fertility of
   varieties

4 THE view generally entertained by naturalists is that species,
   when intercrossed, have been specially endowed with the quality of sterility, in order to prevent the confusion of all organic
   forms.

4:e with sterility, in order to prevent their confusion.

4:f view commonly entertained

5 This view certainly seems at first probable, for species within the
   same country could hardly have kept distinct had they been
   capable of crossing freely.

5:e first highly probable/have been kept/of freely crossing.

5:f species living together could

6 The importance of the fact that hybrids are very generally sterile,
   has, I think, been much underrated by some late writers.

6:d that first crosses between distinct species and hybrids

6:[e]

7 On the theory of natural selection the case is especially important,
   inasmuch as the sterility of hybrids could not possibly be of any
   advantage to them, and therefore could not have been acquired
   by the continued preservation of successive profitable degrees
   of sterility.

7:d important, inasmuch as this sterility can hardly have been increased by the continued preservation of successive, profitable
   degrees of sterility.

7:e The subject is in many ways important for us, more especially as the sterility of species when first crossed, and that of their hybrid offspring, cannot have been acquired by the continued preservation of successive, profitable degrees of sterility.

7:f acquired, as I shall show, by the preservation of successive profitable

8 I hope, however, to be able to show that sterility is not a specially acquired or endowed quality, but is incidental on other acquired differences.

8:c other acquired and little known differences.

8:e But to this subject I shall have to return, and I hope ultimately to show that this sterility is neither a specially acquired nor endowed quality, but is incidental on other acquired and little-known differences in the reproductive systems of the parent-species.

8:f It is an incidental result of differences in the reproductive systems of the parent-species.

9 In treating this subject, two classes of facts, to a large extent fundamentally different, have generally been confounded together; namely, the sterility of two species when first crossed, and the sterility of the hybrids produced from them.

9:f confounded; namely, the sterility of species

10 Pure species have of course their organs of reproduction in a perfect condition, yet when intercrossed they produce either few or no offspring.

11 Hybrids, on the other hand, have their reproductive organs functionally impotent, as may be clearly seen in the state of the male element in both plants and animals; though the organs themselves are perfect in structure, as far as the microscope reveals.

11:d though the formative organs

12 In the first case the two sexual elements which go to form the embryo are perfect; in the second case they are either not at all developed, or are imperfectly developed.

13 This distinction is important, when the cause of the sterility, which is common to the two cases, has to be considered.

14 The distinction has probably been slurred over, owing to the sterility in both cases being looked on as a special endowment, beyond the province of our reasoning powers.

14:f distinction probably has been

15 The fertility of varieties, that is of the forms known or believed to have descended from common parents, when intercrossed, and likewise the fertility of their mongrel offspring, is, on my theory, of equal importance with the sterility of species; for it seems to make a broad and clear distinction between varieties and species.

15:c offspring, is, with reference to my

425

15:f to be descended from common parents, when crossed, and likewise

15.1:d[¶] *Degrees of Sterility.—*

16   First, for the sterility of species when crossed and of their hybrid offspring.

16:d[No ¶]

17   It is impossible to study the several memoirs and works of those two conscientious and admirable observers, Kölreuter and Gärtner, who almost devoted their lives to this subject, without being deeply impressed with the high generality of some degree of sterility.

18   Kölreuter makes the rule universal; but then he cuts the knot, for in ten cases in which he found two forms, considered by most authors as distinct species, quite fertile together, he unhesitatingly ranks them as varieties.

19   Gärtner, also, makes the rule equally universal; and he disputes the entire fertility of Kölreuter's ten cases.

20   But in these and in many other cases, Gärtner is obliged carefully to count the seeds, in order to show that there is any degree of sterility.

21   He always compares the maximum number of seeds produced by two species when crossed and by their hybrid offspring, with the average number produced by both pure parent-species in a state of nature.

21:c parent species

21:d parent-species

21:e when first crossed, and the maximum produced by their

22   But a serious cause of error seems to me to be here introduced: a plant to be hybridised must be castrated, and, what is often more important, must be secluded in order to prevent pollen being brought to it by insects from other plants.

22:d plant, to be hybridised, must be castrated

22:f But causes of serious error here intervene: a plant, to be hybridised

23   Nearly all the plants experimentised on by Gärtner were potted, and apparently were kept in a chamber in his house.

23:d experimented on by Gärtner were potted, and were

24   That these processes are often injurious to the fertility of a plant cannot be doubted; for Gärtner gives in his table about a score of cases of plants which he castrated, and artificially fertilised with their own pollen, and (excluding all cases such as the Leguminosæ, in which there is an acknowledged difficulty in the manipulation) half of these twenty plants had their fertility in some degree impaired.

25   Moreover, as Gärtner during several years repeatedly crossed the

primrose and cowslip, which we have such good reason to believe to be varieties, and only once or twice succeeded in getting fertile seed; as he found the common red and blue pimpernels (Anagallis arvensis and cœrulea), which the best botanists rank as varieties, absolutely sterile together; and as he came to the same conclusion in several other analogous cases; it seems to me that we may well be permitted to doubt whether many other species are really so sterile, when intercrossed, as Gärtner believes.

25:c cases, it

25:d Gärtner repeatedly crossed some forms, such as the common/ rank as varieties, and found them absolutely sterile, we may doubt/intercrossed, as he believed.

25:f many species

26   It is certain, on the one hand, that the sterility of various species when crossed is so different in degree and graduates away so insensibly, and, on the other hand, that the fertility of pure species is so easily affected by various circumstances, that for all practical purposes it is most difficult to say where perfect fertility ends and sterility begins.

27 I think no better evidence of this can be required than that the two most experienced observers who have ever lived, namely, Kölreuter and Gärtner, should have arrived at diametrically opposite conclusions in regard to the very same species.

27:f Gärtner, arrived/to some of the very same forms.

28 It is also most instructive to compare—but I have not space here to enter on details—the evidence advanced by our best botanists on the question whether certain doubtful forms should be ranked as species or varieties, with the evidence from fertility adduced by different hybridisers, or by the same author, from experiments made during different years.

28:f same observer from

29 It can thus be shown that neither sterility nor fertility affords any clear distinction between species and varieties; but that the evidence from this source graduates away, and is doubtful in the same degree as is the evidence derived from other constitutional and structural differences.

29.x-y:f varieties. y The evidence

30   In regard to the sterility of hybrids in successive generations; though Gärtner was enabled to rear some hybrids, carefully guarding them from a cross with either pure parent, for six or seven, and in one case for ten generations, yet he asserts positively that their fertility never increased, but generally greatly decreased.

30:d generally decreased greatly and suddenly.

30:f increases/decreases

427

31 I do not doubt that this is usually the case, and that the fertility often suddenly decreases in the first few generations.

31:d With respect to this decrease, it may first be noticed that, when any deviation in structure or constitution is common to both parents, this is often transmitted in an augmented degree to the offspring; and both sexual elements in hybrid plants are already affected in some degree.

31:f that when

32 Nevertheless I believe that in all these experiments the fertility has been diminished by an independent cause, namely, from close interbreeding.

32:d But I believe in nearly all these cases, that the fertility has been diminished by an independent cause, namely, by too close

32:f believe that their fertility has been diminished in nearly all these cases by an independent

33 I have collected so large a body of facts, showing that close interbreeding lessens fertility, and, on the other hand, that an occasional cross with a distinct individual or variety increases fertility, that I cannot doubt the correctness of this almost universal belief amongst breeders.

33:d showing on the one hand that an occasional cross with a distinct individual or variety increases the vigour and fertility of the offspring, and on the other hand that very close interbreeding lessens their vigour and fertility, that I must admit the correctness

33:f have made so many experiments and collected so many facts/ I cannot doubt the correctness of this conclusion.

34 Hybrids are seldom raised by experimentalists in great numbers; and as the parent-species, or other allied hybrids, generally grow in the same garden, the visits of insects must be carefully prevented during the flowering season: hence hybrids will generally be fertilised during each generation by their own individual pollen; and I am convinced that this would be injurious to their fertility, already lessened by their hybrid origin.

34:d generally have to be fertilised/pollen; and this would probably be injurious

34:f hence hybrids, if left to themselves, will generally be fertilised during each generation by pollen from the same flower; and this

35 I am strengthened in this conviction by a remarkable statement repeatedly made by Gärtner, namely, that if even the less fertile hybrids be artificially fertilised with hybrid pollen of the same kind, their fertility, notwithstanding the frequent ill effects of manipulation, sometimes decidedly increases, and goes on increasing.

35:d effects from manipulation

36 Now, in artificial fertilisation pollen is as often taken by chance

(as I know from my own experience) from the anthers of another flower, as from the anthers of the flower itself which is to be fertilised; so that a cross between two flowers, though probably on the same plant, would be thus effected.

36:d Now, in the process of artificial fertilisation, pollen/probably often on

37 Moreover, whenever complicated experiments are in progress, so careful an observer as Gärtner would have castrated his hybrids, and this would have insured in each generation a cross with the pollen from a distinct flower, either from the same plant or from another plant of the same hybrid nature.

37:b cross with a pollen

37:d with pollen

37:e ensured

38 And thus, the strange fact of the increase of fertility in the successive generations of *artificially fertilised* hybrids may, I believe, be accounted for by close interbreeding having been avoided.

38:d may, as I believe, be accounted for by too close

38:e of an increase/hybrids, in contrast with those spontaneously self-fertilised, may

39 Now let us turn to the results arrived at by the third most experienced hybridiser, namely, the Hon. and Rev. W. Herbert.

39:e by a third

40 He is as emphatic in his conclusion that some hybrids are perfectly fertile—as fertile as the pure parent-species—as are Kölreuter and Gärtner that some degree of sterility between distinct species is a universal law of nature.

41 He experimentised on some of the very same species as did Gärtner.

41:c experimented

42 The difference in their results may, I think, be in part accounted for by Herbert's great horticultural skill, and by his having hothouses at his command.

42:e hot-houses

43 Of his many important statements I will here give only a single one as an example, namely, that "every ovule in a pod of Crinum capense fertilised by C. revolutum produced a plant, which (he says) I never saw to occur in a case of its natural fecundation."

43:e which I

44 So that we here have perfect, or even more than commonly perfect, fertility in a first cross between two distinct species.

44:d perfect or

44:f commonly perfect fertility, in

45   This case of the Crinum leads me to refer to a most singular fact, namely, that there are individual plants, as with certain species of Lobelia, and with all the species of the genus Hippeastrum, which can be far more easily fertilised by the pollen of another and distinct species, than by their own pollen.

45:b plants of certain species of Lobelia and of some other genera, which/own pollen; and all the individuals of nearly all the species of Hippeastrum seem to be in this predicament.

45.x-y:e refer to a singular fact, namely, that individual plants of certain species of Lobelia, Verbascum and Passiflora, can easily be fertilised by pollen from a distinct species, but not by pollen from the same plant, though this pollen can be proved to be perfectly sound by fertilising other plants or species. *y* In the genus Hippeastrum, in Corydalis as shown by Professor Hildebrand, in various orchids as shown by Mr. Scott and Fritz Müller, all the individuals are in this peculiar condition.

46 For these plants have been found to yield seed to the pollen of a distinct species, though quite sterile with their own pollen, notwithstanding that their own pollen was found to be perfectly good, for it fertilised distinct species.

46:[e]

47 So that certain individual plants and all the individuals of certain species can actually be hybridised much more readily than they can be self-fertilised!

47:e So that with some species, certain abnormal individuals, and in other species all the individuals, can actually be hybridised much more readily than they can be fertilised by pollen from the same individual plant!

48 For instance, a bulb of Hippeastrum aulicum produced four flowers; three were fertilised by Herbert with their own pollen, and the fourth was subsequently fertilised by the pollen of a compound hybrid descended from three other and distinct species: the result was that "the ovaries of the three first flowers soon ceased to grow, and after a few days perished entirely, whereas the pod impregnated by the pollen of the hybrid made vigorous growth and rapid progress to maturity, and bore good seed, which vegetated freely."

48:e To give one instance/from three distinct

49 In a letter to me, in 1839, Mr. Herbert told me that he had then tried the experiment during five years, and he continued to try it during several subsequent years, and always with the same result.

49:c letter written in

49:e [freely."] Mr. Herbert tried similar experiments during many years

50 This result has, also, been confirmed by other observers in the

430

case of Hippeastrum with its sub-genera, and in the case of some other genera, as Lobelia, Passiflora and Verbascum.

50:d Lobelia, Verbascum, and especially Passiflora.

50:[e]

51 Although the plants in these experiments appeared perfectly healthy, and although both the ovules and pollen of the same flower were perfectly good with respect to other species, yet as they were functionally imperfect in their mutual self-action, we must infer that the plants were in an unnatural state.

51:e With those plants in which certain individuals alone fail to be fertilised by their own pollen, though they appear quite healthy and although both ovules and pollen are perfectly good with reference to other species, yet they must be in some way in an unnatural condition.

51:[f]

52 Nevertheless these facts show on what slight and mysterious causes the lesser or greater fertility of species when crossed, in comparison with the same species when self-fertilised, sometimes depends.

52:e These cases serve to show/species sometimes

53 The practical experiments of horticulturists, though not made with scientific precision, deserve some notice.

54 It is notorious in how complicated a manner the species of Pelargonium, Fuchsia, Calceolaria, Petunia, Rhododendron, &c., have been crossed, yet many of these hybrids seed freely.

55 For instance, Herbert asserts that a hybrid from Calceolaria integrifolia and plantaginea, species most widely dissimilar in general habit, "reproduced itself as perfectly as if it had been a natural species from the mountains of Chile."

55:f reproduces/Chili."

56 I have taken some pains to ascertain the degree of fertility of some of the complex crosses of Rhododendrons, and I am assured that many of them are perfectly fertile.

57 Mr. C. Noble, for instance, informs me that he raises stocks for grafting from a hybrid between Rhod. Ponticum and Catawbiense, and that this hybrid "seeds as freely as it is possible to imagine."

58 Had hybrids, when fairly treated, gone on decreasing in fertility in each successive generation, as Gärtner believes to be the case, the fact would have been notorious to nurserymen.

58:d treated, always gone/believed

59 Horticulturists raise large beds of the same hybrids, and such alone are fairly treated, for by insect agency the several individuals of the same hybrid variety are allowed to freely cross with each other, and the injurious influence of close interbreeding is thus prevented.

59:f same hybrid, and such/individuals are allowed to cross freely with

60 Any one may readily convince himself of the efficiency of insect-agency by examining the flowers of the more sterile kinds of hybrid rhododendrons, which produce no pollen, for he will find on their stigmas plenty of pollen brought from other flowers.

60:c Rhododendrons

61 In regard to animals, much fewer experiments have been carefully tried than with plants.

62 If our systematic arrangements can be trusted, that is if the genera of animals are as distinct from each other, as are the genera of plants, then we may infer that animals more widely separated in the scale of nature can be more easily crossed than in the case of plants; but the hybrids themselves are, I think, more sterile.

62:c is, if/other as

62:f widely distinct in the scale of nature can be crossed more easily than

63 I doubt whether any case of a perfectly fertile hybrid animal can be considered as thoroughly well authenticated.

63:[f]

64 It should, however, be borne in mind that, owing to few animals breeding freely under confinement, few experiments have been fairly tried: for instance, the canary-bird has been crossed with nine other finches, but as not one of these nine species breeds freely in confinement, we have no right to expect that the first crosses between them and the canary, or that their hybrids, should be perfectly fertile.

64:d but, as

64:f nine distinct species of finches, but, as not one of these breeds

65 Again, with respect to the fertility in successive generations of the more fertile hybrid animals, I hardly know of an instance in which two families of the same hybrid have been raised at the same time from different parents, so as to avoid the ill effects of close interbreeding.

66 On the contrary, brothers and sisters have usually been crossed in each successive generation, in opposition to the constantly repeated admonition of every breeder.

67 And in this case, it is not at all surprising that the inherent sterility in the hybrids should have gone on increasing.

68 If we were to act thus, and pair brothers and sisters in the case of any pure animal, which from any cause had the least tendency to sterility, the breed would assuredly be lost in a very few generations.

68:c lost in a few

69 Although I do not know of any thoroughly well-authenticated cases of perfectly fertile hybrid animals, I have some reason to believe that the hybrids from Cervulus vaginalis and Reevesii, and from Phasianus colchicus with P. torquatus and with P. versicolor are perfectly fertile.

69:c have reason

69:d torquatus, are

69:f Although I know of hardly any

69.1:b There is no doubt that these three pheasants, namely, the common, the true ring-necked, and the Japan, intercross, and are becoming blended together in the woods of several parts of England.

69.1:[d]

69.1.0.1:f M. Quatrefages states that the hybrids from two moths (Bombyx cynthia and arrindia) were proved in Paris to be fertile *inter se* for eight generations.

69.1.1:d From the experiments lately made on a large scale in France, it seems that two such distinct species as the hare and rabbit, when they can be got to breed together, produce offspring almost perfectly fertile.

69.1.1:e It has lately been asserted that/produce almost perfectly fertile offspring; but this statement is as yet very doubtful.

69.1.1:f produce offspring, which are highly fertile when crossed with one of the parent-species.

70 The hybrids from the common and Chinese geese (A. cygnoides), species which are so different that they are generally ranked in distinct genera, have often bred in this country with either pure parent, and in one single instance they have bred *inter se.*

71 This was effected by Mr. Eyton, who raised two hybrids from the same parents but from different hatches; and from these two birds he raised no less than eight hybrids (grandchildren of the pure geese) from one nest.

71:d parents, but

72 In India, however, these cross-bred geese must be far more fertile; for I am assured by two eminently capable judges, namely Mr. Blyth and Capt. Hutton, that whole flocks of these crossed geese are kept in various parts of the country; and as they are kept for profit, where neither pure parent-species exists, they must certainly be highly fertile.

72:d highly or perfectly fertile.

73 A doctrine which originated with Pallas, has been largely accepted by modern naturalists; namely, that most of our domestic animals have descended from two or more aboriginal species, since commingled by intercrossing.

73:b more wild species

73:e The various races of each kind of domesticated animal are quite fertile when crossed together; yet in many cases they are descended from two or more wild species.

73:f With our domesticated animals, the various races when crossed together are quite fertile; yet

74 On this view, the aboriginal species must either at first have produced quite fertile hybrids, or the hybrids must have become in subsequent generations quite fertile under domestication.

74:e From this fact we must conclude either that the aboriginal parent-species produced at first perfectly fertile hybrids, or that the hybrids subsequently reared under domestication became quite fertile.

74:f parent-species at first produced perfectly

75 This latter alternative seems to me the most probable, and I am inclined to believe in its truth, although it rests on no direct evidence.

75:c seems the most

75:d probable, and I have hardly any doubt of its truth

75:e alternative, which was first propounded by Pallas, seems the most probable, and can, indeed, hardly be doubted.

75:f seems by far the most

76 I believe, for instance, that our dogs have descended from several wild stocks; yet, with perhaps the exception of certain indigenous domestic dogs of South America, all are quite **fertile** together; and analogy makes me greatly doubt, whether the several aboriginal species would at first have freely bred together and have produced quite fertile hybrids.

76:d It is, for instance, almost certain that

76:e dogs are descended

76:f together; but analogy

77 So again there is reason to believe that our European and the humped Indian cattle are quite fertile together; but from facts communicated to me by Mr. Blyth, I think they must be considered as distinct species.

77:d again I have lately acquired decisive evidence that the crossed offspring from the Indian humped and common cattle are *inter se* perfectly fertile; and from the observations by Rütimeyer on their important osteological differences, as well as from those by Mr. Blyth on their differences in habits, voice, constitution, &c., these two forms must be regarded as good and distinct species—as good as any in the world.

77:e species as any in the world.

77:f species.

77.1:f The same remarks may be extended to the two chief races of the pig.

78 On this view of the origin of many of our domestic animals, we must either give up the belief of the almost universal sterility of distinct species of animals when crossed; or we must look at sterility, not as an indelible characteristic, but as one capable of being removed by domestication.

78:e According to this

78:f We must, therefore, either give up the belief of the universal sterility of species when crossed; or we must look at this sterility in animals, not

79 Finally, looking to all the ascertained facts on the intercrossing of plants and animals, it may be concluded that some degree of sterility, both in first crosses and in hybrids, is an extremely general result; but that it cannot, under our present state of knowledge, be considered as absolutely universal.

79:d Finally, considering all

80 *Laws governing the Sterility of first Crosses and of Hybrids.—*

80:d [Center] Laws/Hybrids. [Space]

81 We will now consider a little more in detail the circumstances and rules governing the sterility of first crosses and of hybrids.

81:d[¶]

81:f detail the laws governing

82 Our chief object will be to see whether or not the rules indicate that species have specially been endowed with this quality, in order to prevent their crossing and blending together in utter confusion.

82:f not these laws indicate

83 The following rules and conclusions are chiefly drawn up from Gärtner's admirable work on the hybridisation of plants.

83:f following conclusions are drawn up chiefly from

84 I have taken much pains to ascertain how far the rules apply to animals, and considering how scanty our knowledge is in regard to hybrid animals, I have been surprised to find how generally the same rules apply to both kingdoms.

84:d and, considering

84:f far they apply

85 It has been already remarked, that the degree of fertility, both of first crosses and of hybrids, graduates from zero to perfect fertility.

86 It is surprising in how many curious ways this gradation can be shown to exist; but only the barest outline of the facts can here be given.

86:e shown; but

87 When pollen from a plant of one family is placed on the stigma

435

of a plant of a distinct family, it exerts no more influence than so much inorganic dust.

88 From this absolute zero of fertility, the pollen of different species of the same genus applied to the stigma of some one species, yields a perfect gradation in the number of seeds produced, up to nearly complete or even quite complete fertility; and, as we have seen, in certain abnormal cases, even to an excess of fertility, beyond that which the plant's own pollen will produce.

88:e own pollen produces.

88:f species applied to the stigma of some one species of the same genus, yields

89 So in hybrids themselves, there are some which never have produced, and probably never would produce, even with the pollen of either pure parent, a single fertile seed: but in some of these cases a first trace of fertility may be detected, by the pollen of one of the pure parent-species causing the flower of the hybrid to wither earlier than it otherwise would have done; and the early withering of the flower is well known to be a sign of incipient fertilisation.

89:e pollen of the pure parents, a single

90 From this extreme degree of sterility we have self-fertilised hybrids producing a greater and greater number of seeds up to perfect fertility.

91 Hybrids from two species which are very difficult to cross, and which rarely produce any offspring, are generally very sterile; but the parallelism between the difficulty of making a first cross, and the sterility of the hybrids thus produced—two classes of facts which are generally confounded together—is by no means strict.

91:f The hybrids raised from

92 There are many cases, in which two pure species can be united with unusual facility, and produce numerous hybrid-offspring, yet these hybrids are remarkably sterile.

92:e species, as in the genus Verbascum, can

93 On the other hand, there are species which can be crossed very rarely, or with extreme difficulty, but the hybrids, when at last produced, are very fertile.

94 Even within the limits of the same genus, for instance in Dianthus, these two opposite cases occur.

95 The fertility, both of first crosses and of hybrids, is more easily affected by unfavourable conditions, than is the fertility of pure species.

95:f than is that of

96 But the degree of fertility is likewise innately variable; for it is not always the same when the same two species are crossed under the same circumstances, but depends in part upon the

constitution of the individuals which happen to have been chosen for the experiment.

96:f But the fertility of first crosses is likewise innately variable; for it is not always the same in degree when the same two species are crossed under the same circumstances; it depends

97 So it is with hybrids, for their degree of fertility is often found to differ greatly in the several individuals raised from seed out of the same capsule and exposed to exactly the same conditions.

97:c exposed to the same

98 By the term systematic affinity is meant, the resemblance between species in structure and in constitution, more especially in the structure of parts which are of high physiological importance and which differ little in the allied species.

98:f meant, the general resemblance between species in structure and constitution.

99 Now the fertility of first crosses between species, and of the hybrids produced from them, is largely governed by their systematic affinity.

99:f crosses, and

100 This is clearly shown by hybrids never having been raised between species ranked by systematists in distinct families; and on the other hand, by very closely allied species generally uniting with facility.

101 But the correspondence between systematic affinity and the facility of crossing is by no means strict.

102 A multitude of cases could be given of very closely allied species which will not unite, or only with extreme difficulty; and on the other hand of very distinct species which unite with the utmost facility.

103 In the same family there may be a genus, as Dianthus, in which very many species can most readily be crossed; and another genus, as Silene, in which the most persevering efforts have failed to produce between extremely close species a single hybrid.

104 Even within the limits of the same genus, we meet with this same difference; for instance, the many species of Nicotiana have been more largely crossed than the species of almost any other genus; but Gärtner found that N. acuminata, which is not a particularly distinct species, obstinately failed to fertilise, or to be fertilised by, no less than eight other species of Nicotiana.

104:f by no

105 Very many analogous facts could be given.

105:e Many

106 No one has been able to point out what kind, or what amount, of difference in any recognisable character is sufficient to prevent two species crossing.

106:f kind or what amount of difference, in any recognisable character, is

107 It can be shown that plants most widely different in habit and general appearance, and having strongly marked differences in every part of the flower, even in the pollen, in the fruit, and in the cotyledons, can be crossed.

108 Annual and perennial plants, deciduous and evergreen trees, plants inhabiting different stations and fitted for extremely different climates, can often be crossed with ease.

109 By a reciprocal cross between two species, I mean the case, for instance, of a stallion-horse being first crossed with a female-ass, and then a male-ass with a mare: these two species may then be said to have been reciprocally crossed.

109:f instance, of a female-ass being first crossed by a stallion, and then a mare by a male-ass: these

110 There is often the widest possible difference in the facility of making reciprocal crosses.

111 Such cases are highly important, for they prove that the capacity in any two species to cross is often completely independent of their systematic affinity, or of any recognisable difference in their whole organisation.

111:d any difference in their whole organisation, except in their reproductive systems.

111:f affinity, that is of any difference in their structure or constitution, excepting in

112 On the other hand, these cases clearly show that the capacity for crossing is connected with constitutional differences imperceptible by us, and confined to the reproductive system.

112:[d]

113 This difference in the result of reciprocal crosses between the same two species was long ago observed by Kölreuter.

113:d The diversity of result in reciprocal

113:f of the result

114 To give an instance: Mirabilis jalappa can easily be fertilised by the pollen of M. longiflora, and the hybrids thus produced are sufficiently fertile; but Kölreuter tried more than two hundred times, during eight following years, to fertilise reciprocally M. longiflora with the pollen of M. jalappa, and utterly failed.

114:b jalapa/jalapa

115 Several other equally striking cases could be given.

116 Thuret has observed the same fact with certain sea-weeds or Fuci.

117 Gärtner, moreover, found that this difference of facility in

making reciprocal crosses is extremely common in a lesser degree.

118 He has observed it even between forms so closely related (as Matthiola annua and glabra) that many botanists rank them only as varieties.

118:e between closely related forms (as Matthiola annua and glabra) which many botanists rank only

119 It is also a remarkable fact, that hybrids raised from reciprocal crosses, though of course compounded of the very same two species, the one species having first been used as the father and then as the mother, generally differ in fertility in a small, and occasionally in a high degree.

119:d mother, though they rarely differ in external characters, yet generally

120 Several other singular rules could be given from Gärtner: for instance, some species have a remarkable power of crossing with other species; other species of the same genus have a remarkable power of impressing their likeness on their hybrid offspring; but these two powers do not at all necessarily go together.

121 There are certain hybrids which instead of having, as is usual, an intermediate character between their two parents, always closely resemble one of them; and such hybrids, though externally so like one of their pure parent-species, are with rare exceptions extremely sterile.

121:d which, instead

122 So again amongst hybrids which are usually intermediate in structure between their parents, exceptional and abnormal individuals sometimes are born, which closely resemble one of their pure parents; and these hybrids are almost always utterly sterile, even when the other hybrids raised from seed from the same capsule have a considerable degree of fertility.

123 These facts show how completely fertility in the hybrid is independent of its external resemblance to either pure parent.

123:f completely the fertility of a hybrid may be independent

124 Considering the several rules now given, which govern the fertility of first crosses and of hybrids, we see that when forms, which must be considered as good and distinct species, are united, their fertility graduates from zero to perfect fertility, or even to fertility under certain conditions in excess.

125 That their fertility, besides being eminently susceptible to favourable and unfavourable conditions, is innately variable.

126 That it is by no means always the same in degree in the first cross and in the hybrids produced from this cross.

127 That the fertility of hybrids is not related to the degree in which they resemble in external appearance either parent.

128 And lastly, that the facility of making a first cross between any two species is not always governed by their systematic affinity or degree of resemblance to each other.

124 + *5* + *6* + *7* + *8:f* excess; that their/variable; that it/this cross; that the fertility/parent; and

129 This latter statement is clearly proved by reciprocal crosses between the same two species, for according as the one species or the other is used as the father or the mother, there is generally some difference, and occasionally the widest possible difference, in the facility of effecting an union.

129:c by the difference in the result of reciprocal

129:d for, according

130 The hybrids, moreover, produced from reciprocal crosses often differ in fertility.

131 Now do these complex and singular rules indicate that species have been endowed with sterility simply to prevent their becoming confounded in nature?

132 I think not.

133 For why should the sterility be so extremely different in degree, when various species are crossed, all of which we must suppose it would be equally important to keep from blending together?

134 Why should the degree of sterility be innately variable in the individuals of the same species?

135 Why should some species cross with facility, and yet produce very sterile hybrids; and other species cross with extreme difficulty, and yet produce fairly fertile hybrids?

136 Why should there often be so great a difference in the result of a reciprocal cross between the same two species?

137 Why, it may even be asked, has the production of hybrids been permitted? to grant to species the special power of producing hybrids, and then to stop their further propagation by different degrees of sterility, not strictly related to the facility of the first union between their parents, seems to be a strange arrangement.

137.x-y:c permitted? *y* To grant

137.y:e seems a strange

138 The foregoing rules and facts, on the other hand, appear to me clearly to indicate that the sterility both of first crosses and of hybrids is simply incidental or dependent on unknown differences, chiefly in the reproductive systems, of the species which are crossed.

139 The differences being of so peculiar and limited a nature, that, in reciprocal crosses between two species the male sexual element of the one will often freely act on the female sexual element of the other, but not in a reversed direction.

139:c species, the male

138 + 9:d differences in their reproductive systems; the differences

138 + 9:e between the same two

140 It will be advisable to explain a little more fully by an example what I mean by sterility being incidental on other differences, and not a specially endowed quality.

141 As the capacity of one plant to be grafted or budded on another is so entirely unimportant for its welfare in a state of nature, I presume that no one will suppose that this capacity is a *specially* endowed quality, but will admit that it is incidental on differences in the laws of growth of the two plants.

141:c so unimportant

141:f is unimportant for their welfare

142 We can sometimes see the reason why one tree will not take on another, from differences in their rate of growth, in the hardiness of their wood, in the period of the flow or nature of their sap, &c.; but in a multitude of cases we can assign no reason whatever.

143 Great diversity in the size of two plants, one being woody and the other herbaceous, one being evergreen and the other decid-uous, and adaptation to widely different climates, does not always prevent the two grafting together.

143:c climates, do not

144 As in hybridisation, so with grafting, the capacity is limited by systematic affinity, for no one has been able to graft trees together belonging to quite distinct families; and, on the other hand, closely allied species, and varieties of the same species, can usually, but not invariably, be grafted with ease.

144:f to graft together trees belonging

145 But this capacity, as in hybridisation, is by no means absolutely governed by systematic affinity.

146 Although many distinct genera within the same family have been grafted together, in other cases species of the same genus will not take on each other.

147 The pear can be grafted far more readily on the quince, which is ranked as a distinct genus, than on the apple, which is a member of the same genus.

148 Even different varieties of the pear take with different degrees of facility on the quince; so do different varieties of the apricot and peach on certain varieties of the plum.

149 As Gärtner found that there was sometimes an innate differ-ence in different *individuals* of the same two species in crossing; so Sageret believes this to be the case with different individuals of the same two species in being grafted together.

150 As in reciprocal crosses, the facility of effecting an union is

often very far from equal, so it sometimes is in grafting; the common gooseberry, for instance, cannot be grafted on the currant, whereas the currant will take, though with difficulty, on the gooseberry.

151   We have seen that the sterility of hybrids, which have their reproductive organs in an imperfect condition, is a very different case from the difficulty of uniting two pure species, which have their reproductive organs perfect; yet these two distinct cases run to a certain extent parallel.

151:d is a different

151:f distinct classes of cases run to a large extent

152 Something analogous occurs in grafting; for Thouin found that three species of Robinia, which seeded freely on their own roots, and which could be grafted with no great difficulty on another species, when thus grafted were rendered barren.

152:f difficulty on a fourth species

153 On the other hand, certain species of Sorbus, when grafted on other species, yielded twice as much fruit as when on their own roots.

153:f other species yielded

154 We are reminded by this latter fact of the extraordinary case of Hippeastrum, Lobelia, &c., which seeded much more freely when fertilised with the pollen of distinct species, than when self-fertilised with their own pollen.

154:c than when fertilised

154:d Hippeastrum, Passiflora, &c., which seed

154:e pollen of a distinct species, than when fertilised with pollen from the same plant.

155   We thus see, that although there is a clear and fundamental difference between the mere adhesion of grafted stocks, and the union of the male and female elements in the act of reproduction, yet that there is a rude degree of parallelism in the results of grafting and of crossing distinct species.

155:d that, although

155:e and great difference

156 And as we must look at the curious and complex laws governing the facility with which trees can be grafted on each other as incidental on unknown differences in their vegetative systems, so I believe that the still more complex laws governing the facility of first crosses, are incidental on unknown differences, chiefly in their reproductive systems.

156:c crosses are

156:d are incidental on unknown differences in

157 These differences, in both cases, follow to a certain extent, as might have been expected, systematic affinity, by which every

kind of resemblance and dissimilarity between organic beings is attempted to be expressed.

157:f affinity, by which term every

158 The facts by no means seem to me to indicate that the greater or lesser difficulty of either grafting or crossing together various species has been a special endowment; although in the case of crossing, the difficulty is as important for the endurance and stability of specific forms, as in the case of grafting it is unimportant for their welfare.

158:e crossing various

158:f seem to indicate

159 *Causes of the Sterility of first Crosses and of Hybrids.—*

159:d [Center] Origin and Causes/Hybrids. [Space]

159.1:d[¶] At one time it appeared to me probable, as it has to others, that this sterility might have been acquired through natural selection slowly acting on a slightly lessened degree of fertility, which at first spontaneously appeared, like any other variation, in certain individuals of one variety when crossed with another variety.

159.1:e that the sterility of first crosses and of hybrids might have been slowly acquired through the natural selection of slightly lessened degrees of fertility, which spontaneously

159.1:f which, like any other variation, spontaneously appeared in/ with those of another

159.2-3:d For it would clearly be advantageous to two varieties or incipient species, if they could be kept from blending, on the same principle that, when a man is selecting at the same time two varieties, it is necessary that he should keep them separate. *3* In the first place, it may be remarked that distinct regions are often inhabited by groups of species and by single species which when brought together and crossed are found to be more or less sterile; now it could clearly have been of no advantage to such separated species to have been rendered mutually sterile, and consequently this could not have been effected through natural selection; but it may perhaps be argued with truth, that, if a species were rendered sterile with some one compatriot, sterility with other species would probably follow as a necessary contingency.

159.3:e argued, that/would follow

159.3:f that species inhabiting distinct regions are often sterile when crossed; now/if a species was rendered

159.4:d In the second place, it is as much opposed to the theory of natural selection as to that of special creation, that in reciprocal crosses the male element of one form has sometimes been rendered utterly impotent on a second form, whilst at the same time the male element of this second form is enabled freely to fertilise the first form.

443

159.4:e form should be rendered/first form; for this peculiar state of the reproductive system could not possibly be advantageous to either species.

159.4:f is almost as much/should have been rendered/could hardly have been advantageous

159.5:d[¶] But in considering the probability of natural selection having come into action, one great difficulty will be found to lie in the existence of many graduated steps from very slightly lessened fertility to utter and absolute sterility.

159.5:e In considering/action, in rendering species mutually sterile, one/ from slightly lessened fertility to absolute

159.5:f sterile, the greatest difficulty

159.6:d It may be admitted, on the principle above explained, that it would profit an incipient species if it were rendered in some slight degree sterile when crossed with its parent-form or with some other variety; for thus fewer bastardised and deteriorated offspring would be produced to commingle their blood with the newly-forming variety.

159.6:e blood with the new species in process of formation.

159.6:f admitted that it would profit an incipient species, if

159.7-8:d But he who will take the trouble to reflect on the steps by which this first degree of sterility could be increased through natural selection to that high degree which is common with so many species, and which is universal with species which have been differentiated to a generic or family rank, will find the subject extraordinarily complex. *8* After mature reflection it seems to me that this could not have been effected through natural selection; for it could not have been of any direct advantage to an individual animal to breed poorly with another individual of a different variety, and thus to leave few offspring; consequently such individuals could not have been preserved or selected.

159.8:e for it could have been of no direct/breed badly with

159.8:f selection.

159.8.1:e Or take the case of two species which in their present state when crossed, produce few and sterile offspring; now, what is there which could favour the survival of those individuals which happened to be endowed in a slightly higher degree with mutual infertility, and which thus approached by one small step towards absolute sterility?

159.8.1:f Take the case of any two species which, when

159.8.2:e Yet an advance of this kind, if the theory of natural selection be brought to bear, must have incessantly occurred with many species, for a multitude are mutually quite barren.

159.9:d With sterile neuter insects we have reason to believe that modifications in their structure have been slowly accumulated by natural selection, from an advantage having been thus indirectly given to the community to which they belonged over

444

other communities of the same species; but an individual animal, if rendered slightly sterile when crossed with some other variety, would not thus indirectly give any advantage to its nearest relatives or to any other individuals of the same variety, thus leading to their preservation.

159.9:e structure and fertility have been slowly/animal not belonging to a social community, if/not thus itself gain any advantage or indirectly give any advantage to the other

159.10:d From these considerations I infer, as far as animals are concerned, that the various degrees of lessened fertility which occur with species when crossed cannot have been slowly accumulated by means of natural selection.

159.10:[f]

159.11:d[¶] With plants, it is possible that the case may be different.

159.11:e be somewhat different.

159.11:[f]

159.12:d With very many kinds, insects constantly bring pollen from neighbouring plants of the same or of other varieties to the stigma of each flower; and with some this is effected by the wind.

159.12:e With many kinds, insects constantly carry pollen from neighbouring plants to the stigmas of

159.12:[f]

159.13:d Now, if the pollen of any one variety should become by spontaneous variation in ever so slight a degree prepotent over the pollen of other varieties, so that, when deposited by any means on the stigmas of the flowers of its own variety, it obliterated the effects of previously placed pollen of other varieties, this would certainly be an advantage to the variety; for it would thus escape being bastardised and deteriorated in character.

159.13:e of a variety, when deposited on the stigma of the same variety, should/other varieties, this would certainly be an advantage to the variety; for its own pollen would thus obliterate the effects of the pollen of other varieties, and prevent deterioration of character.

159.13:[f]

159.14:d And the more prepotent the pollen could be rendered through natural selection the greater the advantage would be.

159.14:e And the more prepotent the variety's own pollen/selection, the greater

159.14:[f]

159.15:d We know from the researches of Gärtner that prepotency of this kind always accompanies the sterility which follows from crossing distinct species; but we do not know whether prepotency is a consequence of sterility, or sterility a consequence of prepotency.

445

159.15:e that, with species which are mutually sterile, the pollen of each is always prepotent on its own stigma over that of the other species: but we do not know whether this prepotency is a consequence of the mutual sterility, or the sterility a consequence of the prepotency.

159.15:[f]

159.16:d If the latter view be correct, we may infer that, as the prepotency became stronger through natural selection, from being advantageous to a species in process of formation, so the sterility consequent on prepotency would at the same time be augmented; and the final result would be various degrees of sterility, such as actually occur with our existing species when crossed.

159.16:e correct, as the prepotency became/such as occurs with existing species.

159.16:[f]

159.17:d This same view might be extended to animals if the female before each birth received several males, so that the sexual element of the prepotent male of her own variety obliterated all effects from the access of previous males of other varieties; but we have no reason to believe, at least with terrestrial animals, that this is the case; as most males and females pair for each birth, and some few for life.

159.17:e This view/obliterated the effects of the access of previous males belonging to other varieties, but

159.17:[f]

159.18:d[¶] On the whole we may conclude that with animals the sterility of crossed species has not been slowly augmented through natural selection; and as this sterility follows the same general laws in the vegetable as in the animal kingdom, it is improbable, though apparently possible, that crossed plants should have been rendered sterile by a different process from animals.

159.18:e augmented, through/possible, that with plants crossed species should have been rendered sterile by a different process.

159.18:[f]

159.19:d From this consideration, and remembering that species which have never co-existed in the same country, and which therefore could not have profited by having been rendered mutually infertile, yet are sterile when crossed; and bearing in mind that in reciprocal crosses between the same two species there is sometimes the widest difference in the resulting degrees of sterility, we must give up the belief that natural selection has come into play; and we are driven to our former proposition, that the sterility of first crosses, and indirectly of hybrids, is simply incidental on unknown differences in the reproductive systems of the parent-species.

159.19.x-y:e not have received any advantage from having been

446

rendered mutually infertile, yet are generally sterile/play. *y* We are thus driven to our former proposition, namely, that the sterility

159.19.x-y:[f]

159.19.0.1-3:f[¶] But it would be superfluous to discuss this question in detail; for with plants we have conclusive evidence that the sterility of crossed species must be due to some principle, quite independent of natural selection. *2* Both Gärtner and Kölreuter have proved that in genera including numerous species, a series can be formed from species which when crossed yield fewer and fewer seeds, to species which never produce a single seed, but yet are affected by the pollen of certain other species, for the germen swells. *3* It is here manifestly impossible to select the more sterile individuals, which have already ceased to yield seeds; so that this acme of sterility, when the germen alone is affected, cannot have been gained through selection; and from the laws governing the various grades of sterility being so uniform throughout the animal and vegetable kingdoms, we may infer that the cause, whatever it may be, is the same or nearly the same in all cases. [*Space*]

160 We may now look a little closer at the probable causes of the sterility of first crosses and of hybrids.

160:d[¶] now try and look a little closer at the probable nature of these differences, which induce sterility in first crosses, as well as in hybrids.

160:e crosses and in hybrids.

160:f We will now look a little closer at the probable nature of the differences between species which

161 These two cases are fundamentally different, for, as just remarked, in the union of two pure species the male and female sexual elements are perfect, whereas in hybrids they are imperfect.

161:d Pure species and hybrids differ, as already remarked, in the state of their reproductive organs; but from what will presently follow on reciprocally dimorphic and trimorphic plants, it would appear as if some unknown bond or law existed, which causes the young from a union not fully fertile to be themselves more or less infertile.

161:[f]

162 Even in first crosses, the greater or lesser difficulty in effecting a union apparently depends on several distinct causes.

162:c effecting an union

162:d[¶] In the case of first crosses between pure species, the greater or less difficulty in effecting an union and in obtaining offspring apparently

162:f[No¶] In the case of first crosses, the greater

163 There must sometimes be a physical impossibility in the male

element reaching the ovule, as would be the case with a plant having a pistil too long for the pollen-tubes to reach the ovarium.

164 It has also been observed that when pollen of one species is placed on the stigma of a distantly allied species, though the pollen-tubes protrude, they do not penetrate the stigmatic surface.

164:f when the pollen of one

165 Again, the male element may reach the female element, but be incapable of causing an embryo to be developed, as seems to have been the case with some of Thuret's experiments on Fuci.

165:f element but

166 No explanation can be given of these facts, any more than why certain trees cannot be grafted on others.

167 Lastly, an embryo may be developed, and then perish at an early period.

168 This latter alternative has not been sufficiently attended to; but I believe, from observations communicated to me by Mr. Hewitt, who has had great experience in hybridising galli-naceous birds, that the early death of the embryo is a very frequent cause of sterility in first crosses.

168:d hybridising pheasants and fowls, that

168.1:d Mr. Salter has recently given the results of an examination of about 500 eggs produced from various crosses between three species of Gallus and their hybrids; the majority of these eggs had been fertilised; and in the majority of the fertilised eggs, the embryos had either been only partially developed and had then aborted, or had become nearly mature, but the young chickens had been unable to break through the shell.

168.1:e embryos either had been partially

168:f then perished, or

168.2-3:d Of the chickens which were born, more than four-fifths died within the first few days, or at latest weeks, "without any obvious cause, apparently from mere inability to live;" so that from the 500 eggs only twelve chickens were reared. *3* The early death of hybrid embryos probably occurs in like manner with plants; at least it is known that hybrids raised from very distinct species are sometimes weak and dwarfed, and perish at an early age; of which fact Max Wichura has recently given some striking cases with hybrid willows.

168.3:f With plants, hybridised embryos probably often perish in a like manner; at least

168.4:d It may be here worth noticing that in some cases of parthenogenesis, embryos produced from eggs which had not been fertilised, like those produced by the crossing of two distinct species, passed through their early stages of development and then perished; this has been observed by M. Jourdan with the unimpregnated eggs of the silk-moth.

168.4:e from the eggs of silk-moths, which had not been fertilised, passed/perished like the embryos produced by a cross between two distinct species.

168.4:f parthenogenesis, the embryos within the eggs of silk moths which/pass/perish

169 I was at first very unwilling to believe in this view; as hybrids, when once born, are generally healthy and long-lived, as we see in the case of the common mule.

169:d Until becoming acquainted with these facts, I was unwilling to believe in the frequent early death of hybrid embryos; for hybrids

170 Hybrids, however, are differently circumstanced before and after birth: when born and living in a country where their two parents can live, they are generally placed under suitable conditions of life.

170:d parents live

171 But a hybrid partakes of only half of the nature and constitution of its mother, and therefore before birth, as long as it is nourished within its mother's womb or within the egg or seed produced by the mother, it may be exposed to conditions in some degree unsuitable, and consequently be liable to perish at an early period; more especially as all very young beings seem eminently sensitive to injurious or unnatural conditions of life.

171:c womb, or

171:d beings are eminently

171:f mother; it may therefore/by the mother, be exposed

171.1:d But after all, the cause more probably lies in some imperfection in the original act of impregnation, causing the embryo to be imperfectly developed, rather than in the conditions to which it is subsequently exposed.

172 In regard to the sterility of hybrids, in which the sexual elements are imperfectly developed, the case is very different.

172:d is different.

172:f is somewhat different.

173 I have more than once alluded to a large body of facts, which I have collected, showing that when animals and plants are removed from their natural conditions, they are extremely liable to have their reproductive systems seriously affected.

173:d that, when

173:f facts showing

174 This, in fact, is the great bar to the domestication of animals.

175 Between the sterility thus superinduced and that of hybrids, there are many points of similarity.

176 In both cases the sterility is independent of general health, and is often accompanied by excess of size or great luxuriance.

177 In both cases, the sterility occurs in various degrees; in both, the male element is the most liable to be affected; but sometimes the female more than the male.

177:c cases the sterility

178 In both, the tendency goes to a certain extent with systematic affinity, for whole groups of animals and plants are rendered impotent by the same unnatural conditions; and whole groups of species tend to produce sterile hybrids.

179 On the other hand, one species in a group will sometimes resist great changes of conditions with unimpaired fertility; and certain species in a group will produce unusually fertile hybrids.

180 No one can tell, till he tries, whether any particular animal will breed under confinement or any plant seed freely under culture; nor can he tell, till he tries, whether any two species of a genus will produce more or less sterile hybrids.

180:b or any exotic plant

180:d confinement, or

180:f nor can he tell till

181 Lastly, when organic beings are placed during several generations under conditions not natural to them, they are extremely liable to vary, which is due, as I believe, to their reproductive systems having been specially affected, though in a lesser degree than when sterility ensues.

181:e vary, which seems to be partly due to

182 So it is with hybrids, for hybrids in successive generations are eminently liable to vary, as every experimentalist has observed.

182:e hybrids, for their offspring in

183   Thus we see that when organic beings are placed under new and unnatural conditions, and when hybrids are produced by the unnatural crossing of two species, the reproductive system, independently of the general state of health, is affected by sterility in a very similar manner.

183:f affected in

184 In the one case, the conditions of life have been disturbed, though often in so slight a degree as to be inappreciable by us; in the other case, or that of hybrids, the external conditions have remained the same, but the organisation has been disturbed by two different structures and constitutions having been blended into one.

184:f two distinct structures and constitutions, including of course the reproductive systems, having

185 For it is scarcely possible that two organisations should be compounded into one, without some disturbance occurring in the development, or periodical action, or mutual relation of the

different parts and organs one to another, or to the conditions of life.

185:d relations/another or

186 When hybrids are able to breed *inter se*, they transmit to their offspring from generation to generation the same compounded organisation, and hence we need not be surprised that their sterility, though in some degree variable, rarely diminishes.

186:d variable, does not diminish, but is apt to increase; this increase being perhaps intelligible, as before explained, on the principles of inheritance and from too close interbreeding.

186:e diminish; it is even apt to increase, this being generally the result, as before explained, of too

186.1:d The above view of the sterility of hybrids being caused by two different constitutions having been confounded into one has lately been strongly maintained by Max Wichura; but it must be owned that the sterility, so like in every respect to that of hybrids, which affects the illegitimate offspring of dimorphic and trimorphic plants of the same species (as will be immediately described), makes this view rather doubtful.

186.1:e constitutions being confounded/sterility (as will be immediately explained) which affects the offspring of dimorphic and trimorphic plants, when individuals belonging to the same form are united, makes

186.1:f two constitutions being compounded into one has been strongly maintained by Max Wichura.

186.1.1:e It should, however, be borne in mind that the sterility of these plants has been acquired for a special purpose, and may differ in origin from that of hybrids.

186.1.1:[f]

187 It must, however, be confessed that we cannot understand, excepting on vague hypotheses, several facts with respect to the sterility of hybrids; for instance, the unequal fertility of hybrids produced from reciprocal crosses; or the increased sterility in those hybrids which occasionally and exceptionally resemble closely either pure parent.

187:d It must be owned that we cannot understand, on this or any other view, several

187:e must be/on the above or any

187:f It must, however, be

188 Nor do I pretend that the foregoing remarks go to the root of the matter: no explanation is offered why an organism, when placed under unnatural conditions, is rendered sterile.

189 All that I have attempted to show, is that in two cases, in some respects allied, sterility is the common result,—in the one case from the conditions of life having been disturbed, in the other case from the organisation having been disturbed by two organisations having been compounded into one.

189:d organisation or constitution having been disturbed by two organisations being compounded

189:f show is, that

190 It may seem fanciful, but I suspect that a similar parallelism extends to an allied yet very different class of facts.

190:e A similar parallelism apparently extends

190:f parallelism holds good with an allied

191 It is an old and almost universal belief, founded, I think, on a considerable body of evidence, that slight changes in the conditions of life are beneficial to all living things.

191:e founded on

191:f belief founded on a considerable body of evidence, which I have elsewhere given, that

192 We see this acted on by farmers and gardeners in their frequent exchanges of seed, tubers, &c., from one soil or climate to another, and back again.

193 During the convalescence of animals, we plainly see that great benefit is derived from almost any change in the habits of life.

193:e animals, great

193:f in their habits

194 Again, both with plants and animals, there is abundant evidence, that a cross between very distinct individuals of the same species, that is between members of different strains or sub-breeds, gives vigour and fertility to the offspring.

194:c evidence that a cross

195 I believe, indeed, from the facts alluded to in our fourth chapter, that a certain amount of crossing is indispensable even with hermaphrodites; and that close interbreeding continued during several generations between the nearest relations, especially if these be kept under the same conditions of life, always induces weakness and sterility in the progeny.

194 + 5:e between individuals of the same species, which differ to a certain extent, gives vigour and fertility to the offspring; and that close/life, almost always induces weakness and sterility.

194 + 5:f is the clearest evidence/relations, if/always leads to decreased size, weakness, or sterility.

196 Hence it seems that, on the one hand, slight changes in the conditions of life benefit all organic beings, and on the other hand, that slight crosses, that is crosses between the males and females of the same species which have varied and become slightly different, give vigour and fertility to the offspring.

196:f species, which have been subjected to slightly different conditions, or which have slightly varied, give

197 But we have seen that greater changes, or changes of a particular nature, often render organic beings in some degree sterile;

and that greater crosses, that is crosses between males and females which have become widely or specifically different, produce hybrids which are generally sterile in some degree.

197:f But, as we have seen, organic beings long habituated to certain uniform conditions under a state of nature, when subjected, as under confinement, to a considerable change in their conditions, very frequently are rendered more or less sterile; and we know that a cross between two forms, that have become widely or specifically different, produce hybrids which are almost always in some degree sterile.

198 I cannot persuade myself that this parallelism is an accident or an illusion.

198:f I am fully persuaded that this double parallelism is by no means an accident

198.1-2:f He who is able to explain why the elephant and a multitude of other animals are incapable of breeding when kept under only partial confinement in their native country, will be able to explain the primary cause of hybrids being so generally sterile. 2 He will at the same time be able to explain how it is that the races of some of our domesticated animals, which have often been subjected to new and not uniform conditions, are quite fertile together, although they are descended from distinct species, which would probably have been sterile if aboriginally crossed.

199 Both series of facts seem to be connected together by some common but unknown bond, which is essentially related to the principle of life.

199:d life; this principle apparently being that life, as Mr. Herbert Spencer has remarked, depends on, or consists in, the incessant action and reaction of various forces, which, as throughout nature, are always tending towards an equilibrium; and when this tendency is slightly disturbed by any change, the vital forces apparently gain in power. [*Space*]

199:f The above two parallel series/life; this principle, according to Mr. Herbert Spencer, being that life depends/forces gain

199.1:d [*Center*] *Reciprocal Dimorphism and Trimorphism.* [*Space*]

199.2:d[¶] This subject may be here briefly discussed, and will be found to throw considerable light on hybridism.

199.2:e throw some light

199.3:d Several plants belonging to distinct orders present two forms, existing together in about equal numbers, which differ in no respect except in their reproductive organs; one form having a long pistil with short stamens, the other a short pistil with long stamens; both with differently sized pollen-grains.

199.3:e forms, which exist in about equal numbers, and which differ

199.3:f other a short pistil with long stamens; the two having differently

199.4:d With trimorphic plants there are three forms likewise differing in the lengths of their pistils and stamens, in the size and colour of the pollen-grains, and in some other respects; and as in each of the three forms there are two sets of stamens, there are altogether six sets of stamens and three kinds of pistils.

199.4:f sets of stamens, the three forms possess altogether

199.5:d These organs are so proportioned in length to each other, that, in any two of the forms, half the stamens in each stand on a level with the stigma of the third form.

199.5:f that half the stamens in two of the forms stand

199.6:d Now I have shown, and the result has been confirmed by other observers, that, in order to obtain full fertility with these plants, it is necessary that the stigma of the one form should be fertilised by pollen taken from the stamens of corresponding height in the other form.

199.6:f height in another form.

199.7:d So that with dimorphic species two unions, which may be called legitimate, are fully fertile, and two, which may be called illegitimate, are more or less fertile.

199.7:e fertile; and

199.8:d With trimorphic species six unions are legitimate or fully fertile, and twelve are illegitimate or more or less infertile.

199.9-13:d[¶] The infertility which may be observed in various dimorphic and trimorphic plants, when they are illegitimately fertilised, that is by pollen taken from stamens not corresponding in height with the pistil, differs much in degree, up to absolute and utter sterility; just in the same manner as occurs in crossing distinct species. *10* As the degree of sterility in the latter case depends in an eminent degree on the conditions of life being more or less favourable, so I have found it with illegitimate unions. *11* It is well known that if pollen of a distinct species be placed on the stigma of a flower, and its own pollen be afterwards, even after a considerable interval of time, placed on the same stigma, its action is so strongly prepotent that it generally annihilates the effect of the foreign pollen; so it is with the pollen of the several forms of the same species, for legitimate pollen is strongly prepotent over illegitimate pollen, when both are placed on the same stigma. *12* I ascertained this by fertilising several flowers, first illegitimately, and twenty-four hours afterwards legitimately, with pollen taken from a peculiarly coloured variety, and all the seedlings were similarly coloured; this shows that the legitimate pollen, though applied twenty-four hours subsequently, had wholly destroyed or prevented the action of the previously applied illegitimate pollen. *13* Again, as in making reciprocal crosses between the same two species, there is occasionally a great difference in the result, so something analogous occurs with dimorphic plants; for a short-styled cowslip yields more seed when fertilised by the long-styled form, and less seed when fertilised by its own form, than

454

does a long-styled cowslip when fertilised in the two corresponding methods.

199.13:e result, so the same thing occurs with trimorphic plants; for instance, the mid-styled form of Lythrum salicaria was illegitimately fertilised with the greatest ease by pollen from the longer stamens of the short-styled form, and yielded many seeds; but the latter form did not yield a single seed when fertilised by the longer stamens of the mid-styled form.

199.14:d[¶] In all these respects the forms of the same undoubted species when illegitimately united behave in exactly the same manner as do two distinct species when crossed.

199.14:e respects and in others which might have been adduced, the forms

199.14:f respects, and in others which might be added, the forms

199.15-7:d This led me carefully to observe during four years many seedlings, raised from several illegitimate unions. *16* The chief result is that these illegitimate plants, as they may be called, are not fully fertile. *17* It is possible to raise from dimorphic species, both long-styled and short-styled illegitimate plants, and from trimorphic plants all three illegitimate forms; so that these can be properly united in a legitimate manner.

199.17:e forms; these can then be

199.17.x-y:f forms. *y* These

199.18-9:d When this is done, there is no apparent reason why they should not yield as many seeds as did their parents when legitimately fertilised. *19* But such is not the case; they are all infertile, but in various degrees; some being so utterly and incurably sterile that they did not yield during four seasons a single seed or even a seed-capsule.

199.19.x-y:f case. *y* They are all infertile, in/even seed-capsule.

199.20:d These illegitimate plants, which are so sterile, although united with each other in a legitimate manner, may be strictly compared with hybrids when crossed *inter se,* and we all know how sterile these latter generally are.

199.20:e The sterility of these illegitimate plants, when united with each other in a legitimate manner, may be strictly compared with that of hybrids when crossed *inter se.*

199.21:d When on the other hand a hybrid is crossed with either pure parent-species, the sterility is usually much lessened: and so it is when an illegitimate plant is fertilised by a legitimate plant.

199.21:f If, on the other hand, a hybrid

199.22:d In the same manner as the sterility of hybrids does not always run parallel with the difficulty of making the first cross between its two parent-species, so the sterility of certain illegitimate plants was unusually great, whilst the sterility of the union from which they were derived was by no means great.

199.22:e between the two

199.23-4:d With hybrids raised from the same seed-capsule the degree of sterility is innately variable, so it is in a marked manner with illegitimate plants. *24* Lastly, many hybrids are profuse and persistent flowerers, whilst other and more sterile hybrids produce few flowers, and are weak, miserable dwarfs; exactly similar cases occur with the illegitimate offspring of various dimorphic and trimorphic plants.

199.25-6:d[¶] Altogether there is the closest identity in character and behaviour between illegitimate plants and hybrids. *26* It is hardly an exaggeration to maintain that the former are hybrids, but produced within the limits of the same species by the improper union of certain forms, whilst ordinary hybrids are produced from an improper union between so-called distinct species.

199.26:f that illegitimate plants are hybrids, produced within

199.27-8:d We have also already seen that there is the closest similarity in all respects between first illegitimate unions and first crosses between distinct species. *28* All this will perhaps be made more fully apparent by an illustration: we may suppose that a botanist found two well-marked varieties (and such occur) of the long-styled form of the trimorphic Lythrum salicaria, and that he determined to try by crossing whether they were specifically distinct.

199.28:f This will

199.29-31:d He would find that they yielded only about one-fifth of the proper number of seed, and that they behaved in all the other above specified respects as if they had been two distinct species. *30* But to make the case sure, he would raise plants from his supposed hybridised seed, and he would find that the seedlings were miserably dwarfed and utterly sterile, and that they behaved in all other respects like ordinary hybrids. *31* He might then maintain that he had actually proved, in accordance with the common view, that his two varieties were as good and as distinct species as any in the world; but he would be completely mistaken.

199.32:d[¶] The facts now given on dimorphic and trimorphic plants are of importance, because they show us, firstly, that the physiological test of lessened fertility, both in first crosses and in hybrids, is no safe criterion of specific distinction; secondly, because we are thus led to infer, as previously remarked, that there must be some unknown law or bond connecting the infertility both of illegitimate unions and of first crosses, with the infertility of their illegitimate and hybrid offspring; thirdly, because we find, and this seems to me of especial importance, that two or three forms of the same species may exist and may differ in no respect, except in certain characters in their reproductive organs,—such as in the relative lengths of the stamens and pistils, in the size, form, and colour of the pollen-grains, in the structure of the stigma, and in the number and size of the seeds.

199.32:e are important, because they show us, first, that the physio-logical/secondly, because we may conclude that there is some unknown bond which connects the infertility of illegitimate unions with that of their illegitimate offspring, and we are led to extend the same view to first crosses and hybrids; thirdly/ except in their reproductive organs, and yet be sterile when united in certain ways.

199.32:f respect whatever, either in structure or in constitution, relatively to external conditions, and yet

199.33:d With these differences and with no others, either in organ-isation or constitution, between the several forms, which are all hermaphrodites, we find that their illegitimate unions and their illegitimate progeny are more or less sterile, and closely re-semble in a whole series of relations the first unions and the hybrid offspring of distinct species.

199.33:[e]

199.34:d We are thus led to infer that the sterility of species when crossed and of their hybrid progeny is likewise in all probability exclusively due to similar differences confined to their repro-ductive systems.

199.34:[e]

199.34.1:e With dimorphic plants, the unions between the two dis-tinct forms are alone quite fertile, and produce quite fertile offspring, whilst unions between individuals belonging to the same form are more or less sterile; so that the result is exactly the reverse of what occurs with distinct species.

199.34.1:[f]

199.34.2:e With dimorphic plants the resultant sterility is quite in-dependent of any difference in general structure or constitu-tion, for it arises from the union of individuals belonging not only to the same species, but to the same form.

199.34.2:[f]

199.34.3:e It must, therefore, depend on the nature of the sexual elements, which are so adapted to each other, that the male and female elements occurring in the same form do not suit each other, whilst those occurring in the two distinct forms are mu-tually suited to each other.

199.34.3:[f]

199.34.3.1-3:f For we must remember that it is the union of the sexual elements of individuals of the same form, for instance, of two long-styled forms, which results in sterility; whilst it is the union of the sexual elements proper to two distinct forms which is fertile. 2 Hence the case appears at first sight exactly the reverse of what occurs, in the ordinary unions of the indi-viduals of the same species and with crosses between distinct species. 3 It is, however, doubtful whether this is really so; but I will not enlarge on this obscure subject.

199.34.4:e From these considerations, it seems probable that the sterility of distinct species when crossed, and of their hybrid progeny, depends exclusively on the nature of their sexual elements, and not on any general difference in structure or constitution.

199.34.4:f[¶] We may, however, infer as probable from the consideration of dimorphic and trimorphic plants, that the sterility of distinct species when crossed and of their hybrid/any difference in their structure or general constitution.

199.35:d We are indeed led to this same conclusion from considering reciprocal crosses between the same two species, in which the male of one cannot be united, or can be united with great difficulty, with the female of the other species, whilst the converse cross can be effected with perfect facility; for this difference in the facility of making reciprocal crosses and in the fertility of their offspring must be attributed either to the male or to the female element in the one species having been differentiated, with reference to the other sexual element, in a higher degree than in the second species.

199.35:e conclusion by considering reciprocal crosses, in which the male of one species cannot/female of a second species, whilst/ element in the first species having been differentiated, with reference to the sexual elements of the second species in a higher degree than in the converse case.

199.35:f We are also led/facility.

199.36:d That excellent observer, Gärtner, likewise came on general grounds to the same conclusion, namely, that species when crossed are sterile owing to differences confined to their reproductive systems.

199.36:e came to this same/systems. [*Space*]

199.36:f likewise concluded that species

199.37-43:d[¶] Finally, we are naturally led to inquire for what useful end have plants been rendered reciprocally dimorphic and trimorphic? *38* A wide-spread analogy clearly gives us the answer as far as the immediate cause is concerned, namely, to prevent the pollen of each flower acting on the stigma of that flower. *39* We see this effected in a host of flowers by the most curious mechanical contrivances, as I have shown with Orchids, and as could be shown with many plants of many other orders. *40* There are also numerous plants, called dichogamous by C. K. Sprengel, in which the pollen and stigma are never mature at the same time, so that these plants can never fertilise themselves. *41* There are many flowers, which, though they have their stigmas and pollen mature together, and which do not present any obstacle to self-impregnation, yet nevertheless are almost always fertilised by surrounding varieties when growing in the vicinity, as shown by the character of their seedlings. *42* Then, again, we have many flowers with separated **sexes** borne on distinct plants, or on the same, which inevitably prevents self-fertilisation. *43* Lastly, in accordance with the **great** principle prevailing throughout nature, of the same end being gained by the most diversified means, we find in dimorphic and trimorphic plants, in which self-fertilisation is not checked by any of the above-specified means, that this has been effected by the pollen of each flower, and consequently of all the flowers of the same form, having been rendered more or less impotent on their own stigmas; so that its action is easily and wholly obliterated by pollen habitually brought by insects from other individuals and forms of the same species.

199.37-43:[*e*]

199.44-6:d[¶] In searching for the cause of dimorphism and trimorphism in plants, we may, in my opinion, safely go one step further, and conclude that the pollen has been prevented acting on the stigma of the same flower, in order to give vigour to the offspring by leading to the union of two distinct individuals. *45* But on this view it is not a little remarkable that the end has been gained, in the case of dimorphic and trimorphic plants, at the expense of all the plants of the same form being rendered more or less sterile when united, and producing more or less sterile offspring. *46* With respect to the steps by which it is probable that plants have been rendered dimorphic and trimorphic, want of space prevents my entering on the subject; but I will add that there is no special difficulty in this having been effected through variability, through the good gained by the prepotency of one sort of pollen over another, and through the accumulative action of natural selection. [*Space*]

199.44-6:[*e*]

200   *Fertility of Varieties when crossed, and of their Mongrel offspring.—*

200:d [*Center*] *Fertility/Offspring.* [*Space*]

200:f *Fertility/Offspring, not universal.*

201   It may be urged, as a most forcible argument, that there must be some essential distinction between species and varieties, and that there must be some error in all the foregoing remarks, inasmuch as varieties, however much they may differ from each other in external appearance, cross with perfect facility, and yield perfectly fertile offspring.

201:d[¶]

201:e as an overwhelming argument/varieties, inasmuch as the latter, however

202   I fully admit that this is almost invariably the case.

202:c With some exceptions, presently to be given, I fully admit that this is very generally the rule.

202:d is the rule.

203   But if we look to varieties produced under nature, we are immediately involved in hopeless difficulties; for if two hitherto

459

reputed varieties be found in any degree sterile together, they are at once ranked by most naturalists as species.

203:c But the subject is surrounded by difficulties, for looking to varieties produced under nature, if two forms hitherto reputed to be varieties

203:d for, looking

203:e varieties, produced

203:f varieties produced

204 For instance, the blue and red pimpernel, the primrose and cowslip, which are considered by many of our best botanists as varieties, are said by Gärtner not to be quite fertile when crossed, and he consequently ranks them as undoubted species.

204:d pimpernel, which are considered by most botanists

204:f Gärtner to be quite sterile when

205 If we thus argue in a circle, the fertility of all varieties produced under nature will assuredly have to be granted.

206 If we turn to varieties, produced, or supposed to have been produced, under domestication, we are still involved in doubt.

206:f in some doubt.

207 For when it is stated, for instance, that the German Spitz dog unites more easily than other dogs with foxes, or that certain South American indigenous domestic dogs do not readily cross with European dogs, the explanation which will occur to every one, and probably the true one, is that these dogs have descended from several aboriginally distinct species.

207:d dog crosses more easily with the fox than do other dogs, or that certain/readily unite with/from aboriginally

207:e these dogs are descended

207:f that certain/true one, is that they are

208 Nevertheless the perfect fertility of so many domestic varieties, differing widely from each other in appearance, for instance of the pigeon or of the cabbage, is a remarkable fact; more especially when we reflect how many species there are, which, though resembling each other most closely, are utterly sterile when intercrossed.

208:d instance those of the pigeon or the cabbage

208:e pigeon, or of the cabbage

208:f domestic races, differing

209 Several considerations, however, render the fertility of domestic varieties less remarkable than at first appears.

209:c render this fertility/remarkable.

209:e render the fertility

209.0.1:d = 210:[d] in the first place

209.0.1:e In the first place, it may be observed that the amount of external difference between two species is no sure guide to

460

their degree of mutual sterility, so that similar differences in the case of varieties would be no sure guide.

209.0.1.1:e It is almost certain that with species the cause lies exclusively in differences in their sexual constitution.

209.0.1.1:f is certain

209.0.1.2:e Now the conditions to which domesticated animals and cultivated plants have been subjected, have had so little tendency towards modifying the reproductive system in a manner leading to mutual sterility, that we have good grounds for admitting the directly opposite doctrine of Pallas, namely, that such conditions generally eliminate this tendency; so that the domesticated descendants of species, which in their natural state would have been in some degree sterile when crossed, become perfectly fertile together.

209.0.1.2:f Now the varying conditions to which domesticated animals/state probably would

209.0.1.3-4:e With plants, so far is cultivation from giving a tendency towards sterility between distinct species, that in several well-authenticated cases already alluded to, certain plants have been affected in an opposite manner, for they have become self-impotent, whilst still retaining the capacity of fertilising and being fertilised by, other species. *4* If the Pallasian doctrine of the elimination of sterility through long-continued domestication be admitted, and it can hardly be rejected, it becomes in the highest degree improbable that similar circumstances should both induce and eliminate the same tendency; though in certain cases, with species having a peculiar constitution, sterility might occasionally be thus induced.

209.0.1.4:f similar conditions long-continued should likewise induce this tendency/thus caused.

209.0.1.5:e Thus, as I believe, we can understand why with domesticated animals varieties have not been produced which are mutually sterile; and why with plants only a few such cases, immediately to be given, have been observed.

209.0.1.6:e[¶] The real difficulty in our present subject is not, as it appears to me, why domestic varieties have not become mutually infertile when crossed, but why this has so generally occurred with natural varieties as soon as they have been modified in a sufficient and permanent degree to take rank as species.

209.0.1.6:f varieties, as soon as they have been permanently modified in a sufficient degree

209.0.1.7-8:e We are far from precisely knowing the cause; nor is this surprising, seeing how profoundly ignorant we are in regard to the normal and abnormal action of the reproductive system. *8* But we can see that species, owing to their struggle for existence with numerous competitors, must have been exposed to more uniform conditions during long periods of time, than have been domestic varieties; and this may well make a wide difference in the result.

209.0.1.8:f competitors, will have been exposed during long periods of time to more uniform conditions, than have domestic

209.0.1.9:e For we know how commonly wild animals and plants, when taken from their natural conditions and subjected to captivity, are rendered sterile; and the reproductive functions of organic beings which have always lived and been slowly modified under natural conditions would probably in like manner be eminently sensitive to the influence of an unnatural cross.

209.0.1.9:f lived under

209.0.1.10:e Domesticated productions, on the other hand, which, as shown by the mere fact of their domestication, were not originally highly sensitive to changes in their conditions of life, and which can now generally resist with undiminished fertility repeated changes of conditions, might be expected to produce varieties, which would be little liable to have their reproductive powers injuriously affected by the act of crossing with other varieties which had originated in a like manner.

209.1:c In the first place we must remember how ignorant we are regarding the precise cause of sterility, both when species are crossed and when species are removed from their natural conditions.

209.1:d In the second place, we must remember how ignorant we are on the precise causes of sterility, both when species are crossed, and

209.1:[e]

209.2:c On this latter head I have not had space to adduce the many remarkable facts which could have been given; with respect to sterility from crossing, reflect on the difference in the result of reciprocal crosses,—reflect on the singular cases in which a plant can be more easily fertilised by foreign pollen than by its own.

209.2.x-y:d space to give the many remarkable facts which could have been adduced. *y* With respect to sterility from crossing, it is good to reflect on the difference in the result of reciprocal crosses, and on those singular cases in which a plant can be more easily fertilised by pollen from a distinct species than

209.2.x-y:[e]

209.3:c When we think over such cases, and on that of the differently coloured varieties of Verbascum presently to be given, we must feel how ignorant we are, and how little likely it is that we should understand why certain forms are fertile and other forms are sterile when crossed.

209.3:[e]

210 It can, in the first place, be clearly shown that mere external dissimilarity between two species does not determine their greater or lesser degree of sterility when crossed; and we may apply the same rule to domestic varieties.

210:c in the second place

210:[d] = 209.0.1:d

211 In the second place, some eminent naturalists believe that a long course of domestication tends to eliminate sterility in the successive generations of hybrids, which were at first only slightly sterile; and if this be so, we surely ought not to expect to find sterility both appearing and disappearing under nearly the same conditions of life.

211:b hybrids which

211:c In the third place/same domestic conditions

211:d place, there is good evidence for believing that/sterility; and if/same conditions

211:[e]

212 Lastly, and this seems to me by far the most important consideration, new races of animals and plants are produced under domestication by man's methodical and unconscious power of selection, for his own use and pleasure: he neither wishes to select, nor could select, slight differences in the reproductive system, or other constitutional differences correlated with the reproductive system.

212:c domestication chiefly by

212:[e]

212.1:c Domestic productions are less closely adapted to climate and to the other physical conditions of the countries which they inhabit than are those in a state of nature, for they can generally be removed to other and differently constituted countries with entire impunity.

212.1:d with impunity.

212.1:[e]

213 He supplies his several varieties with the same food; treats them in nearly the same manner, and does not wish to alter their general habits of life.

213:c Man supplies/food; he treats

213:[e]

214 Nature acts uniformly and slowly during vast periods of time on the whole organisation, in any way which may be for each creature's own good; and thus she may, either directly, or more probably indirectly, through correlation, modify the reproductive system in the several descendants from any one species.

214:c systems of

214:[e]

215 Seeing this difference in the process of selection, as carried on by man and nature, we need not be surprised at some difference in the result.

215:[e]

216   I have as yet spoken as if the varieties of the same species were invariably fertile when intercrossed.

463

216:c were almost invariably

216:d were invariably

217 But it seems to me impossible to resist the evidence of the existence of a certain amount of sterility in the few following cases, which I will briefly abstract.

217:c it is impossible

218 The evidence is at least as good as that from which we believe in the sterility of a multitude of species.

219 The evidence is, also, derived from hostile witnesses, who in all other cases consider fertility and sterility as safe criterions of specific distinction.

220 Gärtner kept during several years a dwarf kind of maize with yellow seeds, and a tall variety with red seeds, growing near each other in his garden; and although these plants have separated sexes, they never naturally crossed.

220:e red seeds growing

221 He then fertilised thirteen flowers of the one with the pollen of the other; but only a single head produced any seed, and this one head produced only five grains.

221:f one kind with pollen

222 Manipulation in this case could not have been injurious, as the plants have separated sexes.

223 No one, I believe, has suspected that these varieties of maize are distinct species; and it is important to notice that the hybrid plants thus raised were themselves *perfectly* fertile; so that even Gärtner did not venture to consider the two varieties as specifically distinct.

224 Girou de Buzareingues crossed three varieties of gourd, which like the maize has separated sexes, and he asserts that their mutual fertilisation is by so much the less easy as their differences are greater.

225 How far these experiments may be trusted, I know not; but the forms experimentised on, are ranked by Sageret, who mainly founds his classification by the test of infertility, as varieties.

225:c forms experimented on are

225:d varieties, and Naudin has come to the same conclusion.

226 The following case is far more remarkable, and seems at first quite incredible; but it is the result of an astonishing number of experiments made during many years on nine species of Verbascum, by so good an observer and so hostile a witness, as Gärtner: namely, that yellow and white varieties of the same species of Verbascum when intercrossed produce less seed, than do either coloured varieties when fertilised with pollen from their own coloured flowers.

464

226:c witness as

226:d than does either coloured variety when fertilised with pollen from its own

226:e that the yellow and white varieties when crossed produce less seed than the similarly coloured varieties of the same species.

226:f first incredible

227 Moreover, he asserts that when yellow and white varieties of one species are crossed with yellow and white varieties of a *distinct* species, more seed is produced by the crosses between the same coloured flowers, than between those which are differently coloured.

227:b between the similarly coloured

227:d that, when yellow and white varieties of one species are

227.1:e Mr. Scott, also, has experimented on the species and varieties of Verbascum; and although unable to confirm Gärtner's results on the crossing of the distinct species, he finds that the dissimilarly coloured varieties of the same species yield fewer seeds, in the proportion of 86 to 100, than the similarly coloured varieties.

227.1:f Mr. Scott also has

228 Yet these varieties of Verbascum present no other difference besides the mere colour of the flower; and one variety can sometimes be raised from the seed of the other.

228:e varieties differ in no respect except in the colour of their flowers; and/seed of another.

229 From observations which I have made on certain varieties of hollyhock, I am inclined to suspect that they present analogous facts.

229:c From experiments which

229:[d]

230 Kölreuter, whose accuracy has been confirmed by every subsequent observer, has proved the remarkable fact, that one variety of the common tobacco is more fertile, when crossed with a widely distinct species, than are the other varieties.

230:e one particular variety of the common tobacco was more fertile than the other varieties, when crossed with a widely distinct species.

231 He experimentised on five forms, which are commonly reputed to be varieties, and which he tested by the severest trial, namely, by reciprocal crosses, and he found their mongrel offspring perfectly fertile.

231:c experimented

231:f forms which are

232 But one of these five varieties, when used either as father or mother, and crossed with the Nicotiana glutinosa, always

yielded hybrids not so sterile as those which were produced from the four other varieties when crossed with N. glutinosa.

233 Hence the reproductive system of this one variety must have been in some manner and in some degree modified.

234 From these facts; from the great difficulty of ascertaining the infertility of varieties in a state of nature, for a supposed variety if infertile in any degree would generally be ranked as species; from man selecting only external characters in the production of the most distinct domestic varieties, and from not wishing or being able to produce recondite and functional differences in the reproductive system; from these several considerations and facts, I do not think that the very general fertility of varieties can be proved to be of universal occurrence, or to form a fundamental distinction between varieties and species.

234:d supposed variety, if proved to be infertile in any degree, would almost universally be ranked/characters in his domestic/ differences in their reproductive systems; from/that the fertility/between them and species.

234:e From these facts it cannot be maintained that varieties when crossed are invariably quite fertile;—from the great/species;— from man attending only to external characters in his domestic varieties, and from such varieties not having been exposed for a very long period to uniform conditions of life;—from these several considerations we may conclude that fertility does not constitute a fundamental distinction between varieties and species when crossed.

234.x-y:f it can no longer be/fertile. *y* From the great/exposed for very long periods

235 The general fertility of varieties does not seem to me sufficient to overthrow the view which I have taken with respect to the very general, but not invariable, sterility of first crosses and of hybrids, namely, that it is not a special endowment, but is incidental on slowly acquired modifications, more especially in the reproductive systems of the forms which are crossed.

235:c varieties, considering how entirely ignorant we are on the causes of both fertility and sterility, does not seem to me sufficient to overthrow the view taken with respect to the very general, but not invariable, sterility of first crosses between species and of their hybrids, namely

235:d incidental on modifications, slowly impressed, by unknown means, on the reproductive systems of the parent-forms.

235:e The general sterility of crossed species may safely be looked at not as a special acquirement or endowment, but as incidental on changes of an unknown nature in their sexual elements.

235:f at, not

236 *Hybrids and Mongrels compared, independently of their fertility.*—

236:d [Center] Hybrids/fertility. [Space]

237 Independently of the question of fertility, the offspring of species when crossed and of varieties when crossed may be compared in several other respects.

237:d[¶]

237:f species and of varieties when crossed may

238 Gärtner, whose strong wish was to draw a marked line of distinction between species and varieties, could find very few and, as it seems to me, quite unimportant differences between the so-called hybrid offspring of species, and the so-called mongrel offspring of varieties.

238:c draw a distinct line between/few, and, as

239 And, on the other hand, they agree most closely in very many important respects.

239:d in many

240 I shall here discuss this subject with extreme brevity.

241 The most important distinction is, that in the first generation mongrels are more variable than hybrids; but Gärtner admits that hybrids from species which have long been cultivated are often variable in the first generation; and I have myself seen striking instances of this fact.

242 Gärtner further admits that hybrids between very closely allied species are more variable than those from very distinct species; and this shows that the difference in the degree of variability graduates away.

243 When mongrels and the more fertile hybrids are propagated for several generations an extreme amount of variability in their offspring is notorious; but some few cases both of hybrids and mongrels long retaining uniformity of character could be given.

243:e generations, an extreme amount of variability in the offspring in both cases is notorious; but some few instances of both hybrids and mongrels long retaining a uniform character could

244 The variability, however, in the successive generations of mongrels is, perhaps, greater than in hybrids.

245 This greater variability of mongrels than of hybrids does not seem to me at all surprising.

245:e variability in mongrels than in hybrids does not seem at

246 For the parents of mongrels are varieties, and mostly domestic varieties (very few experiments having been tried on natural varieties), and this implies in most cases that there has been recent variability; and therefore we might expect that such variability would often continue and be superadded to that arising from the mere act of crossing.

246:e implies that there has been recent variability, which would often continue and be added to that arising from the act

246:f continue and would augment that

247 The slight degree of variability in hybrids from the first cross or in the first generation, in contrast with their extreme variability in the succeeding generations, is a curious fact and deserves attention.

247:e slight variability of hybrids in the first generation, in contrast with the succeeding

247:f with that in the succeeding

248 For it bears on and corroborates the view which I have taken on the cause of ordinary variability; namely, that it is due to the reproductive system being eminently sensitive to any change in the conditions of life, being thus often rendered either impotent or at least incapable of its proper function of producing offspring identical with the parent-form.

248:e on the view which I have taken of one of the causes of ordinary variability; namely, that the reproductive system from being eminently sensitive to changed conditions of life, fails under these circumstances to perform its proper function of producing offspring identical in all respects with

248:f offspring closely similar in all respects to the parent-form.

249 Now hybrids in the first generation are descended from species (excluding those long cultivated) which have not had their reproductive systems in any way affected, and they are not variable; but hybrids themselves have their reproductive systems seriously affected, and their descendants are highly variable.

249:e long-cultivated

250   But to return to our comparison of mongrels and hybrids: Gärtner states that mongrels are more liable than hybrids to revert to either parent-form; but this, if it be true, is certainly only a difference in degree.

250.1:d Moreover, Gärtner expressly states that hybrids from long-cultivated plants are more subject to reversion than hybrids from species in their natural state; and this probably explains the singular difference in the results arrived at by different observers: thus, Max Wichura doubts whether hybrids ever revert to their parent-forms, and he experimented on uncultivated species of willows; whilst Naudin, on the other hand, insists in the strongest terms on the almost universal tendency to reversion in hybrids, and he experimented chiefly on cultivated plants.

251 Gärtner further insists that when any two species, although closely allied to each other, are crossed with a third species, the hybrids are widely different from each other; whereas if two very distinct varieties of one species are crossed with another species, the hybrids do not differ much.

251:d further states that/whereas, if

252 But this conclusion, as far as I can make out, is founded on a

468

single experiment; and seems directly opposed to the results of several experiments made by Kölreuter.

253 These alone are the unimportant differences, which Gärtner is able to point out, between hybrid and mongrel plants.

253:c out between

253:d Such alone are the unimportant differences which

254 On the other hand, the resemblance in mongrels and in hybrids to their respective parents, more especially in hybrids produced from nearly related species, follows according to Gärtner the same laws.

254:d hand, the degrees and kinds of resemblance/follow

255 When two species are crossed, one has sometimes a prepotent power of impressing its likeness on the hybrid; and so I believe it to be with varieties of plants.

256 With animals one variety certainly often has this prepotent power over another variety.

255.x-y + 256:f hybrid. y So/plants; and with

257 Hybrid plants produced from a reciprocal cross, generally resemble each other closely; and so it is with mongrels from a reciprocal cross.

257:e with mongrel plants from

258 Both hybrids and mongrels can be reduced to either pure parent-form, by repeated crosses in successive generations with either parent.

259 These several remarks are apparently applicable to animals; but the subject is here excessively complicated, partly owing to the existence of secondary sexual characters; but more especially owing to prepotency in transmitting likeness running more strongly in one sex than in the other, both when one species is crossed with another, and when one variety is crossed with another variety.

259:d here much complicated

260 For instance, I think those authors are right, who maintain that the ass has a prepotent power over the horse, so that both the mule and the hinny more resemble the ass than the horse; but that the prepotency runs more strongly in the male-ass than in the female, so that the mule, which is the offspring of the male-ass and mare, is more like an ass, than is the hinny, which is the offspring of the female-ass and stallion.

260:f hinny resemble more closely the ass than the horse; but that the prepotency runs more strongly in the male than in the female ass, so that the mule, which is the offspring of the male ass and mare

261 Much stress has been laid by some authors on the supposed fact, that mongrel animals alone are born closely like one of their parents; but it can be shown that this does sometimes

occur with hybrids; yet I grant much less frequently with hybrids than with mongrels.

261:d that it is only with mongrels that the offspring are not intermediate in character, but closely resemble one of their parents; but this does sometimes occur with hybrids, yet I grant much less frequently with them than with mongrels.

261:f frequently than

262 Looking to the cases which I have collected of cross-bred animals closely resembling one parent, the resemblances seem chiefly confined to characters almost monstrous in their nature, and which have suddenly appeared—such as albinism, melanism, deficiency of tail or horns, or additional fingers and toes; and do not relate to characters which have been slowly acquired by selection.

262:d acquired through selection.

263 Consequently, sudden reversions to the perfect character of either parent would be more likely to occur with mongrels, which are descended from varieties often suddenly produced and semi-monstrous in character, than with hybrids, which are descended from species slowly and naturally produced.

263:f A tendency to sudden reversions to the perfect character of either parent would, also, be much more

264 On the whole I entirely agree with Dr. Prosper Lucas, who, after arranging an enormous body of facts with respect to animals, comes to the conclusion, that the laws of resemblance of the child to its parents are the same, whether the two parents differ much or little from each other, namely in the union of individuals of the same variety, or of different varieties, or of distinct species.

264:d whole, I/differ little or much from
264:f namely, in

265 Laying aside the question of fertility and sterility, in all other respects there seems to be a general and close similarity in the offspring of crossed species, and of crossed varieties.

265:e Independently of the question

266 If we look at species as having been specially created, and at varieties as having been produced by secondary laws, this similarity would be an astonishing fact.

267 But it harmonises perfectly with the view that there is no essential distinction between species and varieties.

268 *Summary of Chapter.—*
268:d [*Center*] *Summary of Chapter.* [*Space*]

269 First crosses between forms sufficiently distinct to be ranked as species, and their hybrids, are very generally, but not universally, sterile.

269:d[¶]

269:f forms, sufficiently

270 The sterility is of all degrees, and is often so slight that the two most careful experimentalists who have ever lived, have come to diametrically opposite conclusions in ranking forms by this test.

270:e that the most careful experimentalists have arrived at diametrically

271 The sterility is innately variable in individuals of the same species, and is eminently susceptible of favourable and unfavourable conditions.

271:e susceptible to the action of

272 The degree of sterility does not strictly follow systematic affinity, but is governed by several curious and complex laws.

273 It is generally different, and sometimes widely different, in reciprocal crosses between the same two species.

274 It is not always equal in degree in a first cross and in the hybrid produced from this cross.

274:d hybrids

275   In the same manner as in grafting trees, the capacity of one species or variety to take on another, is incidental on generally unknown differences in their vegetative systems, so in crossing, the greater or less facility of one species to unite with another, is incidental on unknown differences in their reproductive systems.

275:e incidental on differences, generally of an unknown nature, in their vegetative/another is

276 There is no more reason to think that species have been specially endowed with various degrees of sterility to prevent them crossing and blending in nature, than to think that trees have been specially endowed with various and somewhat analogous degrees of difficulty in being grafted together in order to prevent them becoming inarched in our forests.

276:e prevent their crossing/order to prevent their inarching

276.1:d[¶] The sterility of first crosses and of their hybrid progeny has not, as far as we can judge, been increased through natural selection so as to attain that high degree which is universal with species when rendered widely distinct.

276.1:e been acquired through natural selection.

276.1:f not been

277   The sterility of first crosses between pure species, which have their reproductive systems perfect, seems to depend on several circumstances; in some cases largely on the early death of the embryo.

277:d[*No* ¶] With first crosses between pure species, in which the reproductive system is in a perfect condition, the sterility seems/

embryo, but this apparently depends on some imperfection in the original act of impregnation.

277:e In the case of first crosses it seems/some instances in chief part on the early death of the embryo.

278 The sterility of hybrids, which have their reproductive systems imperfect, and which have had this system and their whole organisation disturbed by being compounded of two distinct species, seems closely allied to that sterility which so frequently affects pure species, when their natural conditions of life have been disturbed.

278:d With hybrids, in which the reproductive system is in an imperfect condition, and in which this system as well as the whole organisation has been disturbed by being compounded from two distinct forms, the sterility apparently is closely allied to that which so frequently affects pure species, when exposed to unnatural conditions of life.

278:e In the case of hybrids, it perhaps depends on their whole organisation having been disturbed by being compounded from two distinct forms; the sterility being closely

278:f it apparently depends/exposed to new and unnatural

278.1:f He who will explain these latter cases will be able to explain the sterility of hybrids.

279 This view is supported by a parallelism of another kind;— namely, that the crossing of forms only slightly different is favourable to the vigour and fertility of their offspring; and that slight changes in the conditions of life are apparently favourable to the vigour and fertility of all organic beings.

279:d kind: namely, that the crossing of forms only slightly differentiated favours the vigour and fertility of their/life apparently add to the vigour

279:e that, first, the crossing/offspring, whilst close interbreeding is injurious; and secondly, that/beings, whilst greater changes are often injurious.

279:f is strongly supported/firstly, slight changes in the conditions of life add to the vigour and fertility of all organic beings; and secondly, that the crossing of forms, which have been exposed to slightly different conditions of life or which have varied, favours the size, vigour, and fertility of their offspring.

279.1:d But the facts given on the sterility of the illegitimate unions of dimorphic and trimorphic plants and of their illegitimate progeny, render it probable that there is some unknown bond connecting in all cases the degree of fertility of first unions with that of their offspring.

279.1:e that some unknown bond in all cases connects the degree

279.1:f The facts/progeny, perhaps render

279.2:d The consideration of these facts on dimorphism, as well as the results of reciprocal crosses, drive us to conclude that in all cases the primary cause of sterility, both in the parents and in the offspring, is confined to differences in their reproductive systems.

279.2:e crosses, clearly leads to the conclusion that the primary cause of the sterility is confined to differences in the sexual elements.

279.2:f sterility of crossed species is confined to differences in their sexual

279.3:d But why in numerous species, descended from a common parent-form, the reproductive system should in all have become more or less modified, leading to their mutual infertility, we do not know in the least; nor whether this has been effected directly, or in correlation with other structural and functional modifications.

279.3:e why, in the case of species, the sexual elements should so generally have become/know.

279.3:f case of distinct species/know; but it seems to stand in some close relation to species having been exposed for long periods of time to nearly uniform conditions of life.

280 It is not surprising that the degree of difficulty in uniting two species, and the degree of sterility of their hybrid-offspring should generally correspond, though due to distinct causes; for both depend on the amount of difference of some kind between the species which are crossed.

280:d[¶] correspond, even if due

280:e that the difficulty in crossing any two species, and the sterility of their hybrid-offspring, should in most cases correspond/difference between

280:f hybrid offspring

281 Nor is it surprising that the facility of effecting a first cross, the fertility of the hybrids produced, and the capacity of being grafted together—though this latter capacity evidently depends on widely different circumstances—should all run, to a certain extent, parallel with the systematic affinity of the forms which are subjected to experiment; for systematic affinity attempts to express all kinds of resemblance between all species.

281:b produced from it, and

281:d cross, and the fertility of the hybrids thus produced, and

281:e forms subjected to experiment; for systematic affinity includes resemblances of all kinds.

282 First crosses between forms known to be varieties, or sufficiently alike to be considered as varieties, and their mongrel offspring, are very generally, but not quite universally, fertile.

282:c not, as is so often falsely stated, universally

282:d often stated

282:e stated, invariably fertile.

283 Nor is this nearly general and perfect fertility surprising, when

473

we remember how liable we are to argue in a circle with respect to varieties in a state of nature; and when we remember that the greater number of varieties have been produced under domestication by the selection of mere external differences, and not of differences in the reproductive system.

283:e Nor is this almost universal and perfect/external differences, and that they have not been long exposed to uniform conditions of life.

283:f when it is remembered how

283.1:d Nor should it be forgotten that long-continued domestication apparently tends to eliminate sterility, and is therefore little likely to induce this same quality.

283.1:e It should also be especially kept in mind, that long-continued domestication tends

284 In all other respects, excluding fertility, there is a close general resemblance between hybrids and mongrels.

284:d Excluding the subject of fertility, in all other respects there is the closest general

284:e Independently of the question of fertility, in all other respects/mongrels, in their variability, in their power of absorbing each other by repeated crosses, and in their inheritance of characters from both parent-forms.

284:f mongrels,—in their variability

285 Finally, then, the facts briefly given in this chapter do not seem to me opposed to, but even rather to support the view, that there is no fundamental distinction between species and varieties.

285:c then, although we are profoundly ignorant in every case of the precise cause of sterility, the facts/support in some respects the view

285:d cause of the sterility of first crosses and of hybrids, the facts/opposed to the view

285:e ignorant of the precise/facts given/opposed to the belief that varieties and species are not fundamentally different.

285:f although we are as ignorant of the precise cause of the sterility of first crosses and of hybrids as we are why animals and plants removed from their natural conditions become sterile, yet the facts given in this chapter do not seem to me opposed to the belief that species aboriginally existed as varieties.

CHAPTER IX.

1:f X.

2 ON THE IMPERFECTION OF THE GEOLOGICAL RECORD.

3 On the absence of intermediate varieties at the present day—On
the nature of extinct intermediate varieties; on their number—
On the vast lapse of time, as inferred from the rate of deposition and of denudation—On the poorness of our palæontological
collections—On the intermittence of geological formations—On
the absence of intermediate varieties in any one formation—
On the sudden appearance of groups of species—On their sudden appearance in the lowest known fossiliferous strata.

3:c collections—On the denudation of granitic areas—On the intermittence

3:e number—On the lapse of time, as inferred from the rate of
denudation and of deposition—On the lapse of time as estimated by years—On the poorness/strata—Antiquity of the
habitable earth.

3:f collections—On the intermittence of geological formations—On
the denudation of granitic areas—On the absence

4 IN the sixth chapter I enumerated the chief objections which
might be justly urged against the views maintained in this
volume.

5 Most of them have now been discussed.

6 One, namely the distinctness of specific forms, and their not being
blended together by innumerable transitional links, is a very
obvious difficulty.

7 I assigned reasons why such links do not commonly occur at the
present day, under the circumstances apparently most favourable for their presence, namely on an extensive and continuous
area with graduated physical conditions.

7:f day under

8 I endeavoured to show, that the life of each species depends in a
more important manner on the presence of other already defined organic forms, than on climate; and, therefore, that the
really governing conditions of life do not graduate away quite
insensibly like heat or moisture.

9 I endeavoured, also, to show that intermediate varieties, from
existing in lesser numbers than the forms which they connect,
will generally be beaten out and exterminated during the
course of further modification and improvement.

10 The main cause, however, of innumerable intermediate links

not now occurring everywhere throughout nature depends on the very process of natural selection, through which new varieties continually take the places of and exterminate their parent-forms.

10:f nature, depends/places of and supplant their

11 But just in proportion as this process of extermination has acted on an enormous scale, so must the number of intermediate varieties, which have formerly existed on the earth, be truly enormous.

11:e existed, be

12 Why then is not every geological formation and every stratum full of such intermediate links?

13 Geology assuredly does not reveal any such finely graduated organic chain; and this, perhaps, is the most obvious and gravest objection which can be urged against my theory.

13:c finely-graduated/obvious and serious objection

13:f against the theory.

14 The explanation lies, as I believe, in the extreme imperfection of the geological record.

15 In the first place it should always be borne in mind what sort of intermediate forms must, on my theory, have formerly existed.

15:e on the theory

16 I have found it difficult, when looking at any two species, to avoid picturing to myself, forms *directly* intermediate between them.

16:e myself forms

17 But this is a wholly false view; we should always look for forms intermediate between each species and a common but unknown progenitor; and the progenitor will generally have differed in some respects from all its modified descendants.

18 To give a simple illustration: the fantail and pouter pigeons have both descended from the rock-pigeon; if we possessed all the intermediate varieties which have ever existed, we should have an extremely close series between both and the rock-pigeon; but we should have no varieties directly intermediate between the fantail and pouter; none, for instance, combining a tail somewhat expanded with a crop somewhat enlarged, the characteristic features of these two breeds.

18:f pigeons are both descended

19 These two breeds, moreover, have become so much modified, that if we had no historical or indirect evidence regarding their origin, it would not have been possible to have determined from a mere comparison of their structure with that of the rock-pigeon, whether they had descended from this species or from some other allied species, such as C. oenas.

476

19:c rock-pigeon (C. livia), whether

19:d that, if/determined, from a mere/rock-pigeon, C. livia, whether

19:f allied form, such

20   So with natural species, if we look to forms very distinct, for instance to the horse and tapir, we have no reason to suppose that links ever existed directly intermediate between them, but between each and an unknown common parent.

20:f links directly intermediate between them ever existed, but

21 The common parent will have had in its whole organisation much general resemblance to the tapir and to the horse; but in some points of structure may have differed considerably from both, even perhaps more than they differ from each other.

22 Hence in all such cases, we should be unable to recognise the parent-form of any two or more species, even if we closely compared the structure of the parent with that of its modified descendants, unless at the same time we had a nearly perfect chain of the intermediate links.

22:d Hence, in

23   It is just possible by my theory, that one of two living forms might have descended from the other; for instance, a horse from a tapir; and in this case *direct* intermediate links will have existed between them.

23:e by the theory

24 But such a case would imply that one form had remained for a very long period unaltered, whilst its descendants had undergone a vast amount of change; and the principle of competition between organism and organism, between child and parent, will render this a very rare event; for in all cases the new and improved forms of life will tend to supplant the old and unimproved forms.

24:b life tend

25   By the theory of natural selection all living species have been connected with the parent-species of each genus, by differences not greater than we see between the varieties of the same species at the present day; and these parent-species, now generally extinct, have in their turn been similarly connected with more ancient species; and so on backwards, always converging to the common ancestor of each great class.

25:f between the natural and domestic varieties/ancient forms; and

26 So that the number of intermediate and transitional links, between all living and extinct species, must have been inconceivably great.

27 But assuredly, if this theory be true, such have lived upon this earth.

27:d upon the earth.

28:d [Center] On the Lapse of Time, as inferred from the rate of Deposition and extent of Denudation. [Space]

29 Independently of our not finding fossil remains of such infinitely numerous connecting links, it may be objected, that time will not have sufficed for so great an amount of organic change, all changes having been effected very slowly through natural selection.

29:d[¶]

29:e time cannot have sufficed

29:f objected that/effected slowly.

30 It is hardly possible for me even to recall to the reader, who may not be a practical geologist, the facts leading the mind feebly to comprehend the lapse of time.

30:e who is not a practical

30:f me to recall to the reader who

31 He who can read Sir Charles Lyell's grand work on the Principles of Geology, which the future historian will recognise as having produced a revolution in natural science, yet does not admit how incomprehensibly vast have been the past periods of time, may at once close this volume.

31:e how vast

31:f science, and yet

32 Not that it suffices to study the Principles of Geology, or to read special treatises by different observers on separate formations, and to mark how each author attempts to give an inadequate idea of the duration of each formation or even each stratum.

32:d even of each

32:f each formation, or

32.1:e We can best gain some idea of past time by knowing the agencies at work, and learning how much of the surface of the land has been denuded, and how much sediment has been deposited.

32.1:f how deeply the surface

32.2:e As Lyell has well remarked, the extent and thickness of our sedimentary formations are the result and the measure of the denudation which the earth's crust has elsewhere undergone.

33 A man must for years examine for himself great piles of superimposed strata, and watch the sea at work grinding down old rocks and making fresh sediment, before he can hope to comprehend anything of the lapse of time, the monuments of which we see around us.

33:e Therefore a man should examine for himself the great piles of superimposed strata, and watch the rivulets bringing down mud, and the waves wearing away the sea-cliffs, in order to comprehend something about the duration of past time, the monuments of which we see all around us.

34 It is good to wander along lines of sea-coast, when formed of moderately hard rocks, and mark the process of degradation.

34:e along the coast

35 The tides in most cases reach the cliffs only for a short time twice a day, and the waves eat into them only when they are charged with sand or pebbles; for there is reason to believe that pure water can effect little or nothing in wearing away rock.

35:b is good evidence that

35:c water effects nothing

36 At last the base of the cliff is undermined, huge fragments fall down, and these remaining fixed, have to be worn away, atom by atom, until reduced in size they can be rolled about by the waves, and then are more quickly ground into pebbles, sand, or mud.

36:c away atom by atom, until reduced in size, they

36:d until, reduced

36:e these, remaining/until after being reduced/then they are

37 But how often do we see along the bases of retreating cliffs rounded boulders, all thickly clothed by marine productions, showing how little they are abraded and how seldom they are rolled about!

38 Moreover, if we follow for a few miles any line of rocky cliff, which is undergoing degradation, we find that it is only here and there, along a short length or round a promontory, that the cliffs are at the present time suffering.

39 The appearance of the surface and the vegetation show that else-where years have elapsed since the waters washed their base.

39.1:e[¶] We have, however, recently learnt from the observations of Ramsay, in the van of excellent observers, of Jukes, Geikie, Croll, and others, that subaerial degradation is a much more important agency than coast-action, or the power of the waves.

39.1:f van of many excellent observers—of Jukes

39.2-3:e The whole surface of the land is exposed to the chemical action of the air and of the rain-water with its dissolved carbonic acid, and in colder countries to frost; the disintegrated matter is carried down even gentle slopes during heavy rain, and to a greater extent than might be supposed, especially in arid districts, by the wind; it is then transported by the streams and rivers, which when rapid deepen their channels, and tritu-rate the fragments. *3* On a rainy day, even in a gently undulat-ting country, we see the effects of subaerial degradation in the muddy rills which flow down each slope.

39.3:f down every slope.

39.4:e Messrs. Ramsay and Whitaker have shown, and the obser-

479

vation is a most striking one, that the great lines of escarpment in the Wealden district and those ranging across England, which formerly were looked at by every one as ancient sea-coasts, cannot have been thus formed, for each line is composed of one and the same formation, whilst our present sea-cliffs are everywhere formed by the intersection of various formations.

39.4:f at as/our sea-cliffs

39.5-6:e This being the case, we are compelled to admit that the escarpments owe their origin in chief part to the rocks of which they are composed having resisted subaerial denudation better than the surrounding surface; this surface consequently has been gradually lowered, with the lines of harder rock left projecting. *6* Nothing impresses the mind with the vast duration of time, according to our ideas of time, more forcibly than the conviction thus gained that subaerial agencies, which apparently have so little power, and which seem to work so slowly, have produced such great results.

39.6:f agencies which

39.7:e[¶] When thus impressed with the slow rate at which the land is worn away through subaerial and littoral action, it is good, in order to appreciate the past duration of time, to consider, on the one hand, the mass of rock which has been removed over many extensive areas, and on the other hand the thickness of our sedimentary formations.

39.7:f masses of rock which have

39.8:e = *52:*[e] struck when

39.9:e = *53:*[e] story is told still more plainly by/cracked, and it/ starts, the surface/down that

39.9:f believe, was slow

39.10:e = *54:*[e][No ¶]

39.11:e = *55:*[e] believes that there/rocks on either side of the crack having

40   He who most closely studies the action of the sea on our shores, will, I believe, be most deeply impressed with the slowness with which rocky coasts are worn away.

40:[e]

40.1:e[¶] On the other hand in all parts of the world the piles of sedimentary strata are of wonderful thickness.

40.1:f hand, in

41 The observations on this head by Hugh Miller, and by that excellent observer Mr. Smith of Jordan Hill, are most impressive.

41:[e]

42 With the mind thus impressed, let any one examine beds of conglomerate many thousand feet in thickness, which, though probably formed at a quicker rate than many other deposits,

yet, from being formed of worn and rounded pebbles, each of which bears the stamp of time, are good to show how slowly the mass has been accumulated.

42.1:b In the Cordillera I estimated one pile of conglomerate at ten thousand feet in thickness.

42.1 + 42:e one mass of conglomerate at ten thousand feet; and although conglomerates have probably been accumulated at a quicker rate than finer sediments, yet from/time, they are good to show how slowly the mass must have been heaped together.

43 Let him remember Lyell's profound remark, that the thickness and extent of sedimentary formations are the result and measure of the degradation which the earth's crust has elsewhere suffered.

43:b Let the observer remember Lyell's profound remark that

43:[e]

44 And what an amount of degradation is implied by the sedimentary deposits of many countries!

44:[e]

45 Professor Ramsay has given me the maximum thickness, in most cases from actual measurement, in a few cases from estimate, of each formation in different parts of Great Britain; and this is the result:—

	Feet.
Palæozoic strata (not including igneous beds) .	57,154
Secondary strata	13,190
Tertiary strata	2,240

—making altogether 72,584 feet; that is, very nearly thirteen and three-quarters British miles.

45:c different

45:e thickness, from actual measurement in most cases, of the successive formations

46 Some of these formations, which are represented in England by thin beds, are thousands of feet in thickness on the Continent.

46:b of the formations

47 Moreover, between each successive formation, we have, in the opinion of most geologists, enormously long blank periods.

47:f geologists, blank periods of enormous length.

48 So that the lofty pile of sedimentary rocks in Britain, gives but an inadequate idea of the time which has elapsed during their accumulation; yet what time this must have consumed!

48:d Britain gives

48:e accumulation.

49 Good observers have estimated that sediment is deposited by the great Mississippi river at the rate of only 600 feet in a hundred thousand years.

49:[e]

50 This estimate may be quite erroneous; yet, considering over what wide spaces very fine sediment is transported by the currents of the sea, the process of accumulation in any one area must be extremely slow.

50:b estimate has no pretension to strict exactness; yet

50:c accumulation over any one extensive area

50:[e]

51 But the amount of denudation which the strata have in many places suffered, independently of the rate of accumulation of the degraded matter, probably offers the best evidence of the lapse of time.

51:[e]

52 I remember having been much struck with the evidence of denudation, when viewing volcanic islands, which have been worn by the waves and pared all round into perpendicular cliffs of one or two thousand feet in height; for the gentle slope of the lava-streams, due to their formerly liquid state, showed at a glance how far the hard, rocky beds had once extended into the open ocean.

52:[e] = 39.8:e

53 The same story is still more plainly told by faults,—those great cracks along which the strata have been upheaved on one side, or thrown down on the other, to the height or depth of thousands of feet; for since the crust cracked, the surface of the land has been so completely planed down by the action of the sea, that no trace of these vast dislocations is externally visible.

53:c crust cracked (it makes no great difference whether the upheaval was sudden, or, as most geologists now believe, was very slow and effected by many starts), the surface

53:d sea that

53:[e] = 39.9:e

54 The Craven fault, for instance, extends for upwards of 30 miles, and along this line the vertical displacement of the strata has varied from 600 to 3000 feet.

54:[e] = 39.10:e

55 Prof. Ramsay has published an account of a downthrow in Anglesea of 2300 feet; and he informs me that he fully believes there is one in Merionethshire of 12,000 feet; yet in these cases there is nothing on the surface to show such prodigious movements; the pile of rocks on the one or other side having been smoothly swept away.

55:c surface of the land to

55:[e] = 39.11:e

56 The consideration of these facts impresses my mind almost in

the same manner as does the vain endeavour to grapple with the idea of eternity.

56:c impresses the mind

56:e these various facts

57 I am tempted to give one other case, the well-known one of the denudation of the Weald.

57:[c]

58 Though it must be admitted that the denudation of the Weald has been a mere trifle, in comparison with that which has removed masses of our palæozoic strata, in parts ten thousand feet in thickness, as shown in Prof. Ramsay's masterly memoir on this subject.

59 Yet it is an admirable lesson to stand on the North Downs and to look at the distant South Downs; for, remembering that at no great distance to the west the northern and southern escarpments meet and close, one can safely picture to oneself the great dome of rocks which must have covered up the Weald within so limited a period as since the latter part of the Chalk formation.

58 + 9:b subject: yet it is an admirable lesson to stand on the intermediate hilly country and look on the one hand at the North Downs, and on the other hand at the South

58 + 9:[c]

60 The distance from the northern to the southern Downs is about 22 miles, and the thickness of the several formations is on an average of about 1100 feet, as I am informed by Prof. Ramsay.

60:[c]

61 But if, as some geologists suppose, a range of older rocks underlies the Weald, on the flanks of which the overlying sedimentary deposits might have accumulated in thinner masses than elsewhere, the above estimate would be erroneous; but this source of doubt probably would not greatly affect the estimate as applied to the western extremity of the district.

61:[c]

62 If, then, we knew the rate at which the sea commonly wears away a line of cliff of any given height, we could measure the time requisite to have denuded the Weald.

62:[c]

63 This, of course, cannot be done; but we may, in order to form some crude notion on the subject, assume that the sea would eat into cliffs 500 feet in height at the rate of one inch in a century.

63:[c]

64 This will at first appear much too small an allowance; but it is the same as if we were to assume a cliff one yard in height to be eaten back along a whole line of coast at the rate of one yard in nearly every twenty-two years.

64:[c]

65 I doubt whether any rock, even as soft as chalk, would yield at this rate excepting on the most exposed coasts; though no doubt the degradation of a lofty cliff would be more rapid from the breakage of the fallen fragments.

65:[c]

66 On the other hand, I do not believe that any line of coast, ten or twenty miles in length, ever suffers degradation at the same time along its whole indented length; and we must remember that almost all strata contain harder layers or nodules, which from long resisting attrition form a breakwater at the base.

66:[c]

66.1:b We may at least confidently believe that no rocky coast 500 feet in height commonly yields at the rate of a foot per century; for this would be the same in amount as a cliff one yard in height retreating twelve yards in twenty-two years; and no one, I think, who has carefully observed the shape of old fallen fragments at the base of cliffs, will admit any near approach to such rapid wearing away.

66.1:[c]

67 Hence, under ordinary circumstances, I conclude that for a cliff 500 feet in height, a denudation of one inch per century for the whole length would be an ample allowance.

67:b circumstances, I should infer that/be a sufficient allowance.

67:[c]

68 At this rate, on the above data, the denudation of the Weald must have required 306,662,400 years; or say three hundred million years.

68:[c]

68.1:b But perhaps it would be safer to allow two or three inches per century, and this would reduce the number of years to one hundred and fifty or one hundred million years.

68.1:[c]

69 The action of fresh water on the gently inclined Wealden district, when upraised, could hardly have been great, but it would somewhat reduce the above estimate.

69:[c]

70 On the other hand, during oscillations of level, which we know this area has undergone, the surface may have existed for millions of years as land, and thus have escaped the action of the sea: when deeply submerged for perhaps equally long periods, it would, likewise, have escaped the action of the coast-waves.

70:[c]

71 So that in all probability a far longer period than 300 million years has elapsed since the latter part of the Secondary period.

71:b that it is not improbable that a longer

484

71:[c]

72   I have made these few remarks because it is highly important
      for us to gain some notion, however imperfect, of the lapse of
      years.

72:c lapse of time.

72:[e]

73 During each of these years, over the whole world, the land and
   the water has been peopled by hosts of living forms.

73:c each year/water have been

73:[e]

74 What an infinite number of generations, which the mind cannot
   grasp, must have succeeded each other in the long roll of years!

74:[e]

75 Now turn to our richest geological museums, and what a paltry
   display we behold!

75:[e] = 76.1:e

75.1-2:e[¶] Nevertheless this impression is partly false. 2 Mr. Croll,
   in a most interesting paper, remarks that we do not err "in
   forming too great a conception of the length of geological
   periods," but in estimating them by years.

75.2:f in an interesting

75.3:e When geologists look at large and complicated phenomena,
   and then at the figures representing several million years, the
   two produce a totally different effect on the mind, and the figures are at once pronounced to be too small.

75.3:f pronounced too

75.4:e But in regard to denudation, Mr. Croll shows, by calculat-
   ing the known amount of sediment annually brought down by
   certain rivers, relatively to the areas of drainage, that 1000 feet
   of rock, disintegrated through subaerial agencies, would thus
   be removed from the mean level of the whole area in the course
   of six million years.

*75.4:f In regard to subaerial denudation/rock, as it became gradu-
   ally disintegrated, would*

75.5:e This seems an astonishing result, and some considerations
   lead to the suspicion that it may be much too large, but even
   if halved or quartered it is still very surprising.

75.5:f be too

75.6-9:e Few of us, however, know what a million really means: Mr.
   Croll gives the following illustration: take a narrow strip of
   paper, 83 feet 4 inches in length, and stretch it along the wall
   of a large hall; then mark off at one end the tenth of an inch.
   7 This tenth of an inch will represent one hundred years, and
   the entire strip a million years. 8 But let it be borne in mind,
   in relation to the subject of this work, what a hundred years
   implies, represented as it is by a measure utterly insignificant

in a hall of the above dimensions. *9* Several eminent breeders, during a single lifetime, have so largely modified some of the higher animals, which propagate their kind much more slowly than most of the lower animals, that they have formed what well deserve to be called new sub-breeds.

75.9:f deserves to be called a new sub-breed.

75.10-2:e Few men have attended with due care to any one strain for more than half a century, so that a hundred years represents the work of two breeders in succession. *11* It is not to be supposed that species in a state of nature ever change so quickly as domestic animals under the guidance of methodical selection. *12* The comparison would be in every way fairer with the results which follow from unconscious selection, that is the preservation of the most useful or beautiful animals, with no intention of modifying the breed; but by this process of unconscious selection, various breeds have been sensibly changed in the course of two or three centuries.

75.12:f with the effects which

75.13-4:e[¶] Species, however, probably change much more slowly, and within the same country only a few change at the same time. *14* This slowness follows from all the inhabitants of the same country being already so well adapted to each other that new places in the polity of nature do not occur until after long intervals, when changes of some kind in the physical conditions or through immigration have occurred; and individual differences or variations of the right nature, by which some of the inhabitants might be better fitted to their new places under the altered circumstances, might not at once occur.

75.14.x-y:f other, that/intervals, due to the occurrence of physical changes of some kind, or through the immigration of new forms. *y* Moreover variations or individual differences of the right/circumstances, would not always occur at once.

75.15:e According to the standard of years we have no means of determining how long a period it takes to modify a species.

75.15:f Unfortunately we have no means of determining, according to the standard of years, how long a period it takes to modify a species; but to the subject of time we must return.

75.16-9:e Mr. Croll judging from the amount of heat-energy in the sun and from the date which he assigns to the last glacial epoch, estimates that only sixty million years have elapsed since the deposition of the first Cambrian formation. *17* This appears a very short period for so many and such great mutations in the forms of life, as have certainly since occurred. *18* It is admitted that many of the elements in the calculation are more or less doubtful, and Sir W. Thomson gives a wide margin to the possible age of the habitable world. *19* But as we have seen, we cannot comprehend what the figures 60,000,000 really imply; and during this, or perhaps a longer roll of years, the land and the waters have everywhere teemed with living creatures, all exposed to the struggle for life and undergoing change.

75.16-9:[f]

76 *On the poorness of our Palæontological collections.—*

76:d [Center] On the Poorness/Collections. [Space]

76.1:e = 75:[e][¶] Now let us turn

77 That our palæontological collections are very imperfect, is admitted by every one.

77:d[¶]

77:e[No ¶] our collections are very imperfect is

78 The remark of that admirable palæontologist, the late Edward Forbes, should not be forgotten, namely, that numbers of our fossil species are known and named from single and often broken specimens, or from a few specimens collected on some one spot.

78:e palæontologist, Edward

78:f should never be forgotten, namely, that very many fossil

79 Only a small portion of the surface of the earth has been geologically explored, and no part with sufficient care, as the important discoveries made every year in Europe prove.

80 No organism wholly soft can be preserved.

81 Shells and bones will decay and disappear when left on the bottom of the sea, where sediment is not accumulating.

81:f bones decay

82 I believe we are continually taking a most erroneous view, when we tacitly admit to ourselves that sediment is being deposited over nearly the whole bed of the sea, at a rate sufficiently quick to embed and preserve fossil remains.

82:e we often take an erroneous

82:f We probably take a quite erroneous view, when we assume that

83 Throughout an enormously large proportion of the ocean, the bright blue tint of the water bespeaks its purity.

84 The many cases on record of a formation conformably covered, after an enormous interval of time, by another and later formation, without the underlying bed having suffered in the interval any wear and tear, seem explicable only on the view of the bottom of the sea not rarely lying for ages in an unaltered condition.

84:e after an immense interval of time

85 The remains which do become embedded, if in sand or gravel, will when the beds are upraised generally be dissolved by the percolation of rain-water.

85:c rain-water charged with carbonic acid.

85:f will, when the beds are upraised, generally

86 I suspect that but few of the very many animals which live on the beach between high and low watermark are preserved.

86:c Some of the many kinds of animals/water mark seem to be rarely preserved.

87 For instance, the several species of the Chthamalinæ (a sub-family of sessile cirripedes) coat the rocks all over the world in infinite numbers: they are all strictly littoral, with the exception of a single Mediterranean species, which inhabits deep water and has been found fossil in Sicily, whereas not one other species has hitherto been found in any tertiary formation: yet it is now known that the genus Chthamalus existed during the chalk period.

87:c water, and

87:e and this has been found fossil

87:f is known

88 The molluscan genus Chiton offers a partially analogous case.

88:[d]

88.1:e Lastly, many great deposits requiring a vast length of time for their accumulation, are entirely destitute of organic remains, without our being able to assign any reason: one of the most striking instances is that of the Flysch formation, which consists of shale and sandstone, several thousand, occasionally even six thousand feet, in thickness, and extending for at least 300 miles from Vienna to Switzerland; and although this great mass has been most carefully searched, no fossils, except a few vegetable remains, have been found.

89   With respect to the terrestrial productions which lived during the Secondary and Palæozoic periods, it is superfluous to state that our evidence from fossil remains is fragmentary in an extreme degree.

89:f evidence is

90 For instance, not a land shell is known belonging to either of these vast periods, with one exception discovered by Sir C. Lyell in the carboniferous strata of North America.

90:b with the exception of one species discovered by Sir C. Lyell and Dr. Dawson in the carboniferous strata of North America, of which shell several specimens have now been collected.

90:c which shell above a hundred specimens

90:e shell until quite recently was known

90:f instance, until recently not a land shell was/America; but now land-shells have been found in the lias.

91 In regard to mammiferous remains, a single glance at the historical table published in the Supplement to Lyell's Manual, will bring home the truth, how accidental and rare is their preservation, far better than pages of detail.

91:d in Lyell's Manual will

91:f remains, a glance

92 Nor is their rarity surprising, when we remember how large a proportion of the bones of tertiary mammals have been discovered either in caves or in lacustrine deposits; and that not a cave or true lacustrine bed is known belonging to the age of our secondary or palæozoic formations.

93 But the imperfection in the geological record mainly results from another and more important cause than any of the foregoing; namely, from the several formations being separated from each other by wide intervals of time.

93:c record largely results

93.1:c This doctrine has been most emphatically admitted by many geologists and palæontologists, who, like E. Forbes, entirely disbelieve in the change of species.

93.1:f been emphatically

94 When we see the formations tabulated in written works, or when we follow them in nature, it is difficult to avoid believing that they are closely consecutive.

95 But we know, for instance, from Sir R. Murchison's great work on Russia, what wide gaps there are in that country between the superimposed formations; so it is in North America, and in many other parts of the world.

96 The most skilful geologist, if his attention had been exclusively confined to these large territories, would never have suspected that during the periods which were blank and barren in his own country, great piles of sediment, charged with new and peculiar forms of life, had elsewhere been accumulated.

96:d been confined exclusively to/that, during

97 And if in each separate territory, hardly any idea can be formed of the length of time which has elapsed between the consecutive formations, we may infer that this could nowhere be ascertained.

97:d if, in

98 The frequent and great changes in the mineralogical composition of consecutive formations, generally implying great changes in the geography of the surrounding lands, whence the sediment has been derived, accords with the belief of vast intervals of time having elapsed between each formation.

98:f sediment was derived, accord

99 But we can, I think, see why the geological formations of each region are almost invariably intermittent; that is, have not followed each other in close sequence.

99:f We

100 Scarcely any fact struck me more when examining many hundred miles of the South American coasts, which have been upraised several hundred feet within the recent period, than the

absence of any recent deposits sufficiently extensive to last for even a short geological period.

101 Along the whole west coast, which is inhabited by a peculiar marine fauna, tertiary beds are so scantily developed, that no record of several successive and peculiar marine faunas will probably be preserved to a distant age.

101:b so poorly developed

102 A little reflection will explain why along the rising coast of the western side of South America, no extensive formations with recent or tertiary remains can anywhere be found, though the supply of sediment must for ages have been great, from the enormous degradation of the coast-rocks and from muddy streams entering the sea.

102:d why, along

103 The explanation, no doubt, is, that the littoral and sub-littoral deposits are continually worn away, as soon as they are brought up by the slow and gradual rising of the land within the grinding action of the coast-waves.

104 We may, I think, safely conclude that sediment must be accumulated in extremely thick, solid, or extensive masses, in order to withstand the incessant action of the waves, when first upraised and during subsequent oscillations of level.

104:e think, conclude/during successive oscillations of level, as well as the subsequent subaerial degradation.

105 Such thick and extensive accumulations of sediment may be formed in two ways; either, in profound depths of the sea, in which case, judging from the researches of E. Forbes, we may conclude that the bottom will be inhabited by extremely few animals, and the mass when upraised will give a most imperfect record of the forms of life which then existed; or, sediment may be accumulated to any thickness and extent over a shallow bottom, if it continue slowly to subside.

105.x:c either in profound/by few animals, but it will not be, as we at last know from the telegraphic soundings, barren of life; consequently the mass/which existed during the period of deposition.

105.x:d as we now know from telegraphic and other deep soundings

105.x:e case the bottom will not be inhabited by so many and such varied forms of life, as the more shallow seas; and the mass when upraised will give an imperfect record of the organisms which existed throughout the world during the period of its accumulation.

105.x:f existed in the neighbourhood during

105.y:c Or

105.y:e Or, sediment may be deposited to any

106 In this latter case, as long as the rate of subsidence and supply of sediment nearly balance each other, the sea will remain shal-

low and favourable for life, and thus a fossiliferous formation thick enough, when upraised, to resist any amount of degradation, may be formed.

106:b thus a rich fossiliferous

106:d formation, thick/resist almost any

106:e and the supply/for many and varied forms, and/resist a large amount of denudation, may

107 I am convinced that all our ancient formations, which are rich in fossils, have thus been formed during subsidence.

107:c that nearly all/are throughout the greater part of their thickness *rich in fossils*

108 Since publishing my views on this subject in 1845, I have watched the progress of Geology, and have been surprised to note how author after author, in treating of this or that great formation, has come to the conclusion that it was accumulated during subsidence.

109 I may add, that the only ancient tertiary formation on the west coast of South America, which has been bulky enough to resist such degradation as it has as yet suffered, but which will hardly last to a distant geological age, was certainly deposited during a downward oscillation of level, and thus gained considerable thickness.

109:e age, was deposited

110 All geological facts tell us plainly that each area has undergone numerous slow oscillations of level, and apparently these oscillations have affected wide spaces.

111 Consequently formations rich in fossils and sufficiently thick and extensive to resist subsequent degradation, may have been formed over wide spaces during periods of subsidence, but only where the supply of sediment was sufficient to keep the sea shallow and to embed and preserve the remains before they had time to decay.

111:f Consequently, formations/degradation, will have

112 On the other hand, as long as the bed of the sea remained stationary, *thick* deposits could not have been accumulated in the shallow parts, which are the most favourable to life.

112:f remains stationary, *thick* deposits cannot have

113 Still less could this have happened during the alternate periods of elevation; or, to speak more accurately, the beds which were then accumulated will have been destroyed by being upraised and brought within the limits of the coast-action.

113:c will generally have

113:f less can this

114 Thus the geological record will almost necessarily be rendered intermittent.

114:[c]

115 I feel much confidence in the truth of these views, for they are in strict accordance with the general principles inculcated by Sir C. Lyell; and E. Forbes independently arrived at a similar conclusion.

115:b Forbes subsequently but independently

115:[c]

115.1-2:c[¶] These remarks apply chiefly to littoral and sub-littoral deposits. 2 In the case of an extensive and shallow sea, such as that within a large part of the Malay Archipelago, where the depth varies from 30 or 40 to 60 fathoms, a widely extended formation might be formed during a period of elevation and yet not suffer excessively from denudation during its slow up-heaval; but the thickness of the formation could not be great, for owing to the elevatory movement it would be less than the depth, supposed to be shallow; the deposit would not generally be much consolidated, nor would it be capped by overlying formations, so that it would run a good chance of being worn away during subsequent oscillations of level.

115.2:d elevation, and/than the depth in which it was formed; nor would the deposit be much consolidated, nor be capped/away by atmospheric degradation and by the action of the sea during

115.3:c It has been suggested by Mr. Hopkins, that if one part of the area, after rising and before being denuded, subsided, the deposit formed during the rising movement, though not thick, might become protected by fresh accumulations, and thus be preserved for an extremely long period,—a consideration which I formerly overlooked.

115.3:d has, however, been suggested/might afterwards become/for a long period.

115.4:c[¶] Mr. Hopkins, in commenting on this subject, states that he believes the entire destruction of any sedimentary bed of considerable horizontal extent to have been of rare occurrence.

115.4:d Mr. Hopkins also expresses his belief that sedimentary beds of considerable horizontal extent have rarely been completely destroyed.

115.5-7:c My remarks apply solely to beds rich in fossils: I have admitted that sediment accumulated in extremely thick, solid, or extensive masses would escape denudation. 6 The point in question is, whether widely extended formations, rich in fossils, and of sufficient thickness to last for a long period, would be formed except during periods of subsidence? 7 My impression is that this has rarely been the case.

115.5-7:[d]

115.8:c As the subject of complete denudation has been broached by Mr. Hopkins, I may remark that all geologists, excepting the few who believe that they see in the metamorphic schists and plutonic rocks the heated primordial nucleus of the globe, will probably admit that rocks of this nature must have been largely denuded.

115.8:d But all geologists, excepting the few who believe that our present metamorphic schists and plutonic rocks once formed the primordial/been denuded on an enormous scale.

115.8:e will admit that these latter rocks have

115.8:f been stript of their covering to an enormous extent.

115.9:c For it is scarcely possible that these rocks should have been solidified and crystallized in a naked condition; but if the metamorphic action occurred at profound depths of the ocean, the former mantle may not have been thick.

115.9:d rocks could have been solidified and crystallized whilst uncovered; but/former protecting mantle may not have been very thick.

115.9:e that such rocks/mantle of rock may

115.10:c Admitting then that such rocks as gneiss, mica-schist, granite, diorite, &c., were once necessarily covered up, how can we account for the extensive and naked areas of such rocks in many parts of the world, except on the belief that they have subsequently been completely denuded of all overlying strata?

115.10:d account for the naked and extensive areas

115.10:e then that gneiss

115.11-2:c That such extensive areas do exist cannot be doubted: the granite region of Parime is described by Humboldt as being at least nineteen times as large as Switzerland. *12* South of the Amazon Boué colours an area composed of such rocks as equal to that of Spain, France, Italy, part of Germany, and the British Islands, all conjoined.

115.12:d Amazon, Boué colours an area composed of rocks of this nature as

115.13:c This region has not been carefully explored, but from the concurrent testimony of travellers, the granite area must be very large: thus, Von Eschwege gives a detailed section of these rocks, stretching from Rio de Janeiro for 260 geographical miles inland in a straight line; and I travelled for 150 miles in another direction and saw nothing but granitic rocks.

115.13:e area is very/direction, and

115.14:c Numerous specimens, collected along the whole coast from near Rio Janeiro to the mouth of the Plata, a distance of 1100 geographical miles, were shown me, and they all belonged to this class.

115.14:d were examined by me

115.15-6:c Inland, along the whole northern bank of the Plata I saw, besides modern tertiary beds, only one small patch of slightly metamorphosed rock, which alone could have formed a part of the original capping of the granitic series. *16* Turning to a well-known region, namely, to the United States and Canada, as shown in Professor H. D. Rogers's beautiful map, I have estimated the areas by cutting out and weighing the paper, and I find that the metamorphic (excluding "the semi-metamorphic") and granitic rocks exceed, in the proportion of 19 to 12·5, not only the true coal measures, well known to be here developed in extraordinary force, but likewise the Umbral series, which together compose the whole newer Palæozoic formation.

115.16:d 12·5, the whole of the newer Palæozoic formations.

115.17:c In many regions the metamorphic and granitic areas would be greatly increased in size, if we could remove all the sedimentary beds which rest unconformably on them, and which at the line of junction have not been metamorphosed, showing that they could not have formed part of the original mantle under which the granitic rocks were crystallized.

115.17:d would be seen to be much more widely extended, if all the sedimentary beds were removed which rest unconformably on them, and which could not have formed part of the original mantle under which they were crystallized.

115.17:f would be found much more widely extended than they appear to be, if

115.18:c Hence it is probable that in some parts of the world whole formations, marking at least sub-stages in the several successive geological epochs, have been completely denuded, with not a wreck left behind.

115.18:e formations have

116 One remark is here worth a passing notice.

117 During periods of elevation the area of the land and of the adjoining shoal parts of the sea will be increased, and new stations will often be formed;—all circumstances most favourable, as previously explained, for the formation of new varieties and species; but during such periods there will generally be a blank in the geological record.

117:c circumstances favourable

118 On the other hand, during subsidence, the inhabited area and number of inhabitants will decrease (excepting the productions on the shores of a continent when first broken up into an archipelago), and consequently during subsidence, though there will be much extinction, fewer new varieties or species will be formed; and it is during these very periods of subsidence, that our great deposits rich in fossils have been accumulated.

118:d excepting on/accumulated. [*Space*]

118:e extinction, few new/that the deposits which are richest

119 Nature may almost be said to have guarded against the frequent discovery of her transitional or linking forms.

119:c her fine transitional

119:[d]

119.1:d [*Center*] On the Absence of Numerous Intermediate Varieties in any one Single Formation. [*Space*]

119.1:f any Single

120   From the foregoing considerations it cannot be doubted that the geological record, viewed as a whole, is extremely imperfect; but if we confine our attention to any one formation, it becomes more difficult to understand, why we do not therein find closely graduated varieties between the allied species which lived at its commencement and at its close.

120:c From these several considerations/becomes much more

120:d considerations, it cannot

120:f understand why

121 Some cases are on record of the same species presenting distinct varieties in the upper and lower parts of the same formation, but, as they are rare, they may be here passed over.

121:d Several cases/they are not common, they

121:e presenting varieties/formation: thus, Trautschold gives a number of instances with Ammonites; and Hilgendorf has described a most curious case of ten graduated forms of Planorbis multiformis in the successive beds of a fresh-water formation in Switzerland.

122 Although each formation has indisputably required a vast number of years for its deposition, I can see several reasons why each should not include a graduated series of links between the species which then lived; but I can by no means pretend to assign due proportional weight to the following considerations.

122:d which lived at its commencement and close; but I cannot assign

122:e deposition, several reasons can be given why each should not commonly include

123   Although each formation may mark a very long lapse of years, each perhaps is short compared with the period requisite to change one species into another.

123:d years, each probably is

124 I am aware that two palæontologists, whose opinions are worthy of much deference, namely Bronn and Woodward, have concluded that the average duration of each formation is twice or thrice as long as the average duration of specific forms.

125 But insuperable difficulties, as it seems to me, prevent us coming to any just conclusion on this head.

125:d us from coming

126 When we see a species first appearing in the middle of any formation, it would be rash in the extreme to infer that it had not elsewhere previously existed.

127 So again when we find a species disappearing before the uppermost layers have been deposited, it would be equally rash to suppose that it then became wholly extinct.

127:d became extinct.

127:e before the last layers

128 We forget how small the area of Europe is compared with the rest of the world; nor have the several stages of the same formation throughout Europe been correlated with perfect accuracy.

129 With marine animals of all kinds, we may safely infer a large amount of migration during climatal and other changes; and when we see a species first appearing in any formation, the probability is that it only then first immigrated into that area.

129:f We may safely infer that with marine animals of all kinds there has been a large amount of migration due to climatal

130 It is well known, for instance, that several species appeared somewhat earlier in the palæozoic beds of North America than in those of Europe; time having apparently been required for their migration from the American to the European seas.

131 In examining the latest deposits of various quarters of the world, it has everywhere been noted, that some few still existing species are common in the deposit, but have become extinct in the immediately surrounding sea; or, conversely, that some are now abundant in the neighbouring sea, but are rare or absent in this particular deposit.

131:d deposits in various

132 It is an excellent lesson to reflect on the ascertained amount of migration of the inhabitants of Europe during the Glacial period, which forms only a part of one whole geological period; and likewise to reflect on the great changes of level, on the inordinately great change of climate, on the prodigious lapse of time, all included within this same glacial period.

132:c during the glacial period, which

132:e likewise to reflect on the changes of level, on the extreme change of climate, and on the great lapse

132:f glacial epoch, which

133 Yet it may be doubted whether in any quarter of the world, sedimentary deposits, *including fossil remains,* have gone on accumulating within the same area during the whole of this period.

133:d whether, in

134 It is not, for instance, probable that sediment was deposited during the whole of the glacial period near the mouth of the Mississippi, within that limit of depth at which marine animals can flourish; for we know what vast geographical changes occurred in other parts of America during this space of time.

134:e can best flourish: for we know that great geographical

135 When such beds as were deposited in shallow water near the mouth of the Mississippi during some part of the glacial period shall have been upraised, organic remains will probably first

496

appear and disappear at different levels, owing to the migration of species and to geographical changes.

135:e migrations

136 And in the distant future, a geologist examining these beds, might be tempted to conclude that the average duration of life of the embedded fossils had been less than that of the glacial period, instead of having been really far greater, that is extending from before the glacial epoch to the present day.

136:c is, extending

136:d geologist, examining these beds, would be

137 In order to get a perfect gradation between two forms in the upper and lower parts of the same formation, the deposit must have gone on accumulating for a very long period, in order to have given sufficient time for the slow process of variation; hence the deposit will generally have to be a very thick one; and the species undergoing modification will have had to live on the same area throughout this whole time.

137:c period, so that there may have been time sufficient for

137:d deposit will have to go on continuously accumulating during a very long period, so that there may be time/hence the deposit will have to be a very thick one; and the species undergoing modification will have to live in the same district throughout

137:e process of modification; hence/undergoing change will

137:f deposit must have gone on continuously accumulating during a long period, sufficient for the slow process of modification; hence the deposit must be a very thick one; and the species undergoing change must have lived in the same district throughout the whole

138 But we have seen that a thick fossiliferous formation can only be accumulated during a period of subsidence; and to keep the depth approximately the same, which is necessary in order to enable the same species to live on the same space, the supply of sediment must nearly have counterbalanced the amount of subsidence.

138:c thick formation, fossiliferous throughout its thickness, can/ necessary that the same species may live

138:d can accumulate only during/necessary that the same marine species/nearly counterbalance

138:f its entire thickness

139 But this same movement of subsidence will often tend to sink the area whence the sediment is derived, and thus diminish the supply whilst the downward movement continues.

139:d tend to submerge the area

139:e will tend

139:f supply, whilst

140 In fact, this nearly exact balancing between the supply of sediment and the amount of subsidence is probably a rare contingency; for it has been observed by more than one palæontologist, that very thick deposits are usually barren of organic remains, except near their upper or lower limits.

141 It would seem that each separate formation, like the whole pile of formations in any country, has generally been intermittent in its accumulation.

142 When we see, as is so often the case, a formation composed of beds of different mineralogical composition, we may reasonably suspect that the process of deposition has been much interrupted, as a change in the currents of the sea and a supply of sediment of a different nature will generally have been due to geographical changes requiring much time.

142:f beds of widely different/been more or less interrupted.

143 Nor will the closest inspection of a formation give any idea of the time which its deposition has consumed.

143:f give us any idea of the length of time which its deposition may have consumed.

144 Many instances could be given of beds only a few feet in thickness, representing formations, elsewhere thousands of feet in thickness, and which must have required an enormous period for their accumulation; yet no one ignorant of this fact would have suspected the vast lapse of time represented by the thinner formation.

144:f formations, which are elsewhere/would have even suspected

145 Many cases could be given of the lower beds of a formation having been upraised, denuded, submerged, and then re-covered by the upper beds of the same formation,—facts, showing what wide, yet easily overlooked, intervals have occurred in its accumulation.

146 In other cases we have the plainest evidence in great fossilised trees, still standing upright as they grew, of many long intervals of time and changes of level during the process of deposition, which would never even have been suspected, had not the trees chanced to have been preserved: thus, Messrs. Lyell and Dawson found carboniferous beds 1400 feet thick in Nova Scotia, with ancient root-bearing strata, one above the other, at no less than sixty-eight different levels.

146:b thus Messrs.

146:d thus Sir C. Lyell and Dr. Dawson

146:f would not have been suspected, had not the trees been/other at

147 Hence, when the same species occur at the bottom, middle, and top of a formation, the probability is that they have not lived on the same spot during the whole period of deposition, but

have disappeared and reappeared, perhaps many times, during the same geological period.

147:f occurs/that it has not/but has disappeared

148 So that if such species were to undergo a considerable amount of modification during any one geological period, a section would not probably include all the fine intermediate gradations which must on my theory have existed between them, but abrupt, though perhaps very slight, changes of form.

148:c perhaps slight

148:d that, if

148:e not include/on our theory

148:f Consequently if it were to undergo a considerable amount of modification during the deposition of any one geological formation, a section/existed, but

149 It is all-important to remember that naturalists have no golden rule by which to distinguish species and varieties; they grant some little variability to each species, but when they meet with a somewhat greater amount of difference between any two forms, they rank both as species, unless they are enabled to connect them together by close intermediate gradations.

149:d together by the closest intermediate

150 And this from the reasons just assigned we can seldom hope to effect in any one geological section.

149 + 150:f gradations; and this, from the reasons just assigned, we

151 Supposing B and C to be two species, and a third, A, to be found in an underlying bed; even if A were strictly intermediate between B and C, it would simply be ranked as a third and distinct species, unless at the same time it could be most closely connected with either one or both forms by intermediate varieties.

151:c underlying and older bed

151:e found in an older and underlying bed

151:f could be closely connected by intermediate varieties with either one or both forms.

152 Nor should it be forgotten, as before explained, that A might be the actual progenitor of B and C, and yet might not at all necessarily be strictly intermediate between them in all points of structure.

152:d not necessarily/all respects.

152:f yet would not

153 So that we might obtain the parent-species and its several modified descendants from the lower and upper beds of a formation, and unless we obtained numerous transitional gradations, we should not recognise their relationship, and should consequently be compelled to rank them all as distinct species.

153:c their blood-relationship

153:e of the same formation/them as distinct

153:f consequently rank

154   It is notorious on what excessively slight differences many palæontologists have founded their species; and they do this the more readily if the specimens come from different sub-stages of the same formation.

155 Some experienced conchologists are now sinking many of the very fine species of D'Orbigny and others into the rank of varieties; and on this view we do find the kind of evidence of change which on my theory we ought to find.

155:e on the theory

155.1:c Look again at the later tertiary deposits, which include many shells believed by the majority of naturalists to be identical with existing species; but some excellent naturalists, as Agassiz and Pictet, maintain that all these tertiary species are specifically distinct, though the distinction is admitted to be very slight; so that here, unless we believe that these eminent naturalists have been misled by their imaginations and that these late tertiary species really present no difference whatever from their living representatives, or unless we believe that the great majority of naturalists are wrong and that the tertiary species are all truly distinct from the recent, we have evidence of a very general slight modification of form of the kind required.

155.1:d imaginations, and that these/evidence of frequent occurrence of slight modifications of the kind

155.1:e evidence of the frequent

155.1:f or unless we admit, in opposition to the judgment of most naturalists, that these tertiary

156 Moreover, if we look to rather wider intervals, namely, to distinct but consecutive stages of the same great formation, we find that the embedded fossils, though almost universally ranked as specifically different, yet are far more closely allied to each other than are the species found in more widely separated formations; but to this subject I shall have to return in the following chapter.

156:c If/intervals of time, namely/related to each/formations; so that here again we have undoubted evidence of change, though not strictly of variation, in the direction required by my theory; but to this latter subject

156:e change in the direction required by the theory

156:f though universally/shall return

157   One other consideration is worth notice: with animals and plants that can propagate rapidly and are not highly locomotive, there is reason to suspect, as we have formerly seen, that their varieties are generally at first local; and that such local varieties do not spread widely and supplant their parent-forms

until they have been modified and perfected in some considerable degree.

157:d With animals and plants that propagate rapidly and do not wander much, there

158 According to this view, the chance of discovering in a formation in any one country all the early stages of transition between any two forms, is small, for the successive changes are supposed to have been local or confined to some one spot.

158:d two such forms

158:f two forms

159 Most marine animals have a wide range; and we have seen that with plants it is those which have the widest range, that oftenest present varieties; so that with shells and other marine animals, it is probably those which have had the widest range, far exceeding the limits of the known geological formations of Europe, which have oftenest given rise, first to local varieties and ultimately to new species; and this again would greatly lessen the chance of our being able to trace the stages of transition in any one geological formation.

159:c other marine animals, it is highly probable that those

159:e other marine animals, it is probable

158:f so that, with/probable that those which had the widest/Europe, have oftenest

159.1:d[¶] It is a more important consideration, clearly leading to the same result, as lately insisted on by Dr. Falconer, namely, that the periods during which species have been undergoing modification, though very long as measured by years, have probably been short in comparison with the periods during which these same species remained without undergoing any change.

159.1:e consideration, leading/period during which each species underwent modification, though long as measured by years, was, from the reasons lately assigned, probably short in comparison with that during which it remained

159.1:f was probably

159.2-3:d We may infer that this has been the case, from there being no inherent tendency in organic beings to become modified or to progress in structure, and from all modifications depending, firstly on long-continued variability, and secondly on changes in the physical conditions of life, or on changes in the habits and structure of competing species, or on the immigration of new forms; and such contingencies will supervene in most cases only after long intervals of time and at a slow rate. 3 These changes, moreover, in the organic and inorganic conditions of life will affect only a limited number of the inhabitants of any one area or country.

159.2-3:[e]

160   It should not be forgotten, that at the present day, with perfect specimens for examination, two forms can seldom be connected by intermediate varieties and thus proved to be the
same species, until many specimens have been collected from
many places; and in the case of fossil species this could rarely
be effected by palæontologists.

160:c varieties, and thus

160:e many specimens are collected

160:f places; and with fossil species this can rarely be done.

161   We shall, perhaps, best perceive the improbability of our being
enabled to connect species by numerous, fine, intermediate,
fossil links, by asking ourselves whether, for instance, geologists at some future period will be able to prove, that our
different breeds of cattle, sheep, horses, and dogs have descended from a single stock or from several aboriginal stocks;
or, again, whether certain sea-shells inhabiting the shores of
North America, which are ranked by some conchologists as
distinct species from their European representatives, and by
other conchologists as only varieties, are really varieties or
are, as it is called, specifically distinct.

161:c intermediate fossil/prove that/really varieties, or

161:e dogs are descended

162   This could be effected only by the future geologist discovering
in a fossil state numerous intermediate gradations; and such
success seems to me improbable in the highest degree.

162:c success is improbable

162:f effected by the future geologist only by his discovering

162.1:c[¶] It has been asserted over and over again, by writers who
believe in the immutability of species, that geology has yielded
no linking forms.

162.1:d geology yields

162.2:c This assertion is entirely erroneous.

162.2:f assertion, as we shall see in the next chapter, is certainly
erroneous.

162.3:c As Mr. Lubbock has recently remarked, "Every species is a
link between other allied forms."

162.3:d As Sir J. Lubbock has remarked

162.4:c We clearly see this if we take a genus having a score of
recent and extinct species and destroy four-fifths of them; for
in this case no one will doubt that the remainder will stand
much more distinct from each other.

162.4:d one doubts

162.4:f If we take a genus having a score of species, recent and
extinct, and destroy four-fifths of them, no

162.5:c If the extreme forms in the genus happen to have been thus
destroyed, the genus itself in most cases will stand more distinct from other allied genera.

162.5:f itself will

162.6:c The camel and the pig, or the horse and the tapir, are now obviously very distinct forms; but if we add the several fossil quadrupeds which have already been discovered to the families including the camel and pig, these forms become joined by links not extremely wide apart.

162.6:[f]

162.7:c The chain of linking forms does not, however, in these cases, or in any case, run straight from the one living form to the other, but takes a circuitous sweep through the forms which lived during long past ages.

162.7:d long-past

162.7:[f]

163   Geological research, though it has added numerous species to existing and extinct genera, and has made the intervals between some few groups less wide than they otherwise would have been, yet has done scarcely anything in breaking down the distinction between species, by connecting them together by numerous, fine, intermediate varieties; and this not having been effected, is probably the gravest and most obvious of all the many objections which may be urged against my views.

163.x:c[No ¶] What geological research has not revealed is the former existence of infinitely numerous gradations, as fine as existing varieties, connecting all known species.

163.x:d revealed, is

163.x:e connecting nearly all our existing species with extinct species.

163.x:f all existing and extinct

163.y:c And this not having been effected by geology is the most obvious of the many

163.y:e But this ought not to be expected; yet this has been repeatedly advanced as a most serious objection against

164   Hence it will be worth while to sum up the foregoing remarks, under an imaginary illustration.

164:c[¶] remarks on the causes of the imperfection of the geological record under

164:e It may be worth

165   The Malay Archipelago is of about the size of Europe from the North Cape to the Mediterranean, and from Britain to Russia; and therefore equals all the geological formations which have been examined with any accuracy, excepting those of the United States of America.

166   I fully agree with Mr. Godwin-Austen, that the present condition of the Malay Archipelago, with its numerous large islands separated by wide and shallow seas, probably represents the former state of Europe, when most of our formations were accumulating.

166:b Europe, whilst most

167 The Malay Archipelago is one of the richest regions of the whole world in organic beings; yet if all the species were to be collected which have ever lived there, how imperfectly would they represent the natural history of the world!

167:f regions in organic

168 But we have every reason to believe that the terrestrial productions of the archipelago would be preserved in an excessively imperfect manner in the formations which we suppose to be there accumulating.

168:e in an extremely imperfect

169 I suspect that not many of the strictly littoral animals, or of those which lived on naked submarine rocks, would be embedded; and those embedded in gravel or sand, would not endure to a distant epoch.

169:d sand would

169:e Not many

170 Wherever sediment did not accumulate on the bed of the sea, or where it did not accumulate at a sufficient rate to protect organic bodies from decay, no remains could be preserved.

171 In our archipelago, I believe that fossiliferous formations could be formed of sufficient thickness to last to an age, as distant in futurity as the secondary formations lie in the past, only during periods of subsidence.

171:b I believe that fossiliferous formations could be formed in the archipelago, of thickness sufficient to last to an age as distant

171:c As the common rule formations rich in fossils would be formed

171:d Formations rich in fossils, and of thickness/past, would generally be formed in the archipelago only

171:e fossils of many kinds, and of thickness

172 These periods of subsidence would be separated from each other by enormous intervals, during which the area would be either stationary or rising; whilst rising, each fossiliferous formation would be destroyed, almost as soon as accumulated, by the incessant coast-action, as we now see on the shores of South America.

172:c whilst rising, the fossiliferous formations on the steeper shores would/America; even throughout the extensive and shallow seas within the archipelago sedimentary beds could not, during the periods of elevation, be accumulated of great thickness, or become capped and protected by subsequent deposits, so as to have a good chance of enduring to an extremely distant future.

172:d intervals of time, during

172:e by immense intervals/beds could hardly be accumulated of

great thickness during the periods of elevation, or become/
enduring to a very distant

172.*x-y:f* America. *y* Even/archipelago, sedimentary

173 During the periods of subsidence there would probably be
much extinction of life; during the periods of elevation, there
would be much variation, but the geological record would then
be least perfect.

173:c subsidence, there would probably

173:f then be less perfect.

174 It may be doubted whether the duration of any one great
period of subsidence over the whole or part of the archipelago,
together with a contemporaneous accumulation of sediment,
would *exceed* the average duration of the same specific forms;
and these contingencies are indispensable for the preservation
of all the transitional gradations between any two or more
species.

175 If such gradations were not fully preserved, transitional varieties would merely appear as so many distinct species.

175:c not all fully

175:e many new and distinct

175:f new, though closely allied species.

176 It is, also, probable that each great period of subsidence would
be interrupted by oscillations of level, and that slight climatal
changes would intervene during such lengthy periods; and in
these cases the inhabitants of the archipelago would have to
migrate, and no closely consecutive record of their modifications could be preserved in any one formation.

176:c It is also probable/archipelago would migrate

177 Very many of the marine inhabitants of the archipelago now
range thousands of miles beyond its confines; and analogy leads
me to believe that it would be chiefly these far-ranging species
which would oftenest produce new varieties; and the varieties
would at first generally be local or confined to one place, but
if possessed of any decided advantage, or when further modified
and improved, they would slowly spread and supplant their
parent-forms.

177:c analogy plainly leads to the belief that

177:d species, though only some of them, which

177:f first be

178 When such varieties returned to their ancient homes, as they
would differ from their former state, in a nearly uniform,
though perhaps extremely slight degree, they would, according
to the principles followed by many palæontologists, be ranked
as new and distinct species.

178:c degree, and as they would be found embedded in slightly
different sub-stages of the same formation, they

178:f state in

179   If then, there be some degree of truth in these remarks, we have no right to expect to find in our geological formations, an infinite number of those fine transitional forms, which on my theory assuredly have connected all the past and present species of the same group into one long and branching chain of life.

179:c If then there/forms which, on my theory, have

179:d find, in

179:e on our theory

180 We ought only to look for a few links, some more closely, some more distantly related to each other; and these links, let them be ever so close, if found in different stages of the same formation, would, by most palæontologists, be ranked as distinct species.

180:c links, and such assuredly we do find—some more distantly, some more closely, related/by many palæontologists

181 But I do not pretend that I should ever have suspected how poor a record of the mutations of life, the best preserved geological section presented, had not the difficulty of our not discovering innumerable transitional links between the species which appeared at the commencement and close of each formation, pressed so hardly on my theory.

181:c life the best

181:d section would present

181:e poor was the record in the best preserved geological sections, had not the absence of innumerable transitional links between the species which lived at the commencement

182   *On the sudden appearance of whole groups of Allied Species.—*

182:c allied

182:d [Center] *On the sudden Appearance of whole Groups of allied Species.* [Space]

183 The abrupt manner in which whole groups of species suddenly appear in certain formations, has been urged by several palæontologists, for instance, by Agassiz, Pictet, and by none more forcibly than by Professor Sedgwick, as a fatal objection to the belief in the transmutation of species.

183:b palæontologists—for/Sedgwick—as

183:c Pictet, and Sedgwick

183:d [¶]

184 If numerous species, belonging to the same genera or families, have really started into life all at once, the fact would be fatal to the theory of descent with slow modification through natural selection.

184:c life at

184:f of evolution through

185 For the development of a group of forms, all of which have descended from some one progenitor, must have been an extremely slow process; and the progenitors must have lived long ages before their modified descendants.

185:f development by this means of/which are descended/long before

186 But we continually over-rate the perfection of the geological record, and falsely infer, because certain genera or families have not been found beneath a certain stage, that they did not exist before that stage.

186.1:c In all cases positive palæontological evidence may be implicitly trusted; negative evidence is worthless, as experience has so often shown.

187 We continually forget how large the world is, compared with the area over which our geological formations have been carefully examined; we forget that groups of species may elsewhere have long existed and have slowly multiplied before they invaded the ancient archipelagoes of Europe and of the United States.

187:c existed, and have slowly multiplied, before/Europe and the United

188 We do not make due allowance for the enormous intervals of time, which have probably elapsed between our consecutive formations,—longer perhaps in some cases than the time required for the accumulation of each formation.

188:c time which have elapsed/in many cases

188:f for the intervals

189 These intervals will have given time for the multiplication of species from some one or some few parent-forms; and in the succeeding formation such species will appear as if suddenly created.

189:c such groups of species

189:f one parent-form; and in the succeeding formation, such groups or species

190   I may here recall a remark formerly made, namely that it might require a long succession of ages to adapt an organism to some new and peculiar line of life, for instance to fly through the air; but that when this had been effected, and a few species had thus acquired a great advantage over other organisms, a comparatively short time would be necessary to produce many divergent forms, which would be able to spread rapidly and widely throughout the world.

190:c namely, that it/instance, to fly through the air; and consequently that the transitional forms would often long remain confined to some one region; but that when this adaptation had once been

190:d but that, when

190:e which would spread

190.1-4:c Professor Pictet, in his excellent Review of this work, in commenting on early transitional forms, and taking birds as an illustration, cannot see how the successive modifications of the anterior limbs of a supposed prototype could possibly have been of any advantage. *2* But look at the penguins of the Southern Ocean; have not these birds their front limbs in this precise intermediate state of "neither true arms nor true wings"? *3* Yet these birds hold their place victoriously in the battle for life; for they exist in infinite numbers and of many kinds. *4* I do not suppose that we here see the real transitional grade through which the wings of birds have passed; but what special difficulty is there in believing that it might profit the modified descendants of the penguin, first to become enabled to flap along the surface of the sea like the logger-headed duck, and ultimately to rise from its surface and glide through the air?

190.4:d grades

191   I will now give a few examples to illustrate these remarks; and to show how liable we are to error in supposing that whole groups of species have suddenly been produced.

191:b remarks, and

191:c illustrate the foregoing remarks

191.1:c Even in so short an interval as that between the first and second editions of Pictet's great work on Palæontology, published in 1844-46 and in 1853-57, the conclusions on the first appearance and disappearance of several groups of animals have been considerably modified; and a third edition would require still further changes.

192   I may recall the well-known fact that in geological treatises, published not many years ago, the great class of mammals was always spoken of as having abruptly come in at the commencement of the tertiary series.

192:c ago, the whole class

192:e ago, mammals were always

193   And now one of the richest known accumulations of fossil mammals belongs to the middle of the secondary series; and one true mammal has been discovered in the new red sandstone at nearly the commencement of this great series.

193:b mammals, for its thickness, belongs

193:d series; and true mammals have been

193:f mammals belongs

194   Cuvier used to urge that no monkey occurred in any tertiary stratum; but now extinct species have been discovered in India, South America, and in Europe even as far back as the eocene stage.

194:d Europe, as far back as the miocene stage.

194.1:b Had it not been for the rare accident of the preservation of footsteps in the new red sandstone of the United States, who would have ventured to suppose that, besides reptiles, no less than at least thirty kinds of birds, some of gigantic size, existed during that period?

194.1:f suppose that no less than at least thirty different bird-like animals, some

194.2-3:b Not a fragment of bone has been discovered in these beds. *3* Notwithstanding that the number of joints shown in the fossil impressions correspond with the number in the several toes of living birds' feet, some authors doubt whether the animals which left the impressions were really birds.

194.3:d corresponds/left these impressions

194.3:[f]

194.4:b Until quite recently these authors might have maintained, and some have maintained, that the whole class of birds came suddenly into existence during an early tertiary period; but now we know, on the authority of Professor Owen (as may be seen in Lyell's 'Manual'), that a bird certainly lived during the deposition of the upper green-sand.

194.4:c Owen, that

194.4:d during the eocene period/greensand; and still more recently, that strange bird, the Archeopteryx, with a long lizard-like tail, bearing a pair of feathers on each joint, and with its wings furnished with two free claws, has been discovered in the oolitic slates of Solenhofen.

194.4:f Not long ago, palæontologists maintained that the whole

194.4.1:d Hardly any recent discovery shows more forcibly than this how little we as yet know of the former inhabitants of the world.

194.4.1:e this, how

195 The most striking case, however, is that of the Whale family; as these animals have huge bones, are marine, and range over the world, the fact of not a single bone of a whale having been discovered in any secondary formation, seemed fully to justify the belief that this great and distinct order had been suddenly produced in the interval between the latest secondary and earliest tertiary formation.

195:[b]

196 But now we may read in the Supplement to Lyell's 'Manual,' published in 1858, clear evidence of the existence of whales in the upper greensand, some time before the close of the secondary period.

196:[b]

197 I may give another instance, which from having passed under my own eyes has much struck me.

197:c which, from/eyes, has

198 In a memoir on Fossil Sessile Cirripedes, I have stated that, from the number of existing and extinct tertiary species; from the extraordinary abundance of the individuals of many species all over the world, from the Arctic regions to the equator, inhabiting various zones of depths from the upper tidal limits to 50 fathoms; from the perfect manner in which specimens are preserved in the oldest tertiary beds; from the ease with which even a fragment of a valve can be recognised; from all these circumstances, I inferred that had sessile cirripedes existed during the secondary periods, they would certainly have been preserved and discovered; and as not one species had been discovered in beds of this age, I concluded that this great group had been suddenly developed at the commencement of the tertiary series.

198:c one species had then been discovered

198:d inferred that, had sessile

198:e stated that, from the large number

198:f Cirripedes, I stated

199 This was a sore trouble to me, adding as I thought one more instance of the abrupt appearance of a great group of species.

199:f I then thought

200 But my work had hardly been published, when a skilful palæontologist, M. Bosquet, sent me a drawing of a perfect specimen of an unmistakeable sessile cirripede, which he had himself extracted from the chalk of Belgium.

201 And, as if to make the case as striking as possible, this sessile cirripede was a Chthamalus, a very common, large, and ubiquitous genus, of which not one specimen has as yet been found even in any tertiary stratum.

201:f this cirripede/one species has

202 Hence we now positively know that sessile cirripedes existed during the secondary period; and these cirripedes might have been the progenitors of our many tertiary and existing species.

202:f Still more recently, a Pyrgoma, a member of a distinct subfamily of sessile cirripedes, has been discovered by Mr. Woodward in the upper chalk; so that we now have abundant evidence of the existence of this group of animals during the secondary period.

203 The case most frequently insisted on by palæontologists of the apparently sudden appearance of a whole group of species, is that of the teleostean fishes, low down in the Chalk period.

203:f down, according to Agassiz, in

204 This group includes the large majority of existing species.

205 Lately, Professor Pictet has carried their existence one sub-stage further back; and some palæontologists believe that certain

much older fishes, of which the affinities are as yet imperfectly known, are really teleostean.

205:f But certain Jurassic and Triassic forms are now commonly admitted to be teleostean; and even some palæozoic forms have thus been classed by one high authority.

206 Assuming, however, that the whole of them did appear, as Agassiz believes, at the commencement of the chalk formation, the fact would certainly be highly remarkable; but I cannot see that it would be an insuperable difficulty on my theory, unless it could likewise be shown that the species of this group appeared suddenly and simultaneously throughout the world at this same period.

206:d Agassiz maintains, at the commencement

206:e insuperable objection to these views, unless

206:f If the teleosteans had really appeared suddenly in the northern hemisphere at the commencement of the chalk formation, the fact would have been highly remarkable; but it would not have formed an insuperable difficulty, unless it could likewise have been shown that at the same period the species were suddenly and simultaneously developed in other quarters of the world.

207 It is almost superfluous to remark that hardly any fossil-fish are known from south of the equator; and by running through Pictet's Palæontology it will be seen that very few species are known from several formations in Europe.

208 Some few families of fish now have a confined range; the teleostean fish might formerly have had a similarly confined range, and after having been largely developed in some one sea, might have spread widely.

208:f teleostean fishes might/sea, have

209 Nor have we any right to suppose that the seas of the world have always been so freely open from south to north as they are at present.

210 Even at this day, if the Malay Archipelago were converted into land, the tropical parts of the Indian Ocean would form a large and perfectly enclosed basin, in which any great group of marine animals might be multiplied; and here they would remain confined, until some of the species became adapted to a cooler climate, and were enabled to double the southern capes of Africa or Australia, and thus reach other and distant seas.

210:f Southern

211 From these and similar considerations, but chiefly from our ignorance of the geology of other countries beyond the confines of Europe and the United States; and from the revolution in our palæontological ideas on many points, which the discoveries of even the last dozen years have effected, it seems to me

to be about as rash in us to dogmatize on the succession of organic beings throughout the world, as it would be for a naturalist to land for five minutes on some one barren point in Australia, and then to discuss the number and range of its productions.

211:e these considerations, from our ignorance/States, and from the revolution in our palæontological knowledge effected by the discoveries of the last dozen years, it seems to me to be about as rash to dogmatize on the succession of organic forms throughout/minutes on a barren

212   *On the sudden appearance of groups of Allied Species in the lowest known fossiliferous strata.—*

212:d [*Center*] *On the sudden Appearance of Groups of allied Species in the lowest known Fossiliferous Strata.* [*Space*]

213   There is another and allied difficulty, which is much graver.

213:d[¶] much more serious.

214   I allude to the manner in which numbers of species of the same group, suddenly appear in the lowest known fossiliferous rocks.

214:d group suddenly

214:e which many species in several of the main divisions of the animal kingdom suddenly

214:f which species belonging to several

215   Most of the arguments which have convinced me that all the existing species of the same group have descended from one progenitor, apply with nearly equal force to the earliest known species.

215:e group are descended from a single progenitor

215:f with equal

216   For instance, I cannot doubt that all the Silurian trilobites have descended from some one crustacean, which must have lived long before the Silurian age, and which probably differed greatly from any known animal.

216:e instance, it cannot be doubted that all the Silurian trilobites are descended

216:f all the Cambrian and Silurian/before the Cambrian age

217   Some of the most ancient Silurian animals, as the Nautilus, Lingula, &c., do not differ much from living species; and it cannot on my theory be supposed, that these old species were the progenitors of all the species of the orders to which they belong, for they do not present characters in any degree intermediate between them.

217:d all the species belonging to the same group which have subsequently appeared, for

217:e on our theory/same groups which have subsequently appeared, for they are not in any degree intermediate in character.

217:f ancient animals

218 If, moreover, they had been the progenitors of these orders, they would almost certainly have been long ago supplanted and exterminated by their numerous and improved descendants.

218:[d]

219 Consequently, if my theory be true, it is indisputable that before the lowest Silurian stratum was deposited, long periods elapsed, as long as, or probably far longer than, the whole interval from the Silurian age to the present day; and that during these vast, yet quite unknown, periods of time, the world swarmed with living creatures.

219:e if the theory/Silurian or Cambrian stratum was deposited long periods elapsed/from the Cambrian age/vast periods the world

219:f lowest Cambrian stratum was deposited, long periods elapsed

219.1:e Here we encounter a formidable objection; for it seems doubtful whether the earth in a fit state for the habitation of living creatures has lasted long enough.

219.1:f earth, in/creatures, has

219.2-3:e Sir W. Thompson concludes that the consolidation of the crust can hardly have occurred less than 20 or more than 400 millions years ago, but probably not less than 98 or more than 200 million years. *3* These very wide limits show how doubtful the data are; and other elements may have to be introduced into the problem.

219.3:f have hereafter to

219.4:e Mr. Croll estimates that about 60 million years have elapsed since the Cambrian period, but this, judging from the small amount of organic change since the commencement of the Glacial epoch, seems a very short time for the many and great mutations of life, which have certainly occurred since the Cambrian formation; and the previous 140 million years can hardly be considered as sufficient for the development of the varied forms of life which certainly existed toward the close of the Cambrian period.

219.4:f epoch, appears a very/forms of life which already existed during the Cambrian period.

219.4.1:f It is, however, probable, as Sir William Thompson insists, that the world at a very early period was subjected to more rapid and violent changes in its physical conditions than those now occurring; and such changes would have tended to induce changes at a corresponding rate in the organisms which then existed.

513

220 To the question why we do not find records of these vast primordial periods, I can give no satisfactory answer.

220:d find richly fossiliferous records

220:e find rich fossiliferous deposits belonging to these assumed earliest periods

220:f periods prior to the Cambrian system, I

221 Several of the most eminent geologists, with Sir R. Murchison at their head, are convinced that we see in the organic remains of the lowest Silurian stratum the dawn of life on this planet.

221:d Several eminent/head, were until recently convinced that we beheld in the organic remains of the lowest Silurian stratum the first dawn

221:e life.

222 Other highly competent judges, as Lyell and the late E. Forbes, dispute this conclusion.

222:d disputed

222:e and E. Forbes, have disputed

223 We should not forget that only a small portion of the world is known with accuracy.

224 M. Barrande has lately added another and lower stage to the Silurian system, abounding with new and peculiar species.

224:d M. Barrande not long ago added

224:e Not long ago M. Barrande added another and lower stage, abounding with new and peculiar species beneath the old Silurian system.

224:f Not very long/species, beneath the then known Silurian system; and now, still lower down in the Lower Cambrian formation, Mr. Hicks has found in South Wales beds rich in trilobites, and containing various molluscs and annelids.

225 Traces of life have been detected in the Longmynd beds beneath Barrande's so-called primordial zone.

225:b beds, beneath

225:d Remnants of several forms have

225:e have also been detected beneath Barrande's so-called primordial zone in the Longmynd group, now divided into two stages, and constituting the Lower Cambrian system.

225:[f]

225.1:e Still more recently, the remarkable discovery has been made by Torell of the remains of monocotyledonous plants in a Swedish formation, corresponding with the Longmynd group; so that terrestrial or fresh-water plants existed several great stages lower down in the series than has hitherto been supposed.

225.1:[f]

226 The presence of phosphatic nodules and bituminous matter in

some of the lowest azoic rocks, probably indicates the former existence of life at these periods.

226:d presence also of phosphatic

226:e indicates life

226.1:d But now within the last year the great discovery of the Eozoon in the Laurentian formation of Canada has been made; and after reading Dr. Carpenter's description of this remarkable fossil, it is impossible to feel any doubt regarding its organic nature.

226.1:e Now the great/made, for after reading Dr. Carpenter's description of this fossil, it is scarcely possible to doubt

226 + 226.1:f The presence of phosphatic nodules and bituminous matter, even in some/periods; and the existence of the Eozoon in the Laurentian formation of Canada is generally admitted.

226.2:d There are three great series of strata beneath the Silurian system in Canada, in the lowest of which the Eozoon was found; and Sir W. Logan states that their "united thickness may possibly far surpass that of all the succeeding rocks, from the base of the palæozoic series to the present time.

226.2.x-y:f found. *y* Sir

226.3:d We are thus carried back to a period so far remote, that the appearance of the so-called Primordial fauna (of Barrande) may by some be considered a comparatively modern event."

226.3:e so remote

226.4:d The Eozoon belongs to the most lowly organised of all classes of animals, but for its class is highly organised; it existed in countless numbers, and, as Dr. Dawson has remarked, certainly preyed on other minute organic beings, which must have lived in great numbers.

226.4:f but is highly organised for its class; it

226.5:d There is also reason to believe that at this enormously remote period plants of some kind existed.

226.5:[e]

226.6:d Thus the words above given, which I wrote in 1859, and which are almost the same with those used by Sir W. Logan, have come true.

226.6:e words, which I wrote in 1859, about the vast periods which had probably elapsed before the Cambrian system, are almost the same with those since used by Sir W. Logan.

226.6:f about the existence of living beings long before the Cambrian period, and which are almost the same with those since used by Sir W. Logan, have proved true.

227 But the difficulty of understanding the absence of vast piles of fossiliferous strata, which on my theory no doubt were somewhere accumulated before the Silurian epoch, is very great.

227:d Notwithstanding these several facts, the difficulty of assigning any good cause for the absence beneath the Silurian formations of vast piles of strata rich in fossils is very great.

227:e Nevertheless the difficulty of assigning any good reason for the absence beneath the Upper Cambrian formations

227:f Nevertheless, the difficulty/absence of vast piles of strata rich in fossils beneath the Cambrian system is

228 If these most ancient beds had been wholly worn away by denudation, or obliterated by metamorphic action, we ought to find only small remnants of the formations next succeeding them in age, and these ought to be very generally in a metamorphosed condition.

228:d If the most ancient beds had been generally worn away by denudation, or if their fossils had been wholly obliterated by metamorphic action, we ought to have found only/ought to have existed almost always in a metamorphosed

228:e It does not seem probable that the most ancient beds have been quite worn away by denudation, or that their fossils have been wholly obliterated by metamorphic action, for if this had been the case we should have found/these would always have existed in a partially metamorphosed

229 But the descriptions which we now possess of the Silurian deposits over immense territories in Russia and in North America, do not support the view, that the older a formation is, the more it has suffered the extremity of denudation and metamorphism.

229:b has always suffered

229:e we possess/has invariably suffered extreme denudation

229:f more invariably it has suffered

230 The case at present must remain inexplicable; and may be truly urged as a valid argument against the views here entertained.

231 To show that it may hereafter receive some explanation, I will give the following hypothesis.

232 From the nature of the organic remains, which do not appear to have inhabited profound depths, in the several formations of Europe and of the United States; and from the amount of sediment, miles in thickness, of which the formations are composed, we may infer that from first to last large islands or tracts of land, whence the sediment was derived, occurred in the neighbourhood of the existing continents of Europe and North America.

232:b remains which

232:e neighbourhood of the now existing

232.1:f This same view has since been maintained by Agassiz and others.

233 But we do not know what was the state of things in the inter-

vals between the successive formations; whether Europe and the United States during these intervals existed as dry land, or as a submarine surface near land, on which sediment was not deposited, or again as the bed of an open and unfathomable sea.

233:b deposited, or as

234   Looking to the existing oceans, which are thrice as extensive as the land, we see them studded with many islands; but not one oceanic island is as yet known to afford even a remnant of any palæozoic or secondary formation.

234:d one truly oceanic island (with the exception of New Zealand, if this can be called a truly oceanic island) is

234:f but hardly one

235   Hence we may perhaps infer, that during the palæozoic and secondary periods, neither continents nor continental islands existed where our oceans now extend; for had they existed there, palæozoic and secondary formations would in all probability have been accumulated from sediment derived from their wear and tear; and would have been at least partially upheaved by the oscillations of level, which we may fairly conclude must have intervened during these enormously long periods.

235:e they existed, palæozoic/tear; and these would/which must

236   If then we may infer anything from these facts, we may infer that where our oceans now extend, oceans have extended from the remotest period of which we have any record; and on the other hand, that where continents now exist, large tracts of land have existed, subjected no doubt to great oscillations of level, since the earliest silurian period.

236:c Silurian

236:d that, where our

236:f since the Cambrian period.

237   The coloured map appended to my volume on Coral Reefs, led me to conclude that the great oceans are still mainly areas of subsidence, the great archipelagoes still areas of oscillations of level, and the continents areas of elevation.

238   But have we any right to assume that things have thus remained from eternity?

238:b remained from the beginning of this world?

238:d But we have no reason to assume that things have thus existed from the beginning of the world.

238:e thus remained from

239   Our continents seem to have been formed by a preponderance, during many oscillations of level, of the force of elevation; but may not the areas of preponderant movement have changed in the lapse of ages?

517

240 At a period immeasurably antecedent to the silurian epoch, continents may have existed where oceans are now spread out; and clear and open oceans may have existed where our continents now stand.

240:c Silurian

240:e period long antecedent

240:f to the Cambrian epoch

241 Nor should we be justified in assuming that if, for instance, the bed of the Pacific Ocean were now converted into a continent, we should there find formations older than the silurian strata, supposing such to have been formerly deposited; for it might well happen that strata which had subsided some miles nearer to the centre of the earth, and which had been pressed on by an enormous weight of superincumbent water, might have undergone far more metamorphic action than strata which have always remained nearer to the surface.

241:c Silurian

241:d find sedimentary formations in a recognisable condition older

241:f older than the Cambrian strata, supposing

242 The immense areas in some parts of the world, for instance in South America, of bare metamorphic rocks, which must have been heated under great pressure, have always seemed to me to require some special explanation; and we may perhaps believe that we see in these large areas, the many formations long anterior to the silurian epoch in a completely metamorphosed condition.

242:c Silurian/metamorphosed but likewise denuded condition.

242:d America, of naked metamorphic

242:e metamorphosed and denuded

242:f anterior to the Cambrian epoch

243   The several difficulties here discussed, namely our not finding in the successive formations infinitely numerous transitional links between the many species which now exist or have existed; the sudden manner in which whole groups of species appear in our European formations; the almost entire absence, as at present known, of fossiliferous formations beneath the Silurian strata, are all undoubtedly of the gravest nature.

243:c namely—that though we find in our geological formations many links between the species which now exist and have existed, we do not find infinitely numerous fine transitional forms closely joining them all together;—the sudden manner in which several whole groups of species first appeared in our European formations;—the almost/strata,—are all undoubtedly of the most serious nature.

243:d that, though/known, of formations rich in fossils beneath

243:e exist and which formerly existed/first appear in/beneath the Cambrian strata

244 We see this in the plainest manner by the fact that all the most eminent palæontologists, namely Cuvier, Owen, Agassiz, Barrande, Falconer, E. Forbes, &c., and all our greatest geologists, as Lyell, Murchison, Sedgwick, &c., have unanimously, often vehemently, maintained the immutability of species.

244:b Cuvier, Agassiz

244:c this in the fact that the most eminent/Barrande, Pictet, Falconer

245 But I have reason to believe that one great authority, Sir Charles Lyell, from further reflexion entertains grave doubts on this subject.

245:d But it is evident from the recent works of Sir Charles Lyell that he now almost gives up this view; and some other great geologists and palæontologists are much shaken in their confidence.

245:e But Sir Charles Lyell now gives the support of his high authority to the opposite side; and most other geologists and palæontologists are much shaken in their former belief.

245:f most geologists

246 I feel how rash it is to differ from these great authorities, to whom, with others, we owe all our knowledge.

246:b these authorities

246:d from the foregoing authorities

246:[e]

247 Those who think the natural geological record in any degree perfect, and who do not attach much weight to the facts and arguments of other kinds given in this volume, will undoubtedly at once reject my theory.

247:d who believe that the geological record is in any degree perfect, will

247:e reject the theory.

248 For my part, following out Lyell's metaphor, I look at the natural geological record, as a history of the world imperfectly kept, and written in a changing dialect; of this history we possess the last volume alone, relating only to two or three countries.

248:e at the geological

248:f record as

249 Of this volume, only here and there a short chapter has been preserved; and of each page, only here and there a few lines.

250 Each word of the slowly-changing language, in which the history is supposed to be written, being more or less different in the interrupted succession of chapters, may represent the appar-

ently abruptly changed forms of life, entombed in our consecutive, but widely separated, formations.

250:c history is written, being more or less different in the successive chapters

250:d language being/represent the forms of life, entombed in our consecutive but widely separated formations, which falsely appear to have been abruptly transformed.

250:e language, more/life, which are entombed in our consecutive formations, and which falsely appear to us to have been abruptly introduced.

251 On this view, the difficulties above discussed are greatly diminished, or even disappear.

CHAPTER X.

1:f XI.

2   ON THE GEOLOGICAL SUCCESSION OF ORGANIC BEINGS.

3 On the slow and successive appearance of new species—On their
different rates of change—Species once lost do not reappear—
Groups of species follow the same general rules in their appearance and disappearance as do single species—On Extinction—
On simultaneous changes in the forms of life throughout the
world—On the affinities of extinct species to each other and to
living species—On the state of development of ancient forms—
On the succession of the same types within the same areas—
Summary of preceding and present chapters.

3:c chapter.

4 LET us now see whether the several facts and rules relating to the
geological succession of organic beings, better accord with the
common view of the immutability of species, or with that of
their slow and gradual modification, through descent and natural selection.

4:e and laws relating

4:f beings accord best with the common/through variation and
natural

5   New species have appeared very slowly, one after another,
both on the land and in the waters.

6 Lyell has shown that it is hardly possible to resist the evidence on
this head in the case of the several tertiary stages; and every
year tends to fill up the blanks between them, and to make the
percentage system of lost and new forms more gradual.

6:e between the stages, and to

6:f make the proportion between the lost and existing forms

7 In some of the most recent beds, though undoubtedly of high
antiquity if measured by years, only one or two species are lost
forms, and only one or two are new forms, having here appeared for the first time, either locally, or, as far as we know,
on the face of the earth.

7:e are extinct, and only one or two are new, having appeared
there for

8 If we may trust the observations of Philippi in Sicily, the successive changes in the marine inhabitants of that island have
been many and most gradual.

8̄:[e]

9 The secondary formations are more broken; but, as Bronn has re-marked, neither the appearance nor disappearance of their many now extinct species has been simultaneous in each separate formation.

9:e of the many extinct species embedded in each formation has been simultaneous.

9:f many species

10 Species of different genera and classes have not changed at the same rate, or in the same degree.

10:f Species belonging to different

11 In the oldest tertiary beds a few living shells may still be found in the midst of a multitude of extinct forms.

11:c In the older tertiary

12 Falconer has given a striking instance of a similar fact, in an existing crocodile associated with many strange and lost mammals and reptiles in the sub-Himalayan deposits.

12:c strange lost

12:e fact, for an existing crocodile is associated with many lost

13 The Silurian Lingula differs but little from the living species of this genus; whereas most of the other Silurian Molluscs and all the Crustaceans have changed greatly.

14 The productions of the land seem to change at a quicker rate than those of the sea, of which a striking instance has lately been observed in Switzerland.

14:f seem to have changed

15 There is some reason to believe that organisms, considered high in the scale of nature, change more quickly than those that are low: though there are exceptions to this rule.

15:d organisms considered

15:e organisms high in the scale, change

16 The amount of organic change, as Pictet has remarked, does not strictly correspond with the succession of our geological formations; so that between each two consecutive formations, the forms of life have seldom changed in exactly the same degree.

16:e remarked, is not the same in each successive so-called formation.

17 Yet if we compare any but the most closely related formations, all the species will be found to have undergone some change.

18 When a species has once disappeared from the face of the earth, we have reason to believe that the same identical form never reappears.

18:c we have no reason/form ever reappears.

19 The strongest apparent exception to this latter rule, is that of the so-called "colonies" of M. Barrande, which intrude for a period in the midst of an older formation, and then allow the

pre-existing fauna to reappear; but Lyell's explanation, namely, that it is a case of temporary migration from a distinct geographical province, seems to me satisfactory.

19:d rule is that of the so-called

19:f seems satisfactory.

20   These several facts accord well with my theory.

21   I believe in no fixed law of development, causing all the inhabitants of a country to change abruptly, or simultaneously, or to an equal degree.

20 + 1:e with our theory, which includes no/inhabitants of an area to change

22   The process of modification must be extremely slow.

23   The variability of each species is quite independent of that of all others.

22 + 3:e be slow, and will generally affect only a few species at the same time; for the variability

22 + 3:f each species is independent of that

24   Whether such variability be taken advantage of by natural selection, and whether the variations be accumulated to a greater or lesser amount, thus causing a greater or lesser amount of modification in the varying species, depends on many complex contingencies,—on the variability being of a beneficial nature, on the power of intercrossing, on the rate of breeding, on the slowly changing physical conditions of the country, and more especially on the nature of the other inhabitants with which the varying species comes into competition.

24:d intercrossing and on the rate

24:e Whether such variations or individual differences as may arise will be accumulated through natural selection in a greater or less degree, thus causing a greater or less amount of permanent modification, will depend on many complex contingencies—on the variations being of a beneficial nature, on the freedom of intercrossing, on the slowly changing physical conditions of the country, on the immigration of new colonists, and on the nature

25   Hence it is by no means surprising that one species should retain the same identical form much longer than others; or, if changing, that it should change less.

25:e changing, that it should change in a less degree.

25:f changing, should change

26   We see the same fact in geographical distribution; for instance, in the land-shells and coleopterous insects of Madeira having come to differ considerably from their nearest allies on the continent of Europe, whereas the marine shells and birds have remained **unaltered**.

26:e We find similar relations between the inhabitants of distinct countries; for instance, the land-shells and coleopterous insects of Madeira have come

26:f between the existing inhabitants

27 We can perhaps understand the apparently quicker rate of change in terrestrial and in more highly organised productions compared with marine and lower productions, by the more complex relations of the higher beings to their organic and inorganic conditions of life, as explained in a former chapter.

⁀8 When many of the inhabitants of a country have become modified and improved, we can understand, on the principle of competition, and on that of the many all-important relations of organism to organism, that any form which does not become in some degree modified and improved, will be liable to be exterminated.

28:d liable to extermination.

28:e inhabitants of any area have become modified and improved, we can understand, on the principle of competition, and from the all-important relations of organism to organism in the struggle for life, that

28:f which did not become in some degree modified and improved, would be

29 Hence we can see why all the species in the same region do at last, if we look to wide enough intervals of time, become modified; for those which do not change will become extinct.

29:c modified, for

29:e we see/to long enough intervals of time, become modified, for otherwise they would become extinct.

30 In members of the same class the average amount of change, during long and equal periods of time, may, perhaps, be nearly the same; but as the accumulation of long-enduring fossiliferous formations depends on great masses of sediment having been deposited on areas whilst subsiding, our formations have been almost necessarily accumulated at wide and irregularly intermittent intervals; consequently the amount of organic change exhibited by the fossils embedded in consecutive formations is not equal.

30:d intervals of time; consequently

30:e long-enduring formations, rich in fossils, depends on great masses of sediment being deposited on subsiding areas, our formations

30:f accumulation of enduring

31 Each formation, on this view, does not mark a new and complete act of creation, but only an occasional scene, taken almost at hazard, in a slowly changing drama.

31:d in an ever slowly

32 We can clearly understand why a species when once lost

should never reappear, even if the very same conditions of life, organic and inorganic, should recur.

33 For though the offspring of one species might be adapted (and no doubt this has occurred in innumerable instances) to fill the exact place of another species in the economy of nature, and thus supplant it; yet the two forms—the old and the new— would not be identically the same; for both would almost certainly inherit different characters from their distinct progenitors.

33:e fill the place/progenitors, and organisms already differing would vary in a different manner.

33:f progenitors; and

34 For instance, it is just possible, if our fantail-pigeons were all destroyed, that fanciers, by striving during long ages for the same object, might make a new breed hardly distinguishable from our present fantail; but if the parent rock-pigeon were also destroyed, and in nature we have every reason to believe that the parent-form will generally be supplanted and exterminated by its improved offspring, it is quite incredible that a fantail, identical with the existing breed, could be raised from any other species of pigeon, or even from the other well-established races of the domestic pigeon, for the newly-formed fantail would be almost sure to inherit from its new progenitor some slight characteristic differences.

34:e were destroyed, that fanciers might make a new breed hardly distinguishable from the present breed; but if the parent rock-pigeon were likewise destroyed, and under nature we have every reason to believe that parent-forms are generally supplanted and exterminated by their improved offspring, it is incredible/even from any other well-established race of the domestic pigeon, for the successive variations would almost certainly be in some degree different, and the newly-formed variety would probably inherit from its progenitor some characteristic differences.

34:f is possible

35 Groups of species, that is, genera and families, follow the same general rules in their appearance and disappearance as do single species, changing more or less quickly, and in a greater or lesser degree.

36 A group does not reappear after it has once disappeared; or its existence, as long as it lasts, is continuous.

36:d A group, when it has once disappeared, never reappears; that is, its existence

37 I am aware that there are some apparent exceptions to this rule, but the exceptions are surprisingly few, so few, that E. Forbes, Pictet, and Woodward (though all strongly opposed to such views as I maintain) admit its truth: and the rule strictly accords with my theory.

37:c so few that

37:e with the theory.

38 For as all the species of the same group have descended from some one species, it is clear that as long as any species of the group have appeared in the long succession of ages, so long must its members have continuously existed, in order to have generated either new and modified or the same old and unmodified forms.

38:d that, as long as any species of the group have been produced, so long must some of its members have existed, in order to generate either the new and modified, or the old

38:e For all the species of the same group, however long it may have lasted, are the modified descendants of each other, and of some common progenitor.

38:f descendants, one from the other, and all from a common

39 Species of the genus Lingula, for instance, must have continuously existed by an unbroken succession of generations, from the lowest Silurian stratum to the present day.

39:e In the genus Lingula, for instance, the species which have successively appeared at all ages must have been connected by an unbroken series of generations

40 We have seen in the last chapter that the species of a group sometimes falsely appear to have come in abruptly; and I have attempted to give an explanation of this fact, which if true would have been fatal to my views.

40:c abruptly in a body; and

40:d chapter that many species

40:e true would be fatal

40:f that whole groups of species sometimes falsely appear to have been abruptly developed; and I

41 But such cases are certainly exceptional; the general rule being a gradual increase in number, till the group reaches its maximum, and then, sooner or later, it gradually decreases.

41:e number, until the group reaches its maximum, and then, sooner or later, a gradual decrease.

42 If the number of the species of a genus, or the number of the genera of a family, be represented by a vertical line of varying thickness, crossing the successive geological formations in which the species are found, the line will sometimes falsely appear to begin at its lower end, not in a sharp point, but abruptly; it then gradually thickens upwards, sometimes keeping for a space of equal thickness, and ultimately thins out in the upper beds, marking the decrease and final extinction of the species.

42:c thickness, ascending through the successive/upwards, often keeping

42:e species included within a genus, or the number of the genera within a family

42:f keeping of equal thickness for a space, and ultimately

43 This gradual increase in number of the species of a group is strictly conformable with my theory; as the species of the same genus, and the genera of the same family, can increase only slowly and progressively; for the process of modification and the production of a number of allied forms must be slow and gradual,—one species giving rise first to two or three varieties, these being slowly converted into species, which in their turn produce by equally slow steps other species, and so on, like the branching of a great tree from a single stem, till the group becomes **large.**

43:d theory, for the species of the same genus/progressively; the process/forms necessarily being a slow and gradual process,— one species first giving rise to two/other varieties and species

43:e with the theory

44   *On Extinction.—*

44:d [*Center*] *On Extinction.* [*Space*]

45 We have as yet spoken only incidentally of the disappearance of species and of groups of species.

45:d[¶]

46 On the theory of natural selection the extinction of old forms and the production of new and improved forms are intimately connected **together.**

47 The old notion of all the inhabitants of the earth having been swept away at successive periods by catastrophes, is very generally given up, even by those geologists, as Elie de Beaumont, Murchison, Barrande, &c., whose general views would naturally lead them to this conclusion.

47:d away by catastrophes at successive periods is

48 On the contrary, we have every reason to believe, from the study of the tertiary formations, that species and groups of species gradually disappear, one after another, first from one spot, then from another, and finally from the world.

48:c world; in some few cases, however, as by the breaking of an isthmus and the consequent irruption of a multitude of new inhabitants, or by the final subsidence of an island, the extinction may have been comparatively rapid.

48.x-y:e world. *y* In/inhabitants into an adjoining sea, or by the final subsidence of an island, the process of extinction may have been rapid.

49 Both single species and whole groups of species last for very unequal periods; some groups, as we have seen, having endured from the earliest known dawn of life to the present day; some having disappeared before the close of the palæozoic period.

49:e day; some have disappeared

50 No fixed law seems to determine the length of time during which any single species or any single genus endures.

51 There is reason to believe that the complete extinction of the species of a group is generally a slower process than their production: if the appearance and disappearance of a group of species be represented, as before, by a vertical line of varying thickness, the line is found to taper more gradually at its upper end, which marks the progress of extermination, than at its lower end, which marks the first appearance and increase in numbers of the species.

51:d extinction of a whole group of species is generally

51:e that the extinction/if their appearance and disappearance be represented/first appearance and the early increase in number

52 In some cases, however, the extermination of whole groups of beings, as of ammonites towards the close of the secondary period, has been wonderfully sudden.

52:c sudden relatively to that of most other groups.

52:e whole groups, as/sudden.

52:f ammonites, towards

53 The whole subject of the extinction of species has been involved in the most gratuitous mystery.

53:e The extinction of species

54 Some authors have even supposed that as the individual has a definite length of life, so have species a definite duration.

54:d that, as

55 No one I think can have marvelled more at the extinction of species, than I have done.

55:e No one can have marvelled more than I have done at the extinction of species.

56 When I found in La Plata the tooth of a horse embedded with the remains of Mastodon, Megatherium, Toxodon, and other extinct monsters, which all co-existed with still living shells at a very late geological period, I was filled with astonishment; for seeing that the horse, since its introduction by the Spaniards into South America, has run wild over the whole country and has increased in numbers at an unparalleled rate, I asked myself what could so recently have exterminated the former horse under conditions of life apparently so favourable.

56:d for, seeing

57 But how utterly groundless was my astonishment!

57:e But my astonishment was groundless.

58 Professor Owen soon perceived that the tooth, though so like that of the existing horse, belonged to an extinct species.

59 Had this horse been still living, but in some degree rare, no

naturalist would have felt the least surprise at its rarity; for rarity is the attribute of a vast number of species of all classes, in all countries.

60 If we ask ourselves why this or that species is rare, we answer that something is unfavourable in its conditions of life; but what that something is, we can hardly ever tell.

61 On the supposition of the fossil horse still existing as a rare species, we might have felt certain from the analogy of all other mammals, even of the slow-breeding elephant, and from the history of the naturalisation of the domestic horse in South America, that under more favourable conditions it would in a very few years have stocked the whole continent.

61:d certain, from the analogy

62 But we could not have told what the unfavourable conditions were which checked its increase, whether some one or several contingencies, and at what period of the horse's life, and in what degree, they severally acted.

63 If the conditions had gone on, however slowly, becoming less and less favourable, we assuredly should not have perceived the fact, yet the fossil horse would certainly have become rarer and rarer, and finally extinct;—its place being seized on by some more successful competitor.

64   It is most difficult always to remember that the increase of every living being is constantly being checked by unperceived injurious agencies; and that these same unperceived agencies are amply sufficient to cause rarity, and finally extinction.

64:d unperceived hostile agencies

64:e every creature is constantly

64.1-2:c So little is this subject understood, that I have heard surprise repeatedly expressed at such great monsters as the Mastadon and the more ancient Dinosaurians having become extinct; as if mere bodily strength gave victory in the battle of life. 2 Mere size, on the contrary, would in some cases determine quicker extermination from the greater amount of requisite food.

64.2:d determine, as has been remarked by Owen, quicker

64.3-4:c Before man inhabited India or Africa, some cause must have checked the continued increase of the existing elephant. 4 A highly capable judge believes that at the present day insects (as Bruce has likewise described in Abyssinia), from incessantly harassing and weakening the elephant, are one chief check to its increase.

64.4:d judge, Dr. Falconer, believes that it is chiefly insects which, from incessantly harassing and weakening the elephant in India, check its increase; and this was Bruce's conclusion with respect to the African elephant in Abyssinia.

64.5:c It is certain that insects of different kinds, and blood-sucking bats, determine the existence of the larger naturalised quadrupeds in several parts of S. America.

64.5:d insects and blood-sucking bats determine

65 We see in many cases in the more recent tertiary formations, that rarity precedes extinction; and we know that this has been the progress of events with those animals which have been exterminated, either locally or wholly, through man's agency.

65:c[¶]

66 I may repeat what I published in 1845, namely, that to admit that species generally become rare before they become extinct —to feel no surprise at the rarity of a species, and yet to marvel greatly when it ceases to exist, is much the same as to admit that sickness in the individual is the forerunner of death—to feel no surprise at sickness, but when the sick man dies, to wonder and to suspect that he died by some unknown deed of violence.

66:d when the species ceases/but, when/some deed

67 The theory of natural selection is grounded on the belief that each new variety, and ultimately each new species, is produced and maintained by having some advantage over those with which it comes into competition; and the consequent extinction of less-favoured forms almost inevitably follows.

68 It is the same with our domestic productions: when a new and slightly improved variety has been raised, it at first supplants the less improved varieties in the same neighbourhood; when much improved it is transported far and near, like our shorthorn cattle, and takes the place of other breeds in other countries.

68:d productions; when/first generally supplants

68:e first supplants

69 Thus the appearance of new forms and the disappearance of old forms, both natural and artificial, are bound together.

69:d old forms, both those naturally and those artificially produced, are

70 In certain flourishing groups, the number of new specific forms which have been produced within a given time is probably greater than that of the old forms which have been exterminated; but we know that the number of species has not gone on indefinitely increasing, at least during the later geological periods, so that looking to later times we may believe that the production of new forms has caused the extinction of about the same number of old forms.

70:b old specific forms which

70:d In flourishing groups/time has at some periods probably been greater than the number of the old specific/that species have not gone on indefinitely increasing, at least during the later geological epochs, so that, looking to later times, we

71   The competition will generally be most severe, as formerly explained and illustrated by examples, between the forms which are most like each other in all respects.

72 Hence the improved and modified descendants of a species will generally cause the extermination of the parent-species; and if many new forms have been developed from any one species, the nearest allies of that species, *i. e.* the species of the same genus, will be the most liable to extermination.

73 Thus, as I believe, a number of new species descended from one species, that is a new genus, comes to supplant an old genus, belonging to the same family.

74 But it must often have happened that a new species belonging to some one group will have seized on the place occupied by a species belonging to a distinct group, and thus caused its extermination; and if many allied forms be developed from the successful intruder, many will have to yield their places; and it will generally be allied forms, which will suffer from some inherited inferiority in common.

74.x-y:e group has seized/thus have caused its extermination. *y* If/ generally be the allied

75 But whether it be species belonging to the same or to a distinct class, which yield their places to other species which have been modified and improved, a few of the sufferers may often long be preserved, from being fitted to some peculiar line of life, or from inhabiting some distant and isolated station, where they have escaped severe competition.

75:e which have yielded their places to other modified and improved species, a few of the sufferers may often be preserved for a long time, from/they will have

76 For instance, a single species of Trigonia, a great genus of shells in the secondary formations, survives in the Australian seas; and a few members of the great and almost extinct group of Ganoid fishes still inhabit our fresh waters.

76:d instance, some species/survive

77 Therefore the utter extinction of a group is generally, as we have seen, a slower process than its production.

78   With respect to the apparently sudden extermination of whole families or orders, as of Trilobites at the close of the palæozoic period and of Ammonites at the close of the secondary period, we must remember what has been already said on the probable wide intervals of time between our consecutive formations; and in these intervals there may have been much slow extermination.

79 Moreover, when by sudden immigration or by unusually rapid development, many species of a new group have taken possession of a new area, they will have exterminated in a correspondingly rapid manner many of the old inhabitants; and the forms

which thus yield their places will commonly be allied, for they will partake of some inferiority in common.

79:d when, by/area, these will have/partake of the same inferiority

79:e possession of an area, many of the older species will have been exterminated in a correspondingly rapid manner; and

80 Thus, as it seems to me, the manner in which single species and whole groups of species become extinct, accords well with the theory of natural selection.

80:e extinct accords

81 We need not marvel at extinction; if we must marvel, let it be at our presumption in imagining for a moment that we understand the many complex contingencies, on which the existence of each species depends.

81:d our own presumption

81:f contingencies on

82 If we forget for an instant, that each species tends to increase inordinately, and that some check is always in action, yet seldom perceived by us, the whole economy of nature will be utterly obscured.

83 Whenever we can precisely say why this species is more abundant in individuals than that; why this species and not another can be naturalised in a given country; then, and not till then, we may justly feel surprise why we cannot account for the extinction of this particular species or group of species.

83:d of any particular

83:e then, and not until then

84 *On the Forms of Life changing almost simultaneously throughout the World.—*

84:d [Center] World. [Space]

85 Scarcely any palæontological discovery is more striking than the fact, that the forms of life change almost simultaneously throughout the world.

85:d[¶]

86 Thus our European Chalk formation can be recognised in many distant parts of the world, under the most different climates, where not a fragment of the mineral chalk itself can be found; namely, in North America, in equatorial South America, in Tierra del Fuego, at the Cape of Good Hope, and in the peninsula of India.

86:f distant regions, under

87 For at these distant points, the organic remains in certain beds present an unmistakeable degree of resemblance to those of the Chalk.

88 It is not that the same species are met with; for in some cases not one species is identically the same, but they belong to the same

families, genera, and sections of genera, and sometimes are similarly characterised in such trifling points as mere superficial sculpture.

89 Moreover other forms, which are not found in the Chalk of Europe, but which occur in the formations either above or below, are similarly absent at these distant points of the world.

89:d Moreover, other/similarly placed at

89:e Europe but/below, occur in the same order at

90 In the several successive palæozoic formations of Russia, Western Europe and North America, a similar parallelism in the forms of life has been observed by several authors: so it is, according to Lyell, with the several European and North American tertiary deposits.

90:d Europe, and

90:f with the European

91 Even if the few fossil species which are common to the Old and New Worlds be kept wholly out of view, the general parallelism in the successive forms of life, in the stages of the widely separated palæozoic and tertiary periods, would still be manifest, and the several formations could be easily correlated.

91:d stages of the palæozoic and of the tertiary

91:e Worlds were kept/life, in the palæozoic and tertiary stages, would

92   These observations, however, relate to the marine inhabitants of distant parts of the world: we have not sufficient data to judge whether the productions of the land and of fresh water change at distant points in the same parallel manner.

92:e inhabitants of the world/water at distant points change in

93 We may doubt whether they have thus changed: if the Megatherium, Mylodon, Macrauchenia, and Toxodon had been brought to Europe from La Plata, without any information in regard to their geological position, no one would have suspected that they had coexisted with still living sea-shells; but as these anomalous monsters coexisted with the Mastodon and Horse, it might at least have been inferred that they had lived during one of the later tertiary stages.

93:b co-existed/co-existed

93:e with sea-shells all still living; but

94   When the marine forms of life are spoken of as having changed simultaneously throughout the world, it must not be supposed that this expression relates to the same thousandth or hundred-thousandth year, or even that it has a very strict geological sense; for if all the marine animals which live at the present day in Europe, and all those that lived in Europe during the pleistocene period (an enormously remote period as measured by years, including the whole glacial epoch), were to be compared with those now living in South America or in

Australia, the most skilful naturalist would hardly be able to say whether the existing or the pleistocene inhabitants of Europe resembled most closely those of the southern hemisphere.

94:e animals now living in Europe, and/period (a very remote/epoch) were compared with those now existing in South/whether the present or

94:f same year, or to the same century, or even

95 So, again, several highly competent observers believe that the existing productions of the United States are more closely related to those which lived in Europe during certain later tertiary stages, than to those which now live here; and if this be so, it is evident that fossiliferous beds deposited at the present day on the shores of North America would hereafter be liable to be classed with somewhat older European beds.

95:e observers maintain that the existing/late tertiary stages, than to the present inhabitants of Europe; and if this be so, it is evident that fossiliferous beds now being deposited on the shores

96 Nevertheless, looking to a remotely future epoch, there can, I think, be little doubt that all the more modern *marine* formations, namely, the upper pliocene, the pleistocene and strictly modern beds, of Europe, North and South America, and Australia, from containing fossil remains in some degree allied, and from not including those forms which are only found in the older underlying deposits, would be correctly ranked as simultaneous in a geological sense.

96:e can be little/which are found only in

97 The fact of the forms of life changing simultaneously, in the above large sense, at distant parts of the world, has greatly struck those admirable observers, MM. de Verneuil and d'Archiac.

98 After referring to the parallelism of the palæozoic forms of life in various parts of Europe, they add, "If struck by this strange sequence, we turn our attention to North America, and there discover a series of analogous phenomena, it will appear certain that all these modifications of species, their extinction, and the introduction of new ones, cannot be owing to mere changes in marine currents or other causes more or less local and temporary, but depend on general laws which govern the whole animal kingdom."

98:d If, struck

99 M. Barrande has made forcible remarks to precisely the same effect.

100 It is, indeed, quite futile to look to changes of currents, climate, or other physical conditions, as the cause of these great muta-

tions in the forms of life throughout the world, under the most different climates.

101 We must, as Barrande has remarked, look to some special law.

102 We shall see this more clearly when we treat of the present distribution of organic beings, and find how slight is the relation between the physical conditions of various countries, and the nature of their inhabitants.

102:f countries and

103 This great fact of the parallel succession of the forms of life throughout the world, is explicable on the theory of natural selection.

104 New species are formed by new varieties arising, which have some advantage over older forms; and those forms, which are already dominant, or have some advantage over the other forms in their own country, would naturally oftenest give rise to new varieties or incipient species; for these latter must be victorious in a still higher degree in order to be preserved and to survive.

104:d incipient species.

104:e by having some advantage over older forms; and the forms, which/country, would be the most likely to give birth to the greatest number of new varieties

104:f country, give

105 We have distinct evidence on this head, in the plants which are dominant, that is, which are commonest in their own homes, and are most widely diffused, having produced the greatest number of new varieties.

105:c commonest and most widely diffused, compared with other plants within their own homes, having

105:d other less dominant plants, producing

105:e diffused, producing

106 It is also natural that the dominant, varying, and far-spreading species, which already have invaded to a certain extent the territories of other species, should be those which would have the best chance of spreading still further, and of giving rise in new countries to new varieties and species.

106:e countries to other new

106:f which have already invaded

107 The process of diffusion may often be very slow, being dependent on climatal and geographical changes, or on strange accidents, but in the long run the dominant forms will generally succeed in spreading.

107:c accidents, or on the gradual acclimatisation of new species to the various climates through which they must pass, but

107:d they have to pass, but in the course of time the dominant

107:e diffusion would often be very slow, depending on climatal and geographical changes, on strange accidents, and on the gradual/they might have to pass, but in the course of time the dominant forms would generally succeed in spreading and would ultimately prevail.

108 The diffusion would, it is probable, be slower with the terrestrial inhabitants of distinct continents than with the marine inhabitants of the continuous sea.

109 We might therefore expect to find, as we apparently do find, a less strict degree of parallel succession in the productions of the land than of the sea.

109:e as we do find, a less strict degree of parallelism in the succession of the productions of the land than with those of

110 Dominant species spreading from any region might encounter still more dominant species, and then their triumphant course, or even their existence, would cease.

110:[e]

111 We know not at all precisely what are all the conditions most favourable for the multiplication of new and dominant species; but we can, I think, clearly see that a number of individuals, from giving a better chance of the appearance of favourable variations, and that severe competition with many already existing forms, would be highly favourable, as would be the power of spreading into new territories.

111:d of any one new

111:[e]

112 A certain amount of isolation, recurring at long intervals of time, would probably be also favourable, as before explained.

112:[e]

113 One quarter of the world may have been most favourable for the production of new and dominant species on the land, and another for those in the waters of the sea.

113:[e]

114 If two great regions had been for a long period favourably circumstanced in an equal degree, whenever their inhabitants met, the battle would be prolonged and severe; and some from one birthplace and some from the other might be victorious.

114:[e]

115 But in the course of time, the forms dominant in the highest degree, wherever produced, would tend everywhere to prevail.

115:[e]

116 As they prevailed, they would cause the extinction of other and inferior forms; and as these inferior forms would be allied in groups by inheritance, whole groups would tend slowly to disappear; though here and there a single member might long be enabled to survive.

116:[d]

117 Thus, as it seems to me, the parallel, and, taken in a large sense, simultaneous, succession of the same forms of life throughout the world, accords well with the principle of new species having been formed by dominant species spreading widely and varying; the new species thus produced being themselves dominant owing to inheritance, and to having already had some advantage over their parents or over other species; these again spreading, varying, and producing new species.

117:d inheritance and to/other species, and again spreading, varying, and producing other new forms.

117:e themselves dominant, owing to their having had some advantage over their already dominant parents, as well as over other species, and again spreading, varying, and producing new forms.

118 The forms which are beaten and which yield their places to the new and victorious forms, will generally be allied in groups, from inheriting some inferiority in common; and therefore as new and improved groups spread throughout the world, old groups will disappear from the world; and the succession of forms in both ways will everywhere tend to correspond.

118:d The old forms which are/therefore, as/succession of forms will everywhere tend to correspond in their first appearance and final disappearance.

118:e world, old groups disappear from the world; and the succession of forms everywhere tends to correspond both in

119 There is one other remark connected with this subject worth making.

120 I have given my reasons for believing that all our greater fossiliferous formations were deposited during periods of subsidence; and that blank intervals of vast duration occurred during the periods when the bed of the sea was either stationary or rising, and likewise when sediment was not thrown down quickly enough to embed and preserve organic remains.

120:c that most of our greater formations, rich in fossils, were deposited

120:e great/duration, as far as fossils are concerned, occurred

121 During these long and blank intervals I suppose that the inhabitants of each region underwent a considerable amount of modification and extinction, and that there was much migration from other parts of the world.

122 As we have reason to believe that large areas are affected by the same movement, it is probable that strictly contemporaneous formations have often been accumulated over very wide spaces in the same quarter of the world; but we are far from having any right to conclude that this has invariably been the case, and that large areas have invariably been affected by the same movements.

122:e but we are very far

123 When two formations have been deposited in two regions during nearly, but not exactly the same period, we should find in both, from the causes explained in the foregoing paragraphs, the same general succession in the forms of life; but the species would not exactly correspond; for there will have been a little more time in the one region than in the other for modification, extinction, and immigration.

123:d exactly, the same

124   I suspect that cases of this nature have occurred in Europe.

124:b nature occur

125 Mr. Prestwich, in his admirable Memoirs on the eocene deposits of England and France, is able to draw a close general parallelism between the successive stages in the two countries; but when he compares certain stages in England with those in France, although he finds in both a curious accordance in the numbers of the species belonging to the same genera, yet the species themselves differ in a manner very difficult to account for, considering the proximity of the two areas,—unless, indeed, it be assumed that an isthmus separated two seas inhabited by distinct, but contemporaneous, faunas.

125:d for considering

126 Lyell has made similar observations on some of the later tertiary formations.

127 Barrande, also, shows that there is a striking general parallelism in the successive Silurian deposits of Bohemia and Scandinavia; nevertheless he finds a surprising amount of difference in the species.

128 If the several formations in these regions have not been deposited during the same exact periods,—a formation in one region often corresponding with a blank interval in the other, —and if in both regions the species have gone on slowly changing during the accumulation of the several formations and during the long intervals of time between them; in this case, the several formations in the two regions could be arranged in the same order, in accordance with the general succession of the form of life, and the order would falsely appear to be strictly parallel; nevertheless the species would not all be the same in the apparently corresponding stages in the two regions.

128:c forms/not be all the same
128:d case the several

129   *On the Affinities of extinct Species to each other, and to living forms.——*

129:d [*Center*] *Extinct/Living Forms.* [*Space*]

130 Let us now look to the mutual affinities of extinct and living species.

130:d[¶]

131 They all fall into one grand natural system; and this fact is at once explained on the principle of descent.

131:e into a few grand classes; and

131:f All

132 The more ancient any form is, the more, as a general rule, it differs from living forms.

133 But, as Buckland long ago remarked, all fossils can be classed either in still existing groups, or between them.

133:e all extinct species can

133:f remarked, extinct species can all be

134 That the extinct forms of life help to fill up the wide intervals between existing genera, families, and orders, cannot be disputed.

134:e up the intervals

134:f orders, is certainly true; but as this statement has often been ignored or even denied, it may be well to make some remarks on this subject, and to give some instances.

135 For if we confine our attention either to the living or to the extinct alone, the series is far less perfect than if we combine both into one general system.

135:f If/extinct species of the same class, the series

136 With respect to the Vertebrata, whole pages could be filled with striking illustrations from our great palæontologist, Owen, showing how extinct animals fall in between existing groups.

136:d vertebrata

136:e filled with illustrations from Owen

136:f In the writings of Professor Owen we continually meet with the expression of generalised forms, as applied to extinct animals; and in the writings of Agassiz, of prophetic or synthetic types; and these terms imply that such forms are in fact intermediate or connecting links.

137 Cuvier ranked the Ruminants and Pachyderms, as the two most distinct orders of mammals; but Owen has discovered so many fossil links, that he has had to alter the whole classification of these two orders; and has placed certain pachyderms in the same sub-order with ruminants: for example, he dissolves by fine gradations the apparently wide difference between the pig and the camel.

137:e classification, and has

137:[f] = 137.1.1:f

137.1:e Another distinguished palæontologist, M. Gaudry, shows that very many of the fossil mammals discovered by him in Attica connect in the plainest manner existing genera.

137.1:f palæontologist, M. Gaudry, has shown in the most striking manner that many of the fossil mammals discovered by him in

Attica serve to break down the intervals between existing genera.

137.1.1:f = *137:[f]* as two of the most distinct orders of mammals; but so many fossil links have been disentombed that Owen has had to alter/ruminants; for example, he dissolves by gradations the apparently wide interval between

137.1.2-6:f The Ungulata or hoofed quadrupeds are now divided into the even-toed or odd-toed divisions; but the Macrauchenia of S. America connects to a certain extent these two grand divisions. *3* No one will deny that the Hipparion is intermediate between the existing horse and certain older ungulate forms. *4* What a wonderful connecting link in the chain of mammals is the Typotherium from S. America, as the name given to it by Professor Gervais expresses, and which cannot be placed in any existing order. *5* The Sirenia form a very distinct group of mammals, and one of the most remarkable peculiarities in the existing dugong and lamentin is the entire absence of hind limbs, without even a rudiment being left; but the extinct Halitherium had, according to Professor Flower, an ossified thigh-bone "articulated to a well-defined acetabulum in the pelvis," and it thus makes some approach to ordinary hoofed quadrupeds, to which the Sirenia are in other respects allied. *6* The cetaceans or whales are widely different from all other mammals, but the tertiary Zeuglodon and Squalodon, which have been placed by some naturalists in an order by themselves, are considered by Professor Huxley to be undoubtedly cetaceans, "and to constitute connecting links with the aquatic carnivora."

137.2:e Even the wide interval between birds and reptiles has been shown by Professor Huxley to be partially bridged over in the most unexpected manner, by, on the one hand, the ostrich and extinct Archeopteryx, and on the other hand, the Compsognathus, one of the Dinosaurians—that group which includes the most gigantic of all terrestrial reptiles.

137.2:f[¶] by the naturalist just quoted to be partially bridged over in the most unexpected manner, on the one hand, by the ostrich and extinct Archeopteryx, and on the other hand, by the Compsognathus

138 In regard to the Invertebrata, Barrande, and a higher authority could not be named, asserts that he is every day taught that palæozoic animals, though belonging to the same orders, families, or genera with those living at the present day, were not at this early epoch limited in such distinct groups as they now are.

138:b Palæozoic

138:d taught that, although palæozoic animals can certainly be classed under existing groups, yet that at this ancient period these groups were not so distinctly separated from each other as they are at the present time.

138:e Turning to the Invertebrata, Barrande asserts, and a higher authority could not be named, that he is every day taught that, although

139 Some writers have objected to any extinct species or group of species being consider as intermediate between living species or groups.

139:f species, or group of species, being considered as intermediate between any two living species, or groups of species.

140 If by this term it is meant that an extinct form is directly intermediate in all its characters between two living forms, the objection is probably valid.

140:c objection is valid.

140:f living forms or groups, the objection is probably valid.

141 But I apprehend that in a perfectly natural classification many fossil species would have to stand between living species, and some extinct genera between living genera, even between genera belonging to distinct families.

141:e But in a natural classification many fossil species certainly stand

142 The most common case, especially with respect to very distinct groups, such as fish and reptiles, seems to be, that supposing them to be distinguished at the present day from each other by a dozen characters, the ancient members of the same two groups would be distinguished by a somewhat lesser number of characters, so that the two groups, though formerly quite distinct, at that period made some small approach to each other.

142:d that, supposing

142:e members are separated by/quite distinct, made at that period a somewhat nearer approach

142:f day by a score of characters, the ancient members are separated by a somewhat lesser number of characters; so that the two groups formerly made a somewhat nearer approach to each other than they now do.

143 It is a common belief that the more ancient a form is, by so much the more it tends to connect by some of its characters groups now widely separated from each other.

144 This remark no doubt must be restricted to those groups which have undergone much change in the course of geological ages; and it would be difficult to prove the truth of the proposition, for every now and then even a living animal, as the Lepidosiren, is discovered having affinities directed towards very distinct groups.

145 Yet if we compare the older Reptiles and Batrachians, the older Fish, the older Cephalopods, and the eocene Mammals, with the more recent members of the same classes, we must admit that there is some truth in the remark.

145:e there is truth

146 Let us see how far these several facts and inferences accord with the theory of descent with modification.

147 As the subject is somewhat complex, I must request the reader to turn to the diagram in the fourth chapter.

148 We may suppose that the numbered letters represent genera, and the dotted lines diverging from them the species in each genus.

148:f letters in italics represent

149 The diagram is much too simple, too few genera and too few species being given, but this is unimportant for us.

150 The horizontal lines may represent successive geological formations, and all the forms beneath the uppermost line may be considered as extinct.

151 The three existing genera, a^{14}, q^{14}, p^{14}, will form a small family; b^{14} and f^{14} a closely allied family or sub-family; and o^{14}, e^{14}, m^{14}, a third family.

152 These three families, together with the many extinct genera on the several lines of descent diverging from the parent-form A, will form an order; for all will have inherited something in common from their ancient and common progenitor.

152:b parent-form (A), will form
152:f (A) will form/ancient progenitor.

153 On the principle of the continued tendency to divergence of character, which was formerly illustrated by this diagram, the more recent any form is, the more it will generally differ from its ancient progenitor.

154 Hence we can understand the rule that the most ancient fossils differ from existing forms.

155 We must not, however, assume that divergence of character is a necessary contingency; it depends solely on the descendants from a species being thus enabled to seize on many and different places in the economy of nature.

156 Therefore it is quite possible, as we have seen in the case of some Silurian forms, that a species might go on being slightly modified in relation to its slightly altered conditions of life, and yet retain throughout a vast period the same general characteristics.

157 This is represented in the diagram by the letter F^{14}.

158 All the many forms, extinct and recent, descended from A, make, as before remarked, one order; and this order, from the continued effects of extinction and divergence of character, has become divided into several sub-families and families, some of which are supposed to have perished at different periods, and some to have endured to the present day.

158:b from (A), make

159   By looking at the diagram we can see that if many of the extinct forms, supposed to be embedded in the successive formations, were discovered at several points low down in the series, the three existing families on the uppermost line would be rendered less distinct from each other.

159:f forms supposed

160   If, for instance, the genera a^1, a^5, a^{10}, f^8, m^3, m^6, m^9, were disinterred, these three families would be so closely linked together that they probably would have to be united into one great family, in nearly the same manner as has occurred with ruminants and pachyderms.

160:c and certain pachyderms.

161   Yet he who objected to call the extinct genera, which thus linked the living genera of three families together, intermediate in character, would be justified, as they are intermediate, not directly, but only by a long and circuitous course through many widely different forms.

161:d genera which

161:f to consider as intermediate the extinct genera, which thus link together the living genera of three families, would be partly justified, for they

162   If many extinct forms were to be discovered above one of the middle horizontal lines or geological formations—for instance, above No. VI.—but none from beneath this line, then only the two families on the left hand (namely, a^{14}, &c., and b^{14}, &c.) would have to be united into one family; and the two other families (namely, a^{14} to f^{14} now including five genera, and o^{14} to m^{14}) would yet remain distinct.

163   These two families, however, would be less distinct from each other than they were before the discovery of the fossils.

162 + 3:d only two of the families (those on the left hand, a^{14}, &c., and b^{14}, &c.) would have to be united into one; and there would remain two families, which would be less

164   If, for instance, we suppose the existing genera of the two families to differ from each other by a dozen characters, in this case the genera, at the early period marked VI., would differ by a lesser number of characters; for at this early stage of descent they have not diverged in character from the common progenitor of the order, nearly so much as they subsequently diverged.

164:d So again, if the three families formed of eight genera (a^{14} to m^{14}), on the uppermost line, be supposed to differ from each other by half a dozen important characters, then the families which existed at the period marked VI. would certainly have differed from each other by a less number of characters; for they would at this early stage of descent have diverged in a less degree from their common progenitor.

164:f again if/half-a-dozen

165 Thus it comes that ancient and extinct genera are often in some slight degree intermediate in character between their modified descendants, or between their collateral relations.

165:f in a greater or less degree

166 In nature the case will be far more complicated than is represented in the diagram; for the groups will have been more numerous, they will have endured for extremely unequal lengths of time, and will have been modified in various degrees.

166:f Under nature the process will be for/numerous; they

167 As we possess only the last volume of the geological record, and that in a very broken condition, we have no right to expect, except in very rare cases, to fill up wide intervals in the natural system, and thus unite distinct families or orders.

167:d up the wide
167:e except in rare cases
167:f thus to unite

168 All that we have a right to expect, is that those groups, which have within known geological periods undergone much modification, should in the older formations make some slight approach to each other; so that the older members should differ less from each other in some of their characters than do the existing members of the same groups; and this by the concurrent evidence of our best palæontologists seems frequently to be the case.

168:e palæontologists is frequently the case.
168:f expect is, that those groups which have, within known geological periods, undergone

169 Thus, on the theory of descent with modification, the main facts with respect to the mutual affinities of the extinct forms of life to each other and to living forms, seem to me explained in a satisfactory manner.

169:e living forms, are explained

170 And they are wholly inexplicable on any other view.

171 On this same theory, it is evident that the fauna of any great period in the earth's history will be intermediate in general character between that which preceded and that which succeeded it.

171:e any one great
171:f fauna during any

172 Thus, the species which lived at the sixth great stage of descent in the diagram are the modified offspring of those which lived at the fifth stage, and are the parents of those which became still more modified at the seventh stage; hence they could hardly fail to be nearly intermediate in character between the forms of life above and below.

172:d Thus the species

173 We must, however, allow for the entire extinction of some preceding forms, and for the coming in of quite new forms by immigration, and for a large amount of modification, during the long and blank intervals between the successive formations.

173:b forms, and in any one region for the immigration of new forms from other regions, and for a large

173:f modification during

174 Subject to these allowances, the fauna of each geological period undoubtedly is intermediate in character, between the preceding and succeeding faunas.

175 I need give only one instance, namely, the manner in which the fossils of the Devonian system, when this system was first discovered, were at once recognised by palæontologists as intermediate in character between those of the overlying carboniferous, and underlying Silurian system.

176 But each fauna is not necessarily exactly intermediate, as unequal intervals of time have elapsed between consecutive formations.

177 It is no real objection to the truth of the statement, that the fauna of each period as a whole is nearly intermediate in character between the preceding and succeeding faunas, that certain genera offer exceptions to the rule.

177:d statement that the fauna

178 For instance, mastodons and elephants, when arranged by Dr. Falconer in two series, first according to their mutual affinities and then according to their periods of existence, do not accord in arrangement.

178:f instance, the species of mastodons/series,—in the first place according to their mutual affinities, and in the second place according to their periods of existence,—do

179 The species extreme in character are not the oldest, or the most recent; nor are those which are intermediate in character, intermediate in age.

179:d oldest or

180 But supposing for an instant, in this and other such cases, that the record of the first appearance and disappearance of the species was perfect, we have no reason to believe that forms successively produced necessarily endure for corresponding lengths of time: a very ancient form might occasionally last much longer than a form elsewhere subsequently produced, especially in the case of terrestrial productions inhabiting separated districts.

180.x-y:f was complete, which is far from the case, we/time. *y* A very ancient form may occasionally have lasted

181 To compare small things with great: if the principal living and

extinct races of the domestic pigeon were arranged as well as they could be in serial affinity, this arrangement would not closely accord with the order in time of their production, and still less with the order of their disappearance; for the parent rock-pigeon now lives; and many varieties between the rock-pigeon and the carrier have become extinct; and carriers which are extreme in the important character of length of beak originated earlier than short-beaked tumblers, which are at the opposite end of the series in this same respect.

181:c series in this respect.

181:e rock-pigeon still lives

181:f great; if/arranged in serial/production, and even less

182 Closely connected with the statement, that the organic remains from an intermediate formation are in some degree intermediate in character, is the fact, insisted on by all palæontologists, that fossils from two consecutive formations are far more closely related to each other, than are the fossils from two remote formations.

183 Pictet gives as a well-known instance, the general resemblance of the organic remains from the several stages of the chalk formation, though the species are distinct in each stage.

183:b Chalk

184 This fact alone, from its generality, seems to have shaken Professor Pictet in his firm belief in the immutability of species.

185 He who is acquainted with the distribution of existing species over the globe, will not attempt to account for the close resemblance of the distinct species in closely consecutive formations, by the physical conditions of the ancient areas having remained nearly the same.

186 Let it be remembered that the forms of life, at least those inhabiting the sea, have changed almost simultaneously throughout the world, and therefore under the most different climates and conditions.

187 Consider the prodigious vicissitudes of climate during the pleistocene period, which includes the whole glacial period, and note how little the specific forms of the inhabitants of the sea have been affected.

187:e glacial epoch, and

188 On the theory of descent, the full meaning of the fact of fossil remains from closely consecutive formations, though ranked as distinct species, being closely related, is obvious.

188:e full meaning of the fossil remains from closely consecutive formations being closely related, though ranked as distinct species, is

189 As the accumulation of each formation has often been interrupted, and as long blank intervals have intervened between

successive formations, we ought not to expect to find, as I attempted to show in the last chapter, in any one or two formations all the intermediate varieties between the species which appeared at the commencement and close of these periods; but we ought to find after intervals, very long as measured by years, but only moderately long as measured geologically, closely allied forms, or, as they have been called by some authors, representative species; and these we assuredly do find.

189:d or in any two

189:e periods: but

189:f two formations, all

190 We find, in short, such evidence of the slow and scarcely sensible mutation of specific forms, as we have a just right to expect to find.

190:d expect.

190:e have the right

190:f mutations

191 *On the state of Development of Ancient Forms.—*

191:c On/Ancient compared with Living Forms.—

191:d [Center] Forms. [Space]

192 There has been much discussion whether recent forms are more highly developed than ancient.

192:[c]

193 I will not here enter on this subject, for naturalists have not as yet defined to each other's satisfaction what is meant by high and low forms.

193:[c]

193.0.1:c We have seen in the fourth chapter that the degree of differentiation and specialisation of the parts of all organic beings, when come to maturity, is the best standard, as yet suggested, of their degree of perfection or highness.

193.0.1:d[¶] parts in all organic beings, when arrived at maturity

193.0.1:f parts in organic

193.1:b The best definition probably is, that the higher forms have their organs more distinctly specialised for different functions; and as such division of physiological labour seems to be an advantage to each being, natural selection will constantly tend in so far to make the later and more modified forms higher than their early progenitors, or than the slightly modified descendants of such progenitors.

193.1:c We have also seen that as the specialisation of parts and organs is an advantage to each being, so natural selection will constantly tend thus to render the organisation of each being more specialised and perfect, and in this sense higher; not but that it may and will leave many creatures with simple and unimproved structures fitted for simple conditions of life, and

547

in some cases will even degrade or simplify the organisation, yet leaving such degraded beings better fitted for their new walks of life.

193.1:d that, as/will tend to

193.1:f parts is an advantage/may leave many

194 But in one particular sense the more recent forms must, on my theory, be higher than the more ancient; for each new species is formed by having had some advantage in the struggle for life over other and preceding forms.

194:b In a more general sense

194:c In another and more general manner we can see that on the theory of natural selection the more recent forms will tend to be higher than their progenitors; for

194:d manner, new species will become superior to their predecessors; for they will in the struggle for life have to beat all the older forms with which they come into close competition.

194:e they will have to beat in the struggle for life all the older

194:f they have to beat/forms, with

195 If under a nearly similar climate, the eocene inhabitants of one quarter of the world were put into competition with the existing inhabitants of the same or some other quarter, the eocene fauna or flora would certainly be beaten and exterminated; as would a secondary fauna by an eocene, and a palæozoic fauna by a secondary fauna.

195:d We may therefore conclude that if under a nearly similar climate the eocene inhabitants of the world could be put into competition with our existing inhabitants, the former would be

195:e with the existing/exterminated by the latter, as would the secondary by the eocene, and the palæozoic by the secondary forms.

196 I do not doubt that this process of improvement has affected in a marked and sensible manner the organisation of the more recent and victorious forms of life, in comparison with the ancient and beaten forms; but I can see no way of testing this sort of progress.

196.x-y:c So that by this fundamental test of victory in the battle for life, as well as by the standard of the specialisation of organs, modern forms ought on the theory of natural selection to stand higher than ancient forms. *y* Is this the case?

196.x:f ought, on the theory of natural selection, to

196.1:c A large majority of palæontologists would certainly answer in the affirmative; but in my judgment I can, after having read the discussions on this subject by Lyell, and Hooker's views in regard to plants, concur only to a limited extent.

196.1:d I cannot, after having read the discussions on this subject by Lyell, Bronn, and Hooker, look at this conclusion as fully proved, though highly probable.

196.1:e would answer in the affirmative; and I suppose that the answer must be admitted as true, though difficult of full proof.

196.1:f affirmative; and it seems that this answer/of proof.

196.2:c Nevertheless it may be anticipated that the evidence will be rendered more decisive by future geological research.

196.2:[d]

196.2.1:d[¶] It is no valid objection to this conclusion or to the general belief that species in the course of time change, that certain Brachiopods have been but slightly modified from an extremely remote geological period, although no explanation can be given of this fact.

196.2.1:e conclusion, that/geological epoch.

196.2.1:f epoch; and that certain land and fresh-water shells have remained nearly the same, from the time when, as far as is known, they first appeared.

196.2.2:d It is not an insuperable difficulty that Foraminifera have not progressed in organisation, as insisted on by Dr. Carpenter, since that most ancient of all epochs the Laurentian formation of Canada; for some organisms would have to remain fitted for simple conditions of life, and what better for this end than these lowly organised Protozoa?

196.2.2:e not, as insisted on by Dr. Carpenter, progressed in organisation since even the Laurentian epoch; for/what could be better

196.2.3:d It is no great difficulty that fresh-water shells, as Professor Phillips has remarked, have remained almost unaltered from the time when they first appeared to the present day; but in this case we can see that these shells will have been subjected to less severe competition than the molluscs which inhabit the far more extensive area of the sea with its innumerable inhabitants.

196.2.3:e has urged, have remained/day; for these/molluscs inhabiting the more

196.2.3:[f]

196.2.4:d Such objections as the above would be fatal to any view which included advance in organisation as a necessary contingent.

196.2.4:f to my view, if it included

196.2.5:d They would be fatal to my view if Foraminifera, for instance, could be proved to have first come into existence during the Laurentian epoch, or Brachiopods during the lower Silurian formations; for if this were proved, there would not have been time sufficient for the development of these organisms up to the standard which they had then reached.

196.2.5:e would likewise be fatal/Brachiopods during the Cambrian formation; for in this case, there/which they then

196.2.5:f fatal, if the above Foraminifera/or the above Brachiopods/which they had then

196.2.6:d When once advanced up to any given point, there is no necessity on the theory of natural selection for their further continued progress; though they will, during each successive age, have to be slightly modified, so as to hold their places in relation to their changing conditions of life.

196.2.6:f When advanced/necessity, on the theory of natural selection, for/relation to slight changes in their conditions.

196.2.7:d All such objections hinge on the question whether we have any sufficient knowledge of the antiquity of the world and of the periods when the various forms of life first appeared; and this may be boldly disputed.

196.2.7:e we really know how old the world is, and at what periods the various forms of life first appeared; and this may be disputed.

196.2.7:f The foregoing objections/period/may well be

196.3:c[¶] The problem is in many ways excessively intricate.

196.3:d problem whether organization on the whole has advanced is

196.4:c The geological record, at all times imperfect, does not extend far enough back, as I believe, to show with unmistakeable clearness that within the known history of the world organisation has largely advanced.

196.4:f back, to

196.5:c Even at the present day, looking to members of the same class, naturalists are not unanimous which forms are highest: thus, some look at the selaceans or sharks from their approach in some important points of structure to reptiles as the highest fish; others look at the teleosteans as the highest.

196.5:d forms are to be ranked as highest/sharks, from/reptiles, as the highest fish

196.5:f forms ought to be

196.6-7:c The ganoids stand intermediate between the selaceans and teleosteans; the latter at the present day are largely preponderant in number; but formerly selaceans and ganoids alone existed; and in this case, according to the standard of highness chosen, so will it be said that fishes have advanced or retrograded in organisation. 7 To attempt to compare in the scale of highness members of distinct types seem hopeless: who will decide whether a cuttle-fish be higher than a bee—that insect which the great Von Baer believed to be "in fact more highly organised than a fish, although upon another type"?

196.7:f compare members of distinct types in the scale of highness seems

197 Crustaceans, for instance, not the highest in their own class, may have beaten the highest molluscs.

197:c In the complex struggle for life it is quite credible that crustaceans, for instance, not very high in their own class, might beat the cephalopods or highest molluscs; and such crustaceans, though not highly developed, would stand very high in the scale of invertebrate animals if judged by the most decisive of all trials—the law of battle.

197:d crustaceans, not very high in their own class, might beat cephalopods, the highest

197:f animals, if

197.1:c Besides these inherent difficulties in deciding which forms are the most advanced in organisation, we ought not solely to compare the highest members of a class at any two distant periods—though undoubtedly this is one and perhaps the most important element in striking a balance—but we ought to compare all the members, high and low, at the two periods.

197.1:e two periods—though

197.2:c At an ancient epoch the highest and lowest molluscs, namely, cephalopods and brachiopods, swarmed in numbers: at the present time both these orders have been greatly reduced, whereas other orders, intermediate in grade of organisation, have largely increased; consequently some naturalists have maintained that molluscs were formerly more highly developed than at present; but a stronger case can be made out on the other side, by considering the vast reduction at the present day of the lowest molluscs, more especially as the existing cephalopods, though so few in number, are more highly organised than their ancient representatives.

197.2:d organisation, have been largely

197.2:e time both orders are greatly reduced, whilst other orders, intermediate in organisation, have largely increased; consequently some naturalists maintain that/on the opposite side, by considering the vast reduction of the lowest molluscs, and the fact that our existing cephalopods, though few

197.2:f lowest molluscoidal animals, namely, cephalopods and brachiopods, swarmed in numbers; at the present time both groups are greatly reduced, whilst others, intermediate/reduction of brachiopods, and

197.3:c We ought also to consider the relative proportional numbers of the high and low classes in the population of the world at the two periods: if, for instance, at the present day there be fifty thousand kinds of vertebrate animals, and if we knew that at some former period only ten thousand kinds had existed, we ought to look at this increase in number of the highest class, which implies a great displacement of lower forms, as a decided advance in the organisation of the world, whether the higher or the lower vertebrata had thus largely increased.

197.3:d at any two

197.3:e to compare the relative proportional numbers of the high and low classes throughout the world at any two periods: if, for instance, at the present day fifty thousand kinds of vertebrate animals exist, and/kinds existed/number in the highest/organisation of the world.

197.3:f numbers at any two periods of the high and low classes throughout the world: if, for

197.4:c We can thus see how hopelessly difficult it will apparently for ever be to compare with perfect fairness, under such extremely complex relations, the standard of organisation of the imperfectly-known faunas of successive periods of the earth's history.

197.4:e We thus see how hopelessly difficult it is to/periods.

197.5:c[¶] We shall appreciate under one important point of view this difficulty the more clearly, by looking to the case of certain existing faunas and floras.

197.5:e appreciate this difficulty the more clearly, by looking to certain

197.5:f difficulty more

198 From the extraordinary manner in which European productions have recently spread over New Zealand, and have seized on places which must have been previously occupied, we may believe, if all the animals and plants of Great Britain were set free in New Zealand, that in the course of time a multitude of British forms would become thoroughly naturalized there, and would exterminate many of the natives.

198:d we must believe, that if/Zealand, in

198:f occupied by the indigenes, we/Zealand, a multitude of British forms would in the course of time become

199 On the other hand, from what we see now occurring in New Zealand, and from hardly a single inhabitant of the southern hemisphere having become wild in any part of Europe, we may doubt, if all the productions of New Zealand were set free in Great Britain, whether any considerable number would be enabled to seize on places now occupied by our native plants and animals.

199:d from the progress of this displacement in New/may well doubt

199:e from hardly/doubt whether, if/Britain, any

199:f from the fact that hardly a single inhabitant of the southern hemisphere has become

200 Under this point of view, the productions of Great Britain may be said to be higher than those of New Zealand.

200:e Britain stand much higher in the scale than

201 Yet the most skilful naturalist from an examination of the species of the two countries could not have foreseen this result.

201:e naturalist, from

201:f countries, could

202 Agassiz insists that ancient animals resemble to a certain extent the embryos of recent animals of the same classes; or that

the geological succession of extinct forms is in some degree parallel to the embryological development of recent forms.

202:d parallel with the embryological

202:e Agassiz and several other highly competent judges insist/recent animals belonging to the same classes; and that the geological succession of extinct forms is nearly parallel with the embryological development of existing forms.

203 I must follow Pictet and Huxley in thinking that the truth of this doctrine is very far from proved.

203:d is far

203:[e]

204 Yet I fully expect to see it hereafter confirmed, at least in regard to subordinate groups, which have branched off from each other within comparatively recent times.

204:d groups which

204:[e]

205 For this doctrine of Agassiz accords well with the theory of natural selection.

205:d accords admirably well

205:e This view accords admirably well with our theory.

206 In a future chapter I shall attempt to show that the adult differs from its embryo, owing to variations supervening at a not early age, and being inherited at a corresponding age.

206:f variations having supervened at a not early age, and having been inherited

207 This process, whilst it leaves the embryo almost unaltered, continually adds, in the course of successive generations, more and more difference to the adult.

208 Thus the embryo comes to be left as a sort of picture, preserved by nature, of the ancient and less modified condition of each animal.

208:e[No ¶] condition of the animal.

208:f nature, of the former and less modified condition of the species.

209 This view may be true, and yet it may never be capable of full proof.

209:e yet may

209:f of proof.

210 Seeing, for instance, that the oldest known mammals, reptiles, and fish strictly belong to their own proper classes, though some of these old forms are in a slight degree less distinct from each other than are the typical members of the same groups at the present day, it would be vain to look for animals having the common embryological character of the Vertebrata, until beds

far beneath the lowest Silurian strata are discovered—a discovery of which the chance is very small.

210:d beds rich in fossils are discovered far beneath the lowest Silurian stratum—a discovery of which the chance is small.

210:e fishes/their proper

210:f lowest Cambrian strata

211 *On the Succession of the same Types within the same areas, during the later tertiary periods.—*

211:d [*Center*] On/Areas/periods. [*Space*]

211:e Tertiary

212 Mr. Clift many years ago showed that the fossil mammals from the Australian caves were closely allied to the living marsupials of that continent.

212:d[¶]

213 In South America, a similar relationship is manifest, even to an uneducated eye, in the gigantic pieces of armour like those of the armadillo, found in several parts of La Plata; and Professor Owen has shown in the most striking manner that most of the fossil mammals, buried there in such numbers, are related to South American types.

213:d armour, like

214 This relationship is even more clearly seen in the wonderful collection of fossil bones made by MM. Lund and Clausen in the caves of Brazil.

215 I was so much impressed with these facts that I strongly insisted, in 1839 and 1845, on this "law of the succession of types," —on "this wonderful relationship in the same continent between the dead and the living."

216 Professor Owen has subsequently extended the same generalisation to the mammals of the Old World.

217 We see the same law in this author's restorations of the extinct and gigantic birds of New Zealand.

218 We see it also in the birds of the caves of Brazil.

219 Mr. Woodward has shown that the same law holds good with sea-shells, but from the wide distribution of most genera of molluscs, it is not well displayed by them.

219:d but, from

219:f most molluscs

220 Other cases could be added, as the relation between the extinct and living land-shells of Madeira; and between the extinct and living brackish-water shells of the Aralo-Caspian Sea.

221 Now what does this remarkable law of the succession of the same types within the same areas mean?

222 He would be a bold man, who after comparing the present

climate of Australia and of parts of South America under the same latitude, would attempt to account, on the one hand, by dissimilar physical conditions for the dissimilarity of the inhabitants of these two continents, and, on the other hand, by similarity of conditions, for the uniformity of the same types in each during the later tertiary periods.

222:d who, after/conditions, for

222:f man who/hand through dissimilar/continents; and, on the other hand through similarity of conditions, for the uniformity of the same types in each continent during

223 Nor can it be pretended that it is an immutable law that marsupials should have been chiefly or solely produced in Australia; or that Edentata and other American types should have been solely produced in South America.

224 For we know that Europe in ancient times was peopled by numerous marsupials; and I have shown in the publications above alluded to, that in America the law of distribution of terrestrial mammals was formerly different from what it now is.

225 North America formerly partook strongly of the present character of the southern half of the continent; and the southern half was formerly more closely allied, than it is at present, to the northern half.

226 In a similar manner we know from Falconer and Cautley's discoveries, that northern India was formerly more closely related in its mammals to Africa than it is at the present time.

226:d know, from

227 Analogous facts could be given in relation to the distribution of marine animals.

228 On the theory of descent with modification, the great law of the long enduring, but not immutable, succession of the same types within the same areas, is at once explained; for the inhabitants of each quarter of the world will obviously tend to leave in that quarter, during the next succeeding period of time, closely allied though in some degree modified descendants.

229 If the inhabitants of one continent formerly differed greatly from those of another continent, so will their modified descendants still differ in nearly the same manner and degree.

230 But after very long intervals of time and after great geographical changes, permitting much inter-migration, the feebler will yield to the more dominant forms, and there will be nothing immutable in the laws of past and present distribution.

230:e time, and after

230:f in the distribution of organic beings.

231 It may be asked in ridicule, whether I suppose that the megatherium and other allied huge monsters have left behind

them in South America the sloth, armadillo, and anteater, as their degenerate descendants.

231:b America, the sloth

231:e monsters, which formerly lived in South America, have left behind them the sloth

232 This cannot for an instant be admitted.

233 These huge animals have become wholly extinct, and have left no progeny.

234 But in the caves of Brazil, there are many extinct species which are closely allied in size and in other characters to the species still living in South America; and some of these fossils may be the actual progenitors of living species.

234:d may have been the actual progenitors of the living

234:e size and in all other/may be

234:f may have been

235 It must not be forgotten that, on my theory, all the species of the same genus have descended from some one species; so that if six genera, each having eight species, be found in one geological formation, and in the next succeeding formation there be six other allied or representative genera with the same number of species, then we may conclude that only one species of each of the six older genera has left modified descendants, constituting the six new genera.

236 The other seven species of the old genera have all died out and have left no progeny.

235 + 6:d genus are the descendants of some one species; so that, if/formation, and in a succeeding/conclude that generally only one species of each of the older genera has left modified descendants, which constitute the several species of the new genera; the other seven species of each of the old genera having died out and left

235 + 6:e on our theory/representative genera each with/seven species of each old genus having

235 + 6:f constitute the new genera containing the several species; the other

237 Or, which would probably be a far commoner case, two or three species of two or three alone of the six older genera will have been the parents of the six new genera; the other old species and the other whole genera having become utterly extinct.

237:b whole old genera

237:d Or, and this probably would be a far commoner case, two or three species of only two or three of the six older genera will have been the parents of the new

237:e probably will be a far commoner case, two or three species in two or three alone of the six older genera will be the parents of the new genera: the other species and the other whole

238 In failing orders, with the genera and species decreasing in numbers, as apparently is the case of the Edentata of South America, still fewer genera and species will have left modified blood-descendants.

238:d will leave

238:e as is the case with the Edentata

238:f numbers as

239 *Summary of the preceding and present Chapters.—*

239:c Chapter.

239:d [Center] Chapter. [Space]

239:f Chapters.

240 I have attempted to show that the geological record is extremely imperfect; that only a small portion of the globe has been geologically explored with care; that only certain classes of organic beings have been largely preserved in a fossil state; that the number both of specimens and of species, preserved in our museums, is absolutely as nothing compared with the incalculable number of generations which must have passed away even during a single formation; that, owing to subsidence being necessary for the accumulation of fossiliferous deposits thick enough to resist future degradation, enormous intervals of time have elapsed between the successive formations; that there has probably been more extinction during the periods of subsidence, and more variation during the periods of elevation, and during the latter the record will have been least perfectly kept; that each single formation has not been continuously deposited; that the duration of each formation is, perhaps, short compared with the average duration of specific forms; that migration has played an important part in the first appearance of new forms in any one area and formation; that widely ranging species are those which have varied most, and have oftenest given rise to new species; and that varieties have at first often been local.

240:c subsidence being almost necessary for the accumulation of deposits rich in fossils and thick/between most of our successive

240:d duration of each formation is, probably, short/varied most frequently, and have oftenest given rise to new species; that varieties have at first been local; and lastly, although each species must have passed through numerous transitional stages, it is probable that the periods, during which each underwent modification, though many and long as measured by years, have been short in comparison with the periods during which each remained in an unchanged condition.

240:e compared with the number/rich in fossil species of very many kinds and thick enough to resist future degradation, great intervals of time must have elapsed

557

240:f rich in fossil species of many kinds, and thick enough to out-last future

241 All these causes taken conjointly, must have tended to make the geological record extremely imperfect, and will to a large extent explain why we do not find interminable varieties, connecting together all the extinct and existing forms of life by the finest graduated steps.

241:c These causes taken conjointly, must have made the geological/why—though we do find many links between the members of the same group—we/together all extinct and existing forms by

241:d conjointly, will to a large extent explain why—though we do find many links between the species of the same

241:e links—we

241.1:d It should also be constantly borne in mind that any linking varieties between two or more forms, which might be found, would be ranked, unless the whole chain could be perfectly restored, as so many new and distinct species; for it is not pretended that we have any sure criterion by which species and varieties can be discriminated.

241.1:f linking variety between two forms/as a new

242 He who rejects these views on the nature of the geological record, will rightly reject my whole theory.

242:d on the imperfection of

242:e reject the whole

242:f rejects this view of the imperfection

243 For he may ask in vain where are the numberless transitional links which must formerly have connected the closely allied or representative species, found in the several stages of the same great formation.

243:d found in the successive stages

244 He may disbelieve in the enormous intervals of time which have elapsed between our consecutive formations; he may overlook how important a part migration must have played, when the formations of any one great region alone, as that of Europe, are considered; he may urge the apparent, but often falsely apparent, sudden coming in of whole groups of species.

244:e in the immense intervals/migration has played

244:f which must have elapsed/region, as those of Europe

245 He may ask where are the remains of those infinitely numerous organisms which must have existed long before the first bed of the Silurian system was deposited: I can answer this latter question only hypothetically, by saying that as far as we can see, where our oceans now extend they have for an enormous period extended, and where our oscillating continents now stand they have stood ever since the Silurian epoch; but that long before that period, the world may have presented a wholly different

aspect; and that the older continents, formed of formations older than any known to us, may now all be in a metamorphosed condition, or may lie buried under the ocean.

245.x:c deposited?

245.x:e before the Cambrian system

245.y:c I can

245.y:d We now know that animals, and probably plants, lived at an epoch immensely remote, long anterior to the primordial zone of the Silurian system, but I can answer the above question only by supposing that where our oceans now extend they have extended for an enormous period, and where our oscillating continents now stand they have stood since the commencement of the Silurian epoch; but that, long before that period, the world presented a widely different/us, exist now only as remnants in a metamorphosed condition, or lie wholly buried

245.y:e that at least one animal did then exist; but I/of the Cambrian system; but that, long before that epoch, the world/lie still buried

245.y:f answer this last question

246   Passing from these difficulties, all the other great leading facts in palæontology seem to me simply to follow on the theory of descent with modification through natural selection.

246:d difficulties, the other

246:f palæontology agree admirably with the theory of descent with modification through variation and natural

247   We can thus understand how it is that new species come in slowly and successively; how species of different classes do not necessarily change together, or at the same rate, or in the same degree; yet in the long run that all undergo modification to some extent.

248   The extinction of old forms is the almost inevitable consequence of the production of new forms.

249   We can understand why when a species has once disappeared it never reappears.

250   Groups of species increase in numbers slowly, and endure for unequal periods of time; for the process of modification is necessarily slow, and depends on many complex contingencies.

251   The dominant species of the larger dominant groups tend to leave many modified descendants, and thus new sub-groups and groups are formed.

251:e species belonging to large and dominant groups tend to leave many modified descendants, which form new sub-groups and groups.

252   As these are formed, the species of the less vigorous groups, from their inferiority inherited from a common progenitor,

559

tend to become extinct together, and to leave no modified off-spring on the face of the earth.

253 But the utter extinction of a whole group of species may often be a very slow process, from the survival of a few descendants, lingering in protected and isolated situations.

253:d species has sometimes been a slow

254 When a group has once wholly disappeared, it does not reappear; for the link of generation has been broken.

255 We can understand how the spreading of the dominant forms of life, which are those that oftenest vary, will in the long run tend to people the world with allied, but modified, descendants; and these will generally succeed in taking the places of those groups of species which are their inferiors in the struggle for existence.

255:e how the dominant forms which spread widely and yield the greatest number of varieties will tend/succeed in displacing the groups which

255:f how it is that dominant/varieties tend

256 Hence, after long intervals of time, the productions of the world will appear to have changed simultaneously.

256:e world appear

257 We can understand how it is that all the forms of life, ancient and recent, make together one grand system; for all are connected by generation.

257:e together a few grand classes; for all are at least thus far connected

257:f classes.

258 We can understand, from the continued tendency to divergence of character, why the more ancient a form is, the more it generally differs from those now living.

259 Why ancient and extinct forms often tend to fill up gaps between existing forms, sometimes blending two groups previously classed as distinct into one; but more commonly only bringing them a little closer together.

258 + 9:d living; why

258 + 9:e groups, previously classed as distinct, into one; but more commonly bringing them only a little

260 The more ancient a form is, the more often, apparently, it displays characters in some degree intermediate between groups now distinct; for the more ancient a form is, the more nearly it will be related to, and consequently resemble, the common progenitor of groups, since become widely divergent.

260:e often it stands in

261 Extinct forms are seldom directly intermediate between exist-

ing forms; but are intermediate only by a long and circuitous course through many extinct and very different forms.

261:c many other extinct and different

261:e through other

262 We can clearly see why the organic remains of closely consecutive formations are more closely allied to each other, than are those of remote formations; for the forms are more closely linked together by generation: we can clearly see why the remains of an intermediate formation are intermediate in character.

262.x-y:e are closely allied; for they are closely linked together by generation. *y* We

263 The inhabitants of each successive period in the world's history have beaten their predecessors in the race for life, and are, in so far, higher in the scale of nature; and this may account for that vague yet ill-defined sentiment, felt by many palæontologists, that organisation on the whole has progressed.

263:c history must have beaten/nature, and their structure will generally have become more specialised; and this may account for that ill-defined yet common sentiment

263:e structure has generally become more specialised; and this may account for the common belief held by

263:f inhabitants of the world at each successive period in its history have beaten/scale, and their

264 If it should hereafter be proved that ancient animals resemble to a certain extent the embryos of more recent animals of the same class, the fact will be intelligible.

264:e Extinct and ancient animals resemble to a certain extent the embryos of the more recent animals belonging to the same classes, and this wonderful fact receives a simple explanation according to our views.

265 The succession of the same types of structure within the same areas during the later geological periods ceases to be mysterious, and is simply explained by inheritance.

265:e is intelligible on the principle of inheritance.

266 If then the geological record be as imperfect as I believe it to be, and it may at least be asserted that the record cannot be proved to be much more perfect, the main objections to the theory of natural selection are greatly diminished or disappear.

266:e imperfect as many believe, and

267 On the other hand, all the chief laws of palæontology plainly proclaim, as it seems to me, that species have been produced by ordinary generation: old forms having been supplanted by new and improved forms of life, produced by the laws of variation still acting round us, and preserved by Natural Selection.

267:e life, the products of Variation and the Survival of the Fittest.

CHAPTER XI.

1:f XII.

GEOGRAPHICAL DISTRIBUTION.

3 Present distribution cannot be accounted for by differences in physical conditions—Importance of barriers—Affinity of the productions of the same continent—Centres of creation—Means of dispersal, by changes of climate and of the level of the land, and by occasional means—Dispersal during the Glacial period co-extensive with the world.

3:d dispersal by changes

3:e period—Alternate Glacial periods in the North and South.

4 IN considering the distribution of organic beings over the face of the globe, the first great fact which strikes us is, that neither the similarity nor the dissimilarity of the inhabitants of various regions can be accounted for by their climatal and other physical conditions.

4:f be wholly accounted for by climatal

5 Of late, almost every author who has studied the subject has come to this conclusion.

6 The case of America alone would almost suffice to prove its truth: for if we exclude the northern parts where the circumpolar land is almost continuous, all authors agree that one of the most fundamental divisions in geographical distribution is that between the New and Old Worlds; yet if we travel over the vast American continent, from the central parts of the United States to its extreme southern point, we meet with the most diversified conditions; the most humid districts, arid deserts, lofty mountains, grassy plains, forests, marshes, lakes, and great rivers, under almost every temperature.

6:e conditions; humid

6:f exclude the arctic and northern temperate parts, all

7 There is hardly a climate or condition in the Old World which cannot be paralleled in the New—at least as closely as the same species generally require; for it is a most rare case to find a group of organisms confined to any small spot, having conditions peculiar in only a slight degree; for instance, small areas in the Old World could be pointed out hotter than any in the New World, yet these are not inhabited by a peculiar fauna or flora.

7:c areas in the Old World can be

7.x-y:e require. *y* No doubt small areas can be pointed out in the

Old World hotter than any in the New World, but these are not inhabited by a fauna different from that of the surrounding districts; for it is very rare to find a group of organisms confined to a small area, having conditions peculiar in only a slight degree.

7.y:f it is rare/area, of which the conditions are peculiar

8 Notwithstanding this parallelism in the conditions of the Old and New Worlds, how widely different are their living productions!

8:e this general parallelism

9   In the southern hemisphere, if we compare large tracts of land in Australia, South Africa, and western South America, between latitudes 25° and 35°, we shall find parts extremely similar in all their conditions, yet it would not be possible to point out three faunas and floras more utterly dissimilar.

10 Or again we may compare the productions of South America south of lat. 35° with those north of 25°, which consequently inhabit a considerably different climate, and they will be found incomparably more closely related to each other, than they are to the productions of Australia or Africa under nearly the same climate.

10:c other than

10:d Or again, we may compare those north of 25°, which consequently are separated by a space of ten degrees of latitude and live under a considerably different climate, yet they are incomparably

10:e latitude and are exposed to considerably different conditions, yet

10:f Or, again

11 Analogous facts could be given with respect to the inhabitants of the sea.

12   A second great fact which strikes us in our general review is, that barriers of any kind, or obstacles to free migration, are related in a close and important manner to the differences between the productions of various regions.

13 We see this in the great difference of nearly all the terrestrial productions of the New and Old Worlds, excepting in the northern parts, where the land almost joins, and where, under a slightly different climate, there might have been free migration for the northern temperate forms, as there now is for the strictly arctic productions.

13:f difference in nearly

14 We see the same fact in the great difference between the inhabitants of Australia, Africa, and South America under the same latitude: for these countries are almost as much isolated from each other as is possible.

14:f latitude; for

15 On each continent, also, we see the same fact; for on the opposite sides of lofty and continuous mountain-ranges, and of great
deserts, and sometimes even of large rivers, we find different
productions; though as mountain-chains, deserts, &c., are not as
impassable, or likely to have endured so long as the oceans separating continents, the differences are very inferior in degree to
those characteristic of distinct continents.

15:d long, as

15:e mountain-ranges, of great deserts, and even of large

16   Turning to the sea, we find the same law.

17 No two marine faunas are more distinct, with hardly a fish,
shell, or crab in common, than those of the eastern and western
shores of South and Central America; yet these great faunas
are separated only by the narrow, but impassable, isthmus of
Panama.

17:e The marine inhabitants of the eastern and western shores of
South America are very distinct, with extremely few fishes,
shells, or crabs in common; but Dr. Günther has recently
shown that on opposite sides of the isthmus of Panama, about
thirty per cent. of the fishes are the same; and this fact has led
naturalists to believe that the isthmus was formerly open.

17:f few shells, crustacea or echinodermata in common; but Dr.
Günther has recently shown that about thirty per cent. of the
fishes are the same on the opposite sides of the isthmus of Panama; and

18 Westward of the shores of America, a wide space of open ocean
extends, with not an island as a halting-place for emigrants;
here we have a barrier of another kind, and as soon as this is
passed we meet in the eastern islands of the Pacific, with another and totally distinct fauna.

18:c Pacific with

19 So that here three marine faunas range far northward and southward, in parallel lines not far from each other, under corresponding climates; but from being separated from each other
by impassable barriers, either of land or open sea, they are
wholly distinct.

19:e that three/southward in/are almost wholly

20 On the other hand, proceeding still further westward from the
eastern islands of the tropical parts of the Pacific, we encounter
no impassable barriers, and we have innumerable islands as
halting-places, until after travelling over a hemisphere we come
to the shores of Africa; and over this vast space we meet with
no well-defined and distinct marine faunas.

20:b halting-places, or continuous coasts, until

20:c farther

20:f hemisphere, we come

21 Although hardly one shell, crab or fish is common to the above-

named three approximate faunas of Eastern and Western America and the eastern Pacific islands, yet many fish range from the Pacific into the Indian Ocean, and many shells are common to the eastern islands of the Pacific and the eastern shores of Africa, on almost exactly opposite meridians of longitude.

21:d crab, or

21:e Although so few shells, crabs, or fishes are common to the above-named

21:f few marine animals are/many fishes range/Africa on

22   A third great fact, partly included in the foregoing statements, is the affinity of the productions of the same continent or sea, though the species themselves are distinct at different points and stations.

22:e statement/continent or of the same sea

23 It is a law of the widest generality, and every continent offers innumerable instances.

24 Nevertheless the naturalist in travelling, for instance, from north to south never fails to be struck by the manner in which successive groups of beings, specifically distinct, yet clearly related, replace each other.

24:d naturalist, in travelling

24:e distinct, though nearly related

25 He hears from closely allied, yet distinct kinds of birds, notes nearly similar, and sees their nests similarly constructed, but not quite alike, with eggs coloured in nearly the same manner.

26 The plains near the Straits of Magellan are inhabited by one species of Rhea (American ostrich), and northward the plains of La Plata by another species of the same genus; and not by a true ostrich or emu, like those found in Africa and Australia under the same latitude.

26:e those inhabiting Africa

27 On these same plains of La Plata, we see the agouti and bizcacha, animals having nearly the same habits as our hares and rabbits and belonging to the same order of Rodents, but they plainly display an American type of structure.

28 We ascend the lofty peaks of the Cordillera and we find an alpine species of bizcacha; we look to the waters, and we do not find the beaver or musk-rat, but the coypu and capybara, rodents of the American type.

28:d Cordillera, and we find an alpine

28:e rodents of the S. American

29 Innumerable other instances could be given.

30 If we look to the islands off the American shore, however much they may differ in geological structure, the inhabitants, though they may be all peculiar species, are essentially American.

30:e inhabitants are essentially American, though they may be all peculiar species.

31 We may look back to past ages, as shown in the last chapter, and we find American types then prevalent on the American continent and in the American seas.

32 We see in these facts some deep organic bond, prevailing throughout space and time, over the same areas of land and water, and independent of their physical conditions.

32:e bond, throughout/independent of physical

32:f water, independently

33 The naturalist must feel little curiosity, who is not led to inquire what this bond is.

33:e must be dull, who

34   This bond, on my theory, is simply inheritance, that cause which alone, as far as we positively know, produces organisms quite like, or, as we see in the case of varieties nearly like each other.

34:b varieties, nearly

34:e bond is/like each other, or, as we see in the case of varieties, nearly alike.

35 The dissimilarity of the inhabitants of different regions may be attributed to modification through natural selection, and in a quite subordinate degree to the direct influence of different physical conditions.

35:e in a subordinate degree to the definite influence

35:f through variation and natural selection, and probably in

36 The degree of dissimilarity will depend on the migration of the more dominant forms of life from one region into another having been effected with more or less ease, at periods more or less remote;—on the nature and number of the former immigrants;—and on their action and reaction, in their mutual struggles for life;—the relation of organism to organism being, as I have already often remarked, the most important of all relations.

36:e been prevented more or less effectually, at/immigrants;—and on the action of the inhabitants on each other in leading to the preservation of different modifications; the relation of organism to organism in the struggle for life being

36:f degrees/been more or less effectually prevented, at

37 Thus the high importance of barriers comes into play by checking migration; as does time for the slow process of modification through natural selection.

38 Widely-ranging species, abounding in individuals, which have already triumphed over many competitors in their own widely-extended homes will have the best chance of seizing on new places, when they spread into new countries.

38:c homes, will

39 In their new homes they will be exposed to new conditions, and will frequently undergo further modification and improvement; and thus they will become still further victorious, and will produce groups of modified descendants.

40 On this principle of inheritance with modification, we can understand how it is that sections of genera, whole genera, and even families are confined to the same areas, as is so commonly and notoriously the case.

40:c families, are

40:e families are

40:f families, are

41   I believe, as was remarked in the last chapter, in no law of necessary development.

41:f There is no evidence, as was remarked in the last chapter, of the existence of any law

42 As the variability of each species is an independent property, and will be taken advantage of by natural selection, only so far as it profits the individual in its complex struggle for life, so the degree of modification in different species will be no uniform quantity.

42:e profits each individual in its complex struggle for life, so the amount of modification

43 If, for instance, a number of species, which stand in direct competition with each other, migrate in a body into a new and afterwards isolated country, they will be little liable to modification; for neither migration nor isolation in themselves can do anything.

43:d If a number of species, after having long competed with each other in their old home, were to migrate in a body into a new and afterwards isolated country, they would be

43:e can effect anything.

43:f themselves effect

44 These principles come into play only by bringing organisms into new relations with each other, and in a lesser degree with the surrounding physical conditions.

45 As we have seen in the last chapter that some forms have retained nearly the same character from an enormously remote geological period, so certain species have migrated over vast spaces, and have not become greatly modified.

45:c greatly or at all modified.

46   On these views, it is obvious, that the several species of the same genus, though inhabiting the most distant quarters of the world, must originally have proceeded from the same source, as they have descended from the same progenitor.

46:c obvious that

46:e According to these/they are descended

47 In the case of those species, which have undergone during whole geological periods but little modification, there is not much difficulty in believing that they may have migrated from the same region; for during the vast geographical and climatal changes which will have supervened since ancient times, almost any amount of migration is possible.

47:e changes which have supervened

47:f periods little

48 But in many other cases, in which we have reason to believe that the species of a genus have been produced within comparatively recent times, there is great difficulty on this head.

49 It is also obvious that the individuals of the same species, though now inhabiting distant and isolated regions, must have proceeded from one spot, where their parents were first produced: for, as explained in the last chapter, it is incredible that individuals identically the same should ever have been produced through natural selection from parents specifically distinct.

49:c should have

49:f as has been explained, it

49.1:d[¶] *Single Centres of supposed Creation.—*

50 We are thus brought to the question which has been largely discussed by naturalists, namely, whether species have been created at one or more points of the earth's surface.

50:d[No ¶]

51 Undoubtedly there are very many cases of extreme difficulty, in understanding how the same species could possibly have migrated from some one point to the several distant and isolated points, where now found.

51:c difficulty in

51:d are many

52 Nevertheless the simplicity of the view that each species was first produced within a single region captivates the mind.

53 He who rejects it, rejects the *vera causa* of ordinary generation with subsequent migration, and calls in the agency of a miracle.

54 It is universally admitted, that in most cases the area inhabited by a species is continuous; and when a plant or animal inhabits two points so distant from each other, or with an interval of such a nature, that the space could not be easily passed over by migration, the fact is given as something remarkable and exceptional.

54:f not have been easily

55 The capacity of migrating across the sea is more distinctly limited in terrestrial mammals, than perhaps in any other organic beings; and, accordingly, we find no inexplicable cases of the same mammal inhabiting distant points of the world.

55:e The incapacity of migrating across a wide sea is more clear in the case of terrestrial mammals, than perhaps with any/inexplicable instances of the same mammals

56 No geologist will feel any difficulty in such cases as Great Britain having been formerly united to Europe, and consequently possessing the same quadrupeds.

56:e geologist feels any difficulty in Great Britain possessing the same quadrupeds with the rest of Europe, for they were no doubt once united.

57 But if the same species can be produced at two separate points, why do we not find a single mammal common to Europe and Australia or South America?

58 The conditions of life are nearly the same, so that a multitude of European animals and plants have become naturalised in America and Australia; and some of the aboriginal plants are identically the same at these distant points of the northern and southern hemispheres?

59 The answer, as I believe, is, that mammals have not been able to migrate, whereas some plants, from their varied means of dispersal, have migrated across the vast and broken interspace.

59:e across the wide and broken interspaces.

60 The great and striking influence which barriers of every kind have had on distribution, is intelligible only on the view that the great majority of species have been produced on one side alone, and have not been able to migrate to the other side.

60:e influence of barriers of all kinds, is/side, and have not been able to migrate to the opposite side.

61 Some few families, many sub-families, very many genera, and a still greater number of sections of genera are confined to a single region; and it has been observed by several naturalists, that the most natural genera, or those genera in which the species are most closely related to each other, are generally local, or confined to one area.

61:c local, or if they have a wide range that their range is continuous.

61:d sections of genera, are confined to a single/other, are generally confined to the same country, or

62 What a strange anomaly it would be, if, when coming one step lower in the series, to the individuals of the same species, a directly opposite rule prevailed; and species were not local, but had been produced in two or more distinct areas!

62:c more quite distinct

62:d it would be if a directly opposite rule were to prevail, when we go down one step lower in the series, namely, to the individuals of the same species, and these had not been, at least at first, confined to some one region!

62:e be, if

63　Hence it seems to me, as it has to many other naturalists, that the view of each species having been produced in one area alone, and having subsequently migrated from that area as far as its powers of migration and subsistence under past and present conditions permitted, is the most probable.

64　Undoubtedly many cases occur, in which we cannot explain how the same species could have passed from one point to the other.

65　But the geographical and climatal changes, which have certainly occurred within recent geological times, must have interrupted or rendered discontinuous the formerly continuous range of many species.

65:e must have rendered

66　So that we are reduced to consider whether the exceptions to continuity of range are so numerous and of so grave a nature, that we ought to give up the belief, rendered probable by general considerations, that each species has been produced within one area, and has migrated thence as far as it could.

67　It would be hopelessly tedious to discuss all the exceptional cases of the same species, now living at distant and separated points; nor do I for a moment pretend that any explanation could be offered of many such cases.

67:e many instances.

67:f points, nor

68　But after some preliminary remarks, I will discuss a few of the most striking classes of facts; namely, the existence of the same species on the summits of distant mountain-ranges, and at distant points in the arctic and antarctic regions; and secondly (in the following chapter), the wide distribution of fresh-water productions; and thirdly, the occurrence of the same terrestrial species on islands and on the mainland, though separated by hundreds of miles of open sea.

68:f islands and on the nearest mainland

69　If the existence of the same species at distant and isolated points of the earth's surface, can in many instances be explained on the view of each species having migrated from a single birthplace; then, considering our ignorance with respect to former climatal and geographical changes and various occasional means of transport, the belief that this has been the universal law, seems to me incomparably the safest.

69:e changes and to the various occasional means of transport, the belief that a single birthplace is the law

70　In discussing this subject, we shall be enabled at the same time to consider a point equally important for us, namely, whether the several distinct species of a genus, which on my theory have all descended from a common progenitor, can have migrated (undergoing modification during some part of their migration) from the area inhabited by their progenitor.

70:e several species of a genus, which must on the theory all be descended from a common progenitor, can have migrated, undergoing modification during their migration, from some one area.

70:f on our theory

71 If it can be shown to be almost invariably the case, that a region, of which most of its inhabitants are closely related to, or belong to the same genera with the species of a second region, has probably received at some former period immigrants from this other region, my theory will be strengthened; for we can clearly understand, on the principle of modification, why the inhabitants of a region should be related to those of another region, whence it has been stocked.

71:d with, the species

71:e When most of the species inhabiting one region are different from those of another region, but are closely allied or belong to the same genera, if in all such cases it can be shown that there probably has been at some former period migration from the one region to the other, our general view will be much strengthened; for the explanation is obvious on the principle of descent with modification.

71:f If, when/another region, though closely allied to them, it can be shown that migration from the one region to the other has probably occurred at some former period, our

72 A volcanic island, for instance, upheaved and formed at the distance of a few hundreds of miles from a continent, would probably receive from it in the course of time a few colonists, and their descendants, though modified, would still be plainly related by inheritance to the inhabitants of the continent.

72:c inhabitants of that continent.

72:e be related

73 Cases of this nature are common, and are, as we shall hereafter more fully see, inexplicable on the theory of independent creation.

73:e hereafter see

74 This view of the relation of species in one region to those in another, does not differ much (by substituting the word variety for species) from that lately advanced in an ingenious paper by Mr. Wallace, in which he concludes, that "every species has come into existence coincident both in space and time with a pre-existing closely allied species."

74:c concludes that

74:e relation of the species of one region to those of another, does not differ much from that advanced by Mr. Wallace, who concludes

75 And I now know from correspondence, that this coincidence he attributes to generation with modification.

75:*e* know that he attributes this coincidence to descent with

75:*f* And it is now well known

76   The previous remarks on "single and multiple centres of creation" do not directly bear on another allied question,— namely whether all the individuals of the same species have descended from a single pair, or single hermaphrodite, or whether, as some authors suppose, from many individuals simultaneously created.

76:*c* namely, whether

76:*e* The discussion on "single and multiple centres of creation" does not/species are descended

76:*f* The question of single or multiple centres of creation differs from another though allied

77 With those organic beings which never intercross (if such exist), the species, on my theory, must have descended from a succession of improved varieties, which will never have blended with other individuals or varieties, but will have supplanted each other; so that, at each successive stage of modification and improvement, all the individuals of each variety will have descended from a single parent.

77:*d* With organic/exist), each species/varieties, which can never

77:*e* intercross, if such exist, each species must be descended from a succession of modified varieties, which have supplanted each other, but which have never blended with other individuals or varieties of the same species; so/all the individuals of the same variety will be descended

77:*f* varieties, that have supplanted each other, but have never/ modification, all the individuals of the same form will be

78 But in the majority of cases, namely, with all organisms which habitually unite for each birth, or which often intercross, I believe that during the slow process of modification the individuals of the species will have been kept nearly uniform by intercrossing; so that many individuals will have gone on simultaneously changing, and the whole amount of modification will not have been due, at each stage, to descent from a single parent.

78:*d* in the great majority

78:*e* or which occasionally intercross, the individuals of the same species inhabiting the same area will be kept nearly uniform by intercrossing; so that many individuals will go on simultaneously changing, and the whole amount of modification at each stage will not be due to

79 To illustrate what I mean: our English race-horses differ slightly from the horses of every other breed; but they do not owe their difference and superiority to descent from any single pair, but to continued care in selecting and training many individuals during many generations.

79:d differ from/their superiority and difference to descent/during each generation.

79:e their difference and superiority to descent/care in the selecting and training of many

80   Before discussing the three classes of facts, which I have selected as presenting the greatest amount of difficulty on the theory of "single centres of creation," I must say a few words on the means of dispersal.

81   *Means of Dispersal.——*
81:d [*Center*] *Means of Dispersal.* [*Space*]

82 Sir C. Lyell and other authors have ably treated this subject.
82:d[¶]

83 I can give here only the briefest abstract of the more important facts.

84 Change of climate must have had a powerful influence on migration: a region when its climate was different may have been a high road for migration, but now be impassable; I shall, however, presently have to discuss this branch of the subject in some detail.

84:e migration; an impassable region when its climate was different from what it now is, may have been a high road for migration; I

84.x-z:f migration. *y* A region now impassable to certain organisms from the nature of its climate, might have been a high road for migration, when the climate was different. *z* I

85 Changes of level in the land must also have been highly influential: a narrow isthmus now separates two marine faunas; submerge it, or let it formerly have been submerged, and the two faunas will now blend or may formerly have blended: where the sea now extends, land may at a former period have connected islands or possibly even continents together, and thus have allowed terrestrial productions to pass from one to the other.

85.x-y:f blend together, or may formerly have blended. *y* Where

86 No geologist will dispute that great mutations of level, have occurred within the period of existing organisms.

86:b level have

86:f geologist disputes

87 Edward Forbes insisted that all the islands in the Atlantic must recently have been connected with Europe or Africa, and Europe likewise with America.

87:e must have been recently connected

88 Other authors have thus hypothetically bridged over every ocean, and have united almost every island to some mainland.

88:e and united

89 If indeed the arguments used by Forbes are to be trusted, it must be admitted that scarcely a single island exists which has not recently been united to some continent.

90 This view cuts the Gordian knot of the dispersal of the same species to the most distant points, and removes many a difficulty: but to the best of my judgment we are not authorized in admitting such enormous geographical changes within the period of existing species.

90:f difficulty; but

91 It seems to me that we have abundant evidence of great oscillations of level in our continents; but not of such vast changes in their position and extension, as to have united them within the recent period to each other and to the several intervening oceanic islands.

91:e oscillations in the level of the land or sea; but not of such vast changes in the position and extension of our continents, as

92 I freely admit the former existence of many islands, now buried beneath the sea, which may have served as halting places for plants and for many animals during their migration.

92:d halting-places

93 In the coral-producing oceans such sunken islands are now marked, as I believe, by rings of coral or atolls standing over them.

93:c marked by

94 Whenever it is fully admitted, as I believe it will some day be, that each species has proceeded from a single birthplace, and when in the course of time we know something definite about the means of distribution, we shall be enabled to speculate with security on the former extension of the land.

94:d as no doubt it

94:f as it

95 But I do not believe that it will ever be proved that within the recent period continents which are now quite separate, have been continuously, or almost continuously, united with each other, and with the many existing oceanic islands.

95:e period most of our continents which now stand quite

95:f almost continuously united

96 Several facts in distribution,—such as the great difference in the marine faunas on the opposite sides of almost every continent, —the close relation of the tertiary inhabitants of several lands and even seas to their present inhabitants,—a certain degree of relation (as we shall hereafter see) between the distribution of mammals and the depth of the sea,—these and other such facts seem to me opposed to the admission of such prodigious geographical revolutions within the recent period, as are necessitated on the view advanced by Forbes and admitted by his many followers.

96:d present inhabitants,—the degree of affinity between the mammals inhabiting islands with those of the nearest continent, being in part determined (as we shall hereafter see) by the depth of the intervening ocean,—these

96:e are necessary

96:f facts are opposed/his followers.

97 The nature and relative proportions of the inhabitants of oceanic islands likewise seem to me opposed to the belief of their former continuity with continents.

97:f islands are likewise opposed

98 Nor does their almost universally volcanic composition favour the admission that they are the wrecks of sunken continents;—if they had originally existed as mountain-ranges on the land, some at least of the islands would have been formed, like other mountain-summits, of granite, metamorphic schists, old fossiliferous or other such rocks, instead of consisting of mere piles of volcanic matter.

98:d does the almost universally volcanic composition of such islands favour

98:e as continental mountain-ranges, some/fossiliferous and other rocks

99   I must now say a few words on what are called accidental means, but which more properly might be called occasional means of distribution.

99:e properly should be

100 I shall here confine myself to plants.

101 In botanical works, this or that plant is stated to be ill adapted for wide dissemination; but for transport across the sea, the greater or less facilities may be said to be almost wholly unknown.

101:f is often stated to be ill adapted for wide dissemination; but the greater or less facilities for transport across the sea may

102 Until I tried, with Mr. Berkeley's aid, a few experiments, it was not even known how far seeds could resist the injurious action of sea-water.

103 To my surprise I found that out of 87 kinds, 64 germinated after an immersion of 28 days, and a few survived an immersion of 137 days.

103.1:c It deserves notice that certain orders were far more affected than others: nine Leguminosæ were tried, and, with one exception, they resisted the salt-water badly; seven species of the allied orders, Hydrophyllaceæ and Polemoniaceæ, were all killed by a month's immersion.

103.1:f more injured than

104 For convenience sake I chiefly tried small seeds, without the capsule or fruit; and as all of these sank in a few days, they

could not be floated across wide spaces of the sea, whether or not they were injured by the salt-water.

104:b convenience'

104:d not have been floated

105 Afterwards I tried some larger fruits, capsules, &c., and some of these floated for a long time.

106 It is well-known what a difference there is in the buoyancy of green and seasoned timber; and it occurred to me that floods might wash down plants or branches, and that these might be dried on the banks, and then by a fresh rise in the stream be washed into the sea.

106:f floods would often wash into the sea dried plants or branches with seed-capsules or fruit attached to them.

107 Hence I was led to dry stems and branches of 94 plants with ripe fruit, and to place them on sea water.

107:b sea-water.

107:f dry the stems

108 The majority sank quickly, but some which whilst green floated for a very short time, when dried floated much longer; for instance, ripe hazel-nuts sank immediately, but when dried, they floated for 90 days and afterwards when planted they germinated; an asparagus plant with ripe berries floated for 23 days, when dried it floated for 85 days, and the seeds afterwards germinated: the ripe seeds of Helosciadium sank in two days, when dried they floated for above 90 days, and afterwards germinated.

108:b dried they floated for 90 days, and afterwards when planted/ seeds afterwards germinated; the ripe

108:d some, which/asparagus-plant

108:f green, floated for a very

109 Altogether out of the 94 dried plants, 18 floated for above 28 days, and some of the 18 floated for a very much longer period.

109:c Altogether, out

109:f days; and

110 So that as $\frac{64}{87}$ seeds germinated after an immersion of 28 days; and as $\frac{18}{94}$ plants with ripe fruit (but not all the same species as in the foregoing experiment) floated, after being dried, for above 28 days, as far as we may infer anything from these scanty facts, we may conclude that the seeds of $\frac{14}{100}$ plants of any country might be floated by sea-currents during 28 days, and would retain their power of germination.

110:f $\frac{64}{87}$ kinds of seeds germinated/ $\frac{18}{94}$ distinct species with/above 28 days, we may conclude, as far as anything can be inferred from these scanty facts, that the seeds of $\frac{14}{100}$ kinds of plants

111 In Johnston's Physical Atlas, the average rate of the several Atlantic currents is 33 miles per diem (some currents running at the rate of 60 miles per diem); on this average, the seeds

of $\frac{14}{100}$ plants belonging to one country might be floated across 924 miles of sea to another country; and when stranded, if blown to a favourable spot by an inland gale, they would germinate.

111:f another country, and when stranded, if blown by an inland gale to a favourable spot, would

112 Subsequently to my experiments, M. Martens tried similar ones, but in a much better manner, for he placed the seeds in a box in the actual sea, so that they were alternately wet and exposed to the air like really floating plants.

113 He tried 98 seeds, mostly different from mine; but he chose many large fruits and likewise seeds from plants which live near the sea; and this would have favoured the average length of their flotation and of their resistance to the injurious action of the salt-water.

113:f favoured both the average length of their flotation and their resistance

114 On the other hand he did not previously dry the plants or branches with the fruit; and this, as we have seen, would have caused some of them to have floated much longer.

115 The result was that $\frac{18}{98}$ of his seeds floated for 42 days, and were then capable of germination.

115:f seeds of different kinds floated

116 But I do not doubt that plants exposed to the waves would float for a less time than those protected from violent movement as in our experiments.

117 Therefore it would perhaps be safer to assume that the seeds of about $\frac{10}{100}$ plants of a flora, after having been dried, could be floated across a space of sea 900 miles in width, and would then germinate.

118 The fact of the larger fruits often floating longer than the small, is interesting; as plants with large seeds or fruit could hardly be transported by any other means; and Alph. de Candolle has shown that such plants generally have restricted ranges.

118:f or fruit which, as Alph. de Candolle has shown, generally have restricted ranges, could hardly be transported by any other means.

119 But seeds may be occasionally transported in another manner.

119:f Seeds

120 Drift timber is thrown up on most islands, even on those in the midst of the widest oceans; and the natives of the coral-islands in the Pacific, procure stones for their tools, solely from the roots of drifted trees, these stones being a valuable royal tax.

120:d Pacific procure

121 I find on examination, that when irregularly shaped stones are

577

embedded in the roots of trees, small parcels of earth are very frequently enclosed in their interstices and behind them,—so perfectly that not a particle could be washed away in the longest transport: out of one small portion of earth thus *completely* enclosed by wood in an oak about 50 years old, three dicotyledonous plants germinated: I am certain of the accuracy of this observation.

121:f I find that when/earth are frequently/away during the longest/enclosed by the roots of an oak

122 Again, I can show that the carcasses of birds, when floating on the sea, sometimes escape being immediately devoured; and seeds of many kinds in the crops of floating birds long retain their vitality: peas and vetches, for instance, are killed by even a few days' immersion in sea-water; but some taken out of the crops of a pigeon, which had floated on artificial salt-water for 30 days, to my surprise nearly all germinated.

122:e saltwater

122:f devoured: and many kinds of seeds in/artificial sea-water for

123 Living birds can hardly fail to be highly effective agents in the transportation of seeds.

124 I could give many facts showing how frequently birds of many kinds are blown by gales to vast distances across the ocean.

125 We may I think safely assume that under such circumstances their rate of flight would often be 35 miles an hour; and some authors have given a far higher estimate.

125:c We may safely

126 I have never seen an instance of nutritious seeds passing through the intestines of a bird; but hard seeds of fruit will pass uninjured through even the digestive organs of a turkey.

126:b fruit pass

127 In the course of two months, I picked up in my garden 12 kinds of seeds, out of the excrement of small birds, and these seemed perfect, and some of them, which I tried, germinated.

127:c which were tried

128 But the following fact is more important: the crops of birds do not secrete gastric juice, and do not in the least injure, as I know by trial, the germination of seeds; now after a bird has found and devoured a large supply of food, it is positively asserted that all the grains do not pass into the gizzard for 12 or even 18 hours.

128:c twelve or even eighteen

128:d now, after

128:e juice, and do not, as I know by trial, injure in the least the germination

129 A bird in this interval might easily be blown to the distance of 500 miles, and hawks are known to look out for tired birds,

and the contents of their torn crops might thus readily get scattered.

130 Mr. Brent informs me that a friend of his had to give up flying carrier-pigeons from France to England, as the hawks on the English coast destroyed so many on their arrival.

130:[c]

131 Some hawks and owls bolt their prey whole, and after an interval of from twelve to twenty hours, disgorge pellets, which, as I know from experiments made in the Zoological Gardens, include seeds capable of germination.

131:d and, after

132 Some seeds of the oat, wheat, millet, canary, hemp, clover, and beet germinated after having been from twelve to twenty-one hours in the stomachs of different birds of prey; and two seeds of beet grew after having been thus retained for two days and fourteen hours.

133 Fresh-water fish, I find, eat seeds of many land and water plants: fish are frequently devoured by birds, and thus the seeds might be transported from place to place.

134 I forced many kinds of seeds into the stomachs of dead fish, and then gave their bodies to fishing-eagles, storks, and pelicans; these birds after an interval of many hours, either rejected the seeds in pellets or passed them in their excrement; and several of these seeds retained their power of germination.

134:d birds, after

135 Certain seeds, however, were always killed by this process.

135.1-2:e[¶] Locusts are sometimes blown to great distances from the land; I myself caught one 370 miles from the coast of Africa, and have heard of others caught at greater distances. 2 The Rev. R. T. Lowe informs Sir C. Lyell that in November 1844 swarms of locusts visited the island of Madeira.

135.2:f informed

135.3-4:e They were in countless numbers, as thick as the flakes of snow in the heaviest snowstorm, and extended upwards as far as could be seen with a telescope. 4 During two or three days they slowly careered round in the air in an immense ellipse, at least five or six miles in diameter, and at night alighted on the taller trees which were completely coated with them.

135.4:f careered round and round in an immense/trees, which

135.5-6:e They then disappeared over the sea, as suddenly as they had appeared, and have not since visited the island. 6 Now, in parts of Natal it is believed by some of the farmers, though on quite insufficient evidence, that injurious seeds are introduced into their grass-land in the dung left by the great flights of locusts which often visit that country.

135.6:f some farmers, though on insufficient

135.7:e In consequence of this belief Mr. Weale sent me in a letter a small packet of the dried pellets, out of which I extracted under the microscope several seeds, and raised from them seven grass plants, belonging to two species, in two genera.

135.7:f species, of two

135.8:e Hence a swarm of locusts, such as that which visited Madeira, might readily be the means of introducing several kinds of plants into an island lying far from the mainland.

136 Although the beaks and feet of birds are generally quite clean, I can show that earth sometimes adheres to them: in one instance I removed twenty-two grains of dry argillaceous earth from one foot of a partridge, and in this earth there was a pebble quite as large as the seed of a vetch.

136:c clean, earth sometimes adheres to them: in one case I removed sixty-one grains, and in another case twenty-two/partridge, and in the earth

136:e generally clean/pebble as large

136:f from the foot

136.1:e Here is a better case: the leg of a woodcock was sent to me by a friend, with a little cake of dry earth attached to the shank, weighing only nine grains; and this contained a seed of the toad-rush (Juncus bufonius) which germinated and flowered.

136.1.1:f Mr. Swaysland, of Brighton, who during the last forty years has paid close attention to our migratory birds, informs me that he has often shot wagtails (Motacillæ), wheatears, and whinchats (Saxicolæ), on their first arrival on our shores, before they had alighted; and he has several times noticed little cakes of earth attached to their feet.

137 Thus seeds might occasionally be transported to great distances; for many facts could be given showing that soil almost everywhere is charged with seeds.

137:e Many facts could be given showing how the soil is almost everywhere charged

137:f how generally soil is charged

137.1:d I will give one case:—Mr. Newton sent me the leg of a red-legged partridge (Caccabis rufa) which had been wounded and could not fly; round the wounded leg and foot a ball of hard earth had collected, and this when removed weighed six and a half ounces.

137.1:e For instance, Prof. Newton/fly, with a ball of hard earth adhering to it, and weighing six and a half ounces.

137.2:d This earth had been kept for three years, but when broken, watered and placed under a bell glass, no less than 82 plants sprung up from it: these consisted of 12 monocotyledons, including the common oat, and at least one kind of grass, and of 70 dicotyledons, which included, judging from the young leaves, at least three distinct species.

137.2:e which consisted, judging from the young leaves, of at

137.2:f The earth

138 Reflect for a moment on the millions of quails which annually cross the Mediterranean; and can we doubt that the earth adhering to their feet would sometimes include a few minute seeds?

138:d With such facts before us, can we doubt that the many birds which are annually blown by gales across great spaces of ocean, and which annually migrate—for instance, the millions of quails across the Mediterranean—must occasionally transport a few seeds embedded in dirt adhering to their feet?

138:e feet or beaks?

139 But I shall presently have to recur to this subject.

139:e shall have

140 As icebergs are known to be sometimes loaded with earth and stones, and have even carried brushwood, bones, and the nest of a land-bird, I can hardly doubt that they must occasionally have transported seeds from one part to another of the arctic and antarctic regions, as suggested by Lyell; and during the Glacial period from one part of the now temperate regions to another.

140:c land-bird, it can hardly be doubted

140:d occasionally, as suggested by Lyell, have transported seeds from one part to another of the arctic and antarctic regions; and

141 In the Azores, from the large number of the species of plants common to Europe, in comparison with the plants of other oceanic islands nearer to the mainland, and (as remarked by Mr. H. C. Watson) from the somewhat northern character of the flora in comparison with the latitude, I suspected that these islands had been partly stocked by ice-borne seeds, during the Glacial epoch.

141:d of plants common to Europe, in comparison with the species in the other Atlantic islands which stand nearer to the mainland, and (as remarked by Mr. H. C. Watson) from their somewhat northern character in

141:e other islands of the Atlantic, which

141:f species on the other

142 At my request Sir C. Lyell wrote to M. Hartung to inquire whether he had observed erratic boulders on these islands, and he answered that he had found large fragments of granite and other rocks, which do not occur in the archipelago.

143 Hence we may safely infer that icebergs formerly landed their rocky burthens on the shores of these mid-ocean islands, and it is at least possible that they may have brought thither the seeds of northern plants.

143:f thither some few seeds

144   Considering that the several above means of transport, and that several other means, which without doubt remain to be discovered, have been in action year after year, for centuries and tens of thousands of years, it would I think be a marvellous fact if many plants had not thus become widely transported.

144:c Considering that these several means of transport, and that

144:d after year, for tens

144:e transport, and that other

144:f would, I think, be

145   These means of transport are sometimes called accidental, but this is not strictly correct: the currents of the sea are not accidental, nor is the direction of prevalent gales of wind.

146   It should be observed that scarcely any means of transport would carry seeds for very great distances; for seeds do not retain their vitality when exposed for a great length of time to the action of sea-water; nor could they be long carried in the crops or intestines of birds.

146:e distances: for seeds

147   These means, however, would suffice for occasional transport across tracts of sea some hundred miles in breadth, or from island to island, or from a continent to a neighbouring island, but not from one distant continent to another.

148   The floras of distant continents would not by such means become mingled in any great degree; but would remain as distinct as we now see them to be.

148:e mingled; but would remain as distinct as they now are.

149   The currents, from their course, would never bring seeds from North America to Britain, though they might and do bring seeds from the West Indies to our western shores, where, if not killed by so long an immersion in salt-water, they could not endure our climate.

149:e by their very long immersion in salt water

150   Almost every year, one or two land-birds are blown across the whole Atlantic Ocean, from North America to the western shores of Ireland and England; but seeds could be transported by these wanderers only by one means, namely, in dirt sticking to their feet, which is in itself a rare accident.

150:d these rare wanderers only by one means, namely, by dirt

150:e feet or beaks, which

151   Even in this case, how small would·the chance be of a seed falling on favourable soil, and coming to maturity!

151:e small the chance would be

151:f small would be the chance of

152   But it would be a great error to argue that because a well-stocked island, like Great Britain, has not, as far as is known

(and it would be very difficult to prove this), received within the last few centuries, through occasional means of transport, immigrants from Europe, or any other continent, that a poorly-stocked island, though standing more remote from the mainland, would not receive colonists by similar means.

153 I do not doubt that out of twenty seeds or animals transported to an island, even if far less well-stocked than Britain, scarcely more than one would be so well fitted to its new home, as to become naturalised.

153:c Out/Britain, perhaps not more
153:e Out of a hundred seeds
153:f hundred kinds of seeds

154 But this, as it seems to me, is no valid argument against what would be effected by occasional means of transport, during the long lapse of geological time, whilst an island was being upheaved and formed, and before it had become fully stocked with inhabitants.

154:c upheaved, and before
154:f But this is no

155 On almost bare land, with few or no destructive insects or birds living there, nearly every seed, which chanced to arrive, would be sure to germinate and survive.

155:b arrive, if fitted for the climate, would
155:d seed which

156 *Dispersal during the Glacial period.—*
156:d [Center] Period. [Space]

157 The identity of many plants and animals, on mountain-summits, separated from each other by hundreds of miles of lowlands, where the Alpine species could not possibly exist, is one of the most striking cases known of the same species living at distant points, without the apparent possibility of their having migrated from one to the other.

157:d[¶] where Alpine/one point to

158 It is indeed a remarkable fact to see so many of the same plants living on the snowy regions of the Alps or Pyrenees, and in the extreme northern parts of Europe; but it is far more remarkable, that the plants on the White Mountains, in the United States of America, are all the same with those of Labrador, and nearly all the same, as we hear from Asa Gray, with those on the loftiest mountains of Europe.

158:d many plants of the same species living

159 Even as long ago as 1747, such facts led Gmelin to conclude that the same species must have been independently created at several distinct points; and we might have remained in this same belief, had not Agassiz and others called vivid attention

to the Glacial period, which, as we shall immediately see, affords a simple explanation of these facts.

159:f at many distinct

160 We have evidence of almost every conceivable kind, organic and inorganic, that within a very recent geological period, central Europe and North America suffered under an Arctic climate.

160:d that, within/arctic

161 The ruins of a house burnt by fire do not tell their tale more plainly, than do the mountains of Scotland and Wales, with their scored flanks, polished surfaces, and perched boulders, of the icy streams with which their valleys were lately filled.

161:e plainly than

162 So greatly has the climate of Europe changed, that in Northern Italy, gigantic moraines, left by old glaciers, are now clothed by the vine and maize.

163 Throughout a large part of the United States, erratic boulders, and rocks scored by drifted icebergs and coast-ice, plainly reveal a former cold period.

163:e boulders and scored rocks plainly

164 The former influence of the glacial climate on the distribution of the inhabitants of Europe, as explained with remarkable clearness by Edward Forbes, is substantially as follows.

164:e explained by

165 But we shall follow the changes more readily, by supposing a new glacial period to come slowly on, and then pass away, as formerly occurred.

165:e period slowly to come on

166 As the cold came on, and as each more southern zone became fitted for arctic beings and ill-fitted for their former more temperate inhabitants, the latter would be supplanted and arctic productions would take their places.

166:e for the inhabitants of the north, they would take the places of the former inhabitants of the temperate regions.

166:f north, these would

167 The inhabitants of the more temperate regions would at the same time travel southward, unless they were stopped by barriers, in which case they would perish.

167:e The latter, at the same time, would travel further and further southward

168 The mountains would become covered with snow and ice, and their former Alpine inhabitants would descend to the plains.

169 By the time that the cold had reached its maximum, we should have a uniform arctic fauna and flora, covering the central

parts of Europe, as far south as the Alps and Pyrenees, and even stretching into Spain.

169:e should have an arctic

170 The now temperate regions of the United States would likewise be covered by arctic plants and animals, and these would be nearly the same with those of Europe; for the present circumpolar inhabitants, which we suppose to have everywhere travelled southward, are remarkably uniform round the world.

170:f animals and

171 We may suppose that the Glacial period came on a little earlier or later in North America than in Europe, so will the southern migration there have been a little earlier or later; but this will make no difference in the final result.

171:d this makes

171:[e]

172 As the warmth returned, the arctic forms would retreat northward, closely followed up in their retreat by the productions of the more temperate regions.

173 And as the snow melted from the bases of the mountains, the arctic forms would seize on the cleared and thawed ground, always ascending higher and higher, as the warmth increased, whilst their brethren were pursuing their northern journey.

173:d ascending, as the warmth increased and the snow still further disappeared, higher and higher, whilst

174 Hence, when the warmth had fully returned, the same arctic species, which had lately lived in a body together on the lowlands of the Old and New Worlds, would be left isolated on distant mountain-summits (having been exterminated on all lesser heights) and in the arctic regions of both hemispheres.

174:d same species, which had lately lived together in a body on the European and North American lowlands, would be found in the arctic regions of the Old and New Worlds, and isolated on many mountain-summits far distant from each other, having been exterminated on all lesser heights.

174:e would again be/Worlds, and on many isolated mountain-summits far distant from each other.

175 Thus we can understand the identity of many plants at points so immensely remote as on the mountains of the United States and of Europe.

175:f as the mountains of the United States and those of Europe.

176 We can thus also understand the fact that the Alpine plants of each mountain-range are more especially related to the arctic forms living due north or nearly due north of them: for the migration as the cold came on, and the re-migration on the returning warmth, will generally have been due south and north.

176:d for the first migration when the cold/warmth, would generally

177 The Alpine plants, for example, of Scotland, as remarked by Mr. H. C. Watson, and those of the Pyrenees, as remarked by Ramond, are more especially allied to the plants of northern Scandinavia; those of the United States to Labrador; those of the mountains of Siberia to the arctic regions of that country.

178 These views, grounded as they are on the perfectly well-ascertained occurrence of a former Glacial period, seem to me to explain in so satisfactory a manner the present distribution of the Alpine and Arctic productions of Europe and America, that when in other regions we find the same species on distant mountain-summits, we may almost conclude without other evidence, that a colder climate permitted their former migration across the low intervening tracts, since become too warm for their existence.

178:d conclude, without other evidence, that a colder climate formerly permitted their migration across the intervening lowlands, now become

179   If the climate, since the Glacial period, has ever been in any degree warmer than at present (as some geologists in the United States believe to have been the case, chiefly from the distribution of the fossil Gnathodon), then the arctic and temperate productions will at a very late period have marched a little further north, and subsequently have retreated to their present homes; but I have met with no satisfactory evidence with respect to this intercalated slightly warmer period, since the Glacial period.

179:c case), then

179:[d]

180   The arctic forms, during their long southern migration and re-migration northward, will have been exposed to nearly the same climate, and, as is especially to be noticed, they will have kept in a body together; consequently their mutual relations will not have been much disturbed, and, in accordance with the principles inculcated in this volume, they will not have been liable to much modification.

180.x-y:d As the arctic forms moved first southward and afterwards backwards to the north, in unison with the changing climate, they will not have been exposed during their long migrations to any great diversity of temperature, and as they will all have migrated in a body together, their mutual relations will not have been much disturbed. *y* Hence, in accordance with the principles inculcated in this volume, these forms will

180.x:e temperature; and as they all migrated

181 But with our Alpine productions, left isolated from the moment of the returning warmth, first at the bases and ultimately on the summits of the mountains, the case will have been some-

what different; for it is not likely that all the same arctic species will have been left on mountain ranges distant from each other, and have survived there ever since; they will, also, in all probability have become mingled with ancient Alpine species, which must have existed on the mountains before the commencement of the Glacial epoch, and which during its coldest period will have been temporarily driven down to the plains; they will, also, have been exposed to somewhat different climatal influences.

181:c mountain-ranges

181:d with the Alpine productions/mountain-ranges far distant/ probability, have become

181:f they will also in/during the coldest/plains; they will, also, have been subsequently exposed

182 Their mutual relations will thus have been in some degree disturbed; consequently they will have been liable to modification; and this we find has been the case; for if we compare the present Alpine plants and animals of the several great European mountain-ranges, though very many of the species are identically the same, some present varieties, some are ranked as doubtful forms, and some few are distinct yet closely allied or representative species.

182:c forms, and many are

182:d mountain-ranges one with another, though many of the species still remain identically the same, some exist as varieties, some as doubtful forms or sub-species, and some as certainly distinct yet closely allied species representing each other on the several ranges.

182:e species remain

182:f modification; and they have been modified; for/sub-species, and some as distinct

183   In illustrating what, as I believe, actually took place during the Glacial period, I assumed that at its commencement the arctic productions were as uniform round the polar regions as they are at the present day.

183:e In the foregoing illustration I have assumed that at the commencement of our imaginary Glacial period, the arctic

184 But the foregoing remarks on distribution apply not only to strictly arctic forms, but also to many sub-arctic and to some few northern temperate forms, for some of these are the same on the lower mountains and on the plains of North America and Europe; and it may be reasonably asked how I account for the necessary degree of uniformity of the sub-arctic and northern temperate forms round the world, at the commencement of the Glacial period.

184:d lower mountain-slopes and on the plains of North America and Europe; and it may be asked how I account for this degree of uniformity in the sub-arctic and temperate

587

184:e But it is necessary also to include many sub-arctic and some few temperate forms, for some/commencement of the real Glacial

184:f is also necessary to assume that many sub-arctic and some few temperate forms were the same round the world, for some of the species which now exist on the lower mountain-slopes and on the plains of North America and Europe are the same; and it may

185 At the present day, the sub-arctic and northern temperate productions of the Old and New Worlds are separated from each other by the Atlantic Ocean and by the extreme northern part of the Pacific.

185:d by the whole Atlantic Ocean and by the northern

186 During the Glacial period, when the inhabitants of the Old and New Worlds lived further southwards than at present, they must have been still more completely separated by wider spaces of ocean.

186:c farther

186:d than they do at/separated from each other by wider spaces of ocean; so that it may well be asked how the same species could have entered two regions then so widely separated.

186:e entered the two continents then

186:f species could then or previously have entered the two continents.

187 I believe the above difficulty may be surmounted by looking to still earlier changes of climate of an opposite nature.

187:d The explanation, I believe, lies in the nature of the climate before the commencement of the **Glacial period.**

188 We have good reason to believe that during the newer Pliocene period, before the **Glacial epoch, and whilst the majority of the** inhabitants of the world were specifically the same as now, the **climate was warmer than at the present day.**

188:d During this, the newer Pliocene period, when the majority/ now, we have **good reason to believe that the climate**

188:e At this, the newer Pliocene period, the majority/now, and we

189 **Hence we may suppose that the organisms now living under** the climate of latitude 60°, during the Pliocene period lived further north under the Polar Circle, in latitude 66°-67°; and that the strictly arctic productions then lived on the broken land still nearer to the pole.

189:c farther

189:d organisms which now live under latitude 60°/66°-67°; and that the present arctic

189:e 60°, lived during the Pliocene period farther

190 Now if we look at a globe, we shall see that under the Polar

Circle there is almost continuous land from western Europe, through Siberia, to eastern America.

190:d Now, if we look at a terrestrial globe, we see

190:e see under the Polar Circle that there

191 And to this continuity of the circumpolar land, and to the consequent freedom for intermigration under a more favourable climate, I attribute the necessary amount of uniformity in the sub-arctic and northern temperate productions of the Old and New Worlds, at a period anterior to the Glacial epoch.

191:d freedom under a more favourable climate for intermigration, I attribute a considerable degree of uniformity

191:e And this continuity of the circumpolar land, with the consequent freedom under a more favourable climate for intermigration, will account for the supposed uniformity of the sub-arctic and temperate

192 Believing, from reasons before alluded to, that our continents have long remained in nearly the same relative position, though subjected to large, but partial oscillations of level, I am strongly inclined to extend the above view, and to infer that during some earlier and still warmer period, such as the older Pliocene period, a large number of the same plants and animals inhabited the almost continuous circumpolar land; and that these plants and animals, both in the Old and New Worlds, began slowly to migrate southwards as the climate became less warm, long before the commencement of the Glacial period.

192:d some still earlier

192:f subjected to great oscillations

193 We now see, as I believe, their descendants, mostly in a modified condition, in the central parts of Europe and the United States.

194 On this view we can understand the relationship, with very little identity, between the productions of North America and Europe,—a relationship which is most remarkable, considering the distance of the two areas, and their separation by the Atlantic Ocean.

194:d highly remarkable/by the whole Atlantic

194:f relationship with

195 We can further understand the singular fact remarked on by several observers, that the productions of Europe and America during the later tertiary stages were more closely related to each other than they are at the present time; for during these warmer periods the northern parts of the Old and New Worlds will have been almost continuously united by land, serving as a bridge, since rendered impassable by cold, for the inter-migration of their inhabitants.

195:b intermigration

195:d during the latter tertiary

195:e during the later tertiary

195:f observers that

196　During the slowly decreasing warmth of the Pliocene period, as soon as the species in common, which inhabited the New and Old Worlds, migrated south of the Polar Circle, they must have been completely cut off from each other.

196:d Worlds, had migrated south of the Polar Circle, they would have

196:e Worlds, migrated south of the Polar Circle, they would be completely

196:f they will have been completely

197　This separation, as far as the more temperate productions are concerned, took place long ages ago.

197:e concerned, must have taken

198　And as the plants and animals migrated southward, they will have become mingled in the one great region with the native American productions, and have had to compete with them; and in the other great region, with those of the Old World.

198:d they would have been liable to become/productions, and would have

198:e As the plants/would become

198:f they will have become

199　Consequently we have here everything favourable for much modification,—for far more modification than with the Alpine productions, left isolated, within a much more recent period, on the several mountain-ranges and on the arctic lands of the two Worlds.

199:d lands of Europe and N. America.

200　Hence it has come, that when we compare the now living productions of the temperate regions of the New and Old Worlds, we find very few identical species (though Asa Gray has lately shown that more plants are identical than was formerly supposed), but we find in every great class many forms, which some naturalists rank as geographical races, and others as distinct species; and a host of closely allied or representative forms which are ranked by all naturalists as specifically distinct.

201　As on the land, so in the waters of the sea, a slow southern migration of a marine fauna, which during the Pliocene or even a somewhat earlier period, was nearly uniform along the continuous shores of the Polar Circle, will account, on the theory of modification, for many closely allied forms now living in areas completely sundered.

201:d which, during/living in marine areas

202　Thus, I think, we can understand the presence of many existing and tertiary representative forms on the eastern and western shores of temperate North America; and the still more striking case of many closely allied crustaceans (as described in Dana's admirable work), of some fish and other marine animals, in the Mediterranean and in the seas of Japan,—areas now separated by a continent and by nearly a hemisphere of equatorial ocean.

202:d presence of some still existing and of some tertiary closely allied forms/Japan,—these two areas being now completely separated by the breadth of a whole continent and by a wide space of ocean.

202:e striking fact of many

202:f some closely allied, still existing and extinct tertiary forms, on/work), some fish and other marine animals, inhabiting the Mediterranean and the seas/continent and by wide spaces

203   These cases of relationship, without identity, of the inhabitants of seas now disjoined, and likewise of the past and present inhabitants of the temperate lands of North America and Europe, are inexplicable on the theory of creation.

203:d of close relationship in many species either now or formerly inhabiting the seas on the eastern and western shores of North America, the Mediterranean and Japan, and the temperate

203:e in species

204 We cannot say that they have been created alike, in correspondence with the nearly similar physical conditions of the areas; for if we compare, for instance, certain parts of South America with the southern continents of the Old World, we see countries closely corresponding in all their physical conditions, but with their inhabitants utterly dissimilar.

204:d cannot maintain that such species have/America with parts of South Africa or Australia, we see countries closely similar in all their physical conditions, but with inhabitants utterly dissimilar. [Space]

204:f conditions, with their inhabitants

204.1:d [Center] Mundane Glacial Period. [Space]

204.1:e Alternate Glacial Periods of the North and South.

204.1:f Periods in the North

205   But we must return to our more immediate subject, the Glacial period.

205:d subject.

206 I am convinced that Forbes's view may be largely extended.

207 In Europe we have the plainest evidence of the cold period, from the western shores of Britain to the Oural range, and southward to the Pyrenees.

207:d we meet with the plainest evidence of the Glacial period

208 We may infer, from the frozen mammals and nature of the mountain vegetation, that Siberia was similarly affected.

208:b infer from

208.1:d In the Lebanon, according to Dr. Hooker, perpetual snow formerly covered the central axis, and fed glaciers which rolled 4000 feet down its valleys.

208.1.1:f The same observer has recently found great moraines at a low level on the Atlas range in N. Africa.

209 Along the Himalaya, at points 900 miles apart, glaciers have left the marks of their former low descent; and in Sikkim, Dr. Hooker saw maize growing on gigantic ancient moraines.

209:f on ancient and gigantic moraines.

210 South of the equator, we have some direct evidence of former glacial action in New Zealand; and the same plants, found on widely separated mountains in this island, tell the same story.

210:b mountains in that island

210:d Southward of the great continent of Asia, on the opposite side of the equator, we now know, from the excellent researches of Dr. J. Haast and Dr. Hector, that enormous glaciers formerly descended to a low level in New Zealand; and the same plants found by Dr. Hooker on widely separated mountains in this island tell the same story of a former cold period.

210:e Southward of the Asiatic continent, on the opposite/Hector, that immense glaciers

210:f we know/Hector, that in New Zealand immense glaciers formerly descended to a low level; and

211 If one account which has been published can be trusted, we have direct evidence of glacial action in the south-eastern corner of Australia.

211:d From facts lately communicated to me by the Rev. W. B. Clarke, it appears also that there are clear traces of former glacial action on the mountains of the south-eastern

211:e facts communicated/are traces

212 Looking to America; in the northern half, ice-borne fragments of rocks have been observed on the eastern side as far south as lat. 36°-37°, and on the shores of the Pacific, where the climate is now so different, as far south as lat. 46°; erratic boulders have, also, been noticed on the Rocky Mountains.

212.x-y:e side of the continent, as far south as lat. 36°-37°/46°. y Erratic

213 In the Cordillera of Equatorial South America, glaciers once extended far below their present level.

213:c equatorial

213:e Cordillera of South America, nearly under the equator, glaciers

214 In central Chile I was astonished at the structure of a vast mound of detritus, about 800 feet in height, crossing a valley of the Andes; and this I now feel convinced was a gigantic moraine, left far below any existing glacier.

214:c In Central Chile I examined a vast mound of detritus, crossing the Portillo valley, which I now fully believe to have been due to ice-action; but we shall hereafter have valuable information on this subject from Mr. D. Forbes, who informs me that he found in the Cordillera, from lat. 13° to 30° S., at about the height of 12,000 feet, strongly-furrowed rocks, resembling those with which he was familiar in Norway, and likewise great masses of detritus, including grooved pebbles.

214:d detritus with great boulders, crossing

214:e valley, which there can hardly be a doubt once formed a huge moraine; and Mr. D. Forbes informs me that he found in various parts of the Cordillera, from lat. 13° to 30° S., at about the height of 12,000 feet, deeply-furrowed rocks

214.1:c Along this whole space of the Cordillera true glaciers do not now exist even at much more considerable heights.

215 Further south on both sides of the continent, from lat. 41° to the southernmost extremity, we have the clearest evidence of former glacial action, in huge boulders transported far from their parent source.

215:c Farther

215:e in numerous immense boulders

215.1:e[¶] From these several facts, namely from the glacial action having extended all round the northern and southern hemispheres—from the period having been in a geological sense recent in both hemispheres—from its having lasted in both during a great length of time, as may be inferred from the amount of work effected—and lastly from glaciers having recently descended to a low level along the whole line of the Cordillera, it formerly appeared to me that we could not avoid the conclusion that the temperature of the whole world had been simultaneously lowered during the Glacial period.

215.1:f Cordillera, it at one time appeared

215.2:e But now Mr. Croll, in a series of admirable memoirs, has attempted to show that a glacial condition of climate is the result of various physical causes, brought into operation by an increase in the excentricity of the earth's orbit.

215.2:f eccentricity

215.3:e All these causes tend towards the same end; but the most powerful appears to be the influence of the excentricity of the orbit upon oceanic currents.

215.3:f be the indirect influence of the eccentricity

215.4:e It follows from Mr. Croll's researches, that cold periods regularly recur every ten or fifteen thousand years; but that at much longer intervals the cold, owing to certain contingencies, is extremely severe, and lasts for a great length of time.

215.4:f According to Mr. Croll, cold periods regularly recur every ten or fifteen thousand years; and these at long intervals are extremely severe, owing to certain contingencies, of which the most important, as Sir C. Lyell has shown, is the relative position of the land and water.

215.5-7:e Mr. Croll believes that the last great Glacial period occurred about 240,000 years ago, and endured with slight alterations of climate for about 160,000 years. 6 With respect to more ancient Glacial periods, several geologists are convinced from direct evidence that such occurred during the Miocene and Eocene formations, not to mention still more ancient formations. 7 But in relation to our present subject, the most important result arrived at by Mr. Croll is, that whenever the northern hemisphere passes through a cold period, the temperature of the southern hemisphere is actually raised, with the winters rendered much milder, chiefly through changes in the direction of the ocean-currents.

215.7:f But the most important result for us, arrived at by Mr. Croll, is that

215.8:e So conversely it is with the northern hemisphere, when the southern passes through a glacial period.

215.8:f it will be with the northern hemisphere, whilst the southern

215.9:e These conclusions have, as we shall immediately see, a most important bearing on geographical distribution; but I will first give the facts, which demand an explanation.

215.9:f This conclusion throws so much light on geographical distribution that I am strongly inclined to trust in it; but

216   We do not know that the Glacial epoch was strictly simultaneous at these several far distant points on opposite sides of the world.

216:d far-distant

216:[e]

217 But we have good evidence in almost every case, that the epoch was included within the latest geological period.

217:d epoch formed part of the latest

217:[e]

218 We have, also, excellent evidence, that it endured for an enormous time, as measured by years, at each point.

218:[e]

219 The cold may have come on, or have ceased, earlier at one point of the globe than at another, but seeing that it endured for long at each, and that it was contemporaneous in a geological sense, it seems to me probable that it was, during a part at least of the period, actually simultaneous throughout the world.

219:c for a long time at/seems probable

219:d but, seeing

219:[e]

220 Without some distinct evidence to the contrary, we may at least admit as probable that the glacial action was simultane-

ous on the eastern and western sides of North America, in the Cordillera under the equator and under the warmer temperate zones, and on both sides of the southern extremity of the continent.

220:c under the equatorial, tropical, and warmer temperate zones, and on both sides of the southern portion of the continent.

220:[e]

221 If this be admitted, it is difficult to avoid believing that the temperature of the whole world was at this period simultaneously cooler.

221:[e]

222 But it would suffice for my purpose, if the temperature was at the same time lower along certain broad belts of longitude.

222:d temperature were at

222:[e]

223 On this view of the whole world, or at least of broad longitudinal belts, having been simultaneously colder from pole to pole, much light can be thrown on the present distribution of identical and allied species.

223:[e]

224 In America, Dr. Hooker has shown that between forty and fifty of the flowering plants of Tierra del Fuego, forming no inconsiderable part of its scanty flora, are common to Europe, enormously remote as these two points are; and there are many closely allied species.

224:e[¶] In South America, Dr. Hooker has shown that besides many closely allied species, between/to North America and Europe, enormously remote as these areas in opposite hemispheres are from each other.

225 On the lofty mountains of equatorial America a host of peculiar species belonging to European genera occur.

226 On the highest mountains of Brazil, some few European genera were found by Gardner, which do not exist in the wide intervening hot countries.

226:d On the Organ mountains of Brazil, some few European temperate, some antarctic, and some Andean genera were found by Gardner, which do not exist in the low intervening hot countries; and I have been informed that Agassiz has lately discovered plain marks of glacial action on these same mountains.

226:e few temperate European, some Antarctic/countries.

227 So on the Silla of Caraccas the illustrious Humboldt long ago found species belonging to genera characteristic of the Cordillera.

227:e On the Silla

227:f Caraccas, the illustrious

228 On the mountains of Abyssinia, several European forms and some few representatives of the peculiar flora of the Cape of Good Hope occur.

228:c several forms characteristic of Europe and some few representatives of the flora of the Cape

228:d[¶] In Africa, several/occur on the mountains of Abyssinia.

229 At the Cape of Good Hope a very few European species, believed not to have been introduced by man, and on the mountains, some few representative European forms are found, which have not been discovered in the intertropical parts of Africa.

229:c Hope itself a very/mountains, several representative

229:f Hope a very/mountains several

229.1-3:d Dr. Hooker has also lately shown that several of the plants living on the upper parts of the lofty island of Fernando Po and on the neighbouring Cameroon mountains, in the Gulf of Guinea, are closely related to those on the mountains of Abyssinia, and likewise to those of temperate Europe. 2 It now also appears, as I hear from Dr. Hooker, that some of these same temperate plants have been discovered by the Rev. R. T. Lowe on the mountains of the Cape de Verde islands. 3 This extension of the same temperate forms, almost under the equator, across the whole continent of Africa and to the mountains of the Cape de Verde archipelago, is one of the most astonishing facts ever recorded in the distribution of plants.

229.3:g Cape Verde

230 On the Himalaya, and on the isolated mountain-ranges of the peninsula of India, on the heights of Ceylon, and on the volcanic cones of Java, many plants occur, either identically the same or representing each other, and at the same time representing plants of Europe, not found in the intervening hot lowlands.

230:d[¶]

231 A list of the genera collected on the loftier peaks of Java raises a picture of a collection made on a hill in Europe!

231:d genera of plants collected on the loftier

231:e made on a hillock in

231:f Java, raises

232 Still more striking is the fact that southern Australian forms are clearly represented by plants growing on the summits of the mountains of Borneo.

232:d that peculiar southern Australian forms are represented

232:e by certain plants

232:f peculiar Australian

233 Some of these Australian forms, as I hear from Dr. Hooker, extend along the heights of the peninsula of Malacca, and are thinly scattered, on the one hand over India and on the other as far north as Japan.

233:d scattered on the one hand over India, and

234 On the southern mountains of Australia, Dr. F. Müller has discovered several European species; other species, not introduced by man, occur on the lowlands; and a long list can be given, as I am informed by Dr. Hooker, of European genera, found in Australia, but not in the intermediate torrid regions.

235 In the admirable 'Introduction to the Flora of New Zealand,' by Dr. Hooker, analogous and striking facts are given in regard to the plants of that large island.

236 Hence we see that throughout the world, the plants growing on the more lofty mountains, and on the temperate lowlands of the northern and southern hemispheres, are sometimes identically the same; but they are much oftener specifically distinct, though related to each other in a most remarkable manner.

236:d that, throughout the world, plants/but much oftener they are specifically/in a remarkable

236:e see that certain plants growing on the more lofty mountains of the tropics in all parts of the world, and on the temperate plains of the north and south, are either the same identical species or varieties of the same species.

236:f same species

236.1:e It should, however, be observed that these plants are not strictly Arctic forms; for, as Mr. H. C. Watson has remarked, "in receding from polar towards equatorial latitudes, the Alpine or mountain floras really become less and less Arctic."

236.1:f arctic

236.2:e Besides these identical and closely allied forms, many species inhabiting the same widely sundered areas, belong to genera not now found in the intermediate tropical lowlands.

237 This brief abstract applies to plants alone: some strictly analogous facts could be given on the distribution of terrestrial animals.

237:e These brief remarks apply to plants alone; but some few analogous facts could be given in regard to terrestrial

238 In marine productions, similar cases occur; as an example, I may quote a remark by the highest authority, Prof. Dana, that "it is certainly a wonderful fact that New Zealand should have a closer resemblance in its crustacea to Great Britain, its antipode, than to any other part of the world."

238:e cases likewise occur; as an example, I may quote a statement by

239 Sir J. Richardson, also, speaks of the reappearance on the shores of New Zealand, Tasmania, &c., of northern forms of fish.

240 Dr. Hooker informs me that twenty-five species of Algæ are

common to New Zealand and to Europe, but have not been found in the intermediate tropical seas.

240.1:e = 256:[e][¶] From the foregoing facts, namely the presence of temperate forms on the highlands across the whole of equatorial Africa, and along the Peninsula of India to Ceylon and the Malay archipelago, and in a less well-marked manner across the wide expanse of tropical South America, it appears almost certain that at some former period, no doubt during the most severe part of the Glacial period, the lowlands of these great continents were everywhere tenanted under the equator by a considerable number of temperate forms.

240.1:f namely, the presence/India, to/part of a Glacial

240.2:e = 257:[e] At this period the equatorial climate at the level of the sea was probably about the same with that now experienced at the height of from five to six thousand feet under the same latitudes, or perhaps even rather cooler.

240.2:f latitude

240.3:e = 258:[e] During this, the coldest period, the lowlands under the equator must have been clothed with a mingled tropical and temperate vegetation, like that described by Hooker as growing luxuriantly at the height of from four to five thousand feet on the lower slopes of the Himalaya, but with perhaps a still greater preponderance of temperate forms.

240.4:e = 258.1:[e] So again, on the mountainous island

240.4:f again in the mountainous

240.5:e = 258.2:[e]

241   It should be observed that the northern species and forms found in the southern parts of the southern hemisphere, and on the mountain-ranges of the intertropical regions, are not arctic, but belong to the northern temperate zones.

242 As Mr. H. C. Watson has recently remarked, "In receding from polar towards equatorial latitudes, the Alpine or mountain floras really become less and less arctic."

241 + 2:d northern forms found on the mountain-ranges of the intertropical regions and in the southern parts of the southern hemisphere are not arctic, but belong to the temperate zones: as Mr. H. C. Watson has recently remarked, "in

241 + 2:[e]

243 Many of the forms living on the mountains of the warmer regions of the earth and in the southern hemisphere are of doubtful value, being ranked by some naturalists as specifically distinct, by others as varieties; but some are certainly identical, and many, though closely related to northern forms, must be ranked as distinct species.

243:d Of these forms, some few are identical with northern temperate species, or are varieties of them, whilst others are ranked by all naturalists as closely allied to, but specifically distinct from, their northern representatives.

243:[e]

244 Now let us see what light can be thrown on the foregoing facts, on the belief, supported as it is by a large body of geological evidence, that the whole world, or a large part of it, was during the Glacial period simultaneously much colder than at present.

244:d was simultaneously colder during the Glacial period than at present.

244:e see whether Mr. Croll's conclusion that when the northern hemisphere suffered from the extreme cold of the great Glacial period, the southern hemisphere was actually warmer, throws any clear light on the present apparently inexplicable distribution of various organisms in the temperate parts of both hemispheres, and on mountains of the tropics.

245 The Glacial period, as measured by years, must have been very long; and when we remember over what vast spaces some naturalised plants and animals have spread within a few centuries, this period will have been ample for any amount of migration.

245.1:e As the cold became more and more intense, we know that Arctic forms invaded the temperate regions; and, from the facts just given, there can hardly be a doubt that some of the more vigorous, dominant, and widest-spreading temperate forms actually then invaded the equatorial lowlands.

245.1:f dominant and widest-spreading temperate forms invaded

245.2:e The inhabitants of these lowlands would at the same time migrate to the tropical and sub-tropical regions of the south, for the southern hemisphere was at this period warmer.

245.2:f these hot lowlands would at the same time have migrated

245.3:e On the decline of the Glacial period, as both hemispheres gradually recovered their former temperatures, the northern temperate forms living on the lowlands under the equator, would be driven to their former homes or be destroyed, being replaced by the equatorial forms returning from the south.

245.3:f would have been driven to their former homes or have been destroyed

245.4:e Some, however, of the northern temperate forms would almost certainly ascend any adjoining high land, where, if sufficiently lofty, they would long survive, like the Arctic forms on the mountains of Europe.

245.4:f certainly have ascended any adjoining high land, where, if sufficiently lofty, they would have long survived like

245.5:e They might survive, even if the climate was not perfectly fitted for them, for the change of temperature must have been very slow, and plants undoubtedly possess a certain capacity for acclimatisation, as shown by their transmitting to their off-spring different constitutional powers of resisting heat and cold.

245.5:*f* They might have survived

245.6:*e*[¶] In the regular course of events the southern hemisphere would be subjected to a severe Glacial period, with the northern hemisphere rendered warmer; and then the southern temperate forms would in their turn invade the equatorial lowlands.

245.6:*f* would in its turn be/forms would invade

245.7-10:*e* The northern forms which had before been left on the mountains would now descend and mingle with the southern forms. *8* These latter, when the warmth returned, would return to their former homes, leaving some few species on the mountains, and carrying southward with them some of the northern temperate forms which had descended from their mountain fastnesses. *9* Thus, we should have some few species identically the same in the northern and southern temperate zones and on the mountains of the intermediate tropical regions. *10* But the species left during a long time on these mountains or in opposite hemispheres, would have to compete with many new forms and would be exposed to somewhat different physical conditions; hence they would be eminently liable to modification, and would generally now exist as varieties or as representative species; and this is the case.

245.10:*f* mountains, or

245.11:*e* We must, also, bear in mind the occurrence in both hemispheres of former Glacial periods; for these will account, in accordance with the same principles, for the many quite distinct species inhabiting the same widely separated areas, and belonging to genera not now found in the intermediate torrid zones.

246 As the cold came slowly on, all the tropical plants and other productions will have retreated from both sides towards the equator, followed in the rear by the temperate productions, and these by the arctic; but with the latter we are not now concerned.

246:*d* on, the tropical plants and animals will

246:[*e*]

246.1-5:*d* The whole problem of what will have occurred is excessively complex. *2* The probable existence before the Glacial period of a pleistocene equatorial flora and fauna, fitted for a hotter climate than any now existing, must not be overlooked. *3* This old equatorial flora will have been almost wholly destroyed, and the two pleistocene sub-tropical floras, commingled and reduced in number, will then have formed the equatorial flora. *4* There will aso probably have been during the Glacial period great changes in the precise nature of the climate, in the degree of humidity, &c.; and various animals and plants will have migrated in different proportions and at different rates. *5* So that altogether during the Glacial period the inhabitants of the tropics must have been greatly disturbed in all their relations of life.

246.1-5:[e]

247 The tropical plants probably suffered much extinction; how much no one can say; perhaps formerly the tropics supported as many species as we see at the present day crowded together at the Cape of Good Hope, and in parts of temperate Australia.

247:d Hence they will have suffered/we now see crowded

247:[e]

248 As we know that many tropical plants and animals can withstand a considerable amount of cold, many might have escaped extermination during a moderate fall of temperature, more especially by escaping into the warmest spots.

248:b into the lowest, most protected, and warmest districts.

248:d[¶]

248:[e]

248.1:d Nor must it be overlooked that, as the cold will have come on very slowly, it is almost certain that many of the inhabitants of the tropics will have become in some degree acclimatised; in the same manner as the same species of plant when living on lowlands and highlands certainly transmit to their seedlings different constitutional powers of resisting cold.

248.1:[e]

249 But the great fact to bear in mind is, that all tropical productions will have suffered to a certain extent.

249:d Nevertheless, it cannot be denied that all tropical productions will have greatly suffered, and the chief difficulty is to understand how they can have escaped entire annihilation.

249:[e]

250 On the other hand, the temperate productions, after migrating nearer to the equator, though they will have been placed under somewhat new conditions, will have suffered less.

250:[e]

251 And it is certain that many temperate plants, if protected from the inroads of competitors, can withstand a much warmer climate than their own.

251:d climate than that proper to them.

251:[e]

252 Hence, it seems to me possible, bearing in mind that the tropical productions were in a suffering state and could not have presented a firm front against intruders, that a certain number of the more vigorous and dominant temperate forms might have penetrated the native ranks and have reached or even crossed the equator.

252:c state, and could/ranks, and

252:[e]

253 The invasion would, of course, have been greatly favoured by high land, and perhaps by a dry climate; for Dr. Falconer informs me that it is the damp with the heat of the tropics which is so destructive to perennial plants from a temperate climate.

253:[e]

254 On the other hand, the most humid and hottest districts will have afforded an asylum to the tropical natives.

254:d districts would have afforded an asylum for the natives.

254:[e]

255 The mountain-ranges north-west of the Himalaya, and the long line of the Cordillera, seem to have afforded two great lines of invasion: and it is a striking fact, lately communicated to me by Dr. Hooker, that all the flowering plants, about forty-six in number, common to Tierra del Fuego and to Europe still exist in North America, which must have lain on the line of march.

255:d fact, communicated

255:[e]

255.1:d We might of course speculate on the land having been formerly higher than at present in various parts of the tropics, where temperate forms apparently have crossed; but as the lines of migration have been so numerous, such speculations would be rash.

255.1:[e]

256 But I do not doubt that some temperate productions entered and crossed even the *lowlands* of the tropics at the period when the cold was most intense,—when arctic forms had migrated some twenty-five degrees of latitude from their native country and covered the land at the foot of the Pyrenees.

256:c country, and

256:d Hence I am forced to believe that in certain regions, as in India, some/forms in Europe had migrated over at least twenty-five degrees of latitude, and

256:[e] = 240.1:e

257 At this period of extreme cold, I believe that the climate under the equator at the level of the sea was about the same with that now felt there at the height of six or seven thousand feet.

257:d height of from five to six thousand

257:[e] = 240.2:e

258 During this the coldest period, I suppose that large spaces of the tropical lowlands were clothed with a mingled tropical and temperate vegetation, like that now growing with strange luxuriance at the base of the Himalaya, as graphically described by Hooker.

258:c period, large spaces of the tropical lowlands were probably clothed

258:d Himalaya, at the height of four or five thousand feet, as so graphically

258:[e] = 240.3:e

258.1:d So again, on the island of Fernando Po, in the Gulf of Guinea, Mr. Mann found temperate European forms first beginning to appear at the height of about five thousand feet.

258.1:[e] = 240.4:e

258.2:d On the mountains of Panama, at the height of only two thousand feet, Dr. Seemann found the vegetation like that of Mexico, "with forms of the torrid zone harmoniously blended with those of the temperate."

258.2:[e] = 240.5:e

258.3:d So that under certain conditions of climate it is certainly possible that strictly tropical forms might have co-existed for an indefinitely long period mingled with temperate forms.

258.3:[e]

258.4-8:d[¶] At one time I had hoped to find evidence that the tropics in some part of the world had escaped the chilling effects of the Glacial period, and had afforded a safe refuge for the suffering tropical productions. 5 We cannot look to the peninsula of India for such a refuge, as temperate forms have reached nearly all its isolated mountain-ranges, as well as Ceylon; we cannot look to the Malay archipelago, for on the volcanic cones of Java we see European forms, and on the heights of Borneo temperate Australian productions. 6 If we look to Africa, we find that not only some temperate European forms have passed through Abyssinia along the eastern side of the continent to its southern extremity; but we now know that temperate forms have likewise travelled in a transverse direction from the mountains of Abyssinia to Fernando Po, aided perhaps in their march by east and west ranges, which there is some reason to believe traverse the continent. 7 But even granting that some one large tropical region had retained during the Glacial period its full warmth, the supposition would be of no avail, for the tropical forms therein preserved could not have travelled to the other great tropical regions within so short a period as has elapsed since the Glacial epoch. 8 Nor are the tropical productions of the whole world by any means of so uniform a character as to appear to have proceeded from any one harbour of refuge.

258.4-8:[e]

258.9-11:d[¶] The eastern plains of tropical South America apparently have suffered least from the Glacial period; yet even here there are on the mountains of Brazil a few southern and northern temperate and some Andean forms, which it appears must have crossed the continent from the Cordillera; and some forms on the Silla of Caraccas, which must have migrated from the same great mountain-chain. 10 But Mr. Bates, who has studied with such care the insect-fauna of the Guiano-Amazonian

603

region, has argued with much force against any recent refrigeration in this great region; for he shows that it abounds with highly peculiar endemic Lepidopterous forms, thus apparently contradicting the belief in much recent extinction near the equator. *11* How far his facts can be explained on the supposition of the almost entire annihilation during the Glacial period of a pleistocene equatorial fauna adapted for greater heat than any now prevailing, and the formation of the present equatorial fauna by the commingling of two former subtropical faunas, I will not pretend to say.

258.9-11:[e]

259   Thus, as I believe, a considerable number of plants, a few terrestrial animals, and some marine productions, migrated during the Glacial period from the northern and southern temperate zones into the intertropical regions, and some even crossed the equator.

259:d Notwithstanding these several difficulties, we are led to believe that a considerable/period both from the northern and from the southern temperate zones into the intertropical regions, and that some of them even

259:[e]

260 As the warmth returned, these temperate forms would naturally ascend the higher mountains, being exterminated on the lowlands; those which had not reached the equator, would re-migrate northward or southward towards their former homes; but the forms, chiefly northern, which had crossed the equator, would travel still further from their homes into the more temperate latitudes of the opposite hemisphere.

260:b equator would re-migrate

260:c farther

260.x-y:d When the heat returned, these temperate forms will naturally have ascended the higher mountains, being exterminated on the lowlands; and the greater number will have re-migrated northward or southward towards their former homes. *y* But any temperate forms which had reached and crossed the equator would have travelled

260.x-y:[e]

261 Although we have reason to believe from geological evidence that the whole body of arctic shells underwent scarcely any modification during their long southern migration and re-migration northward, the case may have been wholly different with those intruding forms which settled themselves on the intertropical mountains, and in the southern hemisphere.

261:d evidence that the arctic/with the intruding northern forms/ mountains and

261:[e]

262 These being surrounded by strangers will have had to compete

with many new forms of life; and it is probable that selected modifications in their structure, habits, and constitutions will have profited them.

262:d that modifications

262:[e]

263 Thus many of these wanderers, though still plainly related by inheritance to their brethren of the northern or southern hemispheres, now exist in their new homes as well-marked varieties or as distinct species.

263:d brethren in the northern hemisphere

263:[e]

263.1:d So it will have been with intruders from the south.

263.1:[e]

264 It is a remarkable fact, strongly insisted on by Hooker in regard to America, and by Alph. de Candolle in regard to Australia, that many more identical plants and allied forms have apparently migrated from the north to the south, than in a reversed direction.

264:d have migrated

264:e identical or now slightly modified species have

264:f fact strongly/or slightly

265 We see, however, a few southern vegetable forms on the mountains of Borneo and Abyssinia.

265:e southern forms

266 I suspect that this preponderant migration from north to south is due to the greater extent of land in the north, and to the northern forms having existed in their own homes in greater numbers, and having consequently been advanced through natural selection and competition to a higher stage of perfection or dominating power, than the southern forms.

266:d from the north to the south

266:e perfection, or

267 And thus, when they became commingled during the Glacial period, the northern forms were enabled to beat the less powerful southern forms.

267:d forms will have been enabled

267:e when the two sets became commingled in the equatorial regions, during the alternations of the Glacial periods, the northern forms were the more powerful and were able to hold their places on the mountains, and afterwards to migrate southward with the southern forms; but not so the southern in regard to the northern forms.

268 Just in the same manner as we see at the present day, that very many European productions cover the ground in La Plata, and in a lesser degree in Australia, and have to a certain extent beaten the natives; whereas extremely few southern forms have

become naturalised in any part of Europe, though hides, wool, and other objects likely to carry seeds have been largely imported into Europe during the last two or three centuries from La Plata, and during the last thirty or forty years from Australia.

268:d part of the northern hemisphere, though

268:e In the same manner we/La Plata, New Zealand, and to a lesser degree in Australia, and have beaten

268:f manner at the present day, we see that

268.1:d The Neilgherrie mountains in India, however, offer a partial exception; for here, as I hear from Dr. Hooker, Australian forms are rapidly sowing themselves and becoming naturalised.

269 Something of the same kind must have occurred on the intertropical mountains: no doubt before the Glacial period they were stocked with endemic Alpine forms; but these have almost everywhere largely yielded to the more dominant forms, generated in the larger areas and more efficient workshops of the north.

269:d Before the Glacial period, no doubt the intertropical mountains were stocked with endemic Alpine forms; but these have almost everywhere yielded

269:e Before the last great Glacial

270 In many islands the native productions are nearly equalled or even outnumbered by the naturalised; and if the natives have not been actually exterminated, their numbers have been greatly reduced, and this is the first stage towards extinction.

270:d equalled, or even out-numbered, by those which have become naturalised there; and if

270:e outnumbered, by those which have become naturalised; and this is the first stage towards their extinction.

271 A mountain is an island on the land; and the intertropical mountains before the Glacial period must have been completely isolated; and I believe that the productions of these islands on the land yielded to those produced within the larger areas of the north, just in the same way as the productions of real islands have everywhere lately yielded to continental forms, naturalised by man's agency.

271:d that the inhabitants of these islands on the land have yielded/everywhere yielded to continental forms lately naturalised there through man's

271:e Mountains are islands on the land, and their inhabitants have yielded/as the inhabitants of real islands have everywhere yielded and are still yielding to continental forms naturalised through

271.1-2:e[¶] The same principles apply to the distribution of terrestrial animals and of marine productions, in the northern and southern temperate zones, and on the intertropical mountains. 2 When during the height of the Glacial period the ocean-currents were widely different to what they now are, some of the inhabitants of the temperate seas might have reached the equator; of these a few would perhaps at once be able to migrate southward, by keeping to the cooler currents, whilst others might remain and survive in the cooler depths, until the southern hemisphere was in its turn subjected to a glacial climate and permitted of their further progress; in nearly the same manner as, according to Forbes, isolated spaces inhabited by Arctic productions exist to the present day in the deeper parts of the temperate seas.   .

271.2:f When, during the height of the Glacial period, the ocean-currents/in the colder depths until/parts of the northern temperate

272   I am far from supposing that all difficulties are removed on the view here given in regard to the range and affinities of the allied species which live in the northern and southern temperate zones and on the mountains of the intertropical regions.

272:d am very far from supposing that all difficulties in regard to the distribution and affinities of the allied species, which/ regions, are removed on the views above given.

272:e am far/affinities of the identical and allied species, which now live so widely separated in the north and south, and sometimes on the intermediate mountain-ranges, are

272:f all the difficulties

272.1-3:d It is extremely difficult to understand how a vast number of peculiar forms confined to the tropics could have been therein preserved during the coldest part of the Glacial period. 2 The number of forms in Australia, which are related to European temperate forms, but which differ so greatly that it is impossible to believe that they could have been modified since the Glacial period, perhaps indicates some much more ancient cold period, even as far back as the miocene age, in accordance with the recent speculations of certain geologists. 3 So again, as I am informed by Mr. Bates, the strongly marked character of several species of Carabus, inhabiting the southern parts of America, indicates that their common progenitor must have been introduced at some early period; and other analogous facts could be given.

272.1-3:[e]

273 Very many difficulties remain to be solved.

273:[d]

274 I do not pretend to indicate the exact lines and means of migration, or the reason why certain species and not others have migrated; why certain species have been modified and have given rise to new groups of forms, and others have remained unaltered.

274:d The exact lines and means of migration during the recent Glacial period cannot be indicated; nor the reason

274.x-y:e lines of migration cannot be indicated. *y* We cannot say why certain species and not/new forms, whilst others

275 We cannot hope to explain such facts, until we can say why one species and not another becomes naturalised by man's agency in a foreign land; why one ranges twice or thrice as far, and is twice or thrice as common, as another species within their own homes.

275:d land; why one species ranges

276 I have said that many difficulties remain to be solved: some of the most remarkable are stated with admirable clearness by Dr. Hooker in his botanical works on the antarctic regions.

277 These cannot be here discussed.

277:[e]

278 I will only say that as far as regards the occurrence of identical species at points so enormously remote as Kerguelen Land, New Zealand, and Fuegia, I believe that towards the close of the Glacial period, icebergs, as suggested by Lyell, have been largely concerned in their dispersal.

278:d that, as far

276 + 8:e Various special difficulties also remain to be solved: for instance, the occurrence, as shown by Dr. Hooker, of the same plants at points so enormously remote as Kerguelen Land, New Zealand, and Fuegia; but icebergs, as suggested by Lyell, may have been concerned

276 + 8:f solved; for

279 But the existence of several quite distinct species, belonging to genera exclusively confined to the south, at these and other distant points of the southern hemisphere, is, on my theory of descent with modification, a far more remarkable case of difficulty.

279:e The existence, at these and other distant points of the southern hemisphere, of species, which, though distinct, belong to genera exclusively confined to the south, is a more remarkable case.

279:f existence at

280 For some of these species are so distinct, that we cannot suppose that there has been time since the commencement of the Glacial period for their migration, and for their subsequent modification to the necessary degree.

280:e Some/commencement of the last Glacial period for their migration and subsequent

281 The facts seem to me to indicate that peculiar and very distinct species have migrated in radiating lines from some common centre; and I am inclined to look in the southern, as in the

northern hemisphere, to a former and warmer period, before the commencement of the Glacial period, when the antarctic lands, now covered with ice, supported a highly peculiar and isolated flora.

281:e that distinct species belonging to the same genera have migrated in radiating lines from a common/Antarctic

281:f seem to indicate

282 I suspect that before this flora was exterminated by the Glacial epoch, a few forms were widely dispersed to various points of the southern hemisphere by occasional means of transport, and by the aid, as halting-places, of existing and now sunken islands, and perhaps at the commencement of the Glacial period, by icebergs.

282:b islands.

282:d halting-places, of now sunken islands.

282:e It may be suspected that before this flora was exterminated during the last Glacial epoch, a few forms had been already widely

282:f aid as

283 By these means, as I believe, the southern shores of America, Australia, New Zealand have become slightly tinted by the same peculiar forms of vegetable life.

283:b Zealand, have

283:d Australia, and New Zealand, became

283:e Thus the southern shores of America, Australia, and New Zealand, might have become slightly tinted by the same peculiar forms of life.

283:f Zealand, may have

284 Sir C. Lyell in a striking passage has speculated, in language almost identical with mine, on the effects of great alternations of climate on geographical distribution.

284:e climate throughout the world on

285 I believe that the world has recently felt one of his great cycles of change; and that on this view, combined with modification through natural selection, a multitude of facts in the present distribution both of the same and of allied forms of life can be explained.

285:e And we have now seen that Mr. Croll's conclusion that successive Glacial periods in the one hemisphere coincided with warmer periods in the opposite hemisphere, together with the admission of the slow modification of species, explains a multitude of facts in the distribution of the same and of the allied forms of life in all parts of the globe.

285:f coincide

286 The living waters may be said to have flowed during one short period from the north and from the south, and to have crossed

at the equator; but to have flowed with greater force from the north so as to have freely inundated the south.

286:e waters have flowed during certain periods from the north and afterwards from the south, and in both cases have reached the equator; but the stream of life has flowed with greater force from the north than in the opposite direction, and has consequently more freely

286:f during one period from the north and during another from the south, and in both cases have reached the equator: but

287 As the tide leaves its drift in horizontal lines, though rising higher on the shores where the tide rises highest, so have the living waters left their living drift on our mountain-summits, in a line gently rising from the arctic lowlands to a great height under the equator.

287:e lines, rising higher/Arctic lowlands to a great altitude under

288 The various beings thus left stranded may be compared with savage races of man, driven up and surviving in the mountain-fastnesses of almost every land, which serve as a record, full of interest to us, of the former inhabitants of the surrounding lowlands.

288:f mountain fastnesses

CHAPTER XII.

1:f XIII.

2   GEOGRAPHICAL DISTRIBUTION—*continued.*

3 Distribution of fresh-water productions—On the inhabitants of oceanic islands—Absence of Batrachians and of terrestrial Mammals—On the relation of the inhabitants of islands to those of the nearest mainland—On colonisation from the nearest source with subsequent modification—Summary of the last and present chapters.

3:c chapter.

3.1:d [*Center*] *Fresh-water Productions.* [*Space*]

4 As lakes and river-systems are separated from each other by barriers of land, it might have been thought that fresh-water productions would not have ranged widely within the same country, and as the sea is apparently a still more impassable barrier, that they never would have extended to distant countries.

4:d and, as

4:e more formidable barrier

4:f and as

5 But the case is exactly the reverse.

6 Not only have many fresh-water species, belonging to quite different classes, an enormous range, but allied species prevail in a remarkable manner throughout the world.

6:f to different

7 I well remember, when first collecting in the fresh waters of Brazil, feeling much surprise at the similarity of the fresh-water insects, shells, &c., and at the dissimilarity of the surrounding terrestrial beings, compared with those of Britain.

7:f When first collecting in the fresh waters of Brazil, I well remember feeling

8   But this power in fresh-water productions of ranging widely, though so unexpected, can, I think, in most cases be explained by their having become fitted, in a manner highly useful to them, for short and frequent migrations from pond to pond, or from stream to stream; and liability to wide dispersal would follow from this capacity as an almost necessary consequence.

8:f But the wide ranging power of fresh-water productions can/ stream to stream within their own countries; and

9 We can here consider only a few cases.

9:f cases; of these, some of the most difficult to explain are presented by fish.

10 In regard to fish, I believe that the same species never occur in the fresh waters of distant continents.

10:f It was formerly believed that the same fresh-water species never existed on two continents distant from each other.

10.1-3:f But Dr. Günther has lately shown that the Galaxias attenuatus inhabits Tasmania, New Zealand, the Falkland Islands, and the mainland of South America. 2 This is a wonderful case, and probably indicates dispersal from an Antarctic centre during a former warm period. 3 This case, however, is rendered in some degree less surprising by the species of this genus having the power of crossing by some unknown means considerable spaces of open ocean: thus there is one species common to New Zealand and to the Auckland Islands, though separated by a distance of about 230 miles.

11 But on the same continent the species often range widely and almost capriciously; for two river-systems will have some fish in common and some different.

11:f On the same continent fresh-water fish often range widely, and as if capriciously; for in two adjoining river-systems some of the species may be the same, and some wholly different.

12 A few facts seem to favour the possibility of their occasional transport by accidental means; like that of the live fish not rarely dropped by whirlwinds in India, and the vitality of their ova when removed from the water.

12.x-y:f It is probable that they are occasionally transported by what may be called accidental means. *y* Thus fishes still alive are not very rarely dropped at distant points by whirlwinds; and it is known that the ova retain their vitality for a considerable time after removal from the water.

13 But I am inclined to attribute the dispersal of fresh-water fish mainly to slight changes within the recent period in the level of the land, having caused rivers to flow into each other.

13:e to changes in the level of the land within the recent period, having

13:f Their dispersal may, however, be mainly attributed to changes/ period, causing

14 Instances, also, could be given of this having occurred during floods, without any change of level.

15 We have evidence in the loess of the Rhine of considerable changes of level in the land within a very recent geological period, and when the surface was peopled by existing land and fresh-water shells.

15:[e]

16 The wide difference of the fish on opposite sides of continuous mountain-ranges, which from an early period must have parted

river-systems and completely prevented their inosculation, seems to lead to this same conclusion.

16:e have completely prevented the inosculation of the river-systems, seems to lead to the same

16:f sides of most mountain-ranges, which are continuous, and which consequently must from an early period have completely prevented the inosculation of the river-systems on the two sides, leads

17 With respect to allied fresh-water fish occurring at very distant points of the world, no doubt there are many cases which cannot at present be explained: but some fresh-water fish belong to very ancient forms, and in such cases there will have been ample time for great geographical changes, and consequently time and means for much migration.

17:f Some

17.1:f Moreover Dr. Günther has recently been led by several considerations to infer that with fishes the same forms have a long endurance.

18 In the second place, salt-water fish can with care be slowly accustomed to live in fresh water; and, according to Valenciennes, there is hardly a single group of fishes confined exclusively to fresh water, so that we may imagine that a marine member of a fresh-water group might travel far along the shores of the sea, and subsequently become modified and adapted to the fresh waters of a distant land.

18:e group of which all the members are confined exclusively to fresh water, so that a marine species of a fresh-water

18:f salt-water fish can/confined to fresh water, so that a marine species belonging to a fresh-water group might travel far along the shores of the sea, and could, it is probable, become adapted without much difficulty to

19 Some species of fresh-water shells have a very wide range, and allied species, which, on my theory, are descended from a common parent and must have proceeded from a single source, prevail throughout the world.

19:c parent, and

19:e have very wide ranges, and allied species which, on our theory

20 Their distribution at first perplexed me much, as their ova are not likely to be transported by birds, and they are immediately killed by sea water, as are the adults.

20:b sea-water

20:e and are immediately

20:f and the ova, as well as the adults, are immediately killed by sea-water.

21 I could not even understand how some naturalised species have rapidly spread throughout the same country.

21:e have spread rapidly throughout

22 But two facts, which I have observed—and no doubt many others remain to be observed—throw some light on this subject.

22:f and many others no doubt will be discovered—throw

23 When a duck suddenly emerges from a pond covered with duck-weed, I have twice seen these little plants adhering to its back; and it has happened to me, in removing a little duck-weed from one aquarium to another, that I have quite unintentionally stocked the one with fresh-water shells from the other.

23:f When ducks suddenly emerge/to their backs/that I have unintentionally

24 But another agency is perhaps more effectual: I suspended a duck's feet, which might represent those of a bird sleeping in a natural pond, in an aquarium, where many ova of fresh-water shells were hatching; and I found that numbers of the extremely minute and just hatched shells crawled on the feet, and clung to them so firmly that when taken out of the water they could not be jarred off, though at a somewhat more advanced age they would voluntarily drop off.

24:b just-hatched

24:c feet in an aquarium

24:f suspended the feet of a duck in an aquarium

25 These just hatched molluscs, though aquatic in their nature, survived on the duck's feet, in damp air, from twelve to twenty hours; and in this length of time a duck or heron might fly at least six or seven hundred miles, and would be sure to alight on a pool or rivulet, if blown across sea to an oceanic island or to any other distant point.

25:c just-hatched

25:e miles, and if blown across the sea to an oceanic island or to any other distant point would be sure to alight on a pool or rivulet.

25:f island, or to any other distant point, would

26 Sir Charles Lyell also informs me that a Dytiscus has been caught with an Ancylus (a fresh-water shell like a limpet) firmly adhering to it; and a water-beetle of the same family, a Colymbetes, once flew on board the 'Beagle,' when forty-five miles distant from the nearest land: how much farther it might have flown with a favouring gale no one can tell.

26:e Lyell informs/might have been blown with

26:f blown by a favouring

27 With respect to plants, it has long been known what enormous ranges many fresh-water and even marsh-species have, both over continents and to the most remote oceanic islands.

27:c fresh-water, and even marsh-species, have

614

27:f marsh species

28 This is strikingly shown, as remarked by Alph. de Candolle, in large groups of terrestrial plants, which have only a very few aquatic members; for these latter seem immediately to acquire, as if in consequence, a very wide range.

28:f strikingly illustrated, according to Alph. de Candolle, in those large groups of terrestrial plants, which have very few aquatic members; for the latter seem immediately to acquire, as if in consequence, a wide

29 I think favourable means of dispersal explain this fact.

30 I have before mentioned that earth occasionally, though rarely, adheres in some quantity to the feet and beaks of birds.

30:f occasionally adheres

31 Wading birds, which frequent the muddy edges of ponds, if suddenly flushed, would be the most likely to have muddy feet.

32 Birds of this order I can show are the greatest wanderers, and are occasionally found on the most remote and barren islands in the open ocean; they would not be likely to alight on the surface of the sea, so that the dirt would not be washed off their feet; when making land, they would be sure to fly to their natural fresh-water haunts.

32:c order, I can show, are

32:f order wander more than those of any other; and they are occasionally found on the most remote and barren islands of the open/that any dirt on their feet would not be washed off; and when gaining the land

33 I do not believe that botanists are aware how charged the mud of ponds is with seeds: I have tried several little experiments, but will here give only the most striking case: I took in February three table-spoonfuls of mud from three different points, beneath water, on the edge of a little pond; this mud when dry weighed only 6¾ ounces; I kept it covered up in my study for six months, pulling up and counting each plant as it grew; the plants were of many kinds, and were altogether 537 in number; and yet the viscid mud was all contained in a breakfast cup!

33:f seeds; I have/when dried weighed

34 Considering these facts, I think it would be an inexplicable circumstance if water-birds did not transport the seeds of fresh-water plants to vast distances, and if consequently the range of these plants was not very great.

34:e plants to unstocked ponds and streams, situated at very distant points.

35 The same agency may have come into play with the eggs of some of the smaller fresh-water animals.

36 Other and unknown agencies probably have also played a part.

37 I have stated that fresh-water fish eat some kinds of seeds, though they reject many other kinds after having swallowed them; even small fish swallow seeds of moderate size, as of the yellow water-lily and Potamogeton.

38 Herons and other birds, century after century, have gone on daily devouring fish; they then take flight and go to other waters, or are blown across the sea; and we have seen that seeds retain their power of germination, when rejected in pellets or in excrement, many hours afterwards.

38:f rejected many hours afterwards in pellets or in the excrement.

39 When I saw the great size of the seeds of that fine water-lily, the Nelumbium, and remembered Alph. de Candolle's remarks on this plant, I thought that its distribution must remain quite inexplicable; but Audubon states that he found the seeds of the great southern water-lily (probably, according to Dr. Hooker, the Nelumbium luteum) in a heron's stomach; although I do not know the fact, yet analogy makes me believe that a heron flying to another pond and getting a hearty meal of fish, would probably reject from its stomach a pellet containing the seeds of the Nelumbium undigested; or the seeds might be dropped by the bird whilst feeding its young, in the same way as fish are known sometimes to be dropped.

39.x-y:f remarks on the distribution of this plant, I thought that the means of its dispersal must remain inexplicable; but/ stomach. *y* Now this bird must often have flown with its stomach thus well stocked to distant ponds, and then getting a hearty meal of fish, analogy makes me believe that it would have rejected the seeds in a pellet in a fit state for germination.

40 In considering these several means of distribution, it should be remembered that when a pond or stream is first formed, for instance, on a rising islet, it will be unoccupied; and a single seed or egg will have a good chance of succeeding.

41 Although there will always be a struggle for life between the individuals of the species, however few, already occupying any pond, yet as the number of kinds is small, compared with those on the land, the competition will probably be less severe between aquatic than between terrestrial species; consequently an intruder from the waters of a foreign country, would have a better chance of seizing on a place, than in the case of terrestrial colonists.

41:d yet, as

41:e between the inhabitants of the same pond, however few in kind, yet as the number even in a well-stocked pond is small in comparison with the number of species inhabiting an equal area of land, the competition/country will have a better chance of seizing on a new place

41:f competition between them will probably/country would have

42 We should, also, remember that some, perhaps many, fresh-water productions are low in the scale of nature, and that we have reason to believe that such low beings change or become modified less quickly than the high; and this will give longer time than the average for the migration of the same aquatic species.

42:c should also remember

42:e remember that many fresh-water/and we have reason to believe that low/give a longer

42:f that such beings become modified more slowly than the high; and this will give time for the migration of aquatic

43 We should not forget the probability of many species having formerly ranged as continuously as fresh-water productions ever can range, over immense areas, and having subsequently become extinct in intermediate regions.

43:f many fresh-water forms having formerly ranged continuously over immense areas, and then having become extinct at intermediate points.

44 But the wide distribution of fresh-water plants and of the lower animals, whether retaining the same identical form or in some degree modified, I believe mainly depends on the wide dispersal of their seeds and eggs by animals, more especially by fresh-water birds, which have large powers of flight, and naturally travel from one to another and often distant piece of water.

44:e have great powers of flight, and naturally travel from one piece of water to another. [*Space*]

44:f modified, apparently depends in main part on

45 Nature, like a careful gardener, thus takes her seeds from a bed of a particular nature, and drops them in another equally well fitted for them.

45:[e]

46   *On the Inhabitants of Oceanic Islands.—*
46:d [*Center*] *On the Inhabitants of Oceanic Islands.* [*Space*]

47 We now come to the last of the three classes of facts, which I have selected as presenting the greatest amount of difficulty, on the view that all the individuals both of the same and of allied species have descended from a single parent; and therefore have all proceeded from a common birthplace, notwithstanding that in the course of time they have come to inhabit distant points of the globe.

47:d[¶]

47:e difficulty, if we accept the view that not only all the individuals of the same species, wherever found, have migrated from some one area, but that allied species, although now inhabiting the most distant points, have proceeded from a single area,—the birthplace of their early progenitor.

47:f difficulty with respect to distribution, on the view that not only all the individuals of the same species have migrated/progenitors.

48 I have already stated that I cannot honestly admit Forbes's view on continental extensions, which, if legitimately followed out, would lead to the belief that within the recent period all existing islands have been nearly or quite joined to some continent.

48:e belief that all existing islands have been continuously or almost continuously joined to some continent within the recent period.

48:f I have already given my reasons for disbelieving in continental extensions within the period of existing species, on so enormous a scale that all the many islands of the several oceans were thus stocked with their present terrestrial inhabitants.

49 This view would remove many difficulties, but it would not, I think, explain all the facts in regard to insular productions.

49:c not explain

49:f This view removes many difficulties, but it does not accord with all the facts in regard to the productions of islands.

50 In the following remarks I shall not confine myself to the mere question of dispersal; but shall consider some other facts, which bear on the truth of the two theories of independent creation and of descent with modification.

50:f dispersal, but shall consider some other cases bearing

51 The species of all kinds which inhabit oceanic islands are few in number compared with those on equal continental areas: Alph. de Candolle admits this for plants, and Wollaston for insects.

52 If we look to the large size and varied stations of New Zealand, extending over 780 miles of latitude, and compare its flowering plants, only 750 in number, with those on an equal area at the Cape of Good Hope or in Australia, we must, I think, admit that something quite independently of any difference in physical conditions has caused so great a difference in number.

52:d New Zealand, for instance, with its lofty mountains and diversified stations, extending over 780 miles of latitude, together with the outlying islands of Auckland, Campbell, and Chatham, contain altogether only 960 kinds of flowering plants; if we compare this moderate number with the species which swarm over equal areas in south-western Australia or at

52:e South-Western/independently of a difference in the physical

52:f admit that some cause, independently of different physical conditions, has given rise to so

53 Even the uniform county of Cambridge has 847 plants, and the little island of Anglesea 764, but a few ferns and a few intro-

duced plants are included in these numbers, and the comparison in some other respects is not quite fair.

54 We have evidence that the barren island of Ascension aboriginally possessed under half-a-dozen flowering plants; yet many have become naturalised on it, as they have on New Zealand and on every other oceanic island which can be named.

54:d possessed less than half-a-dozen flowering plants; yet many have now become

54:e many species have now

55 In St. Helena there is reason to believe that the naturalised plants and animals have nearly or quite exterminated many native productions.

56 He who admits the doctrine of the creation of each separate species, will have to admit, that a sufficient number of the best adapted plants and animals have not been created on oceanic islands; for man has unintentionally stocked them from various sources far more fully and perfectly than has nature.

56:c to admit that

56:d animals were not created

56:e created for oceanic islands; for man has unintentionally stocked them far more fully and perfectly than did nature.

57 Although in oceanic islands the number of kinds of inhabitants is scanty, the proportion of endemic species (*i. e.* those found nowhere else in the world) is often extremely large.

57:d number of the inhabitants is scanty in kind, the proportion

57:e islands the species are few in number, the proportion of endemic kinds (*i. e.*

58 If we compare, for instance, the number of the endemic land-shells in Madeira, or of the endemic birds in the Galapagos Archipelago, with the number found on any continent, and then compare the area of the islands with that of the continent, we shall see that this is true.

58:e of endemic land-shells in Madeira, or of endemic/island

59 This fact might have been expected on my theory, for, as already explained, species occasionally arriving after long intervals in a new and isolated district, and having to compete with new associates, will be eminently liable to modification, and will often produce groups of modified descendants.

59:d intervals of time in

59:e been theoretically expected, for/associates, would be eminently liable to modification, and would often

60 But it by no means follows, that, because in an island nearly all the species of one class are peculiar, those of another class, or of another section of the same class, are peculiar; and this difference seems to depend on the species which do not become

modified having immigrated with facility and in a body, so that their mutual relations have not been much disturbed.

60:b depend partly on/disturbed; and partly on the frequent arrival of unmodified immigrants from the mother-country, and the consequent intercrossing with them.

60:c follows that, because

60:e on the species which are not modified having immigrated in/ mother-country, with which the insular forms have intercrossed.

60.1:b With respect to the effects of this intercrossing, it should be remembered that the offspring of such crosses would almost certainly gain in vigour; so that even an occasional cross would produce more effect than might at first have been anticipated.

60.1:c first be anticipated.

60.1:e It should be borne in mind that the offspring/might have been anticipated.

60.1:f would certainly

61 Thus in the Galapagos Islands nearly every land-bird, but only two out of the eleven marine birds, are peculiar; and it is obvious that marine birds could arrive at these islands more easily than land-birds.

61:b To give a few examples: in

61:c Islands there are 26 land-birds; of these, 21 (or perhaps 23) are peculiar, whereas of the 11 marine birds only 2 are

61:d these 21

61:e I will give a few illustrations of the foregoing remarks: in/ arrive at these islands much more easily and frequently than

62 Bermuda, on the other hand, which lies at about the same distance from North America as the Galapagos Islands do from South America, and which has a very peculiar soil, does not possess one endemic land-bird; and we know from Mr. J. M. Jones's admirable account of Bermuda, that very many North American birds, during their great annual migrations, visit either periodically or occasionally this island.

62:d many North American birds occasionally visit this

62:e possess a single endemic/occasionally or even frequently visit

63 Madeira does not possess one peculiar bird, and many European and African birds are almost every year blown there, as I am informed by Mr. E. V. Harcourt.

63:d Almost every year, as I am informed by Mr. E. V. Harcourt, many European and African birds are blown to Madeira; this island is inhabited by 99 kinds, of which one alone is peculiar, though very closely related to a European form; and three or four other species are confined to this island and to the Canaries.

64 So that these two islands of Bermuda and Madeira have been stocked by birds, which for long ages have struggled together

in their former homes, and have become mutually adapted to each other; and when settled in their new homes, each kind will have been kept by the others to their proper places and habits, and will consequently have been little liable to modification.

64:d that the two/kept by the others to its proper

64:e that the islands of Bermuda and Madeira have been stocked from the neighbouring continents with birds, which for long ages have struggled together, and become mutually adapted; hence when settled in their new homes, each kind would be kept by the others to its proper place and habits, and would consequently be but little

64.x-y:f Islands/ages have there struggled together, and have become mutually co-adapted. *y* Hence/kind will have been kept/ habits, and will consequently have been but

64.1:b Any tendency to modification will, also, have been checked by intercrossing with the unmodified immigrants from the mother-country.

64.1:c will also have

64.1:e modification would also be

64.1:f modification will also have been/immigrants, often arriving from

65 Madeira, again, is inhabited by a wonderful number of peculiar land-shells, whereas not one species of sea-shell is confined to its shores: now, though we do not know how sea-shells are dispersed, yet we can see that their eggs or larvæ, perhaps attached to seaweed or floating timber, or to the feet of wading-birds, might be transported far more easily than land-shells, across three or four hundred miles of open sea.

65:c Madeira again is inhabited

65:d sea-shell is peculiar to its

65:e transported across three or four hundred miles of open sea far more easily than land-shells.

66 The different orders of insects in Madeira apparently present analogous facts.

66:e insects inhabiting Madeira present nearly similar cases.

66:f nearly parallel cases.

67 Oceanic islands are sometimes deficient in certain classes, and their places are apparently occupied by the other inhabitants; in the Galapagos Islands reptiles, and in New Zealand gigantic wingless birds, take the place of mammals.

67:d deficient in animals of certain whole classes, and their places are occupied by animals belonging to other classes; thus in the Galapagos

67:e occupied by other/take, or recently took, the place

67.1:d Although New Zealand is here spoken of as an oceanic island, it is in some degree doubtful whether it should be so ranked; it is of large size, and is not separated from Australia by a profoundly deep sea: from its geological character and the direction of its mountain-ranges, the Rev. W. B. Clarke has lately maintained that this island, as well as New Caledonia, should be considered as appurtenances of Australia.

67.1:e sea; from

68 In the plants of the Galapagos Islands, Dr. Hooker has shown that the proportional numbers of the different orders are very different from what they are elsewhere.

68:e Turning to plants, Dr. Hooker has shown that in the Galapagos Islands the proportional

69 Such cases are generally accounted for by the physical conditions of the islands; but this explanation seems to me not a little doubtful.

69:d All such/explanation is not

69:e All such differences in number, and the absence of certain whole groups of animals and plants on islands, are generally accounted for by supposed differences in their physical conditions; but

69:f differences in the physical conditions of the islands; but

70 Facility of immigration, I believe, has been at least as important as the nature of the conditions.

70:c immigration seems to have been

70:e been fully as important

71   Many remarkable little facts could be given with respect to the inhabitants of remote islands.

71:e of oceanic islands.

72 For instance, in certain islands not tenanted by mammals, some of the endemic plants have beautifully hooked seeds; yet few relations are more striking than the adaptation of hooked seeds for transportal by the wool and fur of quadrupeds.

72:e by a single mammal/more manifest than that hooked seeds are adapted for transportal in the wool or fur

72:f seeds; yet few relations are more manifest than that hooks serve for the transportal of seeds in

73 This case presents no difficulty on my view, for a hooked seed might be transported to an island by some other means; and the plant then becoming slightly modified, but still retaining its hooked seeds, would form an endemic species, having as useless an appendage as any rudimentary organ,—for instance, as the shrivelled wings under the soldered elytra of many insular beetles.

73:e But a hooked seed might be carried to an island by other means; and the plant then becoming modified would form an endemic species, which might still retain the hooks, which would not form a more useless appendage than the shrivelled

73:f species, still retaining its hooks, which would form a useless appendage like the shrivelled wings under the soldered wing-covers of many

74 Again, islands often possess trees or bushes belonging to orders which elsewhere include only herbaceous species; now trees, as Alph. de Candolle has shown, generally have, whatever the cause may be, confined ranges.

74:e orders elsewhere including

74:f orders which elsewhere include

75 Hence trees would be little likely to reach distant oceanic islands; and an herbaceous plant, though it would have no chance of successfully competing in stature with a fully developed tree, when established on an island and having to compete with herbaceous plants alone, might readily gain an advantage by growing taller and taller and overtopping the other plants.

75:d it might have no chance of successfully competing on a continent with many fully developed trees/advantage over them by growing taller and overtopping them.

75:e plant, which had no chance of successfully competing with the many fully developed trees growing on a continent, might, when established on an island, gain an advantage by growing taller and taller and overtopping the other herbaceous plants.

75:f advantage over other herbaceous plants by growing taller and taller and overtopping them.

76 If so, natural selection would often tend to add to the stature of herbaceous plants when growing on an island, to whatever order they belonged, and thus convert them first into bushes and ultimately into trees.

76:d on oceanic islands/trees. [*Space*]

76:e In this case, natural selection would tend to add to the stature of the plant, to whatever order it belonged, and thus convert it first into a bush and then into a tree.

76:f thus first convert it into a bush

76.1:d [*Center*] *Absence of Batrachians and Terrestrial Mammals on Oceanic Islands.* [*Space*]

77 With respect to the absence of whole orders on oceanic islands, Bory St. Vincent long ago remarked that Batrachians (frogs, toads, newts) have never been found on any of the many islands with which the great oceans are studded.

77:e orders of animals on oceanic/newts) are never found

78 I have taken pains to verify this assertion, and I have found it strictly true.

78:d and have found it strictly true, with the exception of New Zealand, of the Andaman Islands, and perhaps of the Salomon Islands.

78:f it true, with the exception of New Zealand, New Caledonia, the Andaman Islands, and perhaps the Salomon Islands and the Seychelles.

78.1:d But I have already remarked that it is doubtful whether New Zealand ought to be classed as an oceanic island; and this is still more doubtful with respect to the Andaman and Salomon groups.

78.1:f Zealand and New Caledonia ought to be classed as oceanic islands/groups and the Seychelles.

79 I have, however, been assured that a frog exists on the mountains of the great island of New Zealand; but I suspect that this exception (if the information be correct) may be explained through glacial agency.

79:[d]

80 This general absence of frogs, toads, and newts on so many oceanic islands cannot be accounted for by their physical conditions; indeed it seems that islands are peculiarly well fitted for these animals; for frogs have been introduced into Madeira, the Azores, and Mauritius, and have multiplied so as to become a nuisance.

80:f many true oceanic/peculiarly fitted

81 But as these animals and their spawn are known to be immediately killed by sea-water, on my view we can see that there would be great difficulty in their transportal across the sea, and therefore why they do not exist on any oceanic island.

81:c sea-water, there would be great difficulty in their transportal across the sea, and therefore on my view we can see why

81:e therefore we

81:f are immediately killed (with the exception, as far as known, of one Indian species) by sea-water/exist on strictly oceanic

82 But why, on the theory of creation, they should not have been created there, it would be very difficult to explain.

83   Mammals offer another and similar case.

84 I have carefully searched the oldest voyages, but have not finished my search; as yet I have not found a single instance, free from doubt, of a terrestrial mammal (excluding domesticated animals kept by the natives) inhabiting an island situated above 300 miles from a continent or great continental island; and many islands situated at a much less distance are equally barren.

84:e voyages, and as

84:f and have

85 The Falkland Islands, which are inhabited by a wolf-like fox, come nearest to an exception; but this group cannot be considered as oceanic, as it lies on a bank connected with the mainland; moreover, icebergs formerly brought boulders to its western shores, and they may have formerly transported foxes, as so frequently now happens in the arctic regions.

85:c mainland, distant from it about 280 miles; moreover

85:e bank in connection with the mainland at the distance of about/as now frequently happens

86 Yet it cannot be said that small islands will not support small mammals, for they occur in many parts of the world on very small islands, if close to a continent; and hardly an island can be named on which our smaller quadrupeds have not become naturalised and greatly multiplied.

86:e support at least small/islands, when lying close

87 It cannot be said, on the ordinary view of creation, that there has not been time for the creation of mammals; many volcanic islands are sufficiently ancient, as shown by the stupendous degradation which they have suffered and by their tertiary strata: there has also been time for the production of endemic species belonging to other classes; and on continents it is thought that mammals appear and disappear at a quicker rate than other and lower animals.

87:c suffered, and by

87:f continents it is known that new species of mammals

88 Though terrestrial mammals do not occur on oceanic islands, aërial mammals do occur on almost every island.

88:e Although terrestrial

88:f aerial

89 New Zealand possesses two bats found nowhere else in the world: Norfolk Island, the Viti Archipelago, the Bonin Islands, the Caroline and Marianne Archipelagoes, and Mauritius, all possess their peculiar bats.

90 Why, it may be asked, has the supposed creative force produced bats and no other mammals on remote islands?

91 On my view this question can easily be answered; for no terrestrial mammal can be transported across a wide space of sea, but bats can fly across.

92 Bats have been seen wandering by day far over the Atlantic Ocean; and two North American species either regularly or occasionally visit Bermuda, at the distance of 600 miles from the mainland.

93 I hear from Mr. Tomes, who has specially studied this family, that many of the same species have enormous ranges, and are found on continents and on far distant islands.

93:e many species

94 Hence we have only to suppose that such wandering species have been modified through natural selection in their new homes in relation to their new position, and we can understand the presence of endemic bats on islands, with the absence of all terrestrial mammals.

94:e modified in their new/on oceanic islands, with the absence of all other terrestrial

95   Besides the absence of terrestrial mammals in relation to the remoteness of islands from continents, there is also a relation, to a certain extent independent of distance, between the depth of the sea separating an island from the neighbouring mainland, and the presence in both of the same mammiferous species or of allied species in a more or less modified condition.

95:e Another interesting relation exists, namely between the depth of the sea separating islands from each other or from the nearest continents, and the degree of affinity of their mammalian inhabitants.

95:f continent

96   Mr. Windsor Earl has made some striking observations on this head in regard to the great Malay Archipelago, which is traversed near Celebes by a space of deep ocean; and this space separates two widely distinct mammalian faunas.

96:d head, since fully confirmed by Mr. Wallace's admirable researches, in

96:e since greatly extended by Mr./ocean, and this separates

97   On either side the islands are situated on moderately deep submarine banks, and they are inhabited by closely allied or identical quadrupeds.

97:e islands stand on a moderately shallow submarine bank, and these are inhabited by the same or by very closely allied quadrupeds.

97:f these islands are inhabited by the same or by closely

98   No doubt some few anomalies occur in this great archipelago, and there is much difficulty in forming a judgment in some cases owing to the probable naturalisation of certain mammals through man's agency; but we shall soon have much light thrown on the natural history of this archipelago by the admirable zeal and researches of Mr. Wallace.

98:[d]

99   I have not as yet had time to follow up this subject in all other quarters of the world; but as far as I have gone, the relation generally holds good.

99:e all quarters

99:f relation holds

100   We see Britain separated by a shallow channel from Europe, and the mammals are the same on both sides; we meet with analogous facts on many islands separated by similar channels from Australia.

100:e For instance, Britain is separated/sides; and so it is with all the islands near the shores of Australia.

101   The West Indian Islands stand on a deeply submerged bank,

nearly 1000 fathoms in depth, and here we find American forms, but the species and even the genera are distinct.

101:e Islands, on the other hand, stand/are quite distinct.

102 As the amount of modification in all cases depends to a certain degree on the lapse of time, and as during changes of level it is obvious that islands separated by shallow channels are more likely to have been continuously united within a recent period to the mainland than islands separated by deeper channels, we can understand the frequent relation between the depth of the sea and the degree of affinity of the mammalian inhabitants of islands with those of a neighbouring continent,—an inexplicable relation on the view of independent acts of creation.

102:e modification which animals of all kinds undergo, partly depends on the lapse of time, and as islands separated from each other or from the mainland by shallow/period than islands separated by deeper channels, we can understand how it is that a relation exists between the depth of the sea separating two mammalian faunas, and the degree of their affinity,—a relation which is quite inexplicable on the theory of

102:f islands which are separated from/channels, are more/than the islands

103 All the foregoing remarks on the inhabitants of oceanic islands,—namely, the scarcity of kinds—the richness in endemic forms in particular classes or sections of classes,—the absence of whole groups, as of batrachians, and of terrestrial mammals notwithstanding the presence of aërial bats,—the singular proportions of certain orders of plants,—herbaceous forms having been developed into trees, &c.,—seem to me to accord better with the view of occasional means of transport having been largely efficient in the long course of time, than with the view of all our oceanic islands having been formerly connected by continuous land with the nearest continent; for on this latter view the migration would probably have been more complete; and if modification be admitted, all the forms of life would have been more equally modified, in accordance with the paramount importance of the relation of organism to organism.

103:c kinds,—the richness/batrachians and of terrestrial

103:d time than

103:e The foregoing statements in regard to the inhabitants of oceanic islands,—namely, the fewness of the species, with a large proportion consisting of endemic forms—the members of certain groups, and not of other groups in the same class, having been modified—the absence of certain whole orders, as of batrachians and of terrestrial mammals, nowithstanding/with the belief in the efficiency of occasional means of transport, carried on during a long course of time, than with the belief in the former connection of all oceanic islands with the nearest continent; for on this latter view it is probable that the various classes would have immigrated more uniformly, and from the species having entered in a body their mutual relations would not have been much disturbed, and consequently they would have been modified either not at all or in a more equal manner.

103:f groups, but not those of other/consequently they would either have not been modified, or all the species in a more equable manner.

104   I do not deny that there are many and grave difficulties in understanding how several of the inhabitants of the more remote islands, whether still retaining the same specific form or modified since their arrival, could have reached their present homes.

104:d and serious difficulties

104:e how many of/form or subsequently modified, have

105   But the probability of many islands having existed as halting-places, of which not a wreck now remains, must not be overlooked.

105:e of islands

105:f of other islands having once existed

106   I will here give a single instance of one of the cases of difficulty.

106:e I will specify one such difficult case.

106:f one difficult

107   Almost all oceanic islands, even the most isolated and smallest, are inhabited by land-shells, generally by endemic species, but sometimes by species found elsewhere.

108   Dr. Aug. A. Gould has given several interesting cases in regard to the land-shells of the islands of the Pacific.

107 + 8:e elsewhere,—striking instances of which have been given by Dr. A. A. Gould in relation to the Pacific.

109   Now it is notorious that land-shells are very easily killed by salt; their eggs, at least such as I have tried, sink in sea-water and are killed by it.

109:c sea-water, and

109:e are easily killed by sea-water; their eggs, at least such as I have tried, sink in it and are killed.

110   Yet there must be, on my view, some unknown, but highly efficient means for their transportal.

110:e according to our view, some unknown, but occasionally efficient

110:f be some

111   Would the just-hatched young occasionally crawl on and adhere to the feet of birds roosting on the ground, and thus get transported?

111:e young sometimes adhere

112   It occurred to me that land-shells, when hybernating and having a membranous diaphragm over the mouth of the shell,

might be floated in chinks of drifted timber across moderately wide arms of the sea.

113 And I found that several species did in this state withstand un-injured an immersion in sea-water during seven days: one of these shells was the Helix pomatia, and after it had again hy-bernated I put it in sea-water for twenty days, and it perfectly recovered.

113:d put it into sea-water

113:e species in this state withstood uninjured an immersion in sea-water during seven days: one shell, the Helix pomatia, after having been thus treated and again hybernating was put into sea-water for twenty days, and perfectly

113:f And I find that several species in this state withstand

113.1:d During this length of time it might have been carried by a marine current, running at an average rate, to a distance of 660 geographical miles.

113.1:e time the shell might have been carried by a marine current of average swiftness, to

114 As this species has a thick calcareous operculum, I removed it, and when it had formed a new membranous one, I immersed it for fourteen days in sea-water, and it recovered and crawled away: but more experiments are wanted on this head.

114:d As this Helix has a thick/one, I again immersed/away.

114:e sea-water, and again it

114.1-4:d Baron Aucapitaine has recently tried similar experiments: he placed 100 land-shells, belonging to ten species, in a box pierced with holes, and immersed it for a fortnight in the sea. *2* Out of the hundred shells, twenty-seven recovered. *3* The presence of an operculum seems to have been of importance, as out of twelve specimens of Cyclostoma elegans, which is thus furnished, eleven revived. *4* It is remarkable, seeing how well Helix pomatia with me resisted the salt-water, that not one out of fifty-four specimens belonging to four species of Helix tried by Aucapitaine, recovered. [*Space*]

114.4:e well with me the Helix pomatia resisted/recovered. [*No space*]

114.4:f well the Helix pomatia resisted with me the salt-water/ four other species

114.4.1:e It is, however, not at all probable that land-shells have often been thus transported; the feet of birds is a more prob-able method. [*Space*]

114.4.1:f birds offer a more

114.5:d [*Center*] *On the Relations of the Inhabitants of Islands to those of the nearest Mainland.* [*Space*]

115 The most striking and important fact for us in regard to the inhabitants of islands, is their affinity to those of the nearest mainland, without being actually the same species.

115:d to islands, is the affinity of their inhabitants to

115:e us is the affinity of the species which inhabit islands to/same.

116 Numerous instances could be given of this fact.

116:c this law.

116:e given.

117 I will give only one, that of the Galapagos Archipelago, situated under the equator, between 500 and 600 miles from the shores of South America.

117:e The Galapagos Archipelago, situated under the equator, lies at the distance of between

118 Here almost every product of the land and water bears the unmistakeable stamp of the American continent.

118:f and of the water

119 There are twenty-six land birds, and twenty-five of these are ranked by Mr. Gould as distinct species, supposed to have been created here; yet the close affinity of most of these birds to American species in every character, in their habits, gestures, and tones of voice, was manifest.

119:c land-birds, and twenty-one, or, perhaps, twenty-three, of these are ranked as distinct species, and are supposed

119:f land-birds; of these, twenty-one or perhaps twenty-three are ranked as distinct species, and would commonly be assumed to have been here created; yet the close affinity of most of these birds to American species is manifest in every character, in their habits, gestures, and tones of voice.

120 So it is with the other animals, and with nearly all the plants, as shown by Dr. Hooker in his admirable memoir on the Flora of this archipelago.

120:d and with a large proportion of the plants, as shown by Dr. Hooker in his admirable Flora

121 The naturalist, looking at the inhabitants of these volcanic islands in the Pacific, distant several hundred miles from the continent, yet feels that he is standing on American land.

122 Why should this be so? why should the species which are supposed to have been created in the Galapagos Archipelago, and nowhere else, bear so plain a stamp of affinity to those created in America?

122:f so plainly the stamp

123 There is nothing in the conditions of life, in the geological nature of the islands, in their height or climate, or in the proportions in which the several classes are associated together, which resembles closely the conditions of the South American coast: in fact there is a considerable dissimilarity in all these respects.

123:d fact, there

123:e together, which closely resembles the conditions/all the foregoing respects.

123:f all these respects.

124 On the other hand, there is a considerable degree of resemblance in the volcanic nature of the soil, in climate, height, and size of the islands, between the Galapagos and Cape de Verde Archipelagos: but what an entire and absolute difference in their inhabitants!

124:c Archipelagoes

124:d soil, in the climate

124:g Cape Verde

125 The inhabitants of the Cape de Verde Islands are related to those of Africa, like those of the Galapagos to America.

125:g Cape Verde

126 I believe this grand fact can receive no sort of explanation on the ordinary view of independent creation; whereas on the view here maintained, it is obvious that the Galapagos Islands would be likely to receive colonists, whether by occasional means of transport or by formerly continuous land, from America; and the Cape de Verde Islands from Africa; and that such colonists would be liable to modification;—the principle of inheritance still betraying their original birthplace.

126:e Facts such as these, admit of no/America; the Cape/to modification,—the principle

126:f colonists from America, whether by occasional means of transport or (though I do not believe in this doctrine) by formerly continuous land, and the Cape de Verde Islands from Africa; such

126:g Cape Verde

127 Many analogous facts could be given: indeed it is an almost universal rule that the endemic productions of islands are related to those of the nearest continent, or of other near islands.

127:e or of the nearest island.

127:f or of the nearest large island.

128 The exceptions are few, and most of them can be explained.

129 Thus the plants of Kerguelen Land, though standing nearer to Africa than to America, are related, and that very closely, as we know from Dr. Hooker's account, to those of America: but on the view that this island has been mainly stocked by seeds brought with earth and stones on icebergs, drifted by the prevailing currents, this anomaly disappears.

129:e Thus although Kerguelen Land stands nearer to Africa than to America, the plants are

130 New Zealand in its endemic plants is much more closely related to Australia, the nearest mainland, than to any other region: and this is what might have been expected; but it is

also plainly related to South America, which, although the next nearest continent, is so enormously remote, that the fact becomes an anomaly.

131 But this difficulty almost disappears on the view that both New Zealand, South America, and other southern lands were long ago partially stocked from a nearly intermediate though distant point, namely from the antarctic islands, when they were clothed with vegetation, before the commencement of the Glacial period.

131:c that New

131:e and the other southern lands have been partially/vegetation, during a warmer tertiary period, before the commencement of the last Glacial

131:f been stocked in part from

132 The affinity, which, though feeble, I am assured by Dr. Hooker is real, between the flora of the south-western corner of Australia and of the Cape of Good Hope, is a far more remarkable case, and is at present inexplicable: but this affinity is confined to the plants, and will, I do not doubt, be some day explained.

132:c will, no doubt

132:e case; but

132:f doubt, some day be explained.

133 The law which causes the inhabitants of an archipelago, though specifically distinct, to be closely allied to those of the nearest continent, we sometimes see displayed on a small scale, yet in a most interesting manner, within the limits of the same archipelago.

133:e The same law which has determined the relationship between the inhabitants of islands and the nearest mainland, is sometimes displayed on a small scale, but in

134 Thus the several islands of the Galapagos Archipelago are tenanted, as I have elsewhere shown, in a quite marvellous manner, by very closely related species; so that the inhabitants of each separate island, though mostly distinct, are related in an incomparably closer degree to each other than to the inhabitants of any other part of the world.

134:e Thus each separate island of the Galapagos Archipelago is tenanted, and the fact is a marvellous one, by distinct species; but these species are related in a very much closer manner to each other than to the inhabitants of any other quarter of the world.

134:f by many distinct species: but these species are related to each other in a very much closer manner than to the inhabitants of the American continent, or of any

135 And this is just what might have been expected on my view, for the islands are situated so near each other that they would

almost certainly receive immigrants from the same original source, or from each other.

135:e This is what might have been expected, for islands situated so near each other would almost necessarily receive immigrants from the same original source, and from

135:f near to each

136 But this dissimilarity between the endemic inhabitants of the islands may be used as an argument against my views; for it may be asked, how has it happened in the several islands situated within sight of each other, having the same geological nature, the same height, climate, &c., that many of the immigrants should have been differently modified, though only in a small degree.

136:d asked how

136:e But how is it that many of the immigrants have been differently modified, though only in a small degree, in islands/&c.?

137 This long appeared to me a great difficulty: but it arises in chief part from the deeply-seated error of considering the physical conditions of a country as the most important for its inhabitants; whereas it cannot, I think, be disputed that the nature of the other inhabitants, with which each has to compete, is at least as important, and generally a far more important element of success.

137:c inhabitants with

137:e important; whereas it cannot be

137:f nature of the other species with

138 Now if we look to those inhabitants of the Galapagos Archipelago which are found in other parts of the world (laying on one side for the moment the endemic species, which cannot be here fairly included, as we are considering how they have come to be modified since their arrival), we find a considerable amount of difference in the several islands.

138:e to the species which inhabit the Galapagos Archipelago and are likewise found in other parts of the world, we find that they differ considerably in

139 This difference might indeed have been expected on the view of the islands having been stocked by occasional means of transport—a seed, for instance, of one plant having been brought to one island, and that of another plant to another island.

139:e expected if the islands have been stocked/another island, though all proceeding from the same general source.

140 Hence when in former times an immigrant settled on any one or more of the islands, or when it subsequently spread from one island to another, it would undoubtedly be exposed to different conditions of life in the different islands, for it would have to compete with different sets of organisms: a plant, for

instance, would find the best-fitted ground more perfectly occupied by distinct plants in one island than in another, and it
would be exposed to the attacks of somewhat different enemies.

140:e immigrant first settled on one of the islands, or when it subsequently spread from one to another, it would undoubtedly
be exposed to different conditions in the different islands, for
it would have to compete with a different set of organisms: a
plant, for instance, would find the ground best fitted for it
occupied by somewhat different species in the different islands,
and would be exposed

140:f plant for instance

141 If then it varied, natural selection would probably favour different varieties in the different islands.

142 Some species, however, might spread and yet retain the same
character throughout the group, just as we see on continents
some species spreading widely and remaining the same.

142:e see some species spreading widely throughout a continent and

143 The really surprising fact in this case of the Galapagos
Archipelago, and in a lesser degree in some analogous instances,
is that the new species formed in the separate islands have not
quickly spread to the other islands.

143:e analogous cases, is that each new species after being formed
in any one island, did not

143:f not spread quickly to

144 But the islands, though in sight of each other, are separated
by deep arms of the sea, in most cases wider than the British
Channel, and there is no reason to suppose that they have at
any former period been continuously united.

145 The currents of the sea are rapid and sweep across the archipelago, and gales of wind are extraordinarily rare; so that the
islands are far more effectually separated from each other than
they appear to be on a map.

145:e appear on a map.

145:f sweep between the islands, and gales

146 Nevertheless a good many species, both those found in other
parts of the world and those confined to the archipelago, are
common to the several islands, and we may infer from certain
facts that these have probably spread from some one island to
the others.

146:d Nevertheless some of the species/spread from one island

146:e islands; and we may infer from their present manner of distribution, that they have spread

146:f both of those found in other parts of the world and of those
confined

147 But we often take, I think, an erroneous view of the probability

of closely-allied species invading each other's territory, when put into free intercommunication.

148 Undoubtedly if one species has any advantage whatever over another, it will in a very brief time wholly or in part supplant it; but if both are equally well fitted for their own places in nature, both probably will hold their own places and keep separate for almost any length of time.

148:d Undoubtedly, if

148:e advantage over/places, both probably will hold their places

148:f both will probably hold their separate places for

149 Being familiar with the fact that many species, naturalised through man's agency, have spread with astonishing rapidity over new countries, we are apt to infer that most species would thus spread; but we should remember that the forms which become naturalised in new countries are not generally closely allied to the aboriginal inhabitants, but are very distinct species, belonging in a large proportion of cases, as shown by Alph. de Candolle, to distinct genera.

149:e rapidity over wide areas, we are apt/remember that the species which/distinct forms, belonging

150 In the Galapagos Archipelago, many even of the birds, though so well adapted for flying from island to island, are distinct on each; thus there are three closely-allied species of mocking-thrush, each confined to its own island.

150:f to island, differ on the different islands; thus

151 Now let us suppose the mocking-thrush of Chatham Island to be blown to Charles Island, which has its own mocking-thrush: why should it succeed in establishing itself there?

151:c mocking-thrush; why

152 We may safely infer that Charles Island is well stocked with its own species, for annually more eggs are laid there than can possibly be reared; and we may infer that the mocking-thrush peculiar to Charles Island is at least as well fitted for its home as is the species peculiar to Chatham Island.

152:e laid and young birds hatched, than

153 Sir C. Lyell and Mr. Wollaston have communicated to me a remarkable fact bearing on this subject; namely, that Madeira and the adjoining islet of Porto Santo possess many distinct but representative land-shells, some of which live in crevices of stone; and although large quantities of stone are annually transported from Porto Santo to Madeira, yet this latter island has not become colonised by the Porto Santo species: nevertheless both islands have been colonised by some European land-shells, which no doubt had some advantage over the indigenous species.

153:e representative species of land-shells, some/been colonised by European

154 From these considerations I think we need not greatly marvel at the endemic and representatives species, which inhabit the several islands of the Galapagos Archipelago, not having universally spread from island to island.

154:f endemic species which/having all spread

155 In many other instances, as in the several districts of the same continent, pre-occupation has probably played an important part in checking the commingling of species under the same conditions of life.

155:e Pre-occupation has also probably/commingling of the species which inhabit different districts with nearly the same physical conditions on the same continent.

155:f On the same continent, also, preoccupation has probably/ conditions.

156 Thus, the south-east and south-west corners of Australia have nearly the same physical conditions, and are united by continuous land, yet they are inhabited by a vast number of distinct mammals, birds, and plants.

156:f plants; so it is, according to Mr. Bates, with the butterflies and other animals inhabiting the great, open, and continuous valley of the Amazons.

157 The principle which determines the general character of the fauna and flora of oceanic islands, namely, that the inhabitants, when not identically the same, yet are plainly related to the inhabitants of that region whence colonists could most readily have been derived,—the colonists having been subsequently modified and better fitted to their new homes,—is of the widest application throughout nature.

157:e The same principle which governs the general character of the inhabitants of oceanic islands, namely, their relation to the source whence colonists could have been most easily derived, together with their subsequent modification, is

157:f namely, the relation

158 We see this on every mountain, in every lake and marsh.

158:e every mountain-summit, in

159 For Alpine species, excepting in so far as the same forms, chiefly of plants, have spread widely throughout the world during the recent Glacial epoch, are related to those of the surrounding lowlands;—thus we have in South America, Alpine humming-birds, Alpine rodents, Alpine plants, &c., all of strictly American forms, and it is obvious that a mountain, as it became slowly upheaved, would naturally be colonised from the surrounding lowlands.

159:d would be colonised

159:e in as far as the same species have become widely spread during/&c., all strictly belonging to American forms; and

159:f during the Glacial/lowlands; thus

160 So it is with the inhabitants of lakes and marshes, excepting in so far as great facility of transport has given the same general forms to the whole world.

160:d has allowed the same fresh-water productions to prevail throughout the world.

160:e far that great facility of transport has allowed many of the same species to prevail through large portions of the world.

160:f far as great facility of transport has allowed the same forms to prevail throughout large

161 We see this same principle in the blind animals inhabiting the caves of America and of Europe.

161:c in some of the blind

161:d principle in the character of most of the blind

162 Other analogous facts could be given.

163 And it will, I believe, be universally found to be true, that wherever in two regions, let them be ever so distant, many closely-allied or representative species occur, there will likewise be found some identical species, showing, in accordance with the foregoing view, that at some former period there has been intercommunication or migration between the two regions.

163:d found true

164 And wherever many closely-allied species occur, there will be found many forms which some naturalists rank as distinct species, and some as varieties; these doubtful forms showing us the steps in the process of modification.

164:d some as mere varieties/in the progress of

163 + 4:e It/identical species; and wherever/distinct species, and others as mere varieties

163 + 4:f be found universally true

165 This relation between the power and extent of migration of a species, either at the present time or at some former period under different physical conditions, and the existence at remote points of the world of other species allied to it, is shown in another and more general way.

165:e The relation between the power and extent of migration in certain species, either at the present or at some former period, and the existence at remote points of the world of allied species, is

165:f world of closely-allied species, is

166 Mr. Gould remarked to me long ago, that in those genera of birds which range over the world, many of the species have very wide ranges.

167 I can hardly doubt that this rule is generally true, though it would be difficult to prove it.

167:f though difficult of proof.

168 Amongst mammals, we see it strikingly displayed in Bats, and in a lesser degree in the Felidæ and Canidæ.

169 We see it, if we compare the distribution of butterflies and beetles.

169:e We see the same rule in the distribution

170 So it is with most fresh-water productions, in which so many genera range over the world, and many individual species have enormous ranges.

170:e most of the inhabitants of fresh water, for many of the genera in the most distinct classes range over the world, and many of the species

171 It is not meant that in world-ranging genera all the species have a wide range, or even that they have on an *average* a wide range; but only that some of the species range very widely; for the facility with which widely-ranging species vary and give rise to new forms will largely determine their average range.

171:d that all the species in world-ranging genera, but that some of them, range very widely.

171:e all, but that some of the species in the genera which range very widely, have themselves very wide ranges.

171:f species have very wide ranges in the genera which range very widely.

172 For instance, two varieties of the same species inhabit America and Europe, and the species thus has an immense range; but, if the variation had been a little greater, the two varieties would have been ranked as distinct species, and the common range would have been greatly reduced.

172:d Nor is it meant that the species have on an average a very wide range; for this will largely depend on how far the process of modification has gone; for instance/variation had been carried a little further, the two varieties would have been ranked as distinct species, and the range

172:e species in such genera have on an average/Europe, and thus the species has an immense range; but, if variation were to be carried a little further, the two varieties would be ranked as distinct species, and the range would be greatly

172:f distinct species, and their range

173 Still less is it meant, that a species which apparently has the capacity of crossing barriers and ranging widely, as in the case of certain powerfully-winged birds, will necessarily range widely; for we should never forget that to range widely implies not only the power of crossing barriers, but the more important power of being victorious in distant lands in the struggle for life with foreign associates.

173:e that species which have the capacity

174 But on the view of all the species of a genus having descended from a single parent, though now distributed to the most re-

mote points of the world, we ought to find, and I believe as a general rule we do find, that some at least of the species range very widely; for it is necessary that the unmodified parent should range widely, undergoing modification during its diffusion, and should place itself under diverse conditions favourable for the conversion of its offspring, firstly into new varieties and ultimately into new species.

174:d distributed at the most/should have ranged widely, undergoing modification during its diffusion, and should have placed/first

174:e But according to the view that all the species of the same genus, though now distributed at the most remote points of the world, are descended from a single progenitor, we ought/ very widely.

174:f species of a genus, though distributed to the most

175   In considering the wide distribution of certain genera, we should bear in mind that some are extremely ancient, and must have branched off from a common parent at a remote epoch; so that in such cases there will have been ample time for great climatal and geographical changes and for accidents of transport; and consequently for the migration of some of the species into all quarters of the world, where they may have become slightly modified in relation to their new conditions.

175:c and that the species must

175:d some of them are extremely ancient, and that their species will have branched off from a common progenitor at a remote epoch; so that in these cases/transport; consequently/where they will have

175:e We should bear in mind in relation to all organic beings that many genera are of very ancient origin, and the species in this case will have had ample time for dispersal and subsequent modification.

175:f mind that many genera in all classes are of ancient

176   There is, also, some reason to believe from geological evidence that organisms low in the scale within each great class, generally change at a slower rate than the higher forms; and consequently the lower forms will have had a better chance of ranging widely and of still retaining the same specific character.

176:d forms; consequently

176:e There is also reason to believe from geological evidence, that within each great class the lower organisms change at a slower rate than the higher; consequently they will

177   This fact, together with the seeds and eggs of many low forms being very minute and better fitted for distant transportation, probably accounts for a law which has long been observed, and which has lately been admirably discussed by Alph. de Candolle in regard to plants, namely, that the lower any group of organisms is, the more widely it is apt to range.

177:e of almost all lowly organised forms being very minute and better fitted for distant transportal, probably/lately been discussed/organisms stands, the more widely it ranges.

177:f with that of the seeds and eggs of most lowly

178 The relations just discussed,—namely, low and slowly-changing organisms ranging more widely than the high,—some of the species of widely-ranging genera themselves ranging widely, —such facts, as alpine, lacustrine, and marsh productions being related (with the exceptions before specified) to those on the surrounding low lands and dry lands, though these stations are so different—the very close relation of the distinct species which inhabit the islets of the same archipelago,—and especially the striking relation of the inhabitants of each whole archipelago or island to those of the nearest mainland,—are, I think, utterly inexplicable on the ordinary view of the independent creation of each species, but are explicable on the view of colonisation from the nearest and readiest source, together with the subsequent modification and better adaptation of the colonists to their new homes.

178:b from the nearest or readiest

178:e namely, lower organisms ranging more widely than the higher,—some/being generally related to those which live on the surrounding low lands and dry lands,—the striking relationship between the inhabitants of islands and those of the nearest mainland—the still closer relationship of the distinct inhabitants of the islands in the same archipelago—are inexplicable/ explicable if we admit colonisation/subsequent adaptation

179 *Summary of last and present Chapters.—*

179:c Chapter

179:d [*Center*] Summary/Chapter. [*Space*]

179:e of the last

179:f Chapters.

180 In these chapters I have endeavoured to show, that if we make due allowance for our ignorance of the full effects of all the changes of climate and of the level of the land, which have certainly occurred within the recent period, and of other similar changes which may have occurred within the same period; if we remember how profoundly ignorant we are with respect to the many and curious means of occasional transport,—a subject which has hardly ever been properly experimentised on; if we bear in mind how often a species may have ranged continuously over a wide area, and then have become extinct in the intermediate tracts, I think the difficulties in believing that all the individuals of the same species, wherever located, have descended from the same parents, are not insuperable.

180:c experimented/tracts, the difficulties

180:d [¶]

180:e effects of the changes of climate/period, and of other changes which have probably occurred,—if we remember how ignorant we are with respect to the many curious means of occasional transport,—if we bear in mind/tracts,—the difficulty is not insuperable in believing that all the individuals of the same species, wherever found, are descended from common parents.

180:f effects of changes of climate/mind, and this is a very important consideration, how

181 And we are led to this conclusion, which has been arrived at by many naturalists under the designation of single centres of creation, by some general considerations, more especially from the importance of barriers and from the analogical distribution of sub-genera, genera, and families.

181:e creation, by various general/barriers of all kinds, and from

182 With respect to the distinct species of the same genus, which on my theory must have spread from one parent-source; if we make the same allowances as before for our ignorance, and remember that some forms of life change most slowly, enormous periods of time being thus granted for their migration, I do not think that the difficulties are insuperable; though they often are in this case, and in that of the individuals of the same species, extremely grave.

182:b extremely great.

182:d allowance

182:e With respect to distinct species belonging to the same genus, which on our theory/life have changed very slowly, enormous periods of time having been thus granted for their migration, the difficulties are far from insuperable; though in this case, and in that of the individuals of the same species, they are often great.

182:f theory have spread/allowances/case, as in

183 As exemplifying the effects of climatal changes on distribution, I have attempted to show how important has been the influence of the modern Glacial period, which I am fully convinced simultaneously affected the whole world, or at least great meridional belts.

183:c great longitudinal belts.

183:e important a part the Glacial period has played, which affected even the equatorial regions, and which, during the alternations of the cold in the north and south, allowed the productions of opposite hemispheres to mingle, and left some of them stranded in all parts of the world on the mountain-summits.

183:f part the last Glacial/stranded on the mountain-summits in all parts of the world.

184 As showing how diversified are the means of occasional transport, I have discussed at some little length the means of dispersal of fresh-water productions.

185   If the difficulties be not insuperable in admitting that in the long course of time the individuals of the same species, and likewise of allied species, have proceeded from some one source; then I think all the grand leading facts of geographical distribution are explicable on the theory of migration (generally of the more dominant forms of life), together with subsequent modification and the multiplication of new forms.

185:c then all

185:e time all the individuals of the same species, and likewise of the several species belonging to the same genus, have/migration together

186 We can thus understand the high importance of barriers, whether of land or water, which separate our several zoological and botanical provinces.

186:e water, in not only separating, but in apparently forming the several

187 We can thus understand the localisation of sub-genera, genera, and families; and how it is that under different latitudes, for instance in South America, the inhabitants of the plains and mountains, of the forests, marshes, and deserts, are in so mysterious a manner linked together by affinity, and are likewise linked to the extinct beings which formerly inhabited the same continent.

187:d together, and

187:e understand the concentration of related species within the same areas; and how/are linked together in so mysterious a manner, and

188 Bearing in mind that the mutual relations of organism to organism are of the highest importance, we can see why two areas having nearly the same physical conditions should often be inhabited by very different forms of life; for according to the length of time which has elapsed since new inhabitants entered one region; according to the nature of the communication which allowed certain forms and not others to enter, either in greater or lesser numbers; according or not, as those which entered happened to come in more or less direct competition with each other and with the aborigines; and according as the immigrants were capable of varying more or less rapidly, there would ensue in different regions, independently of their physical conditions, infinitely diversified conditions of life,—there would be an almost endless amount of organic action and reaction,—and we should find, as we do find, some groups of beings greatly, and some only slightly modified,—some developed in great force, some existing in scanty numbers—in the different great geographical provinces of the world.

188:b of organism to organism is of the highest

188:c come into more or less direct/numbers—in the several great

188:d new colonists entered one of the regions, or both; according to the nature

188:e ensue in the two or more regions

188:f find some groups/numbers—and this we do find in the several

189   On these same principles, we can understand, as I have endeavoured to show, why oceanic islands should have few inhabitants, but of these a great number should be endemic or peculiar; and why, in relation to the means of migration, one group of beings, even within the same class, should have all its species endemic, and another group should have all its species common to other quarters of the world.

189:e its species peculiar, and another group should have all its species the same with those in other

189:f but that of these, a large proportion should be endemic or/ beings should have all its species peculiar, and another group, even within the same class, should have all its species the same with those in an adjoining quarter

190   We can see why whole groups of organisms, as batrachians and terrestrial mammals, should be absent from oceanic islands, whilst the most isolated islands possess their own peculiar species of aërial mammals or bats.

190:d isolated islands should possess

191   We can see why there should be some relation between the presence of mammals, in a more or less modified condition, and the depth of the sea between an island and the mainland.

191:d condition in islands, and the depth of the sea between such islands

191:e presence, in islands, of mammals, in a more or less modified condition, and the depth

191:f why, in islands, there should be some relation between the presence of mammals

192   We can clearly see why all the inhabitants of an archipelago, though specifically distinct on the several islets, should be closely related to each other, and likewise be related, but less closely, to those of the nearest continent or other source whence immigrants were probably derived.

192:d immigrants had probably been derived.

192:e immigrants might have been derived.

192:f other; and should likewise be related, but less closely, to those of the nearest continent, or

193   We can see why in two areas, however distant from each other, there should be a correlation, in the presence of identical species, of varieties, of doubtful species, and of distinct but representative species.

193:e other, where very closely allied or representative species exist, there should almost always exist some identical species.

643

193:f why, if there exist very closely allied or representative species in two areas, however distant from each other, some identical species will almost always there be found.

194 As the late Edward Forbes often insisted, there is a striking parallelism in the laws of life throughout time and space: the laws governing the succession of forms in past times being nearly the same with those governing at the present time the differences in different areas.

194:f space; the laws

195 We see this in many facts.

196 The endurance of each species and group of species is continuous in time; for the exceptions to the rule are so few, that they may fairly be attributed to our not having as yet discovered in an intermediate deposit the forms which are therein absent, but which occur above and below: so in space, it certainly is the general rule that the area inhabited by a single species, or by a group of species, is continuous; and the exceptions, which are not rare, may, as I have attempted to show, be accounted for by migration at some former period under different conditions or by occasional means of transport, and by the species having become extinct in the intermediate tracts.

196:d occur both above/accounted for by former migrations under different conditions or through occasional means of transport, or by

196:e deposit certain forms which are absent in it, but/or by a group of species, is continuous, and the exceptions/different circumstances, or through

196:f for the apparent exceptions

197 Both in time and space, species and groups of species have their points of maximum development.

198 Groups of species, belonging either to a certain period of time, or to a certain area, are often characterised by trifling characters in common, as of sculpture or colour.

198:d species, living during the same period of time, or living within the same area, are often characterised by trifling features in

199 In looking to the long succession of ages, as in now looking to distant provinces throughout the world, we find that some organisms differ little, whilst others belonging to a different class, or to a different order, or even only to a different family of the same order, differ greatly.

199:d succession of past ages/that certain organisms differ little from each other, whilst others belonging to a different class, or to a different order, or only to a different family of the same order, differ greatly from each other.

199:e that species in certain classes differ little from each other, whilst others in a different class, or only in a different

199:f as in looking/whilst those in another class, or only in a different section of the same order, differ

200 In both time and space the lower members of each class generally change less than the higher; but there are in both cases marked exceptions to the rule.

200:d space the lowly organised members of each class generally change less than the highly organised; but

201 On my theory these several relations throughout time and space are intelligible; for whether we look to the forms of life which have changed during successive ages within the same quarter of the world, or to those which have changed after having migrated into distant quarters, in both cases the forms within each class have been connected by the same bond of ordinary generation; and the more nearly any two forms are related in blood, the nearer they will generally stand to each other in time and space; in both cases the laws of variation have been the same, and modifications have been accumulated by the same power of natural selection.

201:d class are connected by the same bond of ordinary generation; and in both

201:e According to our theory/accumulated by the same means of natural

201:f theory, these/look to the allied forms of life which have changed during successive ages, or to those/cases they are connected

CHAPTER XIII.

1:f XIV.

2   Mutual Affinities of Organic Beings: Morphology:
Embryology: Rudimentary Organs.

3 Classification, groups subordinate to groups—Natural system—
Rules and difficulties in classification, explained on the theory
of descent with modification—Classification of varieties—Descent always used in classification—Analogical or adaptive characters—Affinities, general, complex and radiating—Extinction
separates and defines groups—Morphology, between members
of the same class, between parts of the same individual—
Embryology, laws of, explained by variations not supervening
at an early age, and being inherited at a corresponding age—
Rudimentary organs; their origin explained—Summary.

3.1:d [Center] *Classification.* [Space]

4 From the first dawn of life, all organic beings are found to resemble each other in descending degrees, so that they can be
classed in groups under groups.

4:e From a very remote period in the history of the world organic
beings have resembled

4:f From the most remote period in the history of the world organic
beings have been found to resemble

5 This classification is evidently not arbitrary like the grouping
of the stars in constellations.

5:e is not

6 The existence of groups would have been of simple signification,
if one group had been exclusively fitted to inhabit the land,
and another the water; one to feed on flesh, another on vegetable matter, and so on; but the case is widely different in
nature; for it is notorious how commonly members of even
the same sub-group have different habits.

6:f simple significance, if/different, for

7 In our second and fourth chapters, on Variation and on Natural
Selection, I have attempted to show that it is the widely ranging, the much diffused and common, that is the dominant
species belonging to the larger genera, which vary most.

7:c that within each country it/genera in each class, which

8 The varieties, or incipient species, thus produced ultimately become converted, as I believe, into new and distinct species; and
these, on the principle of inheritance, tend to produce other
new and dominant species.

8:c produced, ultimately

8:d converted into

9 Consequently the groups which are now large, and which generally include many dominant species, tend to go on increasing indefinitely in size.

9:c increasing in

10 I further attempted to show that from the varying descendants of each species trying to occupy as many and as different places as possible in the economy of nature, there is a constant tendency in their characters to diverge.

10:f nature, they constantly tend to diverge in character.

11 This conclusion was supported by looking at the great diversity of the forms of life which, in any small area, come into the closest competition, and by looking to certain facts in naturalisation.

11:e This latter conclusion is supported by observing the great diversity of forms which/and by certain

12 I attempted also to show that there is a constant tendency in the forms which are increasing in number and diverging in character, to supplant and exterminate the less divergent, the less improved, and preceding forms.

12:d exterminate the preceding, less divergent and less improved forms.

12:f is a steady tendency

13 I request the reader to turn to the diagram illustrating the action, as formerly explained, of these several principles; and he will see that the inevitable result is that the modified descendants proceeding from one progenitor become broken up into groups subordinate to groups.

13:f is, that

14 In the diagram each letter on the uppermost line may represent a genus including several species; and all the genera on this line form together one class, for all have descended from one ancient but unseen parent, and, consequently, have inherited something in common.

14:c this upper line/ancient parent

14:e and the whole of the genera along this upper line form together one class, for all are descended from one ancient parent and

15 But the three genera on the left hand have, on this same principle, much in common, and form a sub-family, distinct from that including the next two genera on the right hand, which diverged from a common parent at the fifth stage of descent.

15:f that containing the next

16 These five genera have also much, though less, in common; and they form a family distinct from that including the three genera

647

still further to the right hand, which diverged at a still earlier period.

16:e farther

16:f much in common, though less than when grouped in sub-families; and they form a family distinct from that containing the three/at an earlier

17 And all these genera, descended from (A), form an order distinct from the genera descended from (I).

18 So that we here have many species descended from a single progenitor grouped into genera; and the genera are included in, or subordinate to, sub-families, families, and orders, all united into one class.

18:d are subordinate to sub-families, families, and orders, all united into one great class.

18:e genera; and the genera in sub-families, families, and orders, all in one great class.

18:f and the genera into sub-families, families, and orders, all under one

19 Thus, the grand fact in natural history of the subordination of group under group, which, from its familiarity, does not always sufficiently strike us, is in my judgment fully explained.

19:b judgment explained.

19:d fact of the natural subordination of all organic beings in group under

19:e groups under groups

19:f The grand fact of the natural subordination of organic/judgment thus explained.

19.1:d No doubt organic beings, like all other objects, can be classed in groups in many ways, either artificially by single characters or more naturally by a number of characters.

19.1:e classed in

19.2:d We know, for instance, that minerals and the elemental substances can be thus arranged; in this case there is of course no relation in their classification to genealogical succession, and no cause can be assigned for their falling into groups.

19.2:e relation to genealogical succession, and no cause can at present be

19.2.x-y:f arranged. *y* In this

19.3:d But with organic beings the case is different, and the view above given explains their natural arrangement in group under group; and no other explanation has ever been attempted.

19.3:f given accords with their

20 Naturalists try to arrange the species, genera, and families in each class, on what is called the Natural System.

20:d Naturalists, as we have seen, try

21 But what is meant by this system?

22 Some authors look at it merely as a scheme for arranging together those living objects which are most alike, and for separating those which are most unlike; or as an artificial means for
enunciating, as briefly as possible, general propositions,—that
is, by one sentence to give the characters common, for instance,
to all mammals, by another those common to all carnivora, by
another those common to the dog-genus, and then by adding a
single sentence, a full description is given of each kind of dog.

22:d then, by

22:f artificial method of enunciating

23 The ingenuity and utility of this system are indisputable.

24 But many naturalists think that something more is meant by
the Natural System; they believe that it reveals the plan of the
Creator; but unless it be specified whether order in time or
space, or what else is meant by the plan of the Creator, it seems
to me that nothing is thus added to our knowledge.

24:c space, or both, or

25 Such expressions as that famous one of Linnæus, and which we
often meet within a more or less concealed form, that the characters do not make the genus, but that the genus gives the
characters, seem to imply that something more is included in
our classification, than mere resemblance.

25:c classification than

25:d one by Linnæus

25:f Expressions such as that famous one by Linnæus, which we
often meet within a more or less concealed form, namely, that
the characters do/imply that some deeper bond is included in
our classifications

26 I believe that something more is included; and that propinquity
of descent,—the only known cause of the similarity of organic
beings,—is the bond, hidden as it is by various degrees of modification, which is partially revealed to us by our classifications.

26:c is included, and that propinquity of descent—the only/beings
—is

26:f I believe that this is the case, and that community of descent
—the one known cause of close similarity in organic beings—is
the bond, which though observed by various degrees of modification, is

27 Let us now consider the rules followed in classification, and
the difficulties which are encountered on the view that classification either gives some unknown plan of creation, or is
simply a scheme for enunciating general propositions and of
placing together the forms most like each other.

27:d encountered, on

28 It might have been thought (and was in ancient times thought)
that those parts of the structure which determined the habits

of life, and the general place of each being in the economy of nature, would be of very high importance in classification.

29 Nothing can be more false.

30 No one regards the external similarity of a mouse to a shrew, of a dugong to a whale, of a whale to a fish, as of any importance.

31 These resemblances, though so intimately connected with the whole of life of the being, are ranked as merely "adaptive or analogical characters;" but to the consideration of these re-resemblances we shall have to recur.

31:e shall recur.

32 It may even be given as a general rule, that the less any part of the organisation is concerned with special habits, the more important it becomes for classification.

33 As an instance: Owen, in speaking of the dugong, says, "The generative organs being those which are most remotely related to the habits and food of an animal, I have always regarded as affording very clear indications of its true affinities.

33:d organs, being

34 We are least likely in the modifications of these organs to mistake a merely adaptive for an essential character."

35 So with plants, how remarkable it is that the organs of vegetation, on which their whole life depends, are of little signification, excepting in the first main divisions; whereas the organs of reproduction, with their product the seed, are of paramount importance!

35:e With plants/their nutrition and life depend, are of little signification; whereas/seed and embryo, are

35:f plants how

35.1:e So again in formerly discussing morphological differences which are not physiologically important, we have seen that they are often of the highest service in classification.

35.1:f discussing certain morphological characters which are not functionally important

35.2:e This depends on their constancy throughout many allied groups; and the constancy depends chiefly on any slight deviations of structure in such parts not having been preserved and accumulated by natural selection, which acts only on useful characters.

35.2:f and their constancy depends on any slight deviations not/ only on serviceable characters.

36 We must not, therefore, in classifying, trust to resemblances in parts of the organisation, however important they may be for the welfare of the being in relation to the outer world.

36:[e]

37 Perhaps from this cause it has partly arisen, that almost all naturalists lay the greatest stress on resemblances in organs of high vital or physiological importance.

37:[e]

38 No doubt this view of the classificatory importance of organs which are important is generally, but by no means always, true.

38:[e]

39 But their importance for classification, I believe, depends on their greater constancy throughout large groups of species; and this constancy depends on such organs having generally been subjected to less change in the adaptation of the species to their conditions of life.

39:[e]

40 That the mere physiological importance of an organ does not determine its classificatory value, is almost shown by the one fact, that in allied groups, in which the same organ, as we have every reason to suppose, has nearly the same physiological value, its classificatory value is widely different.

40:e[¶] almost proved by the fact

41 No naturalist can have worked at any group without being struck with this fact; and it has been most fully acknowledged in the writings of almost every author.

41:b been fully
41:f worked long at

42 It will suffice to quote the highest authority, Robert Brown, who in speaking of certain organs in the Proteaceæ, says their generic importance, "like that of all their parts, not only in this but, as I apprehend, in every natural family, is very unequal, and in some cases seems to be entirely lost."

42:c this, but
42:d who, in speaking

43 Again in another work he says, the genera of the Connaraceæ "differ in having one or more ovaria, in the existence or absence of albumen, in the imbricate or valvular æstivation.

44 Any one of these characters singly is frequently of more than generic importance, though here even when all taken together they appear insufficient to separate Cnestis from Connarus."

45 To give an example amongst insects, in one great division of the Hymenoptera, the antennæ, as Westwood has remarked, are most constant in structure; in another division they differ much, and the differences are of quite subordinate value in classification; yet no one probably will say that the antennæ in these two divisions of the same order are of unequal physiological importance.

45:d insects: in one
45:e no one will

46 Any number of instances could be given of the varying importance for classification of the same important organ within the same group of beings.

47 Again, no one will say that rudimentary or atrophied organs are of high psychological or vital importance; yet, undoubtedly, organs in this condition are often of high value in classification.

47:f often of much value

48 No one will dispute that the rudimentary teeth in the upper jaws of young ruminants, and certain rudimentary bones of the leg, are highly serviceable in exhibiting the close affinity between Ruminants and Pachyderms.

48:f between ruminants and pachyderms.

49 Robert Brown has strongly insisted on the fact that the rudimentary florets are of the highest importance in the classification of the Grasses.

49:c that the position of the rudimentary florets is of the highest

50 Numerous instances could be given of characters derived from parts which must be considered of very trifling physiological importance, but which are universally admitted as highly serviceable in the definition of whole groups.

51 For instance, whether or not there is an open passage from the nostrils to the mouth, the only character, according to Owen, which absolutely distinguishes fishes and reptiles—the inflection of the angle of the jaws in Marsupials—the manner in which the wings of insects are folded—mere colour in certain Algæ— mere pubescence on parts of the flower in grasses—the nature of the dermal covering, as hair or feathers, in the Vertebrata.

51:f angle of the lower jaw

52 If the Ornithorhynchus had been covered with feathers instead of hair, this external and trifling character would, I think, have been considered by naturalists as important an aid in determining the degree of affinity of this strange creature to birds and reptiles, as an approach in structure in any one internal and important organ.

52:c would have/internal organ.

52:e as an important aid/birds.

53 The importance, for classification, of trifling characters, mainly depends on their being correlated with several other characters of more or less importance.

53:f with many other

54 The value indeed of an aggregate of characters is very evident in natural history.

55 Hence, as has often been remarked, a species may depart from its allies in several characters, both of high physiological importance and of almost universal prevalence, and yet leave us in no doubt where it should be ranked.

55:*f* importance, and of

56 Hence, also, it has been found, that a classification founded on any single character, however important that may be, has always failed; for no part of the organisation is universally constant.

56:*e* is invariably constant.

57 The importance of an aggregate of characters, even when none are important, alone explains, I think, that saying of Linnæus, that the characters do not give the genus, but the genus gives the characters; for this saying seems founded on an appreciation of many trifling points of resemblance, too slight to be defined.

57:*e* explains the aphorism by Linnæus, namely, that the characters/this seems

57:*f* aphorism enunciated by Linnæus

58 Certain plants, belonging to the Malpighiaceæ, bear perfect and degraded flowers; in the latter, as A. de Jussieu has remarked, "the greater number of the characters proper to the species, to the genus, to the family, to the class, disappear, and thus laugh at our classification."

59 But when Aspicarpa produced in France, during several years, only degraded flowers, departing so wonderfully in a number of the most important points of structure from the proper type of the order, yet M. Richard sagaciously saw, as Jussieu observes, that this genus should still be retained amongst the Malpighiaceæ.

59:*f* When/only these degraded

60 This case seems to me well to illustrate the spirit with which our classifications are sometimes necessarily founded.

60:*e* spirit of our classifications.

60:*f* case well illustrates

61 Practically when naturalists are at work, they do not trouble themselves about the physiological value of the characters which they use in defining a group, or in allocating any particular species.

61:*e* group or

62 If they find a character nearly uniform, and common to a great number of forms, and not common to others, they use it as one of high value; if common to some lesser number, they use it as of subordinate value.

63 This principle has been broadly confessed by some naturalists to be the true one; and by none more clearly than by that excellent botanist, Aug. St. Hilaire.

64 If certain characters are always found correlated with others, though no apparent bond of connexion can be discovered between them, especial value is set on them.

64:e connection

64:f If several trifling characters are always found in combination, though

65 As in most groups of animals, important organs, such as those for propelling the blood, or for aërating it, or those for propagating the race, are found nearly uniform, they are considered as highly serviceable in classification; but in some groups of animals all these, the most important vital organs, are found to offer characters of quite subordinate value.

65:f aerating/some groups all

65.1:d Thus, as Fritz Müller has lately remarked, in the same group of crustaceans, Cypridina is furnished with a heart, whilst in two closely allied genera, namely Cypris and Cytherea, there is no such organ; one species of Cypridina has well-developed branchiæ, whilst another species is destitute of them.

66 We can see why characters derived from the embryo should be of equal importance with those derived from the adult, for our classifications of course include all ages of each species.

66:d classification of course includes

66:e for a natural classification/ages.

67 But it is by no means obvious, on the ordinary view, why the structure of the embryo should be more important for this purpose than that of the adult, which alone plays its full part in the economy of nature.

68 Yet it has been strongly urged by those great naturalists, Milne Edwards and Agassiz, that embryonic characters are the most important of any in the classification of animals; and this doctrine has very generally been admitted as true.

68:d true, though its importance has sometimes been exaggerated.

68:e important of all; and this doctrine has very generally been admitted as true.

68.1:d Thus Fritz Müller has arranged the great class of crustaceans in accordance with their embryological differences, for the sake of showing that such an arrangement is not a natural one.

68.1:e Nevertheless their importance has sometimes been exaggerated; in order to show this, Fritz Müller arranged by the aid of such characters the great class of crustaceans, and the arrangement did not prove a natural one.

68.1:f exaggerated, owing to the adaptive characters of larvæ not having been excluded; in order to show this, Fritz Müller arranged by the aid of such characters alone the great

68.1.1:e But there can be no doubt that characters derived from the embryo are generally of the highest value, not only with animals but with plants.

68.1.1:f that embryonic, excluding larval characters, are of the highest value for classification, not

69 The same fact holds good with flowering plants, of which the two main divisions have been founded on characters derived from the embryo,—on the number and position of the embryonic leaves or cotyledons, and on the mode of development of the plumule and radicle.

69:d The general fact of the importance of embryological characters holds/on differences in the embryo,—on the number and position of the cotyledons

69:e Thus the two main divisions of flowering plants are founded

69:f Thus the main

70 In our discussion on embryology, we shall see why such characters are so valuable, on the view of classification tacitly including the idea of descent.

70:e We shall immediately see why these characters possess so high a value in classification, namely, from the natural system being genealogical in its arrangement.

71 Our classifications are often plainly influenced by chains of affinities.

72 Nothing can be easier than to define a number of characters common to all birds; but in the case of crustaceans, such definition has hitherto been found impossible.

72:f but with crustaceans, any such

73 There are crustaceans at the opposite ends of the series, which have hardly a character in common; yet the species at both ends, from being plainly allied to others, and these to others, and so onwards, can be recognised as unequivocally belonging to this, and to no other class of the Articulata.

74 Geographical distribution has often been used, though perhaps not quite logically, in classification, more especially in very large groups of closely allied forms.

75 Temminck insists on the utility or even necessity of this practice in certain groups of birds; and it has been followed by several entomologists and botanists.

76 Finally, with respect to the comparative value of the various groups of species, such as orders, sub-orders, families, sub-families, and genera, they seem to be, at least at present, almost arbitrary.

77 Several of the best botanists, such as Mr. Bentham and others, have strongly insisted on their arbitrary value.

78 Instances could be given amongst plants and insects, of a group of forms, first ranked by practised naturalists as only a genus, and then raised to the rank of a sub-family or family; and this has been done, not because further research has detected important structural differences, at first overlooked, but because numerous allied species, with slightly different grades of difference, have been subsequently discovered.

78:f group first/species with

79   All the foregoing rules and aids and difficulties in classification are explained, if I do not greatly deceive myself, on the
view that the natural system is founded on descent with modification; that the characters which naturalists consider as showing true affinity between any two or more species, are those
which have been inherited from a common parent, and, in so
far, all true classification is genealogical; that community of
descent is the hidden bond which naturalists have been unconsciously seeking, and not some unknown plan of creation, or
the enunciation of general propositions, and the mere putting
together and separating objects more or less alike.

79:d modification;—that the characters/parent, and in so far, all
true classification is genealogical;—that

79:e Natural System/parent, all true classification being genealogical

79:f classification may be explained

80   But I must explain my meaning more fully.

81 I believe that the *arrangement* of the groups within each class, in
due subordination and relation to the other groups, must be
strictly genealogical in order to be natural; but that the *amount*
of difference in the several branches or groups, though allied
in the same degree in blood to their common progenitor, may
differ greatly, being due to the different degrees of modification
which they have undergone; and this is expressed by the forms
being ranked under different genera, families, sections, or
orders.

81:e to each other, must

82 The reader will best understand what is meant, if he will take
the trouble of referring to the diagram in the fourth chapter.

82:d trouble to refer

83 We will suppose the letters A to L to represent allied genera,
which lived during the Silurian epoch, and these have descended from a species which existed at an unknown anterior
period.

83:e represent during the Silurian epoch allied genera, descended
from some still earlier forms.

83:f represent allied genera existing during the Silurian epoch, and
descended from some still earlier form.

84 Species of three of these genera (A, F, and I) have transmitted
modified descendants to the present day, represented by the
fifteen genera (a^{14} to z^{14}) on the uppermost horizontal line.

84:e In three of these genera (A, F, and I) the species have

84:f I), a species has transmitted

85 Now all these modified descendants from a single species, are
represented as related in blood or descent to the same degree;

656

they may metaphorically be called cousins to the same millionth degree; yet they differ widely and in different degrees from each other.

85:e species, are related

85:f descent in the same degree; they may

86 The forms descended from A, now broken up into two or three families, constitute a distinct order from those descended from I, also broken up into two families.

87 Nor can the existing species, descended from A, be ranked in the same genus with the parent A; or those from I, with the parent I.

88 But the existing genus F[14] may be supposed to have been but slightly modified; and it will then rank with the parent-genus F; just as some few still living organic beings belong to Silurian genera.

88:e living organisms belong

89 So that the amount or value of the differences between organic beings all related to each other in the same degree in blood, has come to be widely different.

89:e between these organic beings which are all

89:f that the comparative value of the differences between these organic beings, which

90 Nevertheless their genealogical *arrangement* remains strictly true, not only at the present time, but at each successive period of descent.

91 All the modified descendants from A will have inherited something in common from their common parent, as will all the descendants from I; so will it be with each subordinate branch of descendants, at each successive period.

91:e successive stage.

92 If, however, we choose to suppose that any of the descendants of A or of I have been so much modified as to have more or less completely lost traces of their parentage, in this case, their places in a natural classification will have been more or less completely lost,—as sometimes seems to have occurred with existing organisms.

92:e we suppose any descendant of A or of I to have been so much modified as to have lost all traces of its parentage, in this case, its place in the natural system will likewise be lost,—as seems to have occurred with some few existing

92:f A, or of I, to have become so/will be lost, as

93 All the descendants of the genus F, along its whole line of descent, are supposed to have been but little modified, and they yet form a single genus.

93:e they form

94 But this genus, though much isolated, will still occupy its proper

intermediate position; for F originally was intermediate in character between A and I, and the several genera descended from these two genera will have inherited to a certain extent their characters.

94:e position.

95 This natural arrangement is shown, as far as is possible on paper, in the diagram, but in much too simple a manner.

95:e shown in the diagram as far as is possible on paper, but

95:f The representation of the groups, as here given in the diagram on a flat surface, is much too simple.

95.1:f The branches ought to have diverged in all directions.

96 If a branching diagram had not been used, and only the names of the groups had been written in a linear series, it would have been still less possible to have given a natural arrangement; and it is notoriously not possible to represent in a series, on a flat surface, the affinities which we discover in nature amongst the beings of the same group.

96:f If the names of the groups had been simply written down in a linear series, the representation would have been still less natural; and

97 Thus, on the view which I hold, the natural system is genealogical in its arrangement, like a pedigree; but the degrees of modification which the different groups have undergone, have to be expressed by ranking them under different so-called genera, sub-families, families, sections, orders, and classes.

97:e but the amount of modification which the different groups have undergone has

97:f Thus, the natural/pedigree: but

98 It may be worth while to illustrate this view of classification, by taking the case of languages.

99 If we possessed a perfect pedigree of mankind, a genealogical arrangement of the races of man would afford the best classification of the various languages now spoken throughout the world; and if all extinct languages, and all intermediate and slowly changing dialects, had to be included, such an arrangement would, I think, be the only possible one.

99:c such an arrangement would be

99:f dialects, were to

100 Yet it might be that some very ancient language had altered little, and had given rise to few new languages, whilst others (owing to the spreading and subsequent isolation and states of civilisation of the several races, descended from a common race) had altered much, and had given rise to many new languages and dialects.

100:e some ancient languages had altered very little and had given rise to few new languages, whilst others had altered much owing to the spreading, isolation and state of civilisation of the several co-descended races, and had thus given rise to many new dialects and languages.

100:f isolation, and state

101 The various degrees of difference in the languages from the same stock, would have to be expressed by groups subordinate to groups; but the proper or even only possible arrangement would still be genealogical; and this would be strictly natural, as it would connect together all languages, extinct and modern, by the closest affinities, and would give the filiation and origin of each tongue.

101:e difference between the languages of the same/extinct and recent, by

101:f even the only

102 In confirmation of this view, let us glance at the classification of varieties, which are believed or known to have descended from one species.

102:e to be descended from a single species.

102:f are known or believed to

103 These are grouped under species, with sub-varieties under varieties; and with our domestic productions, several other grades of difference are requisite, as we have seen with pigeons.

103:e under the species, with the sub-varieties under the varieties; and in some cases, as with domestic pigeons, several other grades of difference are requisite.

103:f as with the domestic pigeon, with several other grades of difference.

104 The origin of the existence of groups subordinate to groups, is the same with varieties as with species, namely, closeness of descent with various degrees of modification.

104:c to groups is

104:[e]

105 Nearly the same rules are followed in classifying varieties, as with species.

105:e followed as in classifying species.

106 Authors have insisted on the necessity of classing varieties on a natural instead of an artificial system; we are cautioned, for instance, not to class two varieties of the pine-apple together, merely because their fruit, though the most important part, happens to be nearly identical; no one puts the swedish and common turnips together, though the esculent and thickened stems are so similar.

106:c Swedish

106:e of arranging varieties on

106:f turnip

107 Whatever part is found to be most constant, is used in classing varieties: thus the great agriculturist Marshall says the horns are very useful for this purpose with cattle, because they are less variable than the shape or colour of the body, &c.; whereas with sheep the horns are much less serviceable, because less constant.

108 In classing varieties, I apprehend if we had a real pedigree, a genealogical classification would be universally preferred; and it has been attempted by some authors.

108:d apprehend that if

108:e attempted in some cases.

109 For we might feel sure, whether there had been more or less modification, the principle of inheritance would keep the forms together which were allied in the greatest number of points.

109:f modification, that the principle

110 In tumbler pigeons, though some sub-varieties differ from the others in the important character of having a longer beak, yet all are kept together from having the common habit of tumbling; but the short-faced breed has nearly or quite lost this habit; nevertheless, without any reasoning or thinking on the subject, these tumblers are kept in the same group, because allied in blood and alike in some other respects.

110:e some of the sub-varieties differ in the important/any thought on the subject

110:f character of the length of the beak

111 If it could be proved that the Hottentot had descended from the Negro, I think he would be classed under the Negro group, however much he might differ in colour and other important characters from negroes.

111:[e]

112 With species in a state of nature, every naturalist has in fact brought descent into his classification; for he includes in his lowest grade, or that of a species, the two sexes; and how enormously these sometimes differ in the most important characters, is known to every naturalist: scarcely a single fact can be predicated in common of the males and hermaphrodites of certain cirripedes, when adult, and yet no one dreams of separating them.

112:d common of the adult males and hermaphrodites of certain cirripedes, and

112:e grade, that of the species

112:f that of species

112.1:d = 116:[d] As soon as the three Orchidean forms, Monachanthus, Myanthus, and Catasetum, which had previously been ranked as three distinct genera, were known to be sometimes produced on the same plant, they were immediately considered as varieties; but now I have been able to show that they really constitute the male, female, and hermaphrodite forms of the same species.

112.1:e varieties; and now I have been able to show that they are the male

113 The naturalist includes as one species the several larval stages of the same individual, however much they may differ from each other and from the adult; as he likewise includes the so-called alternate generations of Steenstrup, which can only in a technical sense be considered as the same individual.

113:e species the various larval/adult, as well as the so-called

114 He includes monsters; he includes varieties, not solely because they closely resemble the parent-form, but because they are descended from it.

114:d monsters and varieties, not because they may closely

114:e not from their partial resemblance to the parent-form

115 He who believes that the cowslip is descended from the primrose, or conversely, ranks them together as a single species, and gives a single definition.

115:[d]

116 As soon as three Orchidean forms (Monochanthus, Myanthus, and Catasetum), which had previously been ranked as three distinct genera, were known to be sometimes produced on the same spike, they were immediately included as a single species.

116:[d] = *112.1:d*

117 But it may be asked, what ought we to do, if it could be proved that one species of kangaroo had been produced, by a long course of modification, from a bear?

117:[b]

118 Ought we to rank this one species with bears, and what should we do with the other species?

118:[b]

119 The supposition is of course preposterous; and I might answer by the *argumentum ad hominem,* and ask what should be done if a perfect kangaroo were seen to come out of the womb of a bear?

119:[b]

120 According to all analogy, it would be ranked with bears; but then assuredly all the other species of the kangaroo family would have to be classed under the bear genus.

120:[b]

121 The whole case is preposterous; for where there has been close descent in common, there will certainly be close resemblance or affinity.

121:[b]

122 As descent has universally been used in classing together the

individuals of the same species, though the males and females and larvæ are sometimes extremely diffierent; and as it has been used in classing varieties which have undergone a certain, and sometimes a considerable amount of modification, may not this same element of descent have been unconsciously used in grouping species under genera, and genera under higher groups, though in these cases the modification has been greater in degree, and has taken a longer time to complete?

122:d cases the modification has been much greater

122:e higher groups, all under the so-called natural system?

123 I believe it has thus been unconsciously used; and only thus can I understand the several rules and guides which have been followed by our best systematists.

123:e has been unconsciously used; and thus only can

124 We have no written pedigrees; we have to make out community of descent by resemblances of any kind.

124:f As we have no written pedigrees, we are forced to trace community

125 Therefore we choose those characters which, as far as we can judge, are the least likely to have been modified in relation to the conditions of life to which each species has been recently exposed.

125:f which are the least likely to have been modified, in

126 Rudimentary structures on this view are as good as, or even sometimes better than, other parts of the organisation.

127 We care not how trifling a character may be—let it be the mere inflection of the angle of the jaw, the manner in which an insect's wing is folded, whether the skin be covered by hair or feathers—if it prevail throughout many and different species, especially those having very different habits of life, it assumes high value; for we can account for its presence in so many forms with such different habits, only by its inheritance from a common parent.

127:e only by inheritance

128 We may err in this respect in regard to single points of structure, but when several characters, let them be ever so trifling, occur together throughout a large group of beings having different habits, we may feel almost sure, on the theory of descent, that these characters have been inherited from a common ancestor.

128:d trifling, concur throughout

129 And we know that such correlated or aggregated characters have especial value in classification.

128 + 9:f ancestor; and we know that such aggregated

130 We can understand why a species or a group of species may depart, in several of its most important characteristics, from its allies, and yet be safely classed with them.

130:f depart from its allies, in several of its most important characteristics, and

131 This may be safely done, and is often done, as long as a sufficient number of characters, let them be ever so unimportant, betrays the hidden bond of community of descent.

132 Let two forms have not a single character in common, yet if these extreme forms are connected together by a chain of intermediate groups, we may at once infer their community of descent, and we put them all into the same class.

132:d groups we may
132:f yet, if/groups, we may

133 As we find organs of high physiological importance—those which serve to preserve life under the most diverse conditions of existence—are generally the most constant, we attach especial value to them; but if these same organs, in another group or section of a group, are found to differ much, we at once value them less in our classification.

134 We shall hereafter, I think, clearly see why embryological characters are of such high classificatory importance.

134:e shall presently see

135 Geographical distribution may sometimes be brought usefully into play in classing large and widely-distributed genera, because all the species of the same genus, inhabiting any distinct and isolated region, have in all probability descended from the same parents.

135:e large genera/region, are in
135.1:d[¶] *Analogical Resemblances.—*

136 We can understand, on these views, the very important distinction between real affinities and analogical or adaptive resemblances.

136:d[No ¶] on the above views

137 Lamarck first called attention to this distinction, and he has been ably followed by Macleay and others.

137:f this subject, and he

138 The resemblance, in the shape of the body and in the fin-like anterior limbs, between the dugong, which is a pachydermatous animal, and the whale, and between both these mammals and fishes, is analogical.

138:d resemblance in the shape
138:f limbs between dugongs and whales, and between these two orders of mammals and fishes, are analogical.
138.1-2:f So is the resemblance between a mouse and a shrew-mouse (Sorex), which belong to different orders; and the still closer resemblance, insisted on by Mr. Mivart, between the mouse and a small marsupial animal (Antechinus) of Australia. 2 These latter resemblances may be accounted for, as it seems to me, by

adaptation for similarly active movements through thickets and herbage, together with concealment from enemies.

139 Amongst insects there are innumerable instances: thus Linnæus, misled by external appearances, actually classed an homopterous insect as a moth.

139:f[¶] innumerable similar instances; thus

140 We see something of the same kind even in our domestic varieties, as in the thickened stems of the common and swedish turnip.

140:c Swedish

140:f even with our domestic varieties, as in the strikingly similar shape of the body in the improved breeds of the Chinese and common pig, which are descended from distinct species; and in the similarly thickened stems of the common and specifically distinct Swedish

141 The resemblance of the greyhound and racehorse is hardly more fanciful than the analogies which have been drawn by some authors between very distinct animals.

141:e between widely distinct

141:f The resemblance between the greyhound and the racehorse/ widely different animals.

142 On my view of characters being of real importance for classification, only in so far as they reveal descent, we can clearly understand why analogical or adaptive character, although of the utmost importance to the welfare of the being, are almost valueless to the systematist.

142:f[¶] On the view

143 For animals, belonging to two most distinct lines of descent, may readily become adapted to similar conditions, and thus assume a close external resemblance; but such resemblances will not reveal—will rather tend to conceal their blood-relationship to their proper lines of descent.

143:d readily have become adapted to similar conditions, and thus have assumed

143:e blood-relationship.

143:f may have become

144 We can also understand the apparent paradox, that the very same characters are analogical when one class or order is compared with another, but give true affinities when the members of the same class or order are compared one with another: thus the shape of the body and fin-like limbs are only analogical when whales are compared with fishes, being adaptations in both classes for swimming through the water; but the shape of the body and fin-like limbs serve as characters exhibiting true affinity between the several members of the whale family; for these cetaceans agree in so many characters, great and small, that we cannot doubt that they have inherited their general shape of body and structure of limbs from a common ancestor.

144:c thus, the shape of the body and fin-like limbs are only

144:d also thus understand/class or one order is compared with another, but give true affinities when the members of the same class or order are compared together: thus

144:f one group is compared with another/same group are compared together/water; but between the several members of the whale family, the shape of the body and the fin-like limbs offer characters exhibiting true affinity; for as these parts are so nearly similar throughout the whole family, we cannot doubt that they have been inherited from a common

145 So it is with fishes.

145.0.1-8:f[¶] Numerous cases could be given of striking resemblances in quite distinct beings between single parts or organs, which have been adapted for the same functions. *2* A good instance is afforded by the close resemblance of the jaws of the dog and Tasmanian wolf or Thylacinus,—animals which are widely sundered in the natural system. *3* But this resemblance is confined to general appearance, as in the prominence of the canines, and in the cutting shape of the molar teeth. *4* For the teeth really differ much: thus the dog has on each side of the upper jaw four pre-molars and only two molars; whilst the Thylacinus has three pre-molars and four molars. *5* The molars also differ much in the two animals in relative size and structure. *6* The adult dentition is preceded by a widely different milk dentition. *7* Any one may of course deny that the teeth in either case have been adapted for tearing flesh, through the natural selection of successive variations; but if this be admitted in the one case, it is unintelligible to me that it should be denied in the other. *8* I am glad to find that so high an authority as Professor Flower has come to this same conclusion.

145.0.9-13:f[¶] The extraordinary cases given in a former chapter, of widely different fishes possessing electric organs,—of widely different insects possessing luminous organs,—and of orchids and asclepiads having pollen-masses with viscid discs, come under this same head of analogical resemblances. *10* But these cases are so wonderful that they were introduced as difficulties or objections to our theory. *11* In all such cases some fundamental difference in the growth or development of the parts, and generally in their matured structure, can be detected. *12* The end gained is the same, but the means, though appearing superficially to be the same, are essentially different. *13* The principle formerly alluded to under the term of *analogical variation* has probably in these cases often come into play; that is, the members of the same class, although only distantly allied, have inherited so much in common in their constitution, that they are apt to vary under similar exciting causes in a similar manner; and this would obviously aid in the acquirement through natural selection of parts or organs, strikingly like each other, independently of their direct inheritance from a common progenitor.

145.0.14-5:f[¶] As species belonging to distinct classes have often been adapted by successive slight modifications to live under nearly similar circumstances,—to inhabit, for instance, the three elements of land, air, and water,—we can perhaps understand how it is that a numerical parallelism has sometimes been observed between the sub-groups of distinct classes. *15* A naturalist, struck with a parallelism of this nature, by arbitrarily raising or sinking the value of the groups in several classes (and all our experience shows that their valuation is as yet arbitrary), could easily extend the parallelism over a wide range; and thus the septenary, quinary, quaternary and ternary classifications have probably arisen.

145.0.16:f[¶] There is another curious class of cases in which close external resemblance does not depend on adaptation to similar habits of life, but has been gained for the sake of protection.

145.1:d[¶] The most remarkable case of analogical resemblance ever recorded, though not dependent on adaptation to similar conditions of life, is that given by Mr. Bates with respect to certain butterflies in the Amazonian region closely mimicking other kinds.

145.1:f[*No* ¶] I allude to the wonderful manner in which certain butterflies imitate, as first described by Mr. Bates, other and quite distinct species.

145.2:d This excellent observer shows that in a district where, for instance, an Ithomia abounds in gaudy swarms, another butterfly, namely, a Leptalis, will often be found mingled in the same flock, so like the Ithomia in every shade and stripe of colour and even in the shape of its wings, that Mr. Bates, with his eyes sharpened by collecting during eleven years, was, though always on his guard, continually deceived.

145.2:e Leptalis, is often found mingled in the same flock, and so closely resembles the Ithomia

145.2:f observer has shown that in some districts of S. America, where/flock; and the latter so closely

145.3:d When the mockers and the mocked are caught and compared they are found to be totally different in essential structure, and to belong not only to distinct genera, but often to distinct families.

145.3:f compared, they are found to be very different

145.4:d If this mimicry had occurred in only one or two instances, it might have been passed over as a strange coincidence.

145.4:e Had this mimicry occurred

145.5:d But travel a hundred miles, more or less, from a district where one Leptalis imitates one Ithomia, and a distinct mocker and mocked, equally close in their resemblance, will be found.

145.5:e But, if we proceed from a district where one Leptalis imitates an Ithomia, another mocking and mocked species belonging to the same genera, equally

145.5:f same two genera, equally close in their resemblance, may be

145.6-7:d Altogether no less than ten genera are enumerated, which include species that imitate other butterflies. *7* The mockers and mocked always inhabit the same region; we never find an imitator living remote from the form which it counterfeits.

145.7:e it imitates.

145.8-9:d The mockers are almost invariably rare insects; the mocked in almost every case abound in swarms. *9* In the same district in which a species of Leptalis closely imitates an Ithomia, there are sometimes other Lepidoptera mimicking the same Ithomia; so that in the same place, species of three genera of butterflies and even moths may be found all closely resembling a species of a fourth genus.

145.9:e even a moth are found all closely resembling a butterfly belonging to a fourth

145.9:f same Ithomia: so

145.10-12:d It deserves especial notice that many of the mimicking forms of the Leptalis, as well as of the mimicked forms, can be shown by a graduated series to be merely varieties of the same species; whilst others are undoubtedly distinct species. *11* But why, it may be asked, are certain forms treated as the mimicked and others as the mimickers? *12* Mr. Bates satisfactorily answers this question, by showing that the form which is imitated keeps the usual dress of the group to which it belongs, whilst the counterfeiters have changed their dress and do not resemble their nearest allies.

145.13:d[¶] We are next led to inquire what reason can possibly be assigned for certain butterflies and moths so often assuming the dress of other and quite distinct forms; why, to the perplexity of naturalists, has nature condescended to the tricks of the stage?

145.13:e of another and quite distinct form

145.13:f can be

145.14:d Mr. Bates has, we cannot doubt, hit on the true explanation.

145.14:e has, no doubt

145.15:d The mocked forms, which always abound in numbers, must habitually escape, to a large extent, destruction, otherwise they could not exist in such swarms; and Mr. Bates never saw them preyed on by birds and certain large insects which attack other butterflies; he suspects that this immunity is owing to a peculiar and offensive odour that they emit.

145.15.x-y:e escape destruction to a large extent, otherwise/butterflies. *y* He has good reason to believe that

145.15.x:f swarms; and a large amount of evidence has now been collected, showing that they are distasteful to birds and other insect-devouring animals.

145.16:d The mocking forms, on the other hand, which inhabit

667

the same district, are comparatively rare, and belong to rare groups; hence they must suffer habitually from some danger, for otherwise, from the number of eggs laid by all butterflies, they would, if not persecuted, in three or four generations swarm over the whole country.

145.16:e hand, that inhabit/would in

145.17:d Now if a member of one of these persecuted and rare groups were to assume a dress so like that of a well-protected species that it continually deceived the practised eyes of an entomologist, it would often deceive predacious birds and insects, and thus escape entire annihilation.

145.17:e escape much destruction.

145.17:f predaceous/thus often escape destruction.

145.18:d It may almost be said that Mr. Bates has witnessed the process by which the mimickers have come so closely to resemble the mimicked; for he shows that some of the forms of Leptalis, whether these be ranked as species or varieties, which mimic so many other butterflies, vary much.

145.18:e Mr. Bates may almost be said to have actually witnessed/ he found that some of the forms of Leptalis which mimic so many other butterflies, varied in an extreme degree.

145.19:d In one district several varieties occur, and of these one alone resembles, to a certain extent, the common Ithomia of the same district.

145.19:e occurred, and of these one alone resembled to

145.20:d In another district there are two or three varieties, one of which is much commoner than the others, and this closely mocks an Ithomia.

145.20:e there were two or three varieties, one of which was much commoner than the others, and this closely mocked another form of Ithomia.

145.21:d From many facts of this nature, Mr. Bates concludes that in every case the Leptalis originally varied; and that, when a variety arose which happened to resemble in some degree any common butterfly inhabiting the same district, this variety, from its resemblance to a flourishing and little-persecuted kind, had a better chance of escaping destruction from predacious birds and insects, and was consequently oftener preserved;— "the less perfect degrees of resemblance being generation after generation eliminated, and only the others left to propagate their kind."

145.21:e From facts of this nature, Mr. Bates concludes that the Leptalis first varies; and when a variety happens to resemble/ kind, has a better/insects, and is consequently

145.21:f predaceous

145.22:d So that here we have an excellent illustration of the principle of natural selection.

145.22:e illustration of natural

145.23:d[¶] Mr. Wallace has recently described several equally strik-ing cases of mimicry in the Lepidoptera of the Malay Archi-pelago, and other cases could be given with other orders of insects.

145.23:e other instances could

145.23:f Messrs. Wallace and Trimen have likewise described sev-eral equally striking cases of imitation in the Lepidoptera of the Malay Archipelago and Africa, and with some other insects.

145.24:d Mr. Wallace has also given one instance of mimicry amongst birds, but we have no such cases with the larger animals.

145.24:e has also described one case of/larger quadrupeds.

145.24:f also detected one such case with birds, but we have none with

145.25:d The much greater frequency of mockery with insects than with other animals, is probably the consequence of their small size; insects cannot defend themselves, excepting indeed the kinds that sting, and I have never heard of an instance of these mocking other insects, though they are mocked: insects cannot escape by flight from the larger animals; hence they are re-duced, like most weak creatures, to trickery and dissimulation.

145.25:e of mimicry with insects than

145.25:f of imitation with insects than/kinds furnished with a sting, and I have never heard of an instance of such kinds mocking other insects, though they are mocked; insects cannot easily escape by flight from the larger animals which prey on them; hence

145.25:g them; therefore, speaking metaphorically, they

145.25.1-3:f[¶] It should be observed that the process of imitation probably never commenced between forms widely dissimilar in colour. 2 But starting with species already somewhat like each other, the closest resemblance, if beneficial, could readily be gained by the above means; and if the imitated form was sub-sequently and gradually modified through any agency, the imi-tating form would be led along the same track, and thus be altered to almost any extent, so that it might ultimately assume an appearance or colouring wholly unlike that of the other members of the family to which it belonged. 3 There is, how-ever, some difficulty on this head, for it is necessary to suppose in some cases that ancient members belonging to several dis-tinct groups, before they had diverged to their present extent, accidentally resembled a member of another and protected group in a sufficient degree to afford some slight protection; this having given the basis for the subsequent acquisition of the most perfect resemblance.

146 As members of distinct classes have often been adapted by successive slight modifications to live under nearly similar circumstances,—to inhabit for instance the three elements of

land, air, and water,—we can perhaps understand how it is that a numerical parallelism has sometimes been observed between the sub-groups in distinct classes.

146:d But to return to more ordinary cases of analogical resemblance: as/inhabit, for instance, the three

146:[f]

147 A naturalist, struck by a parallelism of this nature in any one class, by arbitrarily raising or sinking the value of the groups in other classes (and all our experience shows that this valuation has hitherto been arbitrary), could easily extend the parallelism over a wide range; and thus the septenary, quinary, quaternary, and ternary classifications have probably arisen.

147:e that their valuation is as yet arbitrary

147:[f]

147.1:d[¶] On the Nature of the Affinities connecting Organic Beings.—

148 As the modified descendants of dominant species, belonging to the larger genera, tend to inherit the advantages, which made the groups to which they belong large and their parents dominant, they are almost sure to spread widely, and to seize on more and more places in the economy of nature.

148:c advantages which made

148:d[No ¶]

149 The larger and more dominant groups thus tend to go on increasing in size; and they consequently supplant many smaller and feebler groups.

149:c groups within each class thus

150 Thus we can account for the fact that all organisms, recent and extinct, are included under a few great orders, under still fewer classes, and all in one great natural system.

150:e orders, and under still fewer classes.

151 As showing how few the higher groups are in number, and how widely spread they are throughout the world, the fact is striking, that the discovery of Australia has not added a single insect belonging to a new order; and that in the vegetable kingdom, as I learn from Dr. Hooker, it has added only two or three orders of small size.

151:b new class; and

151:c three families of

151:d added an insect

151:e widely they are spread throughout

151:f striking that the discovery

152 In the chapter on geological succession I attempted to show, on the principle of each group having generally diverged much in character during the long-continued process of modification, how it is that the more ancient forms of life often present

characters in some slight degree intermediate between existing groups.

152:c Geological Succession

152:e some degree

153 A few old and intermediate parent-forms having occasionally transmitted to the present day descendants but little modified, will give to us our so-called osculant or aberrant groups.

153:e Some few of these old and intermediate forms having transmitted to the present day descendants but little modified, constitute our so-called osculant or aberrant species.

153:f As some few of the old and intermediate forms have transmitted to the present day descendants but little modified, these constitute

154 The more aberrant any form is, the greater must be the number of connecting forms which on my theory have been exterminated and utterly lost.

154:e which have

155 And we have some evidence of aberrant forms having suffered severely from extinction, for they are generally represented by extremely few species; and such species as do occur are generally very distinct from each other, which again implies extinction.

155:e aberrant groups having suffered severely from extinction, for they are almost always represented

156 The genera Ornithorhynchus and Lepidosiren, for example, would not have been less aberrant had each been represented by a dozen species instead of by a single one; but such richness in species, as I find after some investigation, does not commonly fall to the lot of aberrant genera.

156:e one, or by one or two.

156:f species, instead of as at present by a single one, or by two or three.

157 We can, I think, account for this fact only by looking at aberrant forms as failing groups conquered by more successful competitors, with a few members preserved by some unusual coincidence of favourable circumstances.

157:e aberrant groups as forms which have been conquered by more successful competitors, with a few members still preserved under unusually favourable conditions.

158 Mr. Waterhouse has remarked that, when a member belonging to one group of animals exhibits an affinity to a quite distinct group, this affinity in most cases is general and not special: thus, according to Mr. Waterhouse, of all Rodents, the bizcacha is most nearly related to Marsupials; but in the points in which it approaches this order, its relations are general, and not to any one marsupial species more than to another.

158:f general, that is, not

159 As the points of affinity of the bizcacha to Marsupials are believed to be real and not merely adaptive, they are due on my theory to inheritance in common.

159:e affinity are believed to be real and not merely adaptive, they must be due in accordance with our view to inheritance from a common progenitor.

159:f As these points

160 Therefore we must suppose either that all Rodents, including the bizcacha, branched off from some very ancient Marsupial, which will have had a character in some degree intermediate with respect to all existing Marsupials; or that both Rodents and Marsupials branched off from a common progenitor, and that both groups have since undergone much modification in divergent directions.

160:d some ancient Marsupials, which will naturally have been more or less intermediate in character with

161 On either view we may suppose that the bizcacha has retained, by inheritance, more of the character of its ancient progenitor than have other Rodents; and therefore it will not be specially related to any one existing Marsupial, but indirectly to all or nearly all Marsupials, from having partially retained the character of their common progenitor, or of an early member of the group.

161:e we must suppose/common progenitor, or of some early

161:f of the characters of its

162 On the other hand, of all Marsupials, as Mr. Waterhouse has remarked, the phascolomys resembles most nearly, not any one species, but the general order of Rodents.

162:e Phascolomys

163 In this case, however, it may be strongly suspected that the resemblance is only analogical, owing to the phascolomys having become adapted to habits like those of a Rodent.

163:e Phascolomys

164 The elder De Candolle has made nearly similar observations on the general nature of the affinities of distinct orders of plants.

164:c distinct families of

165 On the principle of the multiplication and gradual divergence in character of the species descended from a common parent, together with their retention by inheritance of some characters in common, we can understand the excessively complex and radiating affinities by which all the members of the same family or higher group are connected together.

165:e common progenitor, together

166 For the common parent of a whole family of species, now

broken up by extinction into distinct groups and sub-groups, will have transmitted some of its characters, modified in various ways and degrees, to all; and the several species will consequently be related to each other by circuitous lines of affinity of various lengths (as may be seen in the diagram so often referred to), mounting up through many predecessors.

166:d family, now/all the species; and they will

166:e common progenitor of a whole

167 As it is difficult to show the blood-relationship between the numerous kindred of any ancient and noble family, even by the aid of a genealogical tree, and almost impossible to do this without this aid, we can understand the extraordinary difficulty which naturalists have experienced in describing, without the aid of a diagram, the various affinities which they perceive between the many living and extinct members of the same great natural class.

167:e family even

168   Extinction, as we have seen in the fourth chapter, has played an important part in defining and widening the intervals between the several groups in each class.

169 We may thus account even for the distinctness of whole classes from each other—for instance, of birds from all other vertebrate animals—by the belief that many ancient forms of life have been utterly lost, through which the early progenitors of birds were formerly connected with the early progenitors of the other veterbrate classes.

169:d with the early progenitors of the other and then less differentiated vertebrate

169:e with the early progenitors of the other and at that time less

170 There has been less entire extinction of the forms of life which once connected fishes with batrachians.

170:e less complete extinction

170:f There has been much less extinction

171 There has been still less in some other classes, as in that of the Crustacea, for here the most wonderfully diverse forms are still tied together by a long, but broken, chain of affinities.

171:c by long, but broken chains

171:d are still linked together by a long and only partially broken chain

171:f less within some whole classes, for instance the Crustacea

172 Extinction has only separated groups: it has by no means made them; for if every form which has ever lived on this earth were suddenly to reappear, though it would be quite impossible to give definitions by which each group could be distinguished from other groups, as all would blend together by steps as fine as those between the finest existing varieties, nevertheless a

natural classification, or at least a natural arrangement, would be possible.

172:c between existing

172:d all would be blended

172:e separated the groups: it has by no/distinguished, still a natural classification

172:f only defined the groups: it has by no

173 We shall see this by turning to the diagram: the letters, A to L, may represent eleven Silurian genera, some of which have produced large groups of modified descendants.

174 Every intermediate link between these eleven genera and their primordial parent, and every intermediate link in each branch and sub-branch of their descendants, may be supposed to be still alive; and the links to be as fine as those between the finest varieties.

173 + 4:e descendants, with every link in each branch and sub-branch still alive; and the links not greater than those between the finest varieties.

173 + 4:f diagram; the letters/between existing varieties.

175 In this case it would be quite impossible to give any definition by which the several members of the several groups could be distinguished from their more immediate parents; or these parents from their ancient and unknown progenitor.

175:e give definitions/parents and descendants.

176 Yet the natural arrangement in the diagram would still hold good; and, on the principle of inheritance, all the forms descended from A, or from I, would have something in common.

176:e Yet the arrangement in the diagram would still hold good and would be natural; for, on the principle of inheritance, all the forms descended, for instance, from A, would

177 In a tree we can specify this or that branch, though at the actual fork the two unite and blend together.

177:e can distinguish this

178 We could not, as I have said, define the several groups; but we could pick out types, or forms, representing most of the characters of each group, whether large or small, and thus give a general idea of the value of the differences between them.

179 This is what we should be driven to, if we were ever to succeed in collecting all the forms in any class which have lived throughout all time and space.

179:e any one class

180 We shall certainly never succeed in making so perfect a collection: nevertheless, in certain classes, we are tending in this direction; and Milne Edwards has lately insisted, in an able paper, on the high importance of looking to types, whether or

not we can separate and define the groups to which such types belong.

180:c shall assuredly never

180:e Assuredly we shall never/tending towards this end; and Milne

181   Finally, we have seen that natural selection, which results from the struggle for existence, and which almost inevitably induces extinction and divergence of character in the many descendants from one dominant parent-species, explains that great and universal feature in the affinities of all organic beings, namely, their subordination in group under group.

181:e inevitably leads to extinction and divergence of character in the descendants

181:f which follows from the struggle/descendants from any one parent-species

182 We use the element of descent in classing the individuals of both sexes and of all ages, although having few characters in common, under one species; we use descent in classing acknowledged varieties, however different they may be from their parent; and I believe this element of descent is the hidden bond of connexion which naturalists have sought under the term of the Natural System.

182:d having but few

182:e ages under one species, although they may have but few characters in common; we

183 On this idea of the natural system being, in so far as it has been perfected, genealogical in its arrangement, with the grades of difference between the descendants from a common parent, expressed by the terms genera, families, orders, &c., we can understand the rules which we are compelled to follow in our classification.

183:e difference expressed

184 We can understand why we value certain resemblances far more than others; why we are permitted to use rudimentary and useless organs, or others of trifling physiological importance; why, in comparing one group with a distinct group, we summarily reject analogical or adaptive characters, and yet use these same characters within the limits of the same group.

184:e others; why we use rudimentary/importance; why, in finding the relations between one group and another, we/use the same

185 We can clearly see how it is that all living and extinct forms can be grouped together in one great system; and how the several members of each class are connected together by the most complex and radiating lines of affinities.

185:e together within a few great classes; and how

186 We shall never, probably, disentangle the inextricable web of affinities between the members of any one class; but when we have a distinct object in view, and do not look to some un-

known plan of creation, we may hope to make sure but slow progress.

186.1:e[¶] Professor Häckel in his 'Generelle Morphologie' and in several other works, has recently brought his great knowledge and abilities to bear on what he calls phylogeny, or the lines of descent of all organic beings.

186.1:f and in other

186.2:e In drawing up the several series he trusts chiefly to embryological characters, but draws aid from homologous and rudimentary organs, as well as from the successive periods at which the various forms of life first appeared in our geological formations.

186.2:f but receives aid/life are believed to have first

186.3:e He has thus boldly made a great beginning, and shows us how classification will in the future be treated.

187   *Morphology.—*

187:d [*Center*] *Morphology.* [*Space*]

188 We have seen that the members of the same class, independently of their habits of life, resemble each other in the general plan of their organisation.

188:d[¶]

189 This resemblance is often expressed by the term "unity of type;" or by saying that the several parts and organs in the different species of the class are homologous.

190 The whole subject is included under the general name of Morphology.

190:e general term of Morphology.

191 This is the most interesting department of natural history, and may be said to be its very soul.

191:f is one of the most interesting departments of natural history, and may almost be said

192 What can be more curious than that the hand of a man, formed for grasping, that of a mole for digging, the leg of the horse, the paddle of the porpoise, and the wing of the bat, should all be constructed on the same pattern, and should include the same bones, in the same relative positions?

192:b include similar bones

192.1-4:f How curious it is, to give a subordinate though striking instance, that the hind-feet of the kangaroo, which are so well fitted for bounding over the open plains,—those of the climbing, leaf-eating koala, equally well fitted for grasping the branches of trees,—those of the ground-dwelling, insect or root eating, bandicoots,—and those of some other Australian marsupials,—should all be constructed on the same extraordinary type, namely with the bones of the second and third digits extremely slender and enveloped within the same skin, so that they appear like a single toe furnished with two claws. 2 Notwithstanding this similarity of pattern, it is obvious that the hind feet of these several animals are used for as widely different purposes as it is possible to conceive. 3 The case is rendered all the more striking by the American opossums, which follow nearly the same habits of life as some of their Australian relatives, having feet constructed on the ordinary plan. 4 Professor Flower, from whom these statements are taken, remarks in conclusion: "We may call this conformity to type, without getting much nearer to an explanation of the phenomenon;" and he then adds "but is it not powerfully suggestive of true relationship, of inheritance from a common ancestor?"

193 Geoffroy St. Hilaire has insisted strongly on the high importance of relative connexion in homologous organs: the parts may change to almost any extent in form and size, and yet they always remain connected together in the same order.

193:e relative position or connexion in homologous parts; they may differ to almost any extent in form and size, and yet will remain connected together in the same invariable order.

193:f[¶] has strongly insisted on/yet remain

194 We never find, for instance, the bones of the arm and forearm, or of the thigh and leg, transposed.

194:f fore-arm

195 Hence the same names can be given to the homologous bones in widely different animals.

196 We see the same great law in the construction of the mouths of insects: what can be more different than the immensely long spiral proboscis of a sphinx-moth, the curious folded one of a bee or bug, and the great jaws of a beetle?—yet all these organs, serving for such different purposes, are formed by infinitely numerous modifications of an upper lip, mandibles, and two pairs of maxillæ.

196:e such widely different

197 Analogous laws govern the construction of the mouths and limbs of crustaceans.

197:e The same law governs

198 So it is with the flowers of plants.

199 Nothing can be more hopeless than to attempt to explain this similarity of pattern in members of the same class, by utility or by the doctrine of final causes.

200 The hopelessness of the attempt has been expressly admitted by Owen in his most interesting work on the 'Nature of Limbs.'

201 On the ordinary view of the independent creation of each being, we can only say that so it is;—that it has so pleased the Creator to construct each animal and plant.

201:d has pleased the Creator to construct all the animals and

677

plants in each great class on a uniformly regulated plan; but this is not a scientific explanation.

201:e uniform **plan**

202 The explanation is manifest on the theory of the natural selection of successive slight modifications,—each modification being profitable in some way to the modified form, but often affecting by correlation of growth other parts of the organisation.

202:e manifest according to the theory of the selection/correlation other

202:f is to a large extent simple on the theory

203 In changes of this nature, there will be little or no tendency to modify the original pattern, or to transpose parts.

203:e to alter the original pattern, or to transpose the parts.

204 The bones of a limb might be shortened and widened to any extent, and become gradually enveloped in thick membrane, so as to serve as a fin; or a webbed foot might have all its bones, or certain bones, lengthened to any extent, and the membrane connecting them increased to any extent, so as to serve as a wing: yet in all this great amount of modification there will be no tendency to alter the framework of bones or the relative connexion of the several parts.

204:e shortened and flattened to any extent, becoming at the same time enveloped in thick membrane, so as to serve as a fin; or a webbed hand might have all its bones, or certain bones, lengthened to any extent, with the membrane connecting them increased, so as to serve as a wing; yet all this modification would not tend to alter the framework of the bones or the relative connexion of the parts.

204:f yet all these modifications

205 If we suppose that the ancient progenitor, the archetype as it may be called, of all mammals, had its limbs constructed on the existing general pattern, for whatever purpose they served, we can at once perceive the plain signification of the homologous construction of the limbs throughout the whole class.

205:e that an early progenitor,—the archetype as it may be called, —of all/throughout the class.

205:f progenitor—the archetype as it may be called—of all mammals, birds, and reptiles, had

206 So with the mouths of insects, we have only to suppose that their common progenitor had an upper lip, mandibles, and two pair of maxillæ, these parts being perhaps very simple in form; and then natural selection will account for the infinite diversity in structure and function of the mouths of insects.

206:b selection, acting on some originally created form, will

206:c selection will

206:e diversity in the structure and functions

207 Nevertheless, it is conceivable that the general pattern of an organ might become so much obscured as to be finally lost, by the atrophy and ultimately by the complete abortion of certain parts, by the soldering together of other parts, and by the doubling or multiplication of others,—variations which we know to be within the limits of possibility.

207:e by the reduction and ultimately by the complete abortion of certain parts, by the fusion of other parts

208 In the paddles of the extinct gigantic sea-lizards, and in the mouths of certain suctorial crustaceans, the general pattern seems to have been thus to a certain extent obscured.

208:e of the gigantic extinct sea-lizards/seems thus to have been partially obscured.

208:f have become partially

209 There is another and equally curious branch of the present subject; namely, the comparison not of the same part in different members of a class, but of the different parts or organs in the same individual.

209:d members of the same class

209:e branch of our present/same parts or organs in different members

209:f our subject; namely, serial homologies, or the comparison of the different parts or organs in the same individual, and not of the same parts or organs in different members of the same class.

210 Most physiologists believe that the bones of the skull are homologous with—that is correspond in number and in relative connexion with—the elemental parts of a certain number of vertebræ.

210:d with—that is, correspond

210:f homologous—that is, correspond in number and in relative connexion—with the elemental

211 The anterior and posterior limbs in each member of the vertebrate and articulate classes are plainly homologous.

211:d vertebrate classes

211:e in all the higher vertebrate

212 We see the same law in comparing the wonderfully complex jaws and legs in crustaceans.

212:e So it is with the wonderfully complex jaws and legs of crustaceans.

213 It is familiar to almost every one, that in a flower the relative position of the sepals, petals, stamens, and pistils, as well as their intimate structure, are intelligible on the view that they consist of metamorphosed leaves, arranged in a spire.

214 In monstrous plants, we often get direct evidence of the possibility of one organ being transformed into another; and we

can actually see in embryonic crustaceans and in many other animals, and in flowers, that organs, which when mature become extremely different, are at an early stage of growth exactly alike.

214:d see in flowers during their early development, as well as in crustaceans and many other animals during their embryonic states, that

214:e see, during the early or embryonic stages of development in flowers, as well as in crustaceans and many other animals, that/ at first exactly

214:f different are

215 How inexplicable are these facts on the ordinary view of creation!

215:f are the cases of serial homologies on

216 Why should the brain be enclosed in a box composed of such numerous and such extraordinarily shaped pieces of bone?

216:f bone, apparently representing vertebræ?

217 As Owen has remarked, the benefit derived from the yielding of the separate pieces in the act of parturition of mammals, will by no means explain the same construction in the skulls of birds.

217:d birds and reptiles.

218 Why should similar bones have been created in the formation of the wing and leg of a bat, used as they are for such totally different purposes?

218:e created to form the wing

218:f purposes, namely flying and walking?

219 Why should one crustacean, which has an extremely complex mouth formed of many parts, consequently always have fewer legs; or conversely, those with many legs have simpler mouths?

220 Why should the sepals, petals, stamens, and pistils in any individual flower, though fitted for such widely different purposes, be all constructed on the same pattern?

220:e pistils, in each flower

220:f such distinct purposes

221 On the theory of natural selection, we can satisfactorily answer these questions.

221:e can answer

221:f can, to a certain extent, answer

221.1-2:f We need not here consider how the bodies of some animals first became divided into a series of segments, or how they became divided into right and left sides, with corresponding organs, for such questions are almost beyond investigation. 2 It is, however, probable that some serial structures are the result of cells multiplying by division, entailing the multiplication of the parts developed from such cells.

222 In the vertebrata, we see a series of internal vertebræ bearing certain processes and appendages; in the articulata, we see the body divided into a series of segments, bearing external appendages; and in flowering plants, we see a series of successive spiral whorls of leaves.

222:*e* processes; in the articulata, the body/plants, spiral whorls

222:[*f*]

223 An indefinite repetition of the same part or organ is the common characteristic (as Owen has observed) of all low or little-modified forms; therefore we may readily believe that the unknown progenitor of the vertebrata possessed many vertebræ; the unknown progenitor of the articulata, many segments; and the unknown progenitor of flowering plants, many spiral whorls of leaves.

223:*d* characteristic, as Owen has observed, of all low or little modified

223:*e* therefore the unknown progenitor of the vertebrata no doubt possessed/plants, many leaves arranged in one or more spires.

223:*f* It must suffice for our purpose to bear in mind that an indefinite/little specialised forms; therefore the unknown progenitor of the Vertebrata probably possessed/Articulata

224 We have formerly seen that parts many times repeated are eminently liable to vary in number and structure; consequently it is quite probable that natural selection, during a long-continued course of modification, should have seized on a certain number of the primordially similar elements, many times repeated, and have adapted them to the most diverse purposes.

224.*x*:*e* structure.

224.*x*:*f* have also formerly/vary, not only in number, but in form.

225 And as the whole amount of modification will have been effected by slight successive steps, we need not wonder at discovering in such parts or organs, a certain degree of fundamental resemblance, retained by the strong principle of inheritance.

225:*c* organs a certain

225:*d* by successive slight steps

224.*y* + 225:*e* Consequently such parts being already present, and being highly variable, would afford the materials for adaptation to the most different purposes; and they would generally retain through the force of inheritance plain traces of their original or fundamental resemblance.

224.*y* + 225:*f* parts, being already present in considerable numbers, and being highly variable, would naturally afford the materials for adaptation to the most different purposes; yet they would generally retain, through the force of inheritance, plain

224.*y* + 225.*1-2*:*f* They would retain this resemblance all the more, as the variations, which afforded the basis for their subsequent modification through natural selection, would tend from the first to be similar; the parts being at an early stage of growth alike, and being subjected to nearly the same conditions. 2 Such parts, whether more or less modified, unless their common origin became wholly obscured, would be serially homologous.

226   In the great class of molluscs, though we can homologise the parts of one species with those of another and distinct species, we can indicate but few serial homologies; that is, we are seldom enabled to say that one part or organ is homologous with another in the same individual.

226:b those of other and distinct

226:e though it can easily be shown that the parts in distinct species are homologous, but few serial homologies can be indicated; that is, we are seldom enabled to say that one part is

226:f though the parts in distinct species can be shown to be homologous, only a few serial homologies, such as the valves of Chitons, can

227 And we can understand this fact; for in molluscs, even in the lowest members of the class, we do not find nearly so much indefinite repetition of any one part, as we find in the other great classes of the animal and vegetable kingdoms.

227:f part as

227.1-6:f[¶] But morphology is a much more complex subject than it at first appears, as has lately been well shown in a remarkable paper by Mr. E. Ray Lankester, who has drawn an important distinction between certain classes of cases which have all been equally ranked by naturalists as homologous. 2 He proposes to call the structures which resemble each other in distinct animals, owing to their descent from a common progenitor with subsequent modification, *homogenous;* and the resemblances which cannot thus be accounted for, he proposes to call *homoplastic. 3* For instance, he believes that the hearts of birds and mammals are as a whole homogenous,—that is, have been derived from a common progenitor; but that the four cavities of the heart in the two classes are homoplastic,—that is, have been independently developed. *4* Mr. Lankester also adduces the close resemblance of the parts on the right and left sides of the body, and in the successive segments of the same individual animal; and here we have parts commonly called homologous, which bear no relation to the descent of distinct species from a common progenitor. *5* Homoplastic structures are the same with those which I have classed, though in a very imperfect manner, as analogous modifications or resemblances. *6* Their formation may be attributed in part to distinct organisms, or to distinct parts of the same organism, having varied in an analagous manner; and in part to similar modifications, having been preserved for the same general purpose or function,—of which many instances have been given.

228   Naturalists frequently speak of the skull as formed of metamorphosed vertebræ: the jaws of crabs as metamorphosed legs; the stamens and pistils of flowers as metamorphosed leaves; but it would in these cases probably be more correct, as Professor Huxley has remarked, to speak of both skull and vertebræ, both jaws and legs, &c.,—as having been metamorphosed, not one from the other, but from some common element.

228:d &c., as/other in their present state, but from some common and simpler element.

228:e vertebræ; the jaws of crabs/pistils in flowers/would in most cases/other, as they now exist, but

228:f both skull and vertebræ, jaws

229 Naturalists, however, use such language only in a metaphorical sense: they are far from meaning that during a long course of descent, primordial organs of any kind—vertebræ in the one case and legs in the other—have actually been modified into skulls or jaws.

229:d Most naturalists

229:e sense; they/been converted into

230 Yet so strong is the appearance of a modification of this nature having occurred, that naturalists can hardly avoid employing language having this plain signification.

230:d of such modifications having actually occurred

230:e of this having occurred

231 On my view these terms may be used literally; and the wonderful fact of the jaws, for instance, of a crab retaining numerous characters, which they would probably have retained through inheritance, if they had really been metamorphosed during a long course of descent from true legs, or from some simple appendage, is explained.

231:d they probably would have/true though simple legs, is

231:e According to the views here maintained, such language may/metamorphosed from true though extremely simple

231:f literally; and/legs, is in part explained.

232   *Embryology.—*

232:d [Center] *Embryology and Development.*

232:e *Development and Embryology.*

232.1:d[¶] This is one of the most important departments of natural history.

232.1:e important subjects in the whole round of

232.2:d Herein are included the ordinary metamorphoses of insects, with which every one is familiar.

232.2:e The metamorphoses of insects, with which every one is familiar, are generally effected abruptly by a few stages; but the transformations are in reality numerous and gradual, though concealed.

232.3:d These are generally effected somewhat abruptly by a few stages and in a concealed manner; but the transformations are in reality numerous and graduated.

232.3:[e]

232.4:d For instance, Sir J. Lubbock has recently shown that a certain ephemerous insect (Chlöeon) during its development moults above twenty times, and each time undergoes a certain amount of change; in such cases we probably behold the act of metamorphosis in its natural or primary progress.

232.4:e A certain ephemerous insect (Chlöeon) during its development, moults, as shown by Sir J. Lubbock, above/change; and in this case we see the act of metamorphosis performed in a primary and gradual manner.

232.5:d What great changes of structure are effected during the development of some animals is seen in the case of insects, but still more plainly with many crustaceans.

232.5:e Many insects, and especially certain crustaceans, show us what wonderful changes of structure can be effected during development.

232.6:d When, however, we read of the several wonderful cases, recently discovered, of the so-called alternate generations of animals, we come to the climax of developmental transformation.

232.6:e Such changes, however, reach their climax in the so-called alternate generations of some of the lower animals.

232.6:g their acme in

232.7:d What fact can be more astonishing than that a delicate branching coralline, studded with polypi and attached to a submarine rock, should produce, first by budding and then by transverse division, a host of huge floating jelly-fishes; and that these should produce eggs, from which are hatched swimming animalcules, which attach themselves to rocks and become developed into branching corallines; and so on in an endless cycle?

232.7:e It is, for instance, an astonishing fact that a delicate/cycle.

232.8:d Hence it will be seen that I follow those naturalists who look at all cases of alternate generation, as essentially modifications of the process of budding, which may supervene at any stage of development.

232.8:[e]

232.9:d This view of the close connection between alternate generations and ordinary metamorphoses has recently been much strengthened by Wagner's discovery of the larva of a Cecidomyia,—that is of the maggot of a fly,—producing asexually within its body other and similar larvæ; these again repeating the process.

232.9:e The belief in the essential identity of the process of alternate generation and of ordinary metamorphosis has been greatly strengthened by Wagner's discovery of the larva or maggot of a fly, namely the Cecidomyia, producing asexually other and similar larvæ.

232.9:f other larvæ, and these others, which finally are developed into mature males and females, propagating their kind in the ordinary manner by eggs.

232.9.1-5:f[¶] It may be worth notice that when Wagner's remarkable discovery was first announced, I was asked how was it possible to account for the larvæ of this fly having acquired the power of asexual reproduction. 2 As long as the case remained unique no answer could be given. 3 But already Grimm has shown that another fly, a Chironomus, reproduces itself in nearly the same manner, and he believes that this occurs frequently in the Order. 4 It is the pupa, and not the larva, of the Chironomus which has this power; and Grimm further shows that this case, to a certain extent, "unites that of the Cecidomyia with the parthenogenesis of the Coccidæ;"—the term parthenogenesis implying that the mature females of the Coccidæ are capable of producing fertile eggs without the concourse of the male. 5 Certain animals belonging to several classes are now known to have the power of ordinary reproduction at an unusually early age; and we have only to accelerate parthenogenetic reproduction by gradual steps to an earlier and earlier age,—Chironomus showing us an almost exactly intermediate stage, viz., that of the pupa—and we can perhaps account for the marvellous case of the Cecidomyia.

233 It has already been casually remarked that certain organs in the individual, which when mature become widely different and serve for different purposes, are in the embryo exactly alike.

233:d[¶] It has already been remarked that various parts and organs of the same individual animal are during an early embryonic period exactly like each other, but become in the adult state widely different and serve for widely different purposes.

233:e been stated that various parts and organs in the same individual are exactly like each other during an early embryonic period, but in the adult state become widely

233:f parts in the same individual which are exactly alike during an early embryonic period, become widely different and serve for widely different purposes in the adult state.

234 The embryos, also, of distinct animals within the same class are often strikingly similar: a better proof of this cannot be given, than a circumstance mentioned by Agassiz, namely, that having forgotten to ticket the embryo of some vertebrate animal, he cannot now tell whether it be that of a mammal, bird, or reptile.

234:c than a statement made by Von Baer, namely, that "the embryos of mammalia, of birds, lizards, and snakes, probably also of chelonia, are in their earliest stages exceedingly like one another, both as a whole and in the mode of development of their parts; so much so, in fact, that we can often distinguish the embryos only by their size.

234.x:d So again it has already been remarked that the embryos of distinct species and genera within the same class are generally closely similar, but become when fully developed widely dissimilar.

234.x:e has been stated that the embryos of the most distinct species within

234.x:f been shown that generally the embryos of the most distinct species belonging to the same class are closely similar, but become, when fully developed, widely

234.y:d A better proof of this latter fact cannot be given than that by Von

234.y:f than the statement by Von Baer that "the embryos of mammalia

234.1-4:c In my possession are two little embryos in spirit, whose names I have omitted to attach, and at present I am quite unable to say to what class they belong. *2* They may be lizards or small birds, or very young mammalia, so complete is the similarity in the mode of formation of the head and trunk in these animals. *3* The extremities, however, are still absent in these embryos. *4* But even if they had existed in the earliest stage of their development we should learn nothing, for the feet of lizards and mammals, the wings and feet of birds, no less than the hands and feet of man, all arise from the same fundamental form."

235 The vermiform larvæ of moths, flies, beetles, &c., resemble each other much more closely than do the mature insects; but in the case of larvæ, the embryos are active, and have been adapted for special lines of life.

235:d &c., generally resemble/and from having been adapted for special lines of life sometimes differ much from each other.

235:e insects; but in these cases the embryos

235:f The larvæ of most crustaceans, at corresponding stages of development, closely resemble each other, however different the adults may become; and so it is with very many other animals.

236 A trace of the law of embryonic resemblance, sometimes lasts till a rather late age: thus birds of the same genus, and of closely allied genera, often resemble each other in their first and second plumage; as we see in the spotted feathers in the thrush group.

236:d resemblance, occasionally lasts

236:e resemblance occasionally/their immature plumage; as we see in the spotted feathers in the young of the thrush group.

236:f and of allied

237 In the cat tribe, most of the species are striped or spotted in lines; and stripes can be plainly distinguished in the whelp of the lion.

237:c and stripes or spots can/lion and the puma.

237:f species when adult are

238 We occasionally though rarely see something of this kind in plants: thus the embryonic leaves of the ulex or furze, and the first leaves of the phyllodineous acaceas, are pinnate or divided like the ordinary leaves of the leguminosæ.

238:c thus the first leaves of the ulex

238:f of the same kind

239 The points of structure, in which the embryos of widely different animals of the same class resemble each other, often have no direct relation to their conditions of existence.

239:e animals within the same

240 We cannot, for instance, suppose that in the embryos of the vertebrata the peculiar loop-like course of the arteries near the branchial slits are related to similar conditions,—in the young mammal which is nourished in the womb of its mother, in the egg of the bird which is hatched in a nest, and in the spawn of a frog under water.

241 We have no more reason to believe in such a relation, than we have to believe that the same bones in the hand of a man, wing of a bat, and fin of a porpoise, are related to similar conditions of life.

241:c that the similar bones

242 No one will suppose that the stripes on the whelp of a lion, or the spots on the young blackbird, are of any use to these animals, or are related to the conditions to which they are exposed.

242:c No good observer will

242:e No one supposes/are of use to these animals.

242:f are of any use

243 The case, however, is different when an animal during any part of its embryonic career is active, and has to provide for itself.

244 The period of activity may come on earlier or later in life; but whenever it comes on, the adaptation of the larva to its conditions of life is just as perfect and as beautiful as in the adult animal.

244.1:d In how important a manner this has acted, has recently been well shown by Sir J. Lubbock in his remarks on the close similarity of the larvæ of some insects belonging to widely different orders, and on the dissimilarity of the larvæ of other insects belonging to the same order, according to their habits of life.

244.1:e to very different/other insects within the same

245 From such special adaptations, the similarity of the larvæ or active embryos of allied animals is sometimes much obscured; and cases could be given of the larvæ of two species, or of two groups of species, differing quite as much, or even more, from each other than do their adult parents.

245:d From such adaptations, especially when including a division of labour during the different stages of development, as when a larva during one stage has to search for food, and during another stage has to search for a place of attachment, the similarity of the larvæ of allied animals is sometimes greatly obscured/differing much more from each other, than

245:e especially when they imply a division/as when the same larva has during one stage to search for food, and has during another stage to/differing more from each other than do the adults.

245.x-y:f Owing to such adaptations, the similarity of the larvæ of allied animals is sometimes greatly obscured; especially when there is a division of labour during the different stages of development, as when the same larva has during one stage to search for food, and during another stage has to search for a place of attachment. *y* Cases can even be given of the larvæ of allied species, or groups of species, differing

246 In most cases, however, the larvæ, though active, still obey more or less closely the law of common embryonic resemblance.

246:b obey, more or less closely, the law

247 Cirripedes afford a good instance of this: even the illustrious Cuvier did not perceive that a barnacle was, as it certainly is, a crustacean; but a glance at the larva shows this to be the case in an unmistakeable manner.

247:e barnacle was a crustacean; but a glance at the larva shows this to be true in

247:f this; even/crustacean: but a glance at the larva shows this in

248 So again the two main divisions of cirripedes, the pedunculated and sessile, which differ widely in external appearance, have larvæ in all their several stages barely distinguishable.

248:c their stages

248:e sessile, though differing

249 The embryo in the course of development generally rises in organisation: I use this expression, though I am aware that it is hardly possible to define clearly what is meant by the organisation being higher or lower.

249:f organisation; I use

250 But no one probably will dispute that the butterfly is higher than the caterpillar.

251 In some cases, however, the mature animal is generally considered as lower in the scale than the larva, as with certain parasitic crustaceans.

251:e animal must be considered

252 To refer once again to cirripedes: the larvæ in the first stage have three pairs of legs, a very simple single eye, and a prosociformed mouth, with which they feed largely, for they increase much in size.

252:f of locomotive organs, a simple

253 In the second stage, answering to the chrysalis stage of butterflies, they have six pairs of beautifully constructed natatory legs, a pair of magnificent compound eyes, and extremely complex antennæ; but they have a closed and imperfect mouth, and cannot feed: their function at this stage is, to search by their well-developed organs of sense, and to reach by their active powers of swimming, a proper place on which to become attached and to undergo their final metamorphosis.

253:f search out by

254 When this is completed they are fixed for life: their legs are now converted into prehensile organs; they again obtain a well-constructed mouth; but they have no antennæ, and their two eyes are now reconverted into a minute, single, and very simple eye-spot.

254:f single, simple

255 In this last and complete state, cirripedes may be considered as either more highly or more lowly organised than they were in the larval condition.

256 But in some genera the larvæ become developed either into hermaphrodites having the ordinary structure, or into what I have called complemental males: and in the latter, the development has assuredly been retrograde; for the male is a mere sack, which lives for a short time, and is destitute of mouth, stomach, or other organ of importance, excepting for reproduction.

256:c males, and in

256:e organs of importance, excepting those for

256:f developed into hermaphrodites having the ordinary structure, and into what I have called complemental males; and in the latter the development has assuredly been retrograde, for the male is a mere sack, which lives for a short time and is destitute of mouth, stomach, and every other organ

257 We are so much accustomed to see differences in structure between the embryo and the adult, and likewise a close similarity in the embryos of widely different animals within the same

class, that we might be led to look at these facts as necessarily contingent in some manner on growth.

257:d of different/as in some manner necessarily contingent on

257:e see a difference in structure between the embryo and the adult, that we are tempted to look at this difference as

257:f as in some necessary manner contingent

258 But there is no obvious reason why, for instance, the wing of a bat, or the fin of a porpoise, should not have been sketched out with all the parts in proper proportion, as soon as any structure became visible in the embryo.

258:e no reason why, for instance, the wings of a bat, or the fins/ all their parts/visible.

258:f any part became

259 And in some whole groups of animals and in certain members of other groups, the embryo does not at any period differ widely from the adult: thus Owen has remarked in regard to cuttle-fish, "there is no metamorphosis; the cephalopodic character is manifested long before the parts of the embryo are completed;" and again in spiders, "there is nothing worthy to be called a metamorphosis."

259:d completed."

259:e In some/other groups this is the case, and the embryo

259.1:d Land-shells and fresh-water crustaceans are born with their proper forms, whilst the marine members of these two great classes pass through considerable and often great developmental changes.

259.1:e born having their proper form, whilst the marine members of the same two great classes pass through considerable and often great changes during their development.

259.1:f forms

259.2:d Spiders, again, barely undergo any metamorphosis.

260 The larvæ of insects, whether adapted to the most diverse and active habits, or quite inactive, being fed by their parents or placed in the midst of proper nutriment, yet nearly all pass through a similar worm-like stage of development; but in some few cases, as in that of Aphis, if we look to the admirable drawings by Professor Huxley of the development of this insect, we see no trace of the vermiform stage.

260:d With almost all insects, the larvæ, whether adapted to diversified and active habits, or remaining inactive, being placed in the midst of proper nutriment or fed by their parents, yet/ see hardly any trace

260:e The larvæ of most insects pass through a worm-like stage, whether they are active and adapted to diversified habits, or are inactive from being placed in the midst of proper nutriment or from being fed by their parents; but/drawings of the development of this insect, by Professor Huxley, we

260.1:d[¶] In some cases it is only the earlier developmental stages which fail; these apparently having been suppressed.

260.1:e Sometimes it/fail.

260.2:d Thus Fritz Müller has recently made the remarkable discovery that certain shrimp-like crustaceans (allied to Penœus) first appear under the simple nauplius-form, and passing through two or more zoea-stages, and through the mysis-stage, finally acquire their mature structure: now in the whole enormous malacostracan class, to which these crustaceans belong, no other member is as yet known to be first developed under the nauplius-form, though very many appear as zoeas; nevertheless Müller assigns reasons for his belief that all these crustaceans would have appeared as nauplii, if there had been no suppression of development;—or that they were primordially developed under this form.

260.2:e has made/and after passing through two or more zoea-stages, and then through the mysis-stage, finally acquire their mature structure: now in the whole great malacostracan/ though many/suppression of development.

260.2:f malacostracan order, to which/belief, that if there had been no suppression of development, all these crustaceans would have appeared as nauplii.

261   How, then, can we explain these several facts in embryology, —namely the very general, but not universal difference in structure between the embryo and the adult;—of parts in the same individual embryo, which ultimately become very unlike and serve for diverse purposes, being at this early period of growth alike;—of embryos of different species within the same class, generally, but not universally, resembling each other;—of the structure of the embryo not being closely related to its conditions of existence, except when the embryo becomes at any period of life active and has to provide for itself;—of the embryo apparently having sometimes a higher organisation than the mature animal, into which it is developed.

261:d namely, the very general/individual embryo which/at an early/developed?

261:e general, though not universal, difference in structure between the embryo and the adult;—the various parts in the same individual/alike;—the general, but not invariable, resemblance between the embryos or larvæ of the most distinct species in the same class;—the embryo retaining whilst within the egg or womb, structures which are of no service to it, either at that period or later in life; whilst embryos at a later period, or larvæ, which have to provide for their own wants, are perfectly adapted to the surrounding conditions;—and lastly the fact of certain larvæ standing higher in the scale of organisation than the mature animals into which they are developed?

261:f alike;—the common, but/class;—the embryo often retaining/

that or at a later period of life; on the other hand larvæ, which have to provide for their own wants, being perfectly

262 I believe that all these facts can be explained, as follows, on the view of descent with modification.

262:e follows.

263   It is commonly assumed, perhaps from monstrosities often affecting the embryo at a very early period, that slight variations necessarily appear at an equally early period.

263:b embryos

263:c embryo

263:e monstrosities affecting/variations or individual differences necessarily

264 But we have little evidence on this head—indeed the evidence rather points the other way; for it is notorious that breeders of cattle, horses, and various fancy animals, cannot positively tell, until some time after the animal has been born, what its merits or form will ultimately turn out.

264:e We have little evidence on this head, but what we have certainly points the other way; for/after birth, what the merits or form of their young animals will turn

264:f what will be the merits or demerits of their young animals.

265 We see this plainly in our own children; we cannot always tell whether the child will be tall or short, or what its precise features will be.

265:d cannot tell

265:e whether a child

266 The question is not, at what period of life any variation has been caused, but at what period it is fully displayed.

266:e life each variation may have been caused, but at what period the effects are displayed.

267 The cause may have acted, and I believe generally has acted, even before the embryo is formed; and the variation may be due to the male and female sexual elements having been affected by the conditions to which either parent, or their ancestors, have been exposed.

267:e generally has acted, on one or both parents before reproduction.

267:f acted, and I believe often has acted, on one or both parents before the act of generation.

268 Nevertheless an effect thus caused at a very early period, even before the formation of the embryo, may appear late in life; as when an hereditary disease, which appears in old age alone, has been communicated to the offspring from the reproductive element of one parent.

268:[e]

269 Or again, as when the horns of cross-bred cattle have been affected by the shape of the horns of either parent.

269:[e]

270 For the welfare of a very young animal, as long as it remains in its mother's womb, or in the egg, or as long as it is nourished and protected by its parent, it must be quite unimportant whether most of its characters are fully acquired a little earlier or later in life.

270:e It deserves notice that it is of no importance to a very/womb or in the egg/parent, whether most of its characters are acquired

271 It would not signify, for instance, to a bird which obtained its food best by having a long beak, whether or not it assumed a beak of this particular length, as long as it was fed by its parents.

271:e food by having a much-curved beak, whether or not whilst young it possessed a beak of this shape, as long as

271:f beak whether

272 Hence, I conclude, that it is quite possible, that each of the many successive modifications, by which each species has acquired its present structure, may have supervened at a not very early period of life; and some direct evidence from our domestic animals supports this view.

272:[e]

273 But in other cases it is quite possible that each successive modification, or most of them, may have appeared at an extremely early period.

273:[e]

274   I have stated in the first chapter, that there is some evidence to render it probable, that at whatever age any variation first appears in the parent, it tends to reappear at a corresponding age in the offspring.

274:d is a large body of facts rendering it probable

274:e that at whatever age a variation

275 Certain variations can only appear at corresponding ages, for instance, peculiarities in the caterpillar, cocoon, or imago states of the silk-moth; or, again, in the horns of almost full-grown cattle.

275:d ages; for

275:e again, in the full-grown horns of cattle.

276 But further than this, variations which, for all that we can see, might have appeared earlier or later in life, tend to appear at a corresponding age in the offspring and parent.

276:e But variations, which, for all that we can see might have appeared either earlier or later in life, likewise tend

276:f have first appeared

277 I am far from meaning that this is invariably the case; and I could give a good many cases of variations (taking the word in the largest sense) which have supervened at an earlier age in the child than in the parent.

277:e give several exceptional cases

277:f case, and

278 These two principles, if their truth be admitted, will, I believe, explain all the above specified leading facts in embryology.

278:e principles, namely, that slight variations generally appear at a not very early period of life, and are inherited at a corresponding not early period, explain, as I believe, all

279 But first let us look at a few analogous cases in domestic varieties.

279:e look to a few analogous cases in our domestic

280 Some authors who have written on Dogs, maintain that the greyhound and bulldog, though appearing so different, are really varieties most closely allied, and have probably descended from the same wild stock; hence I was curious to see how far their puppies differed from each other: I was told by breeders that they differed just as much as their parents, and this, judging by the eye, seemed almost to be the case; but on actually measuring the old dogs and their six-days old puppies, I found that the puppies had not nearly acquired their full amount of proportional difference.

280:c varieties closely

280:e really closely allied varieties, descended/not acquired nearly their

280:f six-days-old puppies

281 So, again, I was told that the foals of cart and race-horses differed as much as the full-grown animals; and this surprised me greatly, as I think it probable that the difference between these two breeds has been wholly caused by selection under domestication; but having had careful measurements made of the dam and of a three-days old colt of a race and heavy cart-horse, I find that the colts have by no means acquired their full amount of proportional difference.

281:e race-horses—breeds which have been almost wholly formed by selection under domestication—differed as much as the full-grown animals; but having had careful measurements made of the dams and of three-days' old colts of race-horses and heavy cart-horses, I find that this is by no means the case.

281:f three-days-old

282 As the evidence appears to me conclusive, that the several domestic breeds of Pigeon have descended from one wild species, I compared young pigeons of various breeds, within twelve hours after being hatched; I carefully measured the

proportions (but will not here give details) of the beak, width of mouth, length of nostril and of eyelid, size of feet and length of leg, in the wild stock, in pouters, fantails, runts, barbs, dragons, carriers, and tumblers.

282:e As we have conclusive evidence that the breeds of the Pigeon are descended from a single wild species, I compared the young within twelve hours after being hatched; I carefully measured in the wild parent-species, in pouters, fantails, runts, barbs, dragons, carriers, and tumblers, the proportions (but will not here give the details) of the beak, width of mouth, length of nostril and of eyelid, size of feet and length of leg.

282:f measured the proportions (but will not here give the details) of the beak, width of mouth, length of nostril and of eyelid, size of feet and length of leg, in the wild parent-species, in pouters, fantails, runts, barbs, dragons, carriers, and tumblers.

283 Now some of these birds, when mature, differ so extraordinarily in length and form of beak, that they would, I cannot doubt, be ranked in distinct genera, had they been natural productions.

283:d would certainly have been ranked as distinct

283:e differ in so extraordinary a degree in the length and form of beak, and in other characters, that they would certainly be ranked as distinct genera if found in a state of nature.

283:f extraordinary a manner in the length/certainly have been ranked

284 But when the nestling birds of these several breeds were placed in a row, though most of them could be distinguished from each other, yet their proportional differences in the above specified several points were incomparably less than in the full-grown birds.

284:e could just be distinguished, yet the proportional differences in the above specified points

284:f distinguished, the proportional

285 Some characteristic points of difference—for instance, that of the width of mouth—could hardly be detected in the young.

286 But there was one remarkable exception to this rule, for the young of the short-faced tumbler differed from the young of the wild rock-pigeon and of the other breeds, in all its proportions, almost exactly as much as in the adult state.

286:e differed in all its proportions from the young of the wild rock-pigeon and of the other breeds, almost

286:f differed from the young of the wild rock-pigeon and of the other breeds, in almost exactly the same proportions as

287 The two principles above given seem to me to explain these facts in regard to the later embryonic stages of our domestic varieties.

287:d given, namely that variations do not generally supervene at

695

a very early age, and that they are inherited at a corresponding age whatever that may have been, seem to me to explain these several facts regarding the later developmental stages

287:e The above two principles explain these facts.

287:f These facts are explained by the above two principles.

288 Fanciers select their horses, dogs, and pigeons, for breeding, when they are nearly grown up: they are indifferent whether the desired qualities and structures have been acquired earlier or later in life, if the full-grown animal possesses them.

288:e select for breeding their dogs, horses, pigeons, &c., when nearly grown up: they are indifferent whether the desired qualities are acquired

288:f select their dogs, horses, pigeons, &c., for breeding, when

289 And the cases just given, more especially that of pigeons, seem to show that the characteristic differences which give value to each breed, and which have been accumulated by man's selection, have not generally first appeared at an early period of life, and have been inherited by the offspring at a corresponding not early period.

289:e pigeons, show/to the breeds and which have been accumulated by man's selection, have not generally appeared at a very early period of life, and have been inherited at

289:f of the pigeons, show that the characteristic differences which have been accumulated by man's selection, and which give value to his breeds, do not generally appear at a very early period of life, and are inherited

290 But the case of the short-faced tumbler, which when twelve hours old had acquired its proper proportions, proves that this is not the universal rule; for here the characteristic differences must either have appeared at an earlier period than usual, or, if not so, the differences must have been inherited, not at the corresponding, but at an earlier age.

290:e old possessed its proper characters, proves

291 Now let us apply these facts and the above two principles—which latter, though not proved true, can be shown to be in some degree probable—to species in a state of nature.

291:d principles to species

291:e these two principles

292 Let us take a genus of birds, descended on my theory from some one parent-species, and of which the several new species have become modified through natural selection in accordance with their diverse habits.

292:e take a group of birds, descended from some ancient form and modified through natural selection for different habits.

293 Then, from the many slight successive steps of variation having supervened at a rather late age, and having been inherited at

a corresponding age, the young of the new species of our supposed genus will manifestly tend to resemble each other much more closely than do the adults, just as we have seen in the case of pigeons.

293:e successive variations having supervened in the several species at a not early age, and having been inherited at a corresponding age, the young will be left but little modified and will resemble each other much more closely than do the adults,— just as we have seen with the breeds of the pigeon.

293:f will have been but little modified, and they will still resemble

294 We may extend this view to whole families or even classes.

294:e view to widely distinct structures and to whole classes.

295 The fore-limbs, for instance, which served as legs in the parent-species, may become, by a long course of modification, adapted in one descendant to act as hands, in another as paddles, in another as wings; and on the above two principles—namely of each successive modification supervening at a rather late age, and being inherited at a corresponding late age—the fore-limbs in the embryos of the several descendants of the parent-species will still resemble each other closely, for they will not have been modified.

295:b may have become

295:e which once served as legs to a remote progenitor, may have become, through a long/wings; but on the above two principles the fore-limbs will not have been much modified in the embryos of these several forms; although in each the embryonic fore-limb will differ greatly from that in the adult.

295:f although in each form the fore-limb will differ greatly in the adult state.

296 But in each individual new species, the embryonic fore-limbs will differ greatly from the fore-limbs in the mature animal; the limbs in the latter having undergone much modification at a rather late period of life, and having thus been converted into hands, or paddles, or wings.

296:b each of our new

296:[e]

297 Whatever influence long-continued exercise or use on the one hand, and disuse on the other, may have in modifying an organ, such influence will mainly affect the mature animal, which has come to its full powers of activity and has to gain its own living; and the effects thus produced will be inherited at a corresponding mature age.

297:d or disuse may have had in/activity and has had to

297:e influence, moreover, long-continued use or disuse may have had in modifying the limbs or other parts of animals, this will chiefly or solely have affected them when mature and when they had to use their full powers to gain their own living; and the effect thus produced will be transmitted to the offspring at

297:f influence long-continued/parts of any species, this will chiefly or solely have affected it when nearly mature, when it was compelled to use its full powers to gain its own living; and the effects thus produced will have been transmitted to the offspring at a corresponding nearly mature

298 Whereas the young will remain unmodified, or be modified in a lesser degree, by the effects of use and disuse.

298:e Thus the young will not be modified or will be modified in a less degree.

298:f or will be modified only in a slight degree, through the effects of the increased use or disuse of parts.

299   In certain cases the successive steps of variation might supervene, from causes of which we are wholly ignorant, at a very early period of life, or each step might be inherited at an earlier period than that at which it first appeared.

299:e In other cases successive variations may have supervened at a very early period of life, or the steps may have been inherited at an earlier age than that at which they first occurred.

299:f With some animals the successive

300 In either case (as with the short-faced tumbler) the young or embryo would closely resemble the mature parent-form.

300:e case, as we have seen with the short-faced tumbler, the young

300:f either of these cases, the young or embryo will closely resemble the mature parent-form, as we have seen with the short-faced tumbler.

301 We have seen that this is the rule of development in certain whole groups of animals, as with cuttle-fish and spiders, and with a few members of the great class of insects, as with Aphis.

301:d cuttle-fish, land-shells, fresh-water crustaceans, spiders/insects.

301:e And this is the rule of development in certain whole groups or sub-groups, as with cuttle-fish, land-shells, fresh-water crustaceans, spiders, and some members

301:f groups, or in certain sub-groups alone, as with cuttle-fish

302 With respect to the final cause of the young in these cases not undergoing any metamorphosis, or closely resembling their parents from their earliest age, we can see that this would result from the two following contingencies; firstly, from the young, during a course of modification carried on for many generations, having to provide for their own wants at a very early stage of development, and secondly, from their following exactly the same habits of life with their parents; for in this case, it would be indispensable for the existence of the species, that the child should be modified at a very early age in the same manner with its parents, in accordance with their similar habits.

302:b contingencies: firstly

302:d metamorphosis, we/provide at a very early stage of development for their own wants, and secondly, from their following (and this might often be of advantage to a species) exactly

302:e these groups not passing through any metamorphosis, we can see that this would follow from the following contingencies; namely, from the young having to provide at a very early age for their own wants, and from their following the same habits of life with their parents; for in this case, it would be indispensable for their existence that they should be modified in the same manner as their parents.

302:f young in such groups

302.1:d Again, with respect to the singular fact of so many terrestrial and fresh-water animals not undergoing any metamorphosis, whilst the marine members of the same classes pass through various transformations, Fritz Müller has suggested that if an animal during a long succession of generations had to change its habits from living in the sea to living on the land or in fresh-water, it would be a great advantage to its descendants during their modification if they were to lose their metamorphoses; for it is not probable that places well adapted for both the larval and mature stages, under such new and greatly changed habits of life, could be found unoccupied or ill-occupied by other organisms.

302.1:e whilst marine members of the same groups pass/suggested that the process of slowly modifying and adapting an animal to live on the land or in fresh water, instead of in the sea, would be greatly simplified by its not passing through any larval stage; for

302.1:f fact that many terrestrial and fresh-water animals do not undergo/life, would commonly be found unoccupied or ill occupied

302.2:d Therefore the modification of a marine animal into a terrestrial or fresh-water one would generally be much more easily effected, if its metamorphoses were suppressed through the gradual acquirement at an earlier and earlier age of the adult structure.

302.2:e In this case the gradual acquirement at an earlier and earlier age of the adult structure would be favoured by natural selection, and all traces of former metamorphoses would finally be lost.

302.2:f selection; and

303 Some further explanation, however, of the embryo not undergoing any metamorphosis is perhaps requisite.

303:[d]

304 If, on the other hand, it profited the young to follow habits of life in any degree different from those of their parent, and consequently to be constructed in a slightly different manner, then, on the principle of inheritance at corresponding ages, the active young or larvæ might easily be rendered by natural selection different to any conceivable extent from their parents.

304:c might be

304:d[¶] parents and consequently to be constructed in a slightly different manner, or if it profited larvæ already having different habits from their parents to change still further their habits, then, on the principle of inheritance at corresponding ages, the young or the larvæ might be rendered by natural selection more and more different from their parents to any conceivable extent.

304:e young of an animal to follow habits of life slightly different from those of the parent-form, and consequently to be constructed in a slightly different manner, or if it profited a larva already widely different from its parent to change still further, then

304:f constructed on a slightly different plan, or if it profited a larva already different from its

305 Such differences might, also, become correlated with successive stages of development; so that the larvæ, in the first stage, might differ greatly from the larvæ in the second stage, as we have seen to be the case with cirripedes.

305:c larvæ in the first

305:d Differences in the larvæ might, also/might come to differ greatly from the larvæ in the second stage, as is the case with so many animals.

305:e that the larvæ, in the first/case with many

305:f larva/of its development/larva/larva

306 The adult might become fitted for sites or habits, in which organs of locomotion or of the senses, &c., would be useless; and in this case the final metamorphosis would be said to be retrograde.

306:d might also become

306:e case the metamorphosis would be retrograde.

306.1:d[¶] From the remarks above made we can see how by alterations of structure in the young, in conformity with altered habits of life, together with inheritance at corresponding ages, the metamorphoses of certain animals might first have been acquired, and subsequently transmitted to numerous modified descendants.

306.1:e remarks just made we can see how by changes of structure in the young, in conformity with changed habits of life, together with inheritance at corresponding ages, animals in certain cases might come to pass through stages of development, perfectly distinct from their primordial, adult condition.

306.1:f animals might/from the primordial condition of their adult progenitors.

306.2:d Fritz Müller, who has recently discussed this whole subject with much ability, goes so far as to believe that the progenitor of all insects probably resembled an adult insect, and that the caterpillar or maggot, and cocoon or pupal stages, have subsequently been acquired; but from this view many naturalists, for instance Sir J. Lubbock, who has likewise recently discussed this subject, would, it is probable, dissent.

306.2:e ability, believes/maggot stages, as well as the cocoon

306.2:f Most of our best authorities are now convinced that the various larval and pupal stages of insects have thus been acquired through adaptation, and not through inheritance from some ancient form.

306.3:d That certain unusual stages in the metamorphoses of insects have arisen from adaptations to peculiar habits of life can hardly be doubted: thus the first larval form of a certain beetle, the Sitaris, as described by M. Fabre, is a minute, active insect, furnished with six legs, two long antennæ, and four eyes.

306.3:e have been acquired through adaptation to peculiar habits of life, there can hardly be a doubt: thus/is an active, minute insect

306.3.x-y:f The curious case of Sitaris—a beetle which passes through certain unusual stages of development—will illustrate how this might occur. *y* The first larval form is described by M. Fabre, as an active

306.4:d These larvæ are hatched in the nest of a bee; and when the male-bees emerge in the spring from their burrows, which they do before the females, the larvæ spring on them, and afterwards take an early and natural opportunity of crawling on to the female-bees.

306.4:e nests of bees; and when the male-bees emerge from their burrows in the spring, which/afterwards crawl on the females whilst paired with the males.

306.4:f burrows, in/crawl on to the females

306.5:d When the latter lay their eggs, one in each cell, on the surface of the contained honey, the larva leaps on the egg and devours it.

306.5:e As soon as the females lay their eggs on the surface of the honey stored in their cells, the larvæ of the Sitaris leap on the eggs and devour them.

306.5:f female bee deposits her eggs on the surface of the honey stored in the cells

306.6:d It then undergoes a complete change; its eyes disappear; its legs and antennæ become rudimentary, and it feeds on honey; so that it now more closely resembles the ordinary larvæ of insects; ultimately it undergoes further transformations, and finally emerges as a perfect beetle.

306.6:e Afterwards these larvæ undergo a complete change; their eyes disappear; their legs and antennæ become rudimentary, and they feed on honey; so that they now more closely resemble the ordinary larvæ of insects; ultimately they undergo a further transformation, and finally emerge as the perfect

306.6:f Afterwards they undergo

306.7:d Now, if an insect, undergoing transformations like those of the Sitaris, had been the progenitor of the whole great class of insects, the general course of development, and especially that of the first larval stage, would probably have been widely different from what is actually the case; and it should be especially noted that the first larval stage would not have represented the adult condition of any insect.

306.7:e Sitaris, were to become the progenitor of a whole new class of insects, their course of development would probably be widely different from what it now is; and the first larval stage certainly would not represent the former condition of any adult and ancient insect.

306.7:f class of insects, the course of development of the new class would be widely different from that of our existing insects; and the first/ancient form.

306.8:d[¶] On the other hand it is probable that with many groups of animals the earlier larval stages do show us, more or less completely, the form of the ancient and adult progenitor of the whole group.

306.8:e is highly probable that with many animals the embryonic or larval stages show us, more or less completely, the state of the progenitor of the whole group in its adult condition.

306.8:f completely, the condition of the progenitor of the whole group in its adult state.

306.9:d In the enormous class of the Crustacea, forms wonderfully distinct from each other, as the suctorial parasites, cirripedes, entomostraca, and even the malacostraca, appear in their first larval state under a similar nauplius form; and as these larvæ feed and live in the open sea, and are not adapted for any peculiar habits of life, and from other reasons assigned by Fritz Müller, it is probable that an independent adult animal, resembling the nauplius, formerly existed at a remote period, and has subsequently produced, through long-continued modification along several divergent lines of descent, the several above-named great Crustacean groups.

306.9:e In the great class of the Crustacea, forms wonderfully distinct from each other, namely, suctorial/appear at first as larvæ under the nauplius-form; and as these larvæ/nauplius, existed at some very remote period, and subsequently produced, along several

306.9:f probable that at some very remote period an independent adult animal, resembling the Nauplius, existed, and subsequently produced, along several divergent lines of descent, the above-named

306.10:d So again it is probable, from what we know of the embryos of mammals, birds, fishes, and reptiles, that all the members in these four great classes are the modified descendants of some one ancient progenitor, which was furnished in its adult state with branchiæ, had a swim-bladder, four simple limbs, and a long tail fitted for an aquatic life.

306.10:e that these animals are the modified/branchiæ, a swim-bladder, four simple limbs, and a long tail, all fitted

306.10:f four fin-like limbs

307   As all the organic beings, extinct and recent, which have ever lived on this earth have to be classed together, and as all have been connected by the finest gradations, the best, or indeed, if our collections were nearly perfect, the only possible arrangement, would be genealogical.

308   Descent being on my view the hidden bond of connexion which naturalists have been seeking under the term of the natural system.

307 + 8:c by fine gradations/genealogical; descent

307 + 8:d or, if/on this view

307 + 8:e lived, can be arranged within a few great classes; and as all within each class have, according to our theory, formerly been connected together by fine gradations, the best, and, if our collections/being the hidden/Natural System.

307 + 8:f theory, been

309   On this view we can understand how it is that, in the eyes of most naturalists, the structure of the embryo is even more important for classification than that of the adult.

310   For the embryo is the animal in its less modified state; and in so far it reveals the structure of its progenitor.

310:[d]

311   In two groups of animal, however much they may at present differ from each other in structure and habits, if they pass through the same or similar embryonic stages, we may feel assured that they have both descended from the same or nearly similar parents, and are therefore in that degree closely related.

311:b animals

311:d two or more groups/through closely similar embryonic stages, we may feel almost assured that they have descended from the same parent-form, and are therefore closely

311:e may differ/feel assured that they all are descended from one parent-form

311:f habits in their adult condition, if

312   Thus, community in embryonic structure reveals community of descent.

312:d descent; but dissimilarity in embryonic development does not prove discommunity of descent, for in one of two groups all the developmental stages may have been suppressed, or may have been so greatly modified as no longer to be recognised, through adaptations, during the earlier periods of growth, to new habits of life.

312:e groups the developmental/modified through adaptation to new habits of life, as to be no longer recognisable.

313 It will reveal this community of descent, however much the structure of the adult may have been modified and obscured; we have seen, for instance, that cirripedes can at once be recognised by their larvæ as belonging to the great class of crustaceans.

313:d Community of descent will, however, often be revealed, although the structure of the adult may have been greatly modified and thus obscured; we have seen, for instance, that cirripedes, though externally so like shell-fish, can

313:e Even in groups, in which the adults have been modified to an extreme degree, community of origin is often revealed by the structure of the larvæ; we/shell-fish, are at once known by their larvæ to belong

314 As the embryonic state of each species and group of species partially shows us the structure of their less modified ancient progenitors, we can clearly see why ancient and extinct forms of life should resemble the embryos of their descendants,—our existing species.

314:d group of species shows us more or less completely the structure of their less modified ancient progenitors, we can see why ancient and extinct forms of life should resemble the embryos of our existing species, their descendants.

314:e As the structure of the embryo generally shows us more or less plainly the structure of its less modified and ancient progenitor, we can see why ancient and extinct forms so often resemble the embryos of existing species in the same class.

314:f As the embryo often shows us more or less plainly the structure of the less modified and ancient progenitor of the group, we can see why ancient and extinct forms so often resemble in their adult state the embryos of existing species of the same class.

315 Agassiz believes this to be a law of nature; but I am bound to confess that I only hope to see the law hereafter proved true.

315:e be a universal law of nature; and I hope to see it hereafter shown in most cases true.

315:f nature; and we may hope hereafter to see the law proved true.

316 It can be proved true in those cases alone in which the ancient state, now supposed to be represented in many embryos, has not been obliterated, either by the successive variations in a long course of modification having supervened at a very early age, or by the variations having been inherited at an earlier period than that at which they first appeared.

316:b represented in existing embryos

316:c early period of growth, or

316:d true only in those cases in which/has been obliterated neither by the successive variations having supervened at a very early period of growth, nor by

316:e can, however, be proved true only in those cases in which the ancient state has not been wholly obliterated either by successive variations having supervened at a very early period of growth, or by such variations

316:f state of the progenitor of the group has not been wholly obliterated, either/inherited at an earlier age than

317 It should also be borne in mind, that the supposed law of resemblance of ancient forms of life to the embryonic stages of recent forms, may be true, but yet, owing to the geological record not extending far enough back in time, may remain for a long period, or for ever, incapable of demonstration.

317:c that the law may

317.1:e The law will not hold good in those cases in which an ancient form became adapted in its larval state to some special line of life, and transmitted the same larval state to a whole group of descendants; for these in their larval condition will not resemble any ancient form in its adult state.

317.1:f not strictly hold/for such larvæ will not resemble any still more ancient

318 Thus, as it seems to me, the leading facts in embryology, which are second in importance to none in natural history, are explained on the principle of slight modifications not appearing, in the many descendants from some one ancient progenitor, at a very early period in the life of each, though perhaps caused at the earliest, and being inherited at a corresponding not early period.

318:d not having appeared/and having been inherited

318:e second to none in importance, are explained on the principle of modifications in the many descendants from some one ancient progenitor, not having appeared at an early period of life, and having been inherited at a corresponding period.

318:f of variations in the many descendants from some one ancient progenitor, having appeared at a not very early

319 Embryology rises greatly in interest, when we thus look at the embryo as a picture, more or less obscured, of the common parent-form of each great class of animals.

319:d embryo of an animal as a picture, more or less obscured, of the progenitor, either in its adult or larval state, of all the members of the same great class.

319:e we look at the embryo as

320 *Rudimentary, atrophied, or aborted organs.—*

320:b Organs

320:d [Center] Rudimentary, Atrophied, and Aborted Organs.
[Space]

321 Organs or parts in this strange condition, bearing the stamp of inutility, are extremely common throughout nature.

321:d[¶]

321:e common, or even general, throughout

321.1:e It would be difficult to name one of the higher animals in which some part is not in a rudimentary condition.

321.1:f be impossible to name one of the higher animals in which some part or other is

322 For instance, rudimentary mammæ are very general in the males of mammals: I presume that the "bastard-wing" in birds may be safely considered as a digit in a rudimentary state: in very many snakes one lobe of the lungs is rudimentary; in other snakes there are rudiments of the pelvis and hind limbs.

322:d general with male mammals

322:e In the mammalia, for instance, the males always possess rudimentary mammæ; in snakes one lobe of the lungs is rudimentary; in birds the "bastard-wing" may safely be considered as a rudimentary digit, and in not a few species the wings cannot be used for flight or are reduced to a rudiment.

322:f males possess/and in some species the whole wing is so far rudimentary that it cannot be used for flight.

323 Some of the cases of rudimentary organs are extremely curious; for instance, the presence of teeth in fœtal whales, which when grown up have not a tooth in their heads; and the presence of teeth, which never cut through the gums, in the upper jaws of our unborn calves.

323:e What can be more curious than the presence/heads; or the teeth, which never cut through the gums, in the upper jaws of unborn calves?

324 It has even been stated on good authority that rudiments of teeth can be detected in the beaks of certain embryonic birds.

324:[e]

325 Nothing can be plainer than that wings are formed for flight, yet in how many insects do we see wings so reduced in size as to be utterly incapable of flight, and not rarely lying under wing-cases, firmly soldered together!

326 The meaning of rudimentary organs is often quite unmistakeable: for instance there are beetles of the same genus (and even of the same species) resembling each other most closely in all respects, one of which will have full-sized wings, and another mere rudiments of membrane; and here it is impossible to doubt, that the rudiments represent wings.

325 + 6.x-y:e Rudimentary organs declare their origin and plain meaning in various ways. *y* There are beetles belonging to closely allied species, or even to the same identical species, which have either full-sized and perfect wings, or mere minute rudiments of membrance, not rarely lying under wing-covers firmly soldered together; and in this case it

325 + 6.x-y:f organs plainly declare their origin and meaning/

mere rudiments of membrane, which not rarely lie under wing-covers firmly soldered together; and in these cases it

327 Rudimentary organs sometimes retain their potentiality, and are merely not developed: this seems to be the case with the mammæ of male mammals, for many instances are on record of these organs having become well developed in full-grown males, and having secreted milk.

327:e potentiality: this occasionally occurs with the mammæ of male mammals, for they have been known to become well developed, and to secrete milk.

327:f mammals, which have been known to become well developed and

328 So again there are normally four developed and two rudimentary teats in the udders of the genus Bos, but in our domestic cows the two sometimes become developed and give milk.

328:c Bos; but

328:e So again in the udders in the genus Bos, there are normally four developed and two rudimentary teats; but the latter in our domestic cows sometimes become well developed and yield milk.

329 In individual plants of the same species the petals sometimes occur as mere rudiments, and sometimes in a well-developed state.

329:b In plants

329:e In regard to plants the petals are sometimes rudimental, and sometimes well-developed in individuals of the same species.

329:f sometimes rudimentary, and sometimes well-developed in the individuals

330 In plants with separated sexes, the male flowers often have a rudiment of a pistil; and Kölreuter found that by crossing such male plants with an hermaphrodite species, the rudiment of the pistil in the hybrid offspring was much increased in size; and this shows that the rudiment and the perfect pistil are essentially alike in nature.

330:d In some plants with their sexes separated, the male flowers include a rudiment of a pistil; and Kölreuter found that by crossing a species of this kind with another hermaphrodite/ shows how essentially alike in nature the rudiment and the perfect pistil are.

330:e In certain diœcious plants Kölreuter found that by crossing a species, in which the male flowers included a rudiment of a pistil, with an hermaphrodite species, having of course a well-developed pistil, the rudiment in the hybrid offspring was much increased in size; and this clearly shows that the rudimentary and perfect pistils are essentially alike in nature.

330:f certain plants having separated sexes Kölreuter

330.1-4:e An animal may possess various parts in a perfect state, and yet they may in one sense be rudimentary, for they are useless: thus the tadpole of the common Salamander or newt, as Mr. G. H. Lewes remarks, "has gills, and passes its existence in the water; but the Salamander atra, which lives high up among the mountains, brings forth its young full-formed. *2* This animal never lives in the water. *3* Yet if we open a gravid female, we find tadpoles inside her with exquisitely feathered gills; and when placed in water they swim about like the tadpoles of the water-newt. *4* Obviously this aquatic organisation has no reference to the future life of the animal, nor has it any adaptation to its embryonic condition; it has solely reference to ancestral adaptations, it repeats a phase in the development of its progenitors."

331 An organ serving for two purposes, may become rudimentary or utterly aborted for one, even the more important purpose; and remain perfectly efficient for the other.

331:c purpose, and

331:e organ, serving

332 Thus in plants, the office of the pistil is to allow the pollen-tubes to reach the ovules protected in the ovarium at its base.

332:e ovules within the ovarium.

333 The pistil consists of a stigma supported on the style; but in some Compositæ, the male florets, which of course cannot be fecundated, have a pistil, which is in a rudimentary state, for it is not crowned with a stigma; but the style remains well developed, and is clothed with hairs as in other compositæ, for the purpose of brushing the pollen out of the surrounding anthers.

333:c compositæ

333:d surrounding and conjoined anthers.

333:e on a style; but in some Compositæ/have a rudimentary pistil/hairs, in the usual manner, for brushing

333:f developed and is clothed in the usual manner with hairs, which serve to brush

334 Again, an organ may become rudimentary for its proper purpose, and be used for a distinct object: in certain fish the swimbladder seems to be rudimentary for its proper function of giving buoyancy, but has become converted into a nascent breathing organ or lung.

334:b to be nearly rudimentary

334:f distinct one: in certain fishes

335 Other similar instances could be given.

335:f Many similar

335.1:b[¶] Organs, however little developed, if of use, should not be called rudimentary; they cannot properly be said to be in an atrophied condition; they may be called nascent, and may hereafter be developed to any extent by natural selection.

335.1:c rudimentary: they may be called nascent, and may hereafter be developed by natural selection to any further extent.

335.1:e be considered as rudimentary

335.1.x-y:f Useful organs, however little they may be developed, unless we have reason to suppose that they were formerly more highly developed, ought not to be considered as rudimentary. *y* They may be in a nascent condition, and in progress towards further development.

335.2:b Rudimentary organs, on the other hand, are essentially useless, as teeth which never cut through the gums; in a still less developed condition, they would be of still less use.

335.2.x:c gums.

335.2.x:f are either quite useless, such as teeth which never cut through the gums, or almost useless, such as the wings of an ostrich, which serve merely as sails.

335.3:b They cannot, therefore, under their present condition, have been formed by natural selection, which acts solely by the preservation of useful modifications; they have been retained, as we shall see, by inheritance, and relate to a former condition of their possessor.

335.2.y + 3.x:c As when in a still less developed condition they would be of still less use, they cannot under the present state of things have been formed/modifications.

335.2.y + 3.x:d As they would be of still less use, when in a still less developed condition, they

335.2.y + 3.x:e of even less use, when in a still less developed condition, they cannot have been formed through variation and natural selection, which latter acts

335.2.y + 3.x:f As organs in this condition would formerly, when still less developed, have been of even less use than at present, they cannot formerly have been produced through variation and natural selection, which acts

335.3.y:c They relate to a former condition of their possessor, and have been retained, as we shall see, by inheritance.

335.3.y:e former state of things, and have been partially retained by the power of inheritance.

335.3.y:f They have been partially retained by the power of inheritance, and relate to a former state of things.

335.4:b It is difficult to know what are nascent organs; looking to the future, we cannot of course tell how any part will be developed, and whether it is now nascent; looking to the past, creatures with an organ in a nascent condition will generally have been supplanted and exterminated by their successors with the organ in a more perfect and developed condition.

335.4:c what organs are nascent; looking to the future

335.4:d now in a nascent condition; looking/successors with the same organ in a more perfect and developed condition, and consequently will not now exist.

335.4:e with an organ in this condition will generally/perfect state, and consequently will have become long ago extinct.

335.4.x-y:f It is, however, often difficult to distinguish between rudimentary and nascent organs; for we can judge only by analogy whether a part is capable of further development, in which case alone it deserves to be called nascent. *y* Organs in this condition will always be somewhat rare; for beings thus provided will commonly have been supplanted

335.5:b The wing of the penguin is of high service, and acts as a fin; it may, therefore, represent the nascent state of the wings of birds; not that I believe this to be the case; it is more probably a reduced organ, modified for a new function: the wing of the Apteryx is useless, and is truly rudimentary.

335.5:d service, acting/state of the wing; not/Apteryx, on the other hand, is quite useless

335.5:f state of the wing: not

335.5.1:d The simple filamentary limbs of the Lepidosiren apparently are in a nascent state; for, as Owen has recently remarked, they are the "beginnings of organs which attain full functional development in higher vertebrates."

335.5.1:f Owen considers the simple filamentary limbs of the Lepidosiren as the "beginnings/vertebrates;" but, according to the view lately advocated by Dr. Günther, they are probably remnants, consisting of the persistent axis of a fin, with the lateral rays or branches aborted.

335.6:b The mammary glands of the Ornithorhynchus may, perhaps, be considered, in comparison with the udder of a cow, as in a nascent state.

335.6:d may, probably, be considered, in comparison with those of the cow, as in a nascent condition.

335.6:e may be considered, in comparison with the udders of a cow

335.7:b The ovigerous frena of certain cirripedes, which are only slightly developed and which have ceased to give attachment to the ova, are nascent branchiæ.

335.7:f which have ceased to give attachment to the ova and are feebly developed, are

336 Rudimentary organs in the individuals of the same species are very liable to vary in degree of development and in other respects.

336:e organs are very liable to vary in development and in other respects in the individuals of the same species.

336:f organs in the individuals of the same species are very liable to vary in the degree of their development and in other respects.

337 Moreover, in closely allied species, the degree to which the same organ has been rendered rudimentary occasionally differs much.

337:e been reduced occasionally

337:f In closely allied species, also, the extent to

338 This latter fact is well exemplified in the state of the wings of the female moths in certain groups.

338:d wings of female

338:e wings in female moths

338:f wings of female moths belonging to the same family.

339 Rudimentary organs may be utterly aborted; and this implies, that we find in an animal or plant no trace of an organ, which analogy would lead us to expect to find, and which is occasionally found in monstrous individuals of the species.

339:e that in certain animals or plants, parts are entirely absent which analogy would lead us to expect to find, and which are occasionally found in monstrous individuals.

339:f find in them, and

340 Thus in the snapdragon (antirrhinum) we generally do not find a rudiment of a fifth stamen; but this may sometimes be seen.

340:d Thus in some Scrophulariaceæ we rarely find even a rudiment/seen plainly or fully developed.

340:e in most of the Scrophulariaceæ the fifth stamen is utterly aborted; yet we may conclude that a fifth stamen once existed, for a rudiment of it is found in many species of the family, and this rudiment occasionally becomes perfectly developed, as may be seen in the common snap-dragon.

340:f may sometimes be

341 In tracing the homologies of the same part in different members of a class, nothing is more common, or more necessary, than the use and discovery of rudiments.

341:d members of the same class

341:e of any part/necessary, in order fully to understand the relations of the parts, than the discovery

341:f or, in order fully to understand the relations of the parts, more useful than

342 This is well shown in the drawings given by Owen of the bones of the leg of the horse, ox, and rhinoceros.

342:f of the leg-bones of

343 It is an important fact that rudimentary organs, such as teeth in the upper jaws of whales and ruminants, can often be detected in the embryo, but afterwards wholly disappear.

344 It is also, I believe, a universal rule, that a rudimentary part or organ is of greater size relatively to the adjoining parts in the embryo, than in the adult; so that the organ at this early

age is less rudimentary, or even cannot be said to be in any degree rudimentary.

344:e part is of

344:f size in the embryo relatively to the adjoining parts, than

345 Hence, also, a rudimentary organ in the adult, is often said to have retained its embryonic condition.

345:b adult is

345:e Hence rudimentary organs in the adult are often said to have retained their embryonic

346 I have now given the leading facts with respect to rudimentary organs.

347 In reflecting on them, every one must be struck with astonishment: for the same reasoning power which tells us plainly that most parts and organs are exquisitely adapted for certain purposes, tells us with equal plainness that these rudimentary or atrophied organs, are imperfect and useless.

347:c atrophied organs are

347:e us that most

347:f astonishment; for

348 In works on natural history rudimentary organs are generally said to have been created "for the sake of symmetry," or in order "to complete the scheme of nature;" but this seems to me no explanation, merely a re-statement of the fact.

348:d merely an imposing re-statement

348.x-y:e history, rudimentary/nature." *y* But this is not an explanation, merely a re-statement

348.1:e Nor is it consistent with itself: thus the boa-constrictor has rudiments of hind-limbs and of a pelvis, and if it be said that these bones have been retained "to complete the scheme of nature," why, as Professor Weismann asks, have they not been retained by other snakes, which do not possess even a vestige of these same bones?

349 Would it be thought sufficient to say that because planets revolve in elliptic courses round the sun, satellites follow the same course round the planets, for the sake of symmetry, and to complete the scheme of nature?

349:d same course round their planets

349:e What would be thought of an astronomer, who maintained that the satellites revolve in elliptic courses round their planets "for the sake of symmetry," because the planets thus revolve round the sun?

349:f astronomer who

350 An eminent physiologist accounts for the presence of rudimentary organs, by supposing that they serve to excrete matter in excess, or injurious to the system; but can we suppose that the

minute papilla, which often represents the pistil in male flowers, and which is formed merely of cellular tissue, can thus act?

350:e excess, or matter injurious/formed of mere cellular

351 Can we suppose that the formation of rudimentary teeth which are subsequently absorbed, can be of any service to the rapidly growing embryonic calf by the excretion of precious phosphate of lime?

351:c teeth, which

351:e that rudimentary teeth, which are subsequently absorbed, are beneficial to the rapidly growing embryonic calf by removing matter so precious as phosphate

352 When a man's fingers have been amputated, imperfect nails sometimes appear on the stumps: I could as soon believe that these vestiges of nails have appeared, not from unknown laws of growth, but in order to excrete horny matter, as that the rudimentary nails on the fin of the manatee were formed for this purpose.

352:e nails have been known to appear on the stumps, and I could as soon believe that these vestiges of nails have been developed in order to excrete/manatee have been developed for this same purpose.

352:f nails are developed

353 On my view of descent with modification, the origin of rudimentary organs is simple.

353:e On the view

353:f is comparatively simple; and we can understand to a large extent the laws governing their imperfect development.

354 We have plenty of cases of rudimentary organs in our domestic productions,—as the stump of a tail in tailless breeds,—the vestige of an ear in earless breeds,—the reappearance of minute dangling horns in hornless breeds of cattle, more especially, according to Youatt, in young animals,—and the state of the whole flower in the cauliflower.

354:e earless breeds of sheep,—the reappearance

355 We often see rudiments of various parts in monsters.

356 But I doubt whether any of these cases throw light on the origin of rudimentary organs in a state of nature, further than by showing that rudiments can be produced; for I doubt whether species under nature ever undergo abrupt changes.

355 + 6:f in monsters; but/produced; for the balance of evidence clearly indicates that species under nature do not undergo great and abrupt

355 + 6.1:f But we learn from the study of our domestic productions that the disuse of parts leads to their reduced size; and that the result is inherited.

357 I believe that disuse has been the main agency; that it has led

713

in successive generations to the gradual reduction of various organs, until they have become rudimentary,—as in the case of the eyes of animals inhabiting dark caverns, and of the wings of birds inhabiting oceanic islands, which have seldom been forced to take flight, and have ultimately lost the power of flying.

357:e forced by beasts of prey to

357.x-y:f[¶] It appears probable that disuse has been the main agent in rendering organs rudimentary. *y* It would at first lead by slow steps to the more and more complete reduction of a part, until at last it became

358 Again, an organ useful under certain conditions, might become injurious under others, as with the wings of beetles living on small and exposed islands; and in this case natural selection would continue slowly to reduce the organ, until it was rendered harmless and rudimentary.

358:f organ, useful/selection will have aided in reducing

359 Any change in function, which can be effected by insensibly small steps, is within the power of natural selection; so that an organ rendered, during changed habits of life, useless or injurious for one purpose, might easily be modified and used for another purpose.

359:b might be modified

359:e in structure and function/rendered, through changed

359:f effected by small stages, is

360 Or an organ might be retained for one alone of its former functions.

360:e An organ might, also, be retained

361 An organ, when rendered useless, may well be variable, for its variations cannot be checked by natural selection.

361:e organ, originally formed by the aid of natural selection, when/variations can no longer be

361:f Organs, originally formed by the aid of natural selection, when rendered useless may well be variable, for their variations

361.1:f All this agrees well with what we see under nature.

362 At whatever period of life disuse or selection reduces an organ, and this will generally be when the being has come to maturity and to its full powers of action, the principle of inheritance at corresponding ages will reproduce the organ in its reduced state at the same age, and consequently will seldom affect or reduce it in the embryo.

362:d age, but will

362:e life either disuse/maturity and has to exert its/same mature age, but will seldom affect it

362:f Moreover, at whatever/ages will tend to reproduce

363 Thus we can understand the greater relative size of rudimentary organs in the embryo, and their lesser relative size in the adult.

363:e greater size of rudimentary organs in the embryo relatively to its other parts, and

363:f relatively to the adjoining parts

363.1:f If, for instance, the digit of an adult animal was used less and less during many generations, owing to some change of habits, or if an organ or gland was less and less functionally exercised, we may infer that it would become reduced in size in the adult descendants of this animal, but would retain nearly its original standard of development in the embryo.

363.2-6:f[¶] There remains, however, this difficulty. *3* After an organ has ceased being used, and has become in consequence much reduced, how can it be still further reduced in size until the merest vestige is left; and how can it be finally quite obliterated? *4* It is scarcely possible that disuse can go on producing any further effect after the organ has once been rendered functionless. *5* Some additional explanation is here requisite which I cannot give. *6* If, for instance, it could be proved that every part of the organisation tends to vary in a greater degree towards diminution than towards augmentation of size, then we should be able to understand how an organ which has become useless would be rendered, independently of the effects of disuse, rudimentary and would at last be wholly suppressed; for the variations towards diminished size would no longer be checked by natural selection.

364 But if each step of the process of reduction were to be inherited, not at the corresponding age, but at an extremely early period of life (as we have good reason to believe to be possible) the rudimentary part would tend to be wholly lost, and we should have a case of complete abortion.

364:b possible), the rudimentary

364:e at a corresponding age, but at a very early period of life, the rudimentary

364:[f]

365 The principle, also, of economy, explained in a former chapter, by which the materials forming any part or structure, if not useful to the possessor, will be saved as far as is possible, will probably often come into play; and this will tend to cause the entire obliteration of a rudimentary organ.

365:d economy in organisation, explained

365:e of the economy of organisation/possible, may often have come into play, and aided in the entire

365:f The principle of the economy of growth, explained in a former chapter, by which the materials forming any part, if not useful to the possessor, are saved as far as is possible, will perhaps come into play in rendering a useless part rudimentary.

365.1:f But this principle will almost necessarily be confined to the earlier stages of the process of reduction; for we cannot suppose that a minute papilla, for instance, representing in a male flower the pistil of the female flower, and formed merely of cellular tissue, could be further reduced or absorbed for the sake of economising nutriment.

366 As the presence of rudimentary organs is thus due to the tendency in every part of the organisation, which has long existed, to be inherited—we can understand, on the genealogical view of classification, how it is that systematists have found rudimentary parts as useful as, or even sometimes more useful than, parts of high physiological importance.

366:f Finally, as rudimentary organs, by whatever steps they may have been degraded into their present useless condition, are the record of a former state of things, and have been retained solely through the power of inheritance,—we can understand, on the genealogical view of classification, how it is that systematists, in placing organisms in their proper places in the natural system, have often found

367 Rudimentary organs may be compared with the letters in a word, still retained in the spelling, but become useless in the pronunciation, but which serve as a clue in seeking for its derivation.

367:e clue for

368 On the view of descent with modification, we may conclude that the existence of organs in a rudimentary, imperfect, and useless condition, or quite aborted, far from presenting a strange difficulty, as they assuredly do on the ordinary doctrine of creation, might even have been anticipated, and can be accounted for by the laws of inheritance.

368:e anticipated in accordance with the views here explained.

368:f do on the old doctrine

369 *Summary.—*

369:d [Center] Summary. [Space]

370 In this chapter I have attempted to show, that the subordination of group to group in all organisms throughout all time; that the nature of the relationship, by which all living and extinct beings are united by complex, radiating, and circuitous lines of affinities into one grand system; the rules followed and the difficulties encountered by naturalists in their classifications; the value set upon characters, if constant and prevalent, whether of high vital importance, or of the most trifling importance, or, as in rudimentary organs, of no importance; the wide opposition in value between analogical or adaptive characters, and characters of true affinity; and other such rules;— all naturally follow on the view of the common parentage of those forms which are considered by naturalists as allied, together with their modification through natural selection, with its contingencies of extinction and divergence of character.

716

370:d[¶] all organic beings throughout/extinct organisms are united

370:e that the arrangement of all organic beings throughout all time in group under group—that the nature/affinities in a few grand classes,—the rules followed and the difficulties encountered by naturalists in their classifications,—the value set/high or the most trifling importance, or, as with rudimentary organs, of no importance,—the wide/true affinity; and other such rules; —all naturally follow if we admit the common parentage of allied forms, together

370:f in groups under groups—that the nature of the relationships by which/affinities into a few/through variation and natural selection, with the contingencies

371 In considering this view of classification, it should be borne in mind that the element of descent has been universally used in ranking together the sexes, ages, and acknowledged varieties of the same species, however different they may be in structure.

371:d ages, dimorphic states, and

371:f however much they may differ from each other in

372 If we extend the use of this element of descent,—the only certainly known cause of similarity in organic beings,—we shall understand what is meant by the natural system: it is genealogical in its attempted arrangement, with the grades of acquired difference marked by the terms varieties, species, genera, families, orders, and classes.

372:d arrangement, and the grades of acquired difference are marked

372:e descent,—the one certainly/Natural System/terms, varieties

373 On this same view of descent with modification, all the great facts in Morphology become intelligible,—whether we look to the same pattern displayed in the homologous organs, to whatever purpose applied, of the different species of a class; or to the homologous parts constructed on the same pattern in each individual animal and plant.

373:d species in the same class

373:e displayed by the different species of the same class in their homologous organs, to whatever purpose applied; or to the homologous parts in

373:f modification, most of the great/or to the serial and lateral homologies in

374 On the principle of successive slight variations, not necessarily or generally supervening at a very early period of life, and being inherited at a corresponding period, we can understand the great leading facts in Embryology; namely, the resemblance in an individual embryo of the homologous parts, which when matured will become widely different from each other in structure and function; and the resemblance in differ-

ent species of a class of the homologous parts or organs, though fitted in the adult members for purposes as different as possible.

374:c matured become

374:d species of the same class

374:e namely, the close resemblance in the individual embryo of the parts which are homologous, and which when matured become widely different in structure and function; and the resemblance in allied though very distinct species of their homologous parts or organs, though fitted in the adult state for purposes as different as is possible.

374:f understand the leading/function; and the resemblance of the homologous parts or organs in allied though distinct species, though fitted in the adult state for habits as different

375 Larvæ are active embryos, which have become specially modified in relation to their habits of life, through the principle of modifications being inherited at corresponding ages.

375:e have been specially modified in a greater or less degree in relation to their habits of life, with their modifications inherited at a corresponding age.

375:f corresponding early age.

376 On this same principle—and bearing in mind, that when organs are reduced in size, either from disuse or selection, it will generally be at that period of life when the being has to provide for its own wants, and bearing in mind how strong is the principle of inheritance—the occurrence of rudimentary organs and their final abortion, present to us no inexplicable difficulties; on the contrary, their presence might have been even anticipated.

376:d might even have been anticipated.

376:e On these same principles/disuse or through natural selection/ strong is the force of inheritance—the occurrence of rudimentary organs might

377 The importance of embryological characters and of rudimentary organs in classification is intelligible, on the view that an arrangement is only so far natural as it is genealogical.

377:e view that a natural arrangement must be genealogical.

378 Finally, the several classes of facts which have been considered in this chapter, seem to me to proclaim so plainly, that the innumerable species, genera, and families of organic beings, with which this world is peopled, have all descended, each within its own class or group, from common parents, and have all been modified in the course of descent, that I should without hesitation adopt this view, even if it were unsupported by other facts or arguments.

378:e families, with which this world is peopled, are all descended, each/by any other

378:f genera and families/by other

CHAPTER XIV.

RECAPITULATION AND CONCLUSION.

3 Recapitulation of the difficulties on the theory of Natural Selection—Recapitulation of the general and special circumstances in its favour—Causes of the general belief in the immutability of species—How far the theory of natural selection may be extended—Effects of its adoption on the study of Natural history—Concluding remarks.

3:d far the theory of Natural Selection

3:e of the objections to the theory

4 As this whole volume is one long argument, it may be convenient to the reader to have the leading facts and inferences briefly recapitulated.

5   That many and grave objections may be advanced against the theory of descent with modification through natural selection, I do not deny.

5:b and serious objections

5:e through variation and natural

6 I have endeavoured to give to them their full force.

7 Nothing at first can appear more difficult to believe than that the more complex organs and instincts should have been perfected, not by means superior to, though analogous with, human reason, but by the accumulation of innumerable slight variations, each good for the individual possessor.

7:e instincts have

8 Nevertheless, this difficulty, though appearing to our imagination insuperably great, cannot be considered real if we admit the following propositions, namely,—that gradations in the perfection of any organ or instinct, which we may consider, either do now exist or could have existed, each good of its kind,—that all organs and instincts are, in ever so slight a degree, variable, —and, lastly, that there is a struggle for existence leading to the preservation of each profitable deviation of structure or instinct.

8:b instinct which

8:c namely, that all organs and instincts are, in ever so slight a degree, variable—that there is a struggle for existence leading to the preservation of each profitable deviation of structure or instinct—and, lastly, that gradations in the perfection of every organ may have existed, each good of its kind.

8:e all parts of the organisation and instincts offer, at least, individual differences—that there is a struggle for existence leading to the preservation of profitable deviations of structure or instinct —and, lastly, that gradations in the state of perfection of each organ

9 The truth of these propositions cannot, I think, be disputed.

10   It is, no doubt, extremely difficult even to conjecture by what gradations many structures have been perfected, more especially amongst broken and failing groups of organic beings; but we see so many strange gradations in nature, as is proclaimed by the canon, "Natura non facit saltum," that we ought to be extremely cautious in saying that any organ or instinct, or any whole being, could not have arrived at its present state by many graduated steps.

10:b nature, that we

10:c beings, which have suffered much extinction; but

10:e instinct, or the whole structure, could

10:f instinct, or any whole

11 There are, it must be admitted, cases of special difficulty on the theory of natural selection; and one of the most curious of these is the existence of two or three defined castes of workers or sterile females in the same community of ants; but I have attempted to show how this difficulty can be mastered.

11:c how these difficulties

11:e difficulty opposed to the theory/sterile female ants in the same community; but

11:f existence in the same community of two or three defined castes of workers or sterile female ants; but

12   With respect to the almost universal sterility of species when first crossed, which forms so remarkable a contrast with the almost universal fertility of varieties when crossed, I must refer the reader to the recapitulation of the facts given at the end of the eighth chapter, which seem to me conclusively to show that this sterility is no more a special endowment than is the incapacity of two trees to be grafted together; but that it is incidental on constitutional differences in the reproductive systems of the intercrossed species.

12:d two distinct trees/on differences confined to the reproductive

12:f end of the ninth chapter/distinct kinds of trees

13 We see the truth of this conclusion in the vast difference in the result, when the same two species are crossed reciprocally; that is, when one species is first used as the father and then as the mother.

13:d reciprocally,—that/mother: analogy from the consideration of

720

dimorphic and trimorphic plants clearly leads us to the same conclusion, for when the forms are illegitimately united, they yield few or no seed, and their offspring are more or less sterile; and these forms of the same undoubted species differ in no respect from each other except in their reproductive organs and functions.

13:e in the results of crossing the same two species reciprocally,— that/leads to

13.x-y:f mother. *y* Analogy/these forms belong to the same undoubted species, and differ from each other in no respect except

14 The fertility of varieties when intercrossed and of their mongrel offspring cannot be considered as universal; nor is their very general fertility surprising when we remember that it is not likely that either their constitutions or their reproductive systems should have been profoundly modified.

14.x:c Although the fertility/offspring has been asserted by so many authors to be universal, this cannot be considered correct after the facts given on the authority of Gärtner and Kölreuter.

14.x:f considered as quite correct after the facts given on the high authority

14.y:c Nor is the very general fertility of varieties, when crossed, surprising, when

14.y:d that their reproductive

14.y:[f]

15 Moreover, most of the varieties which have been experimentised on have been produced under domestication; and as domestication apparently tends to eliminate sterility, we ought not to expect it also to produce sterility.

15:b as domestication (I do not mean mere confinement) apparently

15:c experimented

15:d confinement) almost certainly tends

15:f Most/eliminate that sterility which, judging from analogy, would have affected the parent-species if intercrossed, we ought not to expect that domestication would likewise induce sterility in their modified descendants when crossed.

15.1:f This elimination of sterility apparently follows from the same cause which allows our domestic animals to breed freely under diversified circumstances; and this again apparently follows from their having been gradually accustomed to frequent changes in their conditions of life.

15.2-10:f[¶] A double and parallel series of facts seems to throw much light on the sterility of species, when first crossed, and of their hybrid offspring. *3* On the one side, there is good reason to believe that slight changes in the conditions of life give vigour and fertility to all organic beings. *4* We know also that a cross between the distinct individuals of the same variety, and between distinct varieties, increases the number of their offspring, and certainly gives to them increased size and vigour. 5 This is chiefly owing to the forms which are crossed having been exposed to somewhat different conditions of life; for I have ascertained by a laborious series of experiments that if all the individuals of the same variety be subjected during several generations to the same conditions, the good derived from crossing is often much diminished or wholly disappears. 6 This is one side of the case. 7 On the other side, we know that species which have long been exposed to nearly uniform conditions, when they are subjected under confinement to new and greatly changed conditions, either perish, or if they survive, are rendered sterile, though retaining perfect health. 8 This does occur, or only in a very slight degree, with our domesticated productions, which have long been exposed to fluctuating conditions. 9 Hence, when we find that hybrids produced by a cross between two distinct species are few in number, owing to their perishing soon after conception or at a very early age, or if surviving that they are rendered more or less sterile, it seems highly probable that this result is due to their having been in fact subjected to a great change in their conditions of life, from being compounded of two distinct organisations. 10 He who will explain in a definite manner why, for instance, an elephant or a fox will not breed under confinement in its native country, whilst the domestic pig or dog will breed freely under the most diversified conditions, will at the same time be able to give a definite answer to the question why two distinct species, when crossed, as well as their hybrid offspring, are generally rendered more or less sterile, whilst two domesticated varieties when crossed and their mongrel offspring are perfectly fertile.

16   The sterility of hybrids is a very different case from that of first crosses, for their reproductive organs are more or less functionally impotent; whereas in first crosses the organs on both sides are in a perfect condition.

16:c that of a first cross, for/whereas in first crosses, the organs of both species are

16:d is a different case from that of a first cross, for the reproductive organs of hybrids are more/species are of course in

16:[f]

17 As we continually see that organisms of all kinds are rendered in some degree sterile from their constitutions having been disturbed by slightly different and new conditions of life, we need not feel surprise at hybrids being in some degree sterile, for their constitutions can hardly fail to have been disturbed from being compounded of two distinct organisations.

17:e from being exposed to slightly changed conditions, we/fail to be disturbed from being compounded of two distinct organisations; but whether this is the true cause of their sterility I will not pretend to decide.

17:[f]

18 This parallelism is supported by another parallel, but directly opposite, class of facts; namely, that the vigour and fertility of all organic beings are increased by slight changes in their conditions of life, and that the offspring of slightly modified forms or varieties acquire from being crossed increased vigour and fertility.

18:c facts, namely

18:d varieties when crossed acquire increased

18:e The above parallelism is supported by another

18:[f]

19 So that, on the one hand, considerable changes in the conditions of life and crosses between greatly modified forms, lessen fertility; and on the other hand, lesser changes in the conditions of life and crosses between less modified forms, increase fertility.

19:e hand, a considerable change in the conditions of life and crosses between greatly

19:[f]

20 Turning to geographical distribution, the difficulties encountered on the theory of descent with modification are grave enough.

20:c are serious enough.

21 All the individuals of the same species, and all the species of the same genus, or even higher group, must have descended from common parents; and therefore, in however distant and isolated parts of the world they are now found, they must in the course of successive generations have passed from some one part to the others.

21:d they may now be found/one point to all the others.

21:e generations have travelled from

21:f group, are descended

22 We are often wholly unable even to conjecture how this could have been effected.

23 Yet, as we have reason to believe that some species have retained the same specific form for very long periods, enormously long as measured by years, too much stress ought not to be laid on the occasional wide diffusion of the same species; for during very long periods of time there will always be a good chance for wide migration by many means.

23:b always have been a good

23:d long periods of time, enormously/during very long periods there

23:e time, immensely long as/always be a good

23:f always have been a good

24 A broken or interrupted range may often be accounted for by the extinction of the species in the intermediate regions.

25 It cannot be denied that we are as yet very ignorant of the full extent of the various climatal and geographical changes which have affected the earth during modern periods; and such changes will obviously have greatly facilitated migration.

25:e such changes may obviously have facilitated

25:f ignorant as to the full/such changes will often have

26 As an example, I have attempted to show how potent has been the influence of the Glacial period on the distribution both of the same and of representative species throughout the world.

26:e distribution of the same and of allied species

27 We are as yet profoundly ignorant of the many occasional means of transport.

28 With respect to distinct species of the same genus inhabiting very distant and isolated regions, as the process of modification has necessarily been slow, all the means of migration will have been possible during a very long period; and consequently the difficulty of the wide diffusion of species of the same genus is in some degree lessened.

28:d inhabiting distant/diffusion of the species

29   As on the theory of natural selection an interminable number of intermediate forms must have existed, linking together all the species in each group by gradations as fine as our present varieties, it may be asked, Why do we not see these linking forms all around us?

29:e As according to the theory

29:f our existing varieties

30 Why are not all organic beings blended together in an inextricable chaos?

31 With respect to existing forms, we should remember that we have no right to expect (excepting in rare cases) to discover *directly* connecting links between them, but only between each and some extinct and supplanted form.

32 Even on a wide area, which has during a long period remained continuous, and of which the climate and other conditions of life change insensibly in going from a district occupied by one species into another district occupied by a closely allied species, we have no just right to expect often to find intermediate varieties in the intermediate zone.

32:e insensibly in proceeding from

32:f of which the climatic and

33 For we have reason to believe that only a few species are undergoing change at any one period; and all changes are slowly effected.

33.x-y:c species of a genus ever undergo change; the other species becoming utterly extinct and leaving no modified progeny. *y* Of the species which do change, only a few within the same country change at the same time; and all modifications are slowly effected.

34 I have also shown that the intermediate varieties which will at first probably exist in the intermediate zones, will be liable to be supplanted by the allied forms on either hand; and the latter, from existing in greater numbers, will generally be modified and improved at a quicker rate than the intermediate varieties, which exist in lesser numbers; so that the intermediate varieties will, in the long run, be supplanted and exterminated.

34:e which probably at first existed in the intermediate zones, would be liable to be supplanted by the allied forms on either hand; for the latter, from existing in greater numbers, would generally be modified and improved at a quicker rate than the intermediate varieties, which existed in lesser numbers; so that the intermediate varieties would, in

35 On this doctrine of the extermination of an infinitude of connecting links, between the living and extinct inhabitants of the world, and at each successive period between the extinct and still older species, why is not every geological formation charged with such links?

36 Why does not every collection of fossil remains afford plain evidence of the gradation and mutation of the forms of life?

37 We meet with no such evidence, and this is the most obvious and forcible of the many objections which may be urged against my theory.

37:c Although geological research has undoubtedly revealed the former existence of many links, bringing numerous forms of life much closer together, it does not yield the infinitely many fine gradations between past and present species required on my theory; and this/against it.

37:e on the theory; and this is the most obvious of the many

38 Why, again, do whole groups of allied species appear, though certainly they often falsely appear, to have come in suddenly on the several geological stages?

38:e though this appearance is often false, to have come in suddenly on the successive geological

39 Why do we not find great piles of strata beneath the Silurian system, stored with the remains of the progenitors of the Silurian groups of fossils?

39:d Although we now know that organic beings appeared on this globe, at a period incalculably remote, long before the lowest bed of the Silurian system was deposited, why do we not find beneath this system great piles of strata stored with the remains of the progenitors of the Silurian fossils?

39:e of the Cambrian system/progenitors of the Cambrian fossils?

40 For certainly on my theory such strata must somewhere have been deposited at these ancient and utterly unknown epochs in the world's history.

40:c For on

40:e on the theory

40:f theory, such/epochs of the world's

41   I can answer these questions and grave objections only on the supposition that the geological record is far more imperfect than most geologists believe.

41:c and objections

42 It cannot be objected that there has not been time sufficient for any amount of organic change; for the lapse of time has been so great as to be utterly inappreciable by the human intellect.

42:[e]

43 The number of specimens in all our museums is absolutely as nothing compared with the countless generations of countless species which certainly have existed.

43:e which have certainly existed.

43.1:c The parent form of any two or more species would not be in all its characters directly intermediate between its modified offspring, any more than the rock-pigeon is directly intermediate in crop and tail between its descendants the pouter and fantail pigeons.

43.1:d parent-form

43.1:f descendants, the pouter

44 We should not be able to recognise a species as the parent of any one or more species if we were to examine them ever so closely, unless we likewise possessed many of the intermediate links between their past or parent and present states; and these many links we could hardly ever expect to discover, owing to the imperfection of the geological record.

44:c of another species if we were to examine both ever/past and present

44:e another and modified species, if we were to examine both ever so closely, unless we possessed most of the intermediate links; and owing to the imperfection of the geological record, we have no just right to expect to find so many links.

44:f examine the two ever

44.1:c = 46:[c] If two or three, or even more linking forms were discovered, they would simply be ranked as so many new species, more especially if found in different geological sub-stages, let their differences be ever so slight.

44.1:f ranked by many naturalists as

45 Numerous existing doubtful forms could be named which are probably varieties; but who will pretend that in future ages so

many fossil links will be discovered, that naturalists will be able to decide, on the common view, whether or not these doubtful forms are varieties?

45:d decide whether or not on the common view these

45:e not these doubtful forms ought to be called varieties?

46 As long as most of the links between any two species are unknown, if any one link or intermediate variety be discovered, it will simply be classed as another and distinct species.

46:[c] = 44.1:c

47 Only a small portion of the world has been geologically explored.

48 Only organic beings of certain classes can be preserved in a fossil condition, at least in any great number.

48.1:d Many species when once formed never undergo any further change, but become extinct without leaving modified descendants; and the periods, during which species have undergone modification, though long as measured by years, have probably been short in comparison with the periods during which they have retained the same form.

48.1:f change but/they retained

49 Widely ranging species vary most, and varieties are often at first local,—both causes rendering the discovery of intermediate links less likely.

49:c local—both

49:d It is the dominant and widely ranging species which vary most frequently and vary/links in any one formation less

50 Local varieties will not spread into other and distant regions until they are considerably modified and improved; and when they do spread, if discovered in a geological formation, they will appear as if suddenly created there, and will be simply classed as new species.

50:e when they have spread, and are discovered in a geological

50:f formation, they appear

51 Most formations have been intermittent in their accumulation; and their duration, I am inclined to believe, has been shorter than the average duration of specific forms.

51:d duration has probably been shorter

52 Successive formations are separated from each other by enormous blank intervals of time; for fossiliferous formations, thick enough to resist future degradation, can be accumulated only where much sediment is deposited on the subsiding bed of the sea.

52:c formations are in most cases separated/fossiliferous formations thick enough to resist future degradation can generally be

52:e by blank intervals of time of great length; for/can as a general rule be

727

53 During the alternate periods of elevation and of stationary level the record will be blank.

53:c be generally blank.

53:e will generally be blank.

54 During these latter periods there will probably be more variability in the forms of life; during periods of subsidence, more extinction.

55 With respect to the absence of fossiliferous formations beneath the lowest Silurian strata, I can only recur to the hypothesis given in the ninth chapter.

55:d absence beneath the lowest Silurian strata of formations rich in fossils of many kinds, I can recur only to

55:e absence of strata rich in fossils beneath the Cambrian formation, I

55:f given in the tenth chapter; namely, that though our continents and oceans have endured for an enormous period in nearly their present relative positions, we have no reason to assume that this has always been the case; consequently formations much older than any now known may lie buried beneath the great oceans.

55.1:f With respect to the lapse of time not having been sufficient since our planet was consolidated for the assumed amount of organic change, and this objection, as urged by Sir William Thompson, is probably one of the gravest as yet advanced, I can only say, firstly, that we do not know at what rate species change as measured by years, and secondly, that many philosophers are not as yet willing to admit that we know enough of the constitution of the universe and of the interior of our globe to speculate with safety on its past duration.

56 That the geological record is imperfect all will admit; but that it is imperfect to the degree which I require, few will be inclined to admit.

56:e degree required by our theory, few

56:f[¶]

57 If we look to long enough intervals of time, geology plainly declares that all species have changed; and they have changed in the manner which my theory requires, for they have changed slowly and in a graduated manner.

57:e that species have all changed; and they have changed in the manner required

57:f required by the theory, for

58 We clearly see this in the fossil remains from consecutive formations invariably being much more closely related to each other, than are the fossils from formations distant from each other in time.

58:e from widely separated formations.

59   Such is the sum of the several chief objections and difficulties which may justly be urged against my theory; and I have now briefly recapitulated the answers and explanations which can be given to them.

59:e against the theory

59:f may be justly urged/which, as far as I can see, may be

60 I have felt these difficulties far too heavily during many years to doubt their weight.

61 But it deserves especial notice that the more important objections relate to questions on which we are confessedly ignorant; nor do we know how ignorant we are.

62 We do not know all the possible transitional gradations between the simplest and the most perfect organs; it cannot be pretended that we know all the varied means of Distribution during the long lapse of years, or that we know how imperfect the Geological Record is.

62:f imperfect is the Geological Record.

63 Grave as these several difficulties are, in my judgment they do not overthrow the theory of descent with modification.

63:b descent from a few created forms with subsequent modification.

63:c few primordial forms

63:d Serious as

63:e several objections are, in my judgment they are not sufficient to overthrow the theory of descent with

63:f they are by no means sufficient

64   Now let us turn to the other side of the argument.

65 Under domestication we see much variability.

66 This seems to be mainly due to the reproductive system being eminently susceptible to changes in the conditions of life; so that this system, when not rendered impotent, fails to reproduce offspring exactly like the parent-form.

66:d be in part due

65 + 6:e variability, caused, or at least excited, by changed conditions of life.

65 + 6:f life; but often in so obscure a manner, that we are tempted to consider the variations as spontaneous.

67 Variability is governed by many complex laws,—by correlation of growth, by use and disuse, and by the direct action of the physical conditions of life.

67:e This variability/correlation, by use and disuse, and by the definite action of the surrounding conditions.

67:f Variability is governed by many complex laws,—by correlated

729

growth, compensation, the increased use and disuse of parts, and the definite

68 There is much difficulty in ascertaining how much modification our domestic productions have undergone; but we may safely infer that the amount has been large, and that modifications can be inherited for long periods.

68:e how largely our domestic productions have been modified; but

69 As long as the conditions of life remain the same, we have reason to believe that a modification, which has already been inherited for many generations, may continue to be inherited for an almost infinite number of generations.

70 On the other hand we have evidence that variability, when it has once come into play, does not wholly cease; for new varieties are still occasionally produced by our most anciently domesticated productions.

70:e not cease under domestication for a very long period; for new varieties are still occasionally produced by our oldest domesticated

70:f period; nor do we know that it ever ceases, for

71 Man does not actually produce variability; he only unintentionally exposes organic beings to new conditions of life, and then nature acts on the organisation, and causes variability.

71:f Variability is not actually caused by man; he/organisation and causes it to vary.

72 But man can and does select the variations given to him by nature, and thus accumulate them in any desired manner.

72:f accumulates

73 He thus adapts animals and plants for his own benefit or pleasure.

74 He may do this methodically, or he may do it unconsciously by preserving the individuals most useful to him at the time, without any thought of altering the breed.

74:c time without

74:d him without

74:e useful or pleasing to him without any intention of

75 It is certain that he can largely influence the character of a breed by selecting, in each successive generation, individual differences so slight as to be quite inappreciable by an uneducated eye.

75:c be inappreciable

75:e inappreciable except by an educated

76 This process of selection has been the great agency in the production of the most distinct and useful domestic breeds.

76:e in the formation of

76:f This unconscious process

77 That many of the breeds produced by man have to a large extent the character of natural species, is shown by the inextricable doubts whether very many of them are varieties or aboriginal species.

77:c aboriginally distinct species.

77:e whether many

77:f many breeds

78 There is no obvious reason why the principles which have acted so efficiently under domestication should not have acted under nature.

78:e not act

78:f no reason/not have acted

79 In the preservation of favoured individuals and races, during the constantly-recurrent Struggle for Existence, we see the most powerful and ever-acting means of selection.

79:e In the survival of favoured/see a powerful and ever-acting form of Selection.

80 The struggle for existence inevitably follows from the high geometrical ratio of increase which is common to all organic beings.

81 This high rate of increase is proved by calculation, by the effects of a succession of peculiar seasons, and by the results of naturalisation, as explained in the third chapter.

81:b calculation,—by the rapid increase of many animals and plants during a succession of peculiar seasons, or when naturalised in a new country.

81:e seasons, and when

81:f in new countries.

82 More individuals are born than can possibly survive.

83 A grain in the balance will determine which individual shall live and which shall die,—which variety or species shall increase in number, and which shall decrease, or finally become extinct.

83:e balance may determine which individuals

84 As the individuals of the same species come in all respects into the closest competition with each other, the struggle will generally be most severe between them; it will be almost equally severe between the varieties of the same species, and next in severity between the species of the same genus.

85 But the struggle will often be very severe between beings most remote in the scale of nature.

85:e On the other hand the struggle will often be very severe between beings remote

85:f be severe

86 The slightest advantage in one being, at any age or during any season, over those with which it comes into competition, or

better adaptation in however slight a degree to the surrounding physical conditions, will turn the balance.

86:e The slightest advantage in certain individuals, at/which they come

86:f will, in the long run, turn

87   With animals having separated sexes there will in most cases be a struggle between the males for possession of the females.

87:d for the possession

87:e will be in most cases a struggle

87:f sexes, there

88   The most vigorous individuals, or those which have most successfully struggled with their conditions of life, will generally leave most progeny.

88:d vigorous males, or

89   But success will often depend on having special weapons or means of defence, or on the charms of the males; and the slightest advantage will lead to victory.

89:d on the males having special weapons or means of defence, or on their charms; and

89:e weapons, or means of defence, or charms; and a slight

90   As geology plainly proclaims that each land has undergone great physical changes, we might have expected that organic beings would have varied under nature, in the same way as they generally have varied under the changed conditions of domestication.

90:e expected to find that organic beings have varied under nature, in the same way as they have varied under domestication.

91   And if there be any variability under nature, it would be an unaccountable fact if natural selection had not come into play.

91:e selection did not

91:f there has been any/selection had not

92   It has often been asserted, but the assertion is quite incapable of proof, that the amount of variation under nature is a strictly limited quantity.

92:e is incapable

93   Man, though acting on external characters alone and often capriciously, can produce within a short period a great result by adding up mere individual differences in his domestic productions; and every one admits that there are at least individual differences in species under nature.

93:e that species present individual differences.

94   But, besides such differences, all naturalists have admitted the existence of varieties, which they think sufficiently distinct to be worthy of record in systematic works.

94:c they have considered sufficiently distinct to be worthy of record in their systematic

94:e naturalists admit that varieties exist, which are considered sufficiently distinct to be worthy of record in systematic

94:f that natural varieties

95 No one can draw any clear distinction between individual differences and slight varieties; or between more plainly marked varieties and sub-species, and species.

95:e one has drawn

96 Let it be observed how naturalists differ in the rank which they assign to the many representative forms in Europe and North America.

96:d On separate continents, and on different parts of the same continent when divided by barriers of any kind, and on the several islands in the same archipelago, what a host of forms exist, which some experienced naturalists rank as mere varieties, others as geographical races or sub-species, and others as distinct, though closely allied species!

96:e kind, and on outlying islands, what a multitude of forms exist, which some experienced naturalists rank as varieties

97   If then we have under nature variability and a powerful agent always ready to act and select, why should we doubt that variations in any way useful to beings, under their excessively complex relations of life, would be preserved, accumulated, and inherited?

97:c have variability as well as a powerful agent always ready to act, why

97:d If then animals and plants do vary, let it be ever so slowly or so little, why should we doubt that variations in some way useful to them under their extremely complex relations of life would occasionally occur, and then be preserved and accumulated by natural selection?

97:e so little or so slowly, why should we doubt that the variations or individual differences, which are in any way beneficial would be preserved and accumulated through natural selection, or the survival of the fittest?

97:f so slightly or slowly, why should not variations or individual differences, which are in any way beneficial, be

98 Why, if man can by patience select variations most useful to himself, should nature fail in selecting variations useful, under changing conditions of life, to her living products?

98:d in preserving or selecting

98:e If man can by patience select variations useful to him, why, under changing and complex conditions of life, should not variations useful to nature's living products often arise, and be preserved or selected?

99 What limit can be put to this power, acting during long ages

and rigidly scrutinising the whole constitution, structure, and habits of each creature,—favouring the good and rejecting the bad?

100 I can see no limit to this power, in slowly and beautifully adapting each form to the most complex relations of life.

101 The theory of natural selection, even if we looked no further than this, seems to me to be in itself probable.

101:c farther

101:f look no farther than this, seems to be in the highest degree probable.

102 I have already recapitulated, as fairly as I could, the opposed difficulties and objections: now let us turn to the special facts and arguments in favour of the theory.

102:e theory. [*Space*]

103 On the view that species are only strongly marked and permanent varieties, and that each species first existed as a variety, we can see why it is that no line of demarcation can be drawn between species, commonly supposed to have been produced by special acts of creation, and varieties which are acknowledged to have been produced by secondary laws.

104 On this same view we can understand how it is that in each region where many species of a genus have been produced, and where they now flourish, these same species should present many varieties; for where the manufactory of species has been active, we might expect, as a general rule, to find it still in action; and this is the case if varieties be incipient species.

104:f in a region

105 Moreover, the species of the larger genera, which afford the greater number of varieties or incipient species, retain to a certain degree the character of varieties; for they differ from each other by a less amount of difference than do the species of smaller genera.

106 The closely allied species also of the larger genera apparently have restricted ranges, and they are clustered in little groups round other species—in which respects they resemble varieties.

106:b ranges, and in their affinities they are

106:d other species—in both of which

106:f both respects resembling

107 These are strange relations on the view of each species having been independently created, but are intelligible if all species first existed as varieties.

107:d view that each species was independently created, but are intelligible if each existed first as a variety.

108 As each species tends by its geometrical ratio of reproduction to increase inordinately in number; and as the modified descendants of each species will be enabled to increase by so much

the more as they become more diversified in habits and structure, so as to be enabled to seize on many and widely different places in the economy of nature, there will be a constant tendency in natural selection to preserve the most divergent offspring of any one species.

108:b become diversified

108:f by as much as they become more diversified in habits and structure, so as to be able to seize

109 Hence during a long-continued course of modification, the slight differences, characteristic of varieties of the same species, tend to be augmented into the greater differences characteristic of species of the same genus.

109:d Hence, during

109:f differences characteristic

110 New and improved varieties will inevitably supplant and exterminate the older, less improved and intermediate varieties; and thus species are rendered to a large extent defined and distinct objects.

110:f improved, and intermediate

111 Dominant species belonging to the larger groups tend to give birth to new and dominant forms; so that each large group tends to become still larger, and at the same time more divergent in character.

111:c groups within each class tend to give

112 But as all groups cannot thus succeed in increasing in size, for the world would not hold them, the more dominant groups beat the less dominant.

112:f thus go on increasing

113 This tendency in the large groups to go on increasing in size and diverging in character, together with the almost inevitable contingency of much extinction, explains the arrangement of all the forms of life, in groups subordinate to groups, all within a few great classes, which we now see everywhere around us, and which has prevailed throughout all time.

113:c classes, which has

113:f with the inevitable/life in

114 This grand fact of the grouping of all organic beings seems to me utterly inexplicable on the theory of creation.

114:c beings is utterly

114:e beings under what is called the Natural System, is

115 As natural selection acts solely by accumulating slight, successive, favourable variations, it can produce no great or sudden modification; it can act only by very short and slow steps.

115:e only by short

116 Hence the canon of "Natura non facit saltum," which every

fresh addition to our knowledge tends to make more strictly
correct, is on this theory simply intelligible.

116:b make truer, is

116:e theory intelligible.

116:f tends to confirm, is

116.1:d We can see why throughout nature the same general end is
gained by an almost infinite diversity of means; for every peculiarity when once acquired is long inherited, and structures
already diversified in many ways have to be adapted for the
same general purpose.

116.1:f means, for/already modified in many different ways

117 We can plainly see why nature is prodigal in variety, though
niggard in innovation.

117:d can, in short, see

118 But why this should be a law of nature if each species has been
independently created, no man can explain.

119 Many other facts are, as it seems to me, explicable on this
theory.

120 How strange it is that a bird, under the form of woodpecker,
should have been created to prey on insects on the ground; that
upland geese, which never or rarely swim, should have been
created with webbed feet; that a thrush should have been created to dive and feed on sub-aquatic insects; and that a petrel
should have been created with habits and structure fitting it for
the life of an auk or grebe! and so on in endless other cases.

120:c auk! and

120:e geese which/feet; that a thrush-like bird should have been
created to

120:f of a woodpecker, should prey/swim, should possess webbed
feet; that a thrush-like bird should dive/petrel should have the
habits

121 But on the view of each species constantly trying to increase in
number, with natural selection always ready to adapt the slowly
varying descendants of each to any unoccupied or ill-occupied
place in nature, these facts cease to be strange, or perhaps might
even have been anticipated.

121:e strange, or might

121.1:d[¶] We can understand how it is that such harmonious
beauty generally prevails throughout nature.

121.1:f We can to a certain extent understand how it is that there is
so much beauty throughout nature; for this may be largely
attributed to the agency of selection.

121.2:d That there are exceptions according to our ideas of beauty,
no one will doubt who will look at some of the venomous
snakes, at some fish, and at certain hideous bats with a distorted
resemblance to the human face.

121.2:f That beauty, according to our sense of it, is not universal, must be admitted by every one who will look at some venomous snakes, at some fishes

121.3:d Sexual selection has given, generally to the males alone but sometimes to both sexes, the most brilliant and beautiful colours, as well as other ornaments, to our birds, butterflies, and a few other animals.

121.3:e given the most brilliant colours and other ornaments to the males, but sometimes to both sexes of many birds

121.3:f colours, elegant patterns, and other ornaments to the males, and sometimes to both sexes of many birds, butterflies, and other animals.

121.4:d It has rendered the voices of many male birds musical to their females, as well as to our ears.

121.4:e With birds it has often rendered the voice of the male musical to the female

121.5:d Flowers and fruit have been rendered conspicuous by gaudy colours in contrast with the green foliage, in order that the flowers might be easily seen, visited, and fertilised by insects, and the fruit have their seeds disseminated by birds.

121.5:e insects, and the seeds

121.5:f by brilliant colours in contrast with the green foliage, in order that the flowers may be

121.5.1:f How it comes that certain colours, sounds, and forms should give pleasure to man and the lower animals,—that is, how the sense of beauty in its simplest form was first acquired,— we do not know any more than how certain odours and flavours were first rendered agreeable.

121.6:d And lastly, some living objects have become beautiful through mere symmetry of growth.

121.6:e Lastly

121.6:[f]

122   As natural selection acts by competition, it adapts the inhabitants of each country only in relation to the degree of perfection of their associates; so that we need feel no surprise at the inhabitants of any one country, although on the ordinary view supposed to have been specially created and adapted for that country, being beaten and supplanted by the naturalised productions from another land.

122:e it renders the inhabitants of each country perfect only in relation to the other inhabitants; so that we need feel no surprise at the species of/been created and specially adapted

122:f it adapts and improves the inhabitants of each country only in relation to their co-inhabitants; so

123 Nor ought we to marvel if all the contrivances in nature be not, as far as we can judge, absolutely perfect; and if some of them be abhorrent to our ideas of fitness.

123:f perfect, as in the case even of the human eye; or if

124 We need not marvel at the sting of the bee causing the bee's own death; at drones being produced in such vast numbers for one single act, and being then slaughtered by their sterile sisters; at the astonishing waste of pollen by our fir-trees; at the instinctive hatred of the queen bee for her own fertile daughters; at ichneumonidæ feeding within the live bodies of caterpillars; and at other such cases.

124:b act, with the great majority slaughtered

124:e bee when used against an enemy often causing/such great numbers for one single act, and being then slaughtered/within the living bodies

124:f bee, when used against an enemy, causing/queen-bee/caterpillars; or at

125 The wonder indeed is, on the theory of natural selection, that more cases of the want of absolute perfection have not been observed.

125:f been detected.

126 The complex and little known laws governing variation are the same, as far as we can see, with the laws which have governed the production of so-called specific forms.

126:e governing acknowledged variations/specific differences.

126:f governing the production of varieties are the same, as far as we can judge, with the laws which have governed the production of distinct species.

127 In both cases physical conditions seem to have produced but little direct effect; yet when varieties enter any zone, they occasionally assume some of the characters of the species proper to that zone.

127.x-y:e produced some direct and definite effect, but how much we cannot say. *y* Thus when varieties enter any new station, they occasionally assume some of the characters proper to the species of that station.

128 In both varieties and species, use and disuse seem to have produced some effect; for it is difficult to resist this conclusion when we look, for instance, at the logger-headed duck, which has wings incapable of flight, in nearly the same condition as in the domestic duck; or when we look at the burrowing tucutucu, which is occasionally blind, and then at certain moles, which are habitually blind and have their eyes covered with skin; or when we look at the blind animals inhabiting the dark caves of America and Europe.

128:e produced a considerable effect; for it is impossible to resist
128:f With both/duck, which has/tucu-tucu

129 In both varieties and species correlation of growth seems to have played a most important part, so that when one part has been modified other parts are necessarily modified.

129:e In varieties and species correlated variation seems to have played an important/other parts have been necessarily

129:f With varieties and species, correlated

130 In both varieties and species reversions to long-lost characters occur.

130:f With both varieties and species, reversions to long-lost characters occasionally occur.

131 How inexplicable on the theory of creation is the occasional appearance of stripes on the shoulder and legs of the several species of the horse-genus and in their hybrids!

131:c is the variable appearance

131:d shoulders/horse-genus and of their

131:e creation is the occasional appearance

132 How simply is this fact explained if we believe that these species have descended from a striped progenitor, in the same manner as the several domestic breeds of pigeon have descended from the blue and barred rock-pigeon!

132:e species are all descended from a striped/of the pigeon are descended

133 On the ordinary view of each species having been independently created, why should the specific characters, or those by which the species of the same genus differ from each other, be more variable than the generic characters in which they all agree?

133:f should specific

134 Why, for instance, should the colour of a flower be more likely to vary in any one species of a genus, if the other species, supposed to have been created independently, have differently coloured flowers, than if all the species of the genus have the same coloured flowers?

134:f other species possess differently coloured flowers, than if all possessed the same

135 If species are only well-marked varieties, of which the characters have become in a high degree permanent, we can understand this fact; for they have already varied since they branched off from a common progenitor in certain characters, by which they have come to be specifically distinct from each other; and therefore these same characters would be more likely still to be variable than the generic characters which have been inherited without change for an enormous period.

135:e other; therefore/likely again to vary than

135:f change for an immense period.

136 It is inexplicable on the theory of creation why a part developed in a very unusual manner in any one species of a genus, and therefore, as we may naturally infer, of great importance to the species, should be eminently liable to variation; but, on my

view, this part has undergone, since the several species branched off from a common progenitor, an unusual amount of variability and modification, and therefore we might expect this part generally to be still variable.

136:d to that species, should/expect the part

136:e but, on our view, this part has undergone since

136:f manner in one species alone of a genus/undergone, since

137 But a part may be developed in the most unusual manner, like the wing of a bat, and yet not be more variable than any other structure, if the part be common to many subordinate forms, that is, if it has been inherited for a very long period; for in this case it will have been rendered constant by long-continued natural selection.

138 Glancing at instincts, marvellous as some are, they offer no greater difficulty than does corporeal structure on the theory of the natural selection of successive, slight, but profitable modifications.

138:f than do corporeal structures

139 We can thus understand why nature moves by graduated steps in endowing different animals of the same class with their several instincts.

140 I have attempted to show how much light the principle of gradation throws on the admirable architectural powers of the hive-bee.

141 Habit no doubt sometimes comes into play in modifying instincts; but it certainly is not indispensable, as we see, in the case of neuter insects, which leave no progeny to inherit the effects of long-continued habit.

141:e see in

141:f doubt often comes

142 On the view of all the species of the same genus having descended from a common parent, and having inherited much in common, we can understand how it is that allied species, when placed under considerably different conditions of life, yet should follow nearly the same instincts; why the thrush of South America, for instance, lines her nest with mud like our British species.

142:d under widely different/thrushes of tropical and temperate South America, for instance, line their nests

142:e yet follow

143 On the view of instincts having been slowly acquired through natural selection we need not marvel at some instincts being apparently not perfect and liable to mistakes, and at many instincts causing other animals to suffer.

143:e selection, we

143:f being not

144 If species be only well-marked and permanent varieties, we can at once see why their crossed offspring should follow the same complex laws in their degrees and kinds of resemblance to their parents,—in being absorbed into each other by successive crosses, and in other such points,—as do the crossed offspring of acknowledged varieties.

145 On the other hand, these would be strange facts if species have been independently created, and varieties have been produced by secondary laws.

145:e This similarity would be a strange fact, if species have been independently created and varieties have been produced through secondary

145:f species had been independently created and varieties had been

146 If we admit that the geological record is imperfect in an extreme degree, then such facts as the record gives, support the theory of descent with modification.

146:d then the facts, which the record does give, strongly support

146:f imperfect to an extreme

147 New species have come on the stage slowly and at successive intervals; and the amount of change, after equal intervals of time, is widely different in different groups.

148 The extinction of species and of whole groups of species, which has played so conspicuous a part in the history of the organic world, almost inevitably follows on the principle of natural selection; for old forms will be supplanted by new and improved forms.

148:e follows from the principle of natural selection; for old forms are supplanted

149 Neither single species nor groups of species reappear when the chain of ordinary generation has once been broken.

149:e generation is once broken.

150 The gradual diffusion of dominant forms, with the slow modification of their descendants, causes the forms of life, after long intervals of time, to appear as if they had changed simultaneously throughout the world.

150:d forms with

150:e forms, with

151 The fact of the fossil remains of each formation being in some degree intermediate in character between the fossils in the formations above and below, is simply explained by their intermediate position in the chain of descent.

152 The grand fact that all extinct organic beings belong to the same system with recent beings, falling either into the same or into intermediate groups, follows from the living and the extinct being the offspring of common parents.

152:e extinct beings can be classed with all recent beings, naturally follows

153 As the groups which have descended from an ancient progenitor have generally diverged in character, the progenitor with its early descendants will often be intermediate in character in comparison with its later descendants; and thus we can see why the more ancient a fossil is, the oftener it stands in some degree intermediate between existing and allied groups.

153:d existing allied groups.

153:e As species have generally diverged in character during their long course of descent and modification, we can understand why it is that the more ancient forms, or early progenitors of each group, so often occupy a position in some degree intermediate between existing groups.

154 Recent forms are generally looked at as being, in some vague sense, higher than ancient and extinct forms; and they are in so far higher as the later and more improved forms have conquered the older and less improved organic beings in the struggle for life.

154:c being, on the whole, higher/life; they will also generally have had their organs more specialised for different functions.

154:d they are higher in so far as

154:e higher in the scale of organisation than ancient forms; and they must be higher, in so/less improved forms in the struggle for life; they have also generally had their

154:f looked upon as being, on

154.1:c This fact is perfectly compatible with numerous beings still retaining a simple and little improved organisation fitted for simple conditions of life; it is likewise compatible with some forms having retrograded in organisation, though becoming under each grade of descent better fitted for their changed and degraded habits of life.

154.1:e retaining simple and but little improved structures, fitted/ organisation, by having become at each stage of descent better fitted for changed

154.1:f better fitted for new and degraded

155 Lastly, the law of the long endurance of allied forms on the same continent,—of marsupials in Australia, of edentata in America, and other such cases,—is intelligible, for within a confined country, the recent and the extinct will naturally be allied by descent.

155:d will be closely allied

155:e Lastly, the wonderful law/for generally within the same country, the existing and the extinct

155:f for within the same country the existing and the extinct

156 Looking to geographical distribution, if we admit that there has been during the long course of ages much migration from

one part of the world to another, owing to former climatal and geographical changes and to the many occasional and unknown means of dispersal, then we can understand, on the theory of descent with modification, most of the great leading facts in Distribution.

157 We can see why there should be so striking a parallelism in the distribution of organic beings throughout space, and in their geological succession throughout time; for in both cases the beings have been connected by the bond of ordinary generation, and the means of modification have been the same.

158 We see the full meaning of the wonderful fact, which must have struck every traveller, namely, that on the same continent, under the most diverse conditions, under heat and cold, on mountain and lowland, on deserts and marshes, most of the inhabitants within each great class are plainly related; for they will generally be descendants of the same progenitors and early colonists.

158:d which has struck/they generally are the descendants

158:e they are the descendants

159 On this same principle of former migration, combined in most cases with modification, we can understand, by the aid of the Glacial period, the identity of some few plants, and the close alliance of many others, on the most distant mountains, under the most different climates; and likewise the close alliance of some of the inhabitants of the sea in the northern and southern temperate zones, though separated by the whole intertropical ocean.

159:d mountains, and in the northern and southern temperate zones; and likewise/temperate latitudes, though

160 Although two areas may present the same physical conditions of life, we need feel no surprise at their inhabitants being widely different, if they have been for a long period completely separated from each other; for as the relation of organism to organism is the most important of all relations, and as the two areas will have received colonists from some third source or from each other, at various periods and in different proportions, the course of modification in the two areas will inevitably be different.

160:c two countries may present physical conditions as closely similar as the same species ever require, we/completely sundered from each/and as the two countries will have received/inevitably have been different.

160:e colonists at various periods and in different proportions, from some other country or from each other, the course

161 On this view of migration, with subsequent modification, we can see why oceanic islands should be inhabited by few species, but of these, that many should be peculiar.

161:d islands are inhabited by only few species, but of these, why many are peculiar or endemic forms.

162 We can clearly see why those animals which cannot cross wide spaces of ocean, as frogs and terrestrial mammals, should not inhabit oceanic islands; and why, on the other hand, new and peculiar species of bats, which can traverse the ocean, should so often be found on islands far distant from any continent.

162:c bats, animals which

162:d We clearly see why species of those groups of animals which cannot/mammals, do not inhabit/ocean, are so often found

162:e species belonging to those groups of animals which cannot cross wide spaces of the ocean, as

162:f are often

163 Such facts as the presence of peculiar species of bats, and the absence of all other mammals, on oceanic islands, are utterly inexplicable on the theory of independent acts of creation.

163:e Such cases as the presence of peculiar species of bats on oceanic islands, and the absence of all other terrestrial mammals, are facts utterly

163:f islands and

164   The existence of closely allied or representative species in any two areas, implies, on the theory of descent with modification, that the same parents formerly inhabited both areas; and we almost invariably find that wherever many closely allied species inhabit two areas, some identical species common to both still exist.

164:c still exist there.

164:e same parent-forms formerly/identical species are still common to both.

164:f both areas: and

165 Wherever many closely allied yet distinct species occur, many doubtful forms and varieties of the same species likewise occur.

165:e occur, doubtful forms and varieties belonging to the same groups likewise

166 It is a rule of high generality that the inhabitants of each area are related to the inhabitants of the nearest source whence immigrants might have been derived.

167 We see this in nearly all the plants and animals of the Galapagos archipelago, of Juan Fernandez, and of the other American islands being related in the most striking manner to the plants and animals of the neighbouring American mainland; and those of the Cape de Verde archipelago and other African islands to the African mainland.

167:e in the striking relation of nearly/islands, to the plants/mainland; and of those of the Cape de Verde archipelago and of the other

744

167:f archipelago, and

168 It must be admitted that these facts receive no explanation on the theory of creation.

169 The fact, as we have seen, that all past and present organic beings constitute one grand natural system, with group subordinate to group, and with extinct groups often falling in between recent groups, is intelligible on the theory of natural selection with its contingencies of extinction and divergence of character.

169:e beings can be arranged within a few great classes, in groups subordinate to groups, and with the extinct groups often falling in between the recent

170 On these same principles we see how it is, that the mutual affinities of the species and genera within each class are so complex and circuitous.

170:e of the forms within

171 We see why certain characters are far more serviceable than others for classification;—why adaptive characters, though of paramount importance to the being, are of hardly any importance in classification; why characters derived from rudimentary parts, though of no service to the being, are often of high classificatory value; and why embryological characters are the most valuable of all.

171:d embryological characters are often the most
171:e beings/beings

172 The real affinities of all organic beings are due to inheritance or community of descent.

172:e beings, in contradistinction to their adaptive resemblances, are

173 The natural system is a genealogical arrangement, in which we have to discover the lines of descent by the most permanent characters, however slight their vital importance may be.

173:e The Natural System is a genealogical arrangement, with the acquired grades of difference, marked by the terms, varieties, species, genera, families, &c.; and we/characters whatever they may be and of however slight vital importance.

174 The framework of bones being the same in the hand of a man, wing of a bat, fin of the porpoise, and leg of the horse,—the same number of vertebræ forming the neck of the giraffe and of the elephant,—and innumerable other such facts, at once explain themselves on the theory of descent with slow and slight successive modifications.

174:c being similar in
174:e The similar framework of bones in

175 The similarity of pattern in the wing and leg of a bat, though used for such different purpose,—in the jaws and legs of a crab,

—in the petals, stamens, and pistils of a flower, is likewise intelligible on the view of the gradual modification of parts or organs, which were alike in the early progenitor of each class.

175:e wing and in the leg of a bat/were aboriginally alike in an early progenitor in each of these classes.

175:f likewise, to a large extent, intelligible

176 On the principle of successive variations not always supervening at an early age, and being inherited at a corresponding not early period of life, we can clearly see why the embryos of mammals, birds, reptiles, and fishes should be so closely alike, and should be so unlike the adult forms.

176:d we clearly/fishes are so closely similar, and are so unlike their adult

176:e similar, and so unlike the adult

176:f fishes should be so closely

177 We may cease marvelling at the embryo of an air-breathing mammal or bird having branchial slits and arteries running in loops, like those in a fish which has to breathe the air dissolved in water, by the aid of well-developed branchiæ.

177:e those of a fish/water by

178 Disuse, aided sometimes by natural selection, will often tend to reduce an organ, when it has become useless by changed habits or under changed conditions of life; and we can clearly understand on this view the meaning of rudimentary organs.

178:e selection, has often reduced organs when they have become useless under changed habits or conditions

178:f selection, will often have reduced organs when rendered useless under changed habits or conditions of life; and we can understand

179 But disuse and selection will generally act on each creature, when it has come to maturity and has to play its full part in the struggle for existence, and will thus have little power of acting on an organ during early life; hence the organ will not be much reduced or rendered rudimentary at this early age.

179:e power on an organ during early life; hence the organ will not be reduced

180 The calf, for instance, has inherited teeth, which never cut through the gums of the upper jaw, from an early progenitor having well-developed teeth; and we may believe, that the teeth in the mature animal were reduced, during successive generations, by disuse or by the tongue and palate having been fitted by natural selection to browse without their aid; whereas in the calf, the teeth have been left untouched by selection or disuse, and on the principle of inheritance at corresponding ages have been inherited from a remote period to the present day.

180:c palate, or lips, having become better fitted

746

180:f were formerly reduced by disuse, owing to the tongue and palate, or lips, having become excellently fitted through natural/left unaffected, and on

181 On the view of each organic being and each separate organ having been specially created, how utterly inexplicable it is that parts, like the teeth in the embryonic calf or like the shrivelled wings under the soldered wing-covers of some beetles, should thus so frequently bear the plain stamp of inutility!

181:d and each of its separate parts having/that organs, like the teeth in the embryonic calf or the shrivelled

181:e being with all its separate/that organs bearing the plain stamp of inutility, such as the teeth/wing-covers of many beetles, should so frequently occur.

181:f each organism with/inexplicable is it that

182 Nature may be said to have taken pains to reveal, by rudimentary organs and by homologous structures, her scheme of modification, which it seems that we wilfully will not understand.

182:e reveal her scheme of modification, by means of rudimentary organs, embryological and homologous structures, but we wilfully will not understand the scheme.

182:f organs, of embryological and homologous structures, but we are too blind to understand her meaning.

183   I have now recapitulated the chief facts and considerations which have thoroughly convinced me that species have changed, and are still slowly changing by the preservation and accumulation of successive slight favourable variations.

183:b species have been modified, during a long course of descent, by the preservation or the natural selection of many successive

183:e recapitulated the facts/descent, chiefly through the natural selection of numerous successive, slight, favourable

183:f descent. *y* This has been effected chiefly/variations; aided in an important manner by the inherited effects of the use and disuse of parts; and in an unimportant manner, that is in relation to adaptive structures, whether past or present, by the direct action of external conditions, and by variations which seem to us in our ignorance to arise spontaneously.

183.0.0.1-4:f It appears that I formerly underrated the frequency and value of these latter forms of variation, as leading to permanent modifications of structure independently of natural selection. 2 But as my conclusions have lately been much misrepresented, and it has been stated that I attribute the modification of species exclusively to natural selection, I may be permitted to remark that in the first edition of this work, and subsequently, I placed in a most conspicuous position—namely, at the close of the Introduction—the following words: "I am convinced that natural selection has been the main but not the exclusive means of modification." *3* This has been of no avail.

4 Great is the power of steady misrepresentation; but the history of science shows that fortunately this power does not long endure.

183.1:b I cannot believe that a false theory would explain, as it seems to me that the theory of natural selection does explain, the several large classes of facts above specified.

183.1:f[¶] It can hardly be supposed that a false theory would explain, in so satisfactory a manner as does the theory of natural selection, the several

183.1.0.1-2:f It has recently been objected that this is an unsafe method of arguing; but it is a method used in judging of the common events of life, and has often been used by the greatest natural philosophers. *2* The undulatory theory of light has thus been arrived at; and the belief in the revolution of the earth on its own axis was until lately supported by hardly any direct evidence.

183.1.1-3:c It is no valid objection that science as yet throws no light on the far higher problem of the essence or origin of life. *2* Who can explain what is the essence of the attraction of gravity? *3* No one now objects to following out the results consequent on this unknown element of attraction; notwithstanding that Leibnitz formerly accused Newton of introducing "occult qualities and miracles into philosophy."

183.2:b I see no good reason why the views given in this volume should shock the religious feelings of any one.

183.2:c[¶]

183.2.1:c It is satisfactory, as showing how transient such impressions are, to remember that the greatest discovery ever made by man, namely, the law of the attraction of gravity, was also attacked by Leibnitz, "as subversive of natural and inferentially of revealed religion."

183.3:b A celebrated author and divine has written to me that "he has gradually learnt to see that it is just as noble a conception of the Deity to believe that He created a few original forms capable of self-development into other and needful forms, as to believe that He required a fresh act of creation to supply the voids caused by the action of His laws."

184 Why, it may be asked, have all the most eminent living naturalists and geologists rejected this view of the mutability of species?

184:b[¶]

184:c have nearly all

184:e asked, until recently did nearly/geologists reject

184:f geologists disbelieve in the mutability

185 It cannot be asserted that organic beings in a state of nature are subject to no variation; it cannot be proved that the amount of variation in the course of long ages is a limited quantity; no clear distinction has been, or can be, drawn between species and well-marked varieties.

186 It cannot be maintained that species when intercrossed are invariably sterile, and varieties invariably fertile; or that sterility is a special endowment and sign of creation.

187 The belief that species were immutable productions was almost unavoidable as long as the history of the world was thought to be of short duration; and now that we have acquired some idea of the lapse of time, we are too apt to assume, without proof, that the geological record is so perfect that it would have afforded us plain evidence of the mutation of species, if they had undergone mutation.

188   But the chief cause of our natural unwillingness to admit that one species has given birth to other and distinct species, is that we are always slow in admitting any great change of which we do not see the intermediate steps.

188:e see the steps.

188:f admitting great changes

189 The difficulty is the same as that felt by so many geologists, when Lyell first insisted that long lines of inland cliffs had been formed, and great valleys excavated, by the slow action of the coast-waves.

189:e by the agencies which we still see at work.

190 The mind cannot possibly grasp the full meaning of the term of a hundred million years; it cannot add up and perceive the full effects of many slight variations, accumulated during an almost infinite number of generations.

190:e term of even ten million

190:f even a million

191   Although I am fully convinced of the truth of the views given in this volume under the form of an abstract, I by no means expect to convince experienced naturalists whose minds are stocked with a multitude of facts all viewed, during a long course of years, from a point of view directly opposite to mine.

192 It is so easy to hide our ignorance under such expressions as the "plan of creation," "unity of design," &c., and to think that we give an explanation when we only restate a fact.

192:f re-state

193 Any one whose disposition leads him to attach more weight to unexplained difficulties than to the explanation of a certain number of facts will certainly reject my theory.

193:e reject the theory.

194 A few naturalists, endowed with much flexibility of mind, and who have already begun to doubt on the immutability of species, may be influenced by this volume; but I look with confidence to the future, to young and rising naturalists, who

will be able to view both sides of the question with impartiality.

194:e doubt the immutability

194:f future,—to

195 Whoever is led to believe that species are mutable will do good service by conscientiously expressing his conviction; for only thus can the load of prejudice by which this subject is overwhelmed be removed.

196 Several eminent naturalists have of late published their belief that a multitude of reputed species in each genus are not real species; but that other species are real, that is, have been independently created.

197 This seems to me a strange conclusion to arrive at.

198 They admit that a multitude of forms, which till lately they themselves thought were special creations, and which are still thus looked at by the majority of naturalists, and which consequently have every external characteristic feature of true species,—they admit that these have been produced by variation, but they refuse to extend the same view to other and very slightly different forms.

198:e have all the external characteristic features/other and slightly

199 Nevertheless they do not pretend that they can define, or even conjecture, which are the created forms of life, and which are those produced by secondary laws.

200 They admit variation as a *vera causa* in one case, they arbitrarily reject it in another, without assigning any distinction in the two cases.

201 The day will come when this will be given as a curious illustration of the blindness of preconceived opinion.

202 These authors seem no more startled at a miraculous act of creation than at an ordinary birth.

203 But do they really believe that at innumerable periods in the earth's history certain elemental atoms have been commanded suddenly to flash into living tissues?

204 Do they believe that at each supposed act of creation one individual or many were produced?

205 Were all the infinitely numerous kinds of animals and plants created as eggs or seed, or as full grown? and in the case of mammals, were they created bearing the false marks of nourishment from the mother's womb?

206 Although naturalists very properly demand a full explanation of every difficulty from those who believe in the mutability of species, on their own side they ignore the whole subject of the first appearance of species in what they consider reverent silence.

206:c Undoubtedly these same questions cannot be answered by those who, under the present state of science, believe in the creation of a few aboriginal forms, or of some one form of life.

206:e who believe in the appearance or creation of only a few forms of life, or of some one form alone.

206:f Undoubtedly some of these

206.1:c It has been asserted by several authors that it is as easy to believe in the creation of a hundred million beings as of one; but Maupertuis' philosophical axiom "of least action" leads the mind more willing to admit the smaller number; and certainly we ought not to believe that innumerable beings within each great class have been created with plain, but deceptive, marks of descent from a single parent.

206.1:e been maintained by

206.1:f creation of a million

206.1.1-6:f[¶] As a record of a former state of things, I have retained in the foregoing paragraphs, and elsewhere, several sentences which imply that naturalists believe in the separate creation of each species; and I have been much censured for having thus expressed myself. *2* But undoubtedly this was the general belief when the first edition of the present work appeared. *3* I formerly spoke to very many naturalists on the subject of evolution, and never once met with any sympathetic agreement. *4* It is probable that some did then believe in evolution, but they were either silent, or expressed themselves so ambiguously that it was not easy to understand their meaning. *5* Now things are wholly changed, and almost every naturalist admits the great principle of evolution. *6* There are, however, some who still think that species have suddenly given birth, through quite unexplained means, to new and totally different forms: but, as I have attempted to show, weighty evidence can be opposed to the admission of great and abrupt modifications. *7* Under a scientific point of view, and as leading to further investigation, but little advantage is gained by believing that new forms are suddenly developed in an inexplicable manner from old and widely different forms, over the old belief in the creation of species from the dust of the earth.

207   It may be asked how far I extend the doctrine of the modification of species.

208   The question is difficult to answer, because the more distinct the forms are which we may consider, by so much the arguments fall away in force.

208:e we consider

208:f arguments in favour of community of descent become fewer in number and less in

209   But some arguments of the greatest weight extend very far.

210   All the members of whole classes can be connected together by

chains of affinities, and all can be classified on the same principle, in groups subordinate to groups.

210:e classes are connected together by a chain

210:f can be classed on

211 Fossil remains sometimes tend to fill up very wide intervals between existing orders.

212 Organs in a rudimentary condition plainly show that an early progenitor had the organ in a fully developed state; and this in some instances necessarily implies an enormous amount of modification in the descendants.

212:e instances implies

212:f[¶] developed condition; and this in some cases implies

213 Throughout whole classes various structures are formed on the same pattern, and at an embryonic age the species closely resemble each other.

213:e at a very early age the embryos closely

214 Therefore I cannot doubt that the theory of descent with modification embraces all the members of the same class.

214:f same great class or kingdom.

215 I believe that animals have descended from at most only four or five progenitors, and plants from an equal or lesser number.

215:e animals are descended

216   Analogy would lead me one step further, namely, to the belief that all animals and plants have descended from some one prototype.

216:c farther

216:e plants are descended

217 But analogy may be a deceitful guide.

218 Nevertheless all living things have much in common, in their chemical composition, their germinal vesicles, their cellular structure, and their laws of growth and reproduction.

218:c common,—in their chemical composition, their cellular structure, their laws of growth, and their liability to injurious influences.

218:f common, in their chemical

219 We see this even in so trifling a circumstance as that the same poison often similarly affects plants and animals; or that the poison secreted by the gall-fly produces monstrous growths on the wild rose or oak-tree.

219:e trifling a fact as

219.1:c In all organic beings the union of a male and female elemental cell seems occasionally to be necessary for the production of a new being.

752

219.1:e With all organic beings sexual reproduction seems to be essentially similar.

219.1:f beings, excepting perhaps some of the very lowest, sexual

219.2:c In all, as far as is at present known, the germinal vesicle is the same.

219.3:c So that every individual organic being starts from a common origin.

219.2 + 3:e With all, as far/same; so that all organisms start

219.4:c If we look even to the two main divisions—namely, to the animal and vegetable kingdoms—certain low forms are so far intermediate in character that naturalists have disputed to which kingdom they should be referred, and, as Professor Asa Gray has remarked, "the spores and other reproductive bodies of many of the lower algæ may claim to have first a characteristically animal, and then an unequivocally vegetable existence."

219.4.x-y:f referred. *y* As

220 Therefore I should infer from analogy that probably all the organic beings which have ever lived on this earth have descended from some one primordial form, into which life was first breathed.

220:b breathed by the Creator.

220:c Therefore, on the principle of natural selection with divergence of character, it does not seem incredible that, from some such low and intermediate form, both animals and plants may have been developed; and, if we admit this, we must admit that all/earth may have/form.

220:e must likewise admit/earth may be descended

220.1:c But this inference is chiefly grounded on analogy, and it is immaterial whether or not it be accepted. [*No space*]

220.1.1:e No doubt it is possible, as Mr. G. H. Lewes has urged, that at the first commencement of life many different forms were evolved; but if so, we may conclude that only a very few have left modified descendants.

220.2:c The case is different with the members of each great class, as the Vertebrata, the Articulata, &c.; for here, as has just been remarked, we have in the laws of homology and embryology, &c., distinct evidence that all have descended from a single parent.

220.2:d as I have just remarked

220.2:e For, as I have recently remarked in regard to the members of each great class, such as the Vertebrata, Articulata, &c., we have distinct evidence in their embryological, homologous and rudimentary structures, that within each class all are descended from a single progenitor.

220.2:f great kingdom, such/within each kingdom all the members are

753

221　When the views entertained in this volume on the origin of species, or when analogous views are generally admitted, we can dimly foresee that there will be a considerable revolution in natural history.

221:b views advanced by me in this volume, and by Mr. Wallace in the Linnean Journal, or when analogous views on the origin of species are

221:f Wallace, or

222　Systematists will be able to pursue their labours as at present; but they will not be incessantly haunted by the shadowy doubt whether this or that form be in essence a species.

222:e form be a true species.

223　This I feel sure, and I speak after experience, will be no slight relief.

223:f This, I feel sure and

224　The endless disputes whether or not some fifty species of British brambles are true species will cease.

224:e are good species

225　Systematists will have only to decide (not that this will be easy) whether any form be sufficiently constant and distinct from other forms, to be capable of definition; and if definable, whether the differences be sufficiently important to deserve a specific name.

226　This latter point will become a far more essential consideration than it is at present; for differences, however slight, between any two forms, if not blended by intermediate gradations, are looked at by most naturalists as sufficient to raise both forms to the rank of species.

227　Hereafter we shall be compelled to acknowledge that the only distinction between species and well-marked varieties is, that the latter are known, or believed, to be connected at the present day by intermediate gradations, whereas species were formerly thus connected.

227:f[¶]

228　Hence, without quite rejecting the consideration of the present existence of intermediate gradations between any two forms, we shall be led to weigh more carefully and to value higher the actual amount of difference between them.

228:b without rejecting

229　It is quite possible that forms now generally acknowledged to be merely varieties may hereafter be thought worthy of specific names, as with the primrose and cowslip; and in this case scientific and common language will come into accordance.

229:d names; and in this

230　In short, we shall have to treat species in the same manner as those naturalists treat genera, who admit that genera are merely artificial combinations made for convenience.

231 This may not be a cheering prospect; but we shall at least be freed from the vain search for the undiscovered and undiscoverable essence of the term species.

232 The other and more general departments of natural history will rise greatly in interest.

233 The terms used by naturalists of affinity, relationship, community of type, paternity, morphology, adaptive characters, rudimentary and aborted organs, &c., will cease to be metaphorical, and will have a plain signification.

233:f naturalists, of affinity

234 When we no longer look at an organic being as a savage looks at a ship, as at something wholly beyond his comprehension; when we regard every production of nature as one which has had a history; when we contemplate every complex structure and instinct as the summing up of many contrivances, each useful to the possessor, nearly in the same way as when we look at any great mechanical invention as the summing up of the labour, the experience, the reason, and even the blunders of numerous workmen; when we thus view each organic being, how far more interesting, I speak from experience, will the study of natural history become!

234:d interesting,—I speak from experience,—will

234:e ship, as something/had a long history; when we contemplate/ possessor, in the same way as any great mechanical invention is the summing/experience,—does the study

235 A grand and almost untrodden field of inquiry will be opened, on the causes and laws of variation, on correlation of growth, on the effects of use and disuse, on the direct action of external conditions, and so forth.

235:e correlation, on the effects

236 The study of domestic productions will rise immensely in value.

237 A new variety raised by man will be a far more important and interesting subject for study than one more species added to the infinitude of already recorded species.

237:b be a more

238 Our classifications will come to be, as far as they can be so made, genealogies; and will then truly give what may be called the plan of creation.

239 The rules for classifying will no doubt become simpler when we have a definite object in view.

240 We possess no pedigrees or armorial bearings; and we have to discover and trace the many diverging lines of descent in our natural genealogies, by characters of any kind which have long been inherited.

241 Rudimentary organs will speak infallibly with respect to the nature of long-lost structures.

242 Species and groups of species, which are called aberrant, and which may fancifully be called living fossils, will aid us in forming a picture of the ancient forms of life.

242:f of species which are

243 Embryology will reveal to us the structure, in some degree obscured, of the prototypes of each great class.

243:d will often reveal

244 When we can feel assured that all the individuals of the same species, and all the closely allied species of most genera, have within a not very remote period descended from one parent, and have migrated from some one birthplace; and when we better know the many means of migration, then, by the light which geology now throws, and will continue to throw, on former changes of climate and of the level of the land, we shall surely be enabled to trace in an admirable manner the former migrations of the inhabitants of the whole world.

244:f birth-place

245 Even at present, by comparing the differences of the inhabitants of the sea on the opposite sides of a continent, and the nature of the various inhabitants of that continent in relation to their apparent means of immigration, some light can be thrown on ancient geography.

245:f differences between the inhabitants of the sea/various inhabitants on that

246 The noble science of Geology loses glory from the extreme imperfection of the record.

247 The crust of the earth with its embedded remains must not be looked at as a well-filled museum, but as a poor collection made at hazard and at rare intervals.

248 The accumulation of each great fossiliferous formation will be recognised as having depended on an unusual concurrence of circumstances, and the blank intervals between the successive stages as having been of vast duration.

249 But we shall be able to gauge with some security the duration of these intervals by a comparison of the preceding and succeeding organic forms.

250 We must be cautious in attempting to correlate as strictly contemporaneous two formations, which include few identical species, by the general succession of their forms of life.

250:e which do not include many identical species, by the general succession of the forms

251 As species are produced and exterminated by slowly acting and still existing causes, and not by miraculous acts of creation and by catastrophies; and as the most important of all causes of

organic change is one which is almost independent of altered and perhaps suddenly altered physical conditions, namely, the mutual relation of organism to organism,—the improvement of one being entailing the improvement or the extermination of others; it follows, that the amount of organic change in the fossils of consecutive formations probably serves as a fair measure of the lapse of actual time.

251:e improvement of one organism entailing

251:f creation; and as the most/measure of the relative, though not actual lapse of time.

252 A number of species, however, keeping in a body might remain for a long period unchanged, whilst within this same period, several of these species, by migrating into new countries and coming into competition with foreign associates, might become modified; so that we must not overrate the accuracy of organic change as a measure of time.

253 During early periods of the earth's history, when the forms of life were probably fewer and simpler, the rate of change was probably slower; and at the first dawn of life, when very few forms of the simplest structure existed, the rate of change may have been slow in an extreme degree.

253:[f]

254 The whole history of the world, as at present known, although of a length quite incomprehensible by us, will hereafter be recognised as a mere fragment of time, compared with the ages which have elapsed since the first creature, the progenitor of innumerable extinct and living descendants, was created.

254:e The history of the world, as at present known, although of immense length, will hereafter be recognised as short, compared with the ages which must have elapsed since the first organic beings, the progenitors of innumerable extinct and living descendants, appeared on the stage.

254:[f]

255 In the distant future I see open fields for far more important researches.

255:f In the future

256 Psychology will be based on a new foundation, that of the necessary acquirement of each mental power and capacity by gradation.

256:f be securely based on the foundation already well laid by Mr. Herbert Spencer, that

257 Light will be thrown on the origin of man and his history.

257:f Much light

258 Authors of the highest eminence seem to be fully satisfied with the view that each species has been independently created.

259 To my mind it accords better with what we know of the laws

impressed on matter by the Creator, that the production and extinction of the past and present inhabitants of the world should have been due to secondary causes, like those determining the birth and death of the individual.

260 When I view all beings not as special creations, but as the lineal descendants of some few beings which lived long before the first bed of the Silurian system was deposited, they seem to me to become ennobled.

260:f bed of the Cambrian system

261 Judging from the past, we may safely infer that not one living species will transmit its unaltered likeness to a distant futurity.

262 And of the species now living very few will transmit progeny of any kind to a far distant futurity; for the manner in which all organic beings are grouped, shows that the greater number of species of each genus, and all the species of many genera, have left no descendants, but have become utterly extinct.

263 We can so far take a prophetic glance into futurity as to foretel that it will be the common and widely-spread species, belonging to the larger and dominant groups, which will ultimately prevail and procreate new and dominant species.

263:c foretell/groups within each class, which

264 As all the living forms of life are the lineal descendants of those which lived long before the Silurian epoch, we may feel certain that the ordinary succession by generation has never once been broken, and that no cataclysm has desolated the whole world.

264:f before the Cambrian epoch

265 Hence we may look with some confidence to a secure future of equally inappreciable length.

265:f of great length.

266 And as natural selection works solely by and for the good of each being, all corporeal and mental endowments will tend to progress towards perfection.

267 It is interesting to contemplate an entangled bank, clothed with many plants of many kinds, with birds singing on the bushes, with various insects flitting about, and with worms crawling through the damp earth, and to reflect that these elaborately constructed forms, so different from each other, and dependent on each other in so complex a manner, have all been produced by laws acting around us.

267:f contemplate a tangled bank

268 These laws, taken in the largest sense, being Growth with Reproduction; Inheritance which is almost implied by reproduction; Variability from the indirect and direct action of the external conditions of life, and from use and disuse; a Ratio of Increase so high as to lead to a Struggle for Life, and as a

758

consequence to Natural Selection, entailing Divergence of Character and the Extinction of less-improved forms.

268:e action of the conditions

268:f disuse: a Ratio

269 Thus, from the war of nature, from famine and death, the most exalted object which we are capable of conceiving, namely, the production of the higher animals, directly follows.

270 There is grandeur in this view of life, with its several powers, having been originally breathed into a few forms or into one; and that, whilst this planet has gone cycling on according to the fixed law of gravity, from so simple a beginning endless forms most beautiful and most wonderful have been, and are being, evolved.

270:b breathed by the Creator into a few

270:f being evolved.

GLOSSARY

PRINCIPAL SCIENTIFIC TERMS USED IN THE
PRESENT VOLUME.* [1]

ABERRANT.—Forms or groups of animals or plants which deviate in important characters from their nearest allies, so as not to be easily included in the same group with them, are said to be aberrant.

ABERRATION (in Optics).—In the refraction of light by a convex lens the rays passing through different parts of the lens are brought to a focus at slightly different distances,—this is called *spherical aberration;* at the same time the coloured rays are separated by the prismatic action of the lens and likewise brought to a focus at different distances,—this is *chromatic aberration.*

ABNORMAL.—Contrary to the general rule.

ABORTED.—An organ is said to be aborted, when its development has been arrested at a very early stage.

ALBINISM.—Albinos are animals in which the usual colouring matters characteristic of the species have not been produced in the skin and its appendages. Albinism is the state of being an albino.

ALGÆ.—A class of plants including the ordinary sea-weeds and the filamentous fresh-water weeds.

ALTERATION OF GENERATIONS.—This term is applied to a peculiar mode of reproduction which prevails among many of the lower animals, in which the egg produces a living form quite different from its parent, but from which the parent-form is reproduced by a process of budding, or by the division of the substance of the first product of the egg.

AMMONITES.—A group of fossil, spiral, chambered shells, allied to the existing pearly Nautilus, but having the partitions between the chambers waved in complicated patterns at their junction with the outer wall of the shell.

ANALOGY.—That resemblance of structures which depends upon similarity of function, as in the wings of insects and birds. Such structures are said to be *analogous,* and to be *analogues* of each other.

ANIMALCULE.—A minute animal: generally applied to those visible only by the microscope.

ANNELIDS.—A class of worms in which the surface of the body exhibits a more or less distinct division into rings or segments, generally provided with appendages for locomotion and with gills. It includes the ordinary marine worms, the earthworms, and the leeches.

ANTENNÆ.—Jointed organs appended to the head in Insects, Crustacea and Centipedes, and not belonging to the mouth.

ANTHERS.—The summits of the stamens of flowers, in which the pollen or fertilising dust is produced.

---

* I am indebted to the kindness of Mr. W. S. Dallas for this Glossary, which has been given because several readers have complained to me that some of the terms used were unintelligible to them. Mr. Dallas has endeavoured to give the explanations of the terms in as popular a form as possible.

[1] [NOTE: This section first appeared in *f*.]

APLACENTALIA, APLACENTATA or Aplacental Mammals. See *Mammalia.*

ARCHETYPAL.—Of or belonging to the Archetypes, or ideal primitive form upon which all the beings of a group seem to be organised.

ARTICULATA.—A great division of the Animal Kingdom characterised generally by having the surface of the body divided into rings, called segments, a greater or less number of which are furnished with jointed legs (such as Insects, Crustaceans and Centipedes).

ASYMMETRICAL.—Having the two sides unlike.

ATROPHIED.—Arrested in development at a very early stage.

BALANUS.—The genus including the common Acorn-shells which live in abundance on the rocks of the sea-coast.

BATRACHIANS.—A class of animals allied to the Reptiles, but undergoing a peculiar metamorphosis, in which the young animal is generally aquatic and breathes by gills. (*Examples,* Frogs, Toads, and Newts.)

BOULDERS.—Large transported blocks of stone generally imbedded in clays or gravels.

BRACHIOPODA.—A class of marine Mollusca, or soft-bodied animals, furnished with a bivalve shell, attached to submarine objects by a stalk which passes through an aperture in one of the valves, and furnished with fringed arms, by the action of which food is carried to the mouth.

BRANCHIÆ.—Gills or organs for respiration in water.

BRANCHIAL.—Pertaining to gills or branchiæ.

CAMBRIAN SYSTEM.—A Series of very ancient Palæozoic rocks, between the Laurentian and the Silurian. Until recently these were regarded as the oldest fossiliferous rocks.

CANIDÆ.—The Dog-family, including the Dog, Wolf, Fox, Jackal, &c.

CARAPACE.—The shell enveloping the anterior part of the body in Crustaceans generally; applied also to the hard shelly pieces of the Cirripedes.

CARBONIFEROUS.—This term is applied to the great formation which includes, among other rocks, the coal-measures. It belongs to the oldest, or Palæozoic, system of formations.

CAUDAL.—Of or belonging to the tail.

CEPHALOPODS.—The highest class of the Mollusca, or soft-bodied animals, characterised by having the mouth surrounded by a greater or less number of fleshy arms or tentacles, which, in most living species, are furnished with sucking-cups. (*Examples,* Cuttle-fish, Nautilus.)

CETACEA.—An order of Mammalia, including the Whales, Dolphins, &c., having the form of the body fish-like, the skin naked, and only the fore-limbs developed.

CHELONIA.—An order of Reptiles including the Turtles, Tortoises, &c.

CIRRIPEDES.—An order of Crustaceans including the Barnacles and Acorn-shells. Their young resemble those of many other Crustaceans in form; but when mature they are always attached to other objects, either directly or by means of a stalk, and their bodies are enclosed by a calcareous shell composed of several pieces, two of which can open to give issue to a bunch of curled, jointed tentacles, which represent the limbs.

COCCUS.—The genus of Insects including the Cochineal. In these the male is a minute, winged fly, and the female generally a motionless, berry-like mass.

COCOON.—A case usually of silky material, in which insects are frequently enveloped during the second or resting-stage (pupa) of their existence. The term "cocoon-stage" is here used as equivalent to "pupa-stage."

CŒLOSPERMOUS.—A term applied to those fruits of the Umbelliferæ which have the seed hollowed on the inner face.

COLEOPTERA.—Beetles, an order of Insects, having a biting mouth and the first pair of wings more or less horny, forming sheaths for the second pair, and usually meeting in a straight line down the middle of the back.

COLUMN.—A peculiar organ in the flowers of Orchids, in which the stamens, style and stigma (or the reproductive parts) are united.

COMPOSITÆ or COMPOSITOUS PLANTS.—Plants in which the inflorescence consists of numerous small flowers (florets) brought together into a dense head, the base of which is enclosed by a common envelope. (*Examples,* the Daisy, Dandelion, &c.)

CONFERVÆ.—The filamentous weeds of fresh water.

CONGLOMERATE.—A rock made up of fragments of rock or pebbles, cemented together by some other material.

COROLLA.—The second envelope of a flower, usually composed of coloured, leaf-like organs (petals), which may be united by their edges either in the basal part or throughout.

CORRELATION.—The normal coincidence of one phenomenon, character, &c., with another.

CORYMB.—A bunch of flowers in which those springing from the lower part of the flower stalk are supported on long stalks so as to be nearly on a level with the upper ones.

COTYLEDONS.—The first or seed-leaves of plants.

CRUSTACEANS.—A class of articulated animals, having the skin of the body generally more or less hardened by the deposition of calcareous matter, breathing by means of gills. (*Examples,* Crab, Lobster, Shrimp, &c.)

CURCULIO.—The old generic term for the Beetles known as Weevils, characterised by their four-jointed feet, and by the head being produced into a sort of beak, upon the sides of which the antennæ are inserted.

CUTANEOUS.—Of or belonging to the skin.

DEGRADATION.—The wearing down of land by the action of the sea or of meteoric agencies.

DENUDATION.—The wearing away of the surface of the land by water.

DEVONIAN SYSTEM or formation.—A series of Palæozoic rocks, including the Old Red Sandstone.

DICOTYLEDONS OR DICOTYLEDONOUS PLANTS.—A class of plants characterised by having two seed-leaves, by the formation of new wood beween the bark and the old wood (exogenous growth) and by the reticulation of the veins of the leaves. The parts of the flowers are generally in multiples of five.

DIFFERENTIATION.—The separation or discrimination of parts or organs which in simpler forms of life are more or less united.

DIMORPHIC.—Having two distinct forms.—Dimorphism is the condition of the appearance of the same species under two dissimilar forms.

DIŒCIOUS.—Having the organs of the sexes upon distinct individuals.

DIORITE.—A peculiar form of Greenstone.

DORSAL.—Of or belonging to the back.

EDENTATA.—A peculiar order of Quadrupeds, characterised by the absence of at least the middle incisor (front) teeth in both jaws. (*Examples,* the Sloths and Armadillos.)

ELYTRA.—The hardened fore-wings of Beetles, serving as sheaths for the membranous hind-wings, which constitute the true organs of flight.

EMBRYO.—The young animal undergoing development within the egg or womb.

EMBRYOLOGY.—The study of the development of the embryo.

ENDEMIC.—Peculiar to a given locality.

ENTOMOSTRACA.—A division of the class Crustacea, having all the segments of the body usually distinct, gills attached to the feet or organs of the mouth, and the feet fringed with fine hairs. They are generally of small size.

EOCENE.—The earliest of the three divisions of the Tertiary epoch of geologists. Rocks of this age contain a small proportion of shells identical with species now living.

EPHEMEROUS INSECTS.—Insects allied to the May-fly.

FAUNA.—The totality of the animals naturally inhabiting a certain country or region, or which have lived during a given geological period.

FELIDÆ.—The Cat-family.

FERAL.—Having become wild from a state of cultivation or domestication.

FLORA.—The totality of the plants growing naturally in a country, or during a given geological period.

FLORETS.—Flowers imperfectly developed in some respects, and collected into a dense spike or head, as in the Grasses, the Dandelion, &c.

FŒTAL.—Of or belonging to the fœtus, or embryo in course of development.

FORAMINIFERA.—A class of animals of very low organisation, and generally of small size, having a jelly-like body, from the surface of which delicate filaments can be given off and retracted for the prehension of external objects, and having a calcareous or sandy shell, usually divided into chambers, and perforated with small apertures.

FOSSILIFEROUS.—Containing fossils.

FOSSORIAL.—Having a faculty of digging. The Fossorial Hymenoptera are a group of Wasp-like Insects, which burrow in sandy soil to make nests for their young.

FRENUM (pl. FRENA).—A small band or fold of skin.

FUNGI (sing. FUNGUS).—A class of cellular plants, of which Mushrooms, Toadstools, and Moulds, are familiar examples.

FURCULA.—The forked bone formed by the union of the collar-bones in many birds, such as the common Fowl.

GALLINACEOUS BIRDS.—An order of Birds of which the common Fowl, Turkey, and Pheasant, are well-known examples.

GALLUS.—The genus of birds which includes the common Fowl.

GANGLION.—A swelling or knot from which nerves are given off as from a centre.

GANOID FISHES.—Fishes covered with peculiar enamelled bony scales. Most of them are extinct.

GERMINAL VESICLE.—A minute vesicle in the eggs of animals, from which the development of the embryo proceeds.

GLACIAL PERIOD.—A period of great cold and of enormous extension of ice upon the surface of the earth. It is believed that glacial periods have occurred repeatedly during the geological history of the earth, but the term is generally applied to the close of the Tertiary epoch, when nearly the whole of Europe was subjected to an arctic climate.

GLAND.—An organ which secretes or separates some peculiar product from the blood or sap of animals or plants.

GLOTTIS.—The opening of the windpipe into the œsophagus or gullet.

GNEISS.—A rock approaching granite in composition, but more or less laminated, and really produced by the alteration of a sedimentary deposit after its consolidation.

GRALLATORES.—The so-called Wading-birds (Storks, Cranes, Snipes, &c.),

which are generally furnished with long legs, bare of feathers above the heel, and have no membranes between the toes.

GRANITE.—A rock consisting essentially of crystals of felspar and mica in a mass of quartz.

HABITAT.—The locality in which a plant or animal naturally lives.

HEMIPTERA.—An order or sub-order of Insects, characterised by the possession of a jointed beak or rostrum, and by having the fore-wings horny in the basal portion and membranous at the extremity, where they cross each other. This group includes the various species of Bugs.

HERMAPHRODITE.—Possessing the organs of both sexes.

HOMOLOGY.—That relation between parts which results from their development from corresponding embryonic parts, either in different animals, as in the case of the arm of man, the fore-leg of a quadruped, and the wing of a bird; or in the same individual, as in the case of the fore and hind legs in quadrupeds, and the segments or rings and their appendages of which the body of a worm, a centipede, &c., is composed. The latter is called *serial homology*. The parts which stand in such a relation to each other are said to be *homologous*, and one such part or organ is called the *homologue* of the other. In different plants the parts of the flower are homologous, and in general these parts are regarded as homologous with leaves.

HOMOPTERA.—An order or sub-order of Insects having (like the Hemiptera) a jointed beak, but in which the fore-wings are either wholly membranous or wholly leathery. The *Cicadæ,* Frog-hoppers, and *Aphides,* are well-known examples.

HYBRID.—The offspring of the union of two distinct species.

HYMENOPTERA.—An order of Insects possessing biting jaws and usually four membranous wings in which there are a few veins. Bees and Wasps are familiar examples of this group.

HYPERTROPHIED.—Excessively developed.

ICHNEUMONIDÆ.—A family of Hymenopterous insects, the members of which lay their eggs in the bodies or eggs of other insects.

IMAGO.—The perfect (generally winged) reproductive state of an insect.

INDIGENES.—The aboriginal animal or vegetable inhabitants of a country or region.

INFLORESCENCE.—The mode of arrangement of the flowers of plants.

INFUSORIA.—A class of microscopic Animalcules, so called from their having originally been observed in infusions of vegetable matters. They consist of a gelatinous material enclosed in a delicate membrane, the whole or part of which is furnished with short vibrating hairs (called cilia), by means of which the animalcules swim through the water or convey the minute particles of their food to the orifice of the mouth.

INSECTIVOROUS.—Feeding on Insects.

INVERTEBRATA, or INVERTEBRATE ANIMALS.—Those animals which do not possess a backbone or spinal column.

LACUNÆ.—Spaces left among the tissues in some of the lower animals, and serving in place of vessels for the circulation of the fluids of the body.

LAMELLATED.—Furnished with lamellæ or little plates.

LARVA (pl. LARVÆ).—The first condition of an insect at its issuing from the egg, when it is usually in the form of a grub, caterpillar, or maggot.

LARYNX.—The upper part of the windpipe opening into the gullet.

LAURENTIAN.—A group of greatly altered and very ancient rocks, which is greatly developed along the course of the St. Laurence, whence the name. It is in these that the earliest known traces of organic bodies have been found.

LEGUMINOSÆ.—An order of plants represented by the common Peas and Beans, having an irregular flower in which one petal stands up like a wing, and the stamens and pistil are enclosed in a sheath formed by two other petals. The fruit is a pod (or legume).

LEMURIDÆ.—A group of four-handed animals, distinct from the Monkeys and approaching the Insectivorous Quadrupeds in some of their characters and habits. Its members have the nostrils curved or twisted, and a claw instead of a nail upon the first finger of the hind hands.

LEPIDOPTERA.—An order of Insects, characterised by the possession of a spiral proboscis, and of four large more or less scaly wings. It includes the well-known Butterflies and Moths.

LITTORAL.—Inhabiting the seashore.

LOESS.—A marly deposit of recent (Post-Tertiary) date, which occupies a great part of the valley of the Rhine.

MALACOSTRACA.—The higher division of the Crustacea, including the ordinary Crabs, Lobsters, Shrimps, &c., together with the Woodlice and Sand-hoppers.

MAMMALIA.—The highest class of animals, including the ordinary hairy quadrupeds, the Whales, and Man, and characterised by the production of living young which are nourished after birth by milk from the teats (*Mammæ, Mammary glands*) of the mother. A striking difference in embryonic development has led to the division of this class into two great groups; in one of these, when the embryo has attained a certain stage, a vascular connection, called the *placenta*, is formed between the embryo and the mother; in the other this is wanting, and the young are produced in a very incomplete state. The former, including the greater part of the class, are called *Placental mammals;* the latter, or *Aplacental mammals,* include the Marsupials and Monotremes *(Ornithorhynchus).*

MAMMIFEROUS.—Having mammæ or teats (see MAMMALIA).

MANDIBLES, in Insects.—The first or uppermost pair of jaws, which are generally solid, horny, biting organs. In Birds the term is applied to both jaws with their horny coverings. In Quadrupeds the mandible is properly the lower jaw.

MARSUPIALS.—An order of Mammalia in which the young are born in a very incomplete state of development, and carried by the mother, while sucking, in a ventral pouch (marsupium), such as the Kangaroos, Opossums, &c. (see MAMMALIA).

MAXILLÆ, in Insects.—The second or lower pair of jaws, which are composed of several joints and furnished with peculiar jointed appendages called palpi, or feelers.

MELANISM.—The opposite of albinism; an undue development of colouring material in the skin and its appendages.

METAMORPHIC ROCKS.—Sedimentary rocks which have undergone alteration, generally by the action of heat, subsequently to their deposition and consolidation.

MOLLUSCA.—One of the great divisions of the Animal Kingdom, including those animals which have a soft body, usually furnished with a shell, and in which the nervous ganglia, or centres, present no definite general arrangement. They are generally known under the denomination of "shell-fish;" the cuttle-fish, and the common snails, whelks, oysters, mussels, and cockles, may serve as examples of them.

MONOCOTYLEDONS, or MONOCOTYLEDONOUS PLANTS.—Plants in which the seed sends up only a single seed-leaf (or cotyledon); characterised by the absence of consecutive layers of wood in the stem (endogenous growth), by the veins of the leaves being generally straight, and by the parts of the flowers being generally in multiples of three. (*Examples*, Grasses, Lilies, Orchids, Palms, &c.)

MORAINES.—The accumulations of fragments of rock brought down by glaciers.

MORPHOLOGY.—The law of form or structure independent of function.

MYSIS-STAGE.—A stage in the development of certain Crustaceans (Prawns), in which they closely resemble the adults of a genus (*Mysis*) belonging to a slightly lower group.

NASCENT.—Commencing development.

NATATORY.—Adapted for the purpose of swimming.

NAUPLIUS-FORM.—The earliest stage in the development of many Crustacea, especially belonging to the lower groups. In this stage the animal has a short body, with indistinct indications of a division into segments, and three pairs of fringed limbs. This form of the common fresh-water *Cyclops* was described as a distinct genus under the name of *Nauplius*.

NEURATION.—The arrangement of the veins or nervures in the wings of Insects.

NEUTERS.—Imperfectly developed females of certain social insects (such as Ants and Bees), which perform all the labours of the community. Hence they are also called *workers*.

NICTITATING MEMBRANE.—A semi-transparent membrane, which can be drawn across the eye in Birds and Reptiles, either to moderate the effects of a strong light or to sweep particles of dust, &c., from the surface of the eye.

OCELLI.—The simple eyes or stemmata of Insects, usually situated on the crown of the head between the great compound eyes.

ŒSOPHAGUS.—The gullet.

OOLITIC.—A great series of secondary rocks, so called from the texture of some of its members, which appear to be made up of a mass of small *egg-like* calcareous bodies.

OPERCULUM.—A calcareous plate employed by many Mollusca to close the aperture of their shell. The *opercular valves* of Cirripedes are those which close the aperture of the shell.

ORBIT.—The bony cavity for the reception of the eye.

ORGANISM.—An organised being, whether plant or animal.

ORTHOSPERMOUS.—A term applied to those fruits of the Umbelliferæ which have the seed straight.

OSCULANT.—Forms or groups apparently intermediate between and connecting other groups are said to be osculant.

OVA.—Eggs.

OVARIUM or OVARY (in plants).—The lower part of the pistil or female organ of the flower, containing the ovules or incipient seeds; by growth after the other organs of the flower have fallen, it usually becomes converted into the fruit.

OVIGEROUS.—Egg-bearing.

OVULES (of plants).—The seeds in the earliest condition.

PACHYDERMS.—A group of Mammalia, so called from their thick skins, and including the Elephant, Rhinoceros, Hippopotamus, &c.

PALÆOZOIC.—The oldest system of fossiliferous rocks.

767

PALPI.—Jointed appendages to some of the organs of the mouth in Insects and Crustacea.

PAPILIONACEÆ.—An order of Plants (see LEGUMINOSÆ).—The flowers of these plants are called *papilionaceous*, or butterfly-like, from the fancied resemblance of the expanded superior petals to the wings of a butterfly.

PARASITE.—An animal or plant living upon or in, and at the expense of, another organism.

PARTHENOGENESIS.—The production of living organisms from unimpregnated eggs or seeds.

PEDUNCULATED.—Supported upon a stem or stalk. The pedunculated oak has its acorns borne upon a footstalk.

PELORIA or PELORISM.—The appearance of regularity of structure in the flowers of plants which normally bear irregular flowers.

PELVIS.—The bony arch to which the hind limbs of vertebrate animals are articulated.

PETALS.—The leaves of the corolla, or second circle of organs in a flower. They are usually of delicate texture and brightly coloured.

PHYLLODINEOUS.—Having flattened, leaf-like twigs or leafstalks instead of true leaves.

PIGMENT.—The colouring material produced generally in the superficial parts of animals. The cells secreting it are called *pigment-cells*.

PINNATE.—Bearing leaflets on each side of a central stalk.

PISTILS.—The female organs of a flower, which occupy a position in the centre of the other floral organs. The pistil is generally divisible into the ovary or germen, the style and the stigma.

PLACENTALIA, PLACENTATA, or Placental Mammals.—See MAMMALIA.

PLANTIGRADES.—Quadrupeds which walk upon the whole sole of the foot, like the Bears.

PLASTIC.—Readily capable of change.

PLEISTOCENE PERIOD.—The latest portion of the Tertiary epoch.

PLUMULE (in plants).—The minute bud between the seed-leaves of newly-germinated plants.

PLUTONIC ROCKS.—Rocks supposed to have been produced by igneous action in the depths of the earth.

POLLEN.—The male element in flowering plants; usually a fine dust produced by the anthers, which, by contact with the stigma effects the fecundation of the seeds. This impregnation is brought about by means of tubes (*pollen-tubes*) which issue from the pollen-grains adhering to the stigma, and penetrate through the tissues until they reach the ovary.

POLYANDROUS (flowers).—Flowers having many stamens.

POLYGAMOUS PLANTS.—Plants in which some flowers are unisexual and others hermaphrodite. The unisexual (male and female) flowers, may be on the same or on different plants.

POLYMORPHIC.—Presenting many forms.

POLYZOARY.—The common structure formed by the cells of the Polyzoa, such as the well-known Sea-mats.

PREHENSILE.—Capable of grasping.

PREPOTENT.—Having a superiority of power.

PRIMARIES.—The feathers forming the tip of the wing of a bird, and inserted upon that part which represents the hand of man.

PROCESSES.—Projecting portions of bones, usually for the attachment of muscles, ligaments, &c.

PROPOLIS.—A resinous material collected by the Hive-Bees from the opening buds of various trees.

PROTEAN.—Exceedingly variable.

PROTOZOA.—The lowest great division of the Animal Kingdom. These animals are composed of a gelatinous material, and show scarcely any trace of distinct organs. The Infusoria, Foraminifera, and Sponges, with some other forms, belong to this division.

PUPA (pl. PUPÆ).—The second stage in the development of an Insect, from which it emerges in the perfect (winged) reproductive form. In most insects the *pupal stage* is passed in perfect repose. The *chrysalis* is the pupal state of Butterflies.

RADICLE.—The minute root of an embryo plant.

RAMUS.—One half of the lower jaw in the Mammalia. The portion which rises to articulate with the skull is called the *ascending ramus*.

RANGE.—The extent of country over which a plant or animal is naturally spread. *Range in time* expresses the distribution of a species or group through the fossiliferous beds of the earth's crust.

RETINA.—The delicate inner coat of the eye, formed by nervous filaments spreading from the optic nerve, and serving for the perception of the impressions produced by light.

RETROGRESSION.—Backward development. When an animal, as it approaches maturity, becomes less perfectly organised than might be expected from its early stages and known relationships, it is said to undergo a *retrograde development* or *metamorphosis*.

RHIZOPODS.—A class of lowly organised animals (Protozoa), having a gelatinous body, the surface of which can be protruded in the form of root-like processes or filaments, which serve for locomotion and the prehension of food. The most important order is that of the Foraminifera.

RODENTS.—The gnawing Mammalia, such as the Rats, Rabbits, and Squirrels. They are especially characterised by the possession of a single pair of chisel-like cutting teeth in each jaw, between which and the grinding teeth there is a great gap.

RUBUS.—The Bramble Genus.

RUDIMENTARY.—Very imperfectly developed.

RUMINANTS.—The group of Quadrupeds which ruminate or chew the cud, such as oxen, sheep, and deer. They have divided hoofs, and are destitute of front teeth in the upper jaw.

SACRAL.—Belonging to the sacrum, or the bone composed usually of two or more united vertebræ to which the sides of the pelvis in vertebrate animals are attached.

SARCODE.—The gelatinous material of which the bodies of the lowest animals (Protozoa) are composed.

SCUTELLÆ.—The horny plates with which the feet of birds are generally more or less covered, especially in front.

SEDIMENTARY FORMATIONS.—Rocks deposited as sediments from water.

SEGMENTS.—The transverse rings of which the body of an articulate animal or Annelid is composed.

SEPALS.—The leaves or segments of the calyx, or outermost envelope of an ordinary flower. They are usually green, but sometimes brightly coloured.

SERRATURES.—Teeth like those of a saw.

SESSILE.—Not supported on a stem or footstalk.

SILURIAN SYSTEM.—A very ancient system of fossiliferous rocks belonging to the earlier part of the Palæozoic series.

SPECIALISATION.—The setting apart of a particular organ for the performance of a particular function.

SPINAL CHORD.—The central portion of the nervous system in the Vertebrata, which descends from the brain through the arches of the verte-bræ, and gives off nearly all the nerves to the various organs of the body.

STAMENS.—The male organs of flowering plants, standing in a circle within the petals. They usually consist of a filament and an anther, the anther being the essential part in which the pollen, or fecundating dust, is formed.

STERNUM.—The breast-bone.

STIGMA.—The apical portion of the pistil in flowering plants.

STIPULES.—Small leafy organs placed at the base of the footstalks of the leaves in many plants.

STYLE.—The middle portion of the perfect pistil, which rises like a column from the ovary and supports the stigma at its summit.

SUBCUTANEOUS.—Situated beneath the skin.

SUCTORIAL.—Adapted for sucking.

SUTURES (in the skull).—The lines of junction of the bones of which the skull is composed.

TARSUS (pl. TARSI).—The jointed feet of articulate animals, such as Insects.

TELEOSTEAN FISHES.—Fishes of the kind familiar to us in the present day, having the skeleton usually completely ossified and the scales horny.

TENTACULA or TENTACLES.—Delicate fleshy organs of prehension or touch possessed by many of the lower animals.

TERTIARY.—The latest geological epoch, immediately preceding the estab-lishment of the present order of things.

TRACHEA.—The wind-pipe or passage for the admission of air to the lungs.

TRIDACTYLE.—Three-fingered, or composed of three movable parts attached to a common base.

TRILOBITES.—A peculiar group of extinct Crustaceans, somewhat resembling the Woodlice in external form, and, like some of them, capable of roll-ing themselves up into a ball. Their remains are found only in the Palæozoic rocks, and most abundantly in those of Silurian age.

TRIMORPHIC.—Presenting three distinct forms.

UMBELLIFERÆ.—An order of plants in which the flowers, which contain five stamens and a pistil with two styles, are supported upon footstalks which spring from the top of the flower stem and spread out like the wires of an umbrella, so as to bring all the flowers in the same head (*umbel*) nearly to the same level. (*Examples*, Parsley and Carrot.)

UNGULATA.—Hoofed quadrupeds.

UNICELLULAR.—Consisting of a single cell.

VASCULAR.—Containing blood-vessels.

VERMIFORM.—Like a worm.

VERTEBRATA: or VERTEBRATE ANIMALS.—The highest division of the animal kingdom, so called from the presence in most cases of a backbone composed of numerous joints or *vertebræ*, which constitutes the centre of the skeleton and at the same time supports and protects the central parts of the nervous system.

WHORLS.—The circles or spiral lines in which the parts of plants are arranged upon the axis of growth.

WORKERS.—See Neuters.

ZOËA-STAGE.—The earliest stage in the development of many of the higher

Crustacea, so called from the name of *Zoëa* applied to these young animals when they were supposed to constitute a peculiar genus.

Zooids.—In many of the lower animals (such as the Corals, Medusæ, &c.) reproduction takes place in two ways, namely, by means of eggs and by a process of budding with or without separation from the parent of the product of the latter, which is often very different from that of the egg. The individuality of the species is represented by the whole of the form produced between two sexual reproductions; and these forms, which are apparently individual animals, have been called *zooids*.

APPENDIX I

STATISTICAL SUMMARY OF VARIANTS

TABLE I

	Sentences Dropped	Rewritten	Added
% of Totals	8	145	39

The total dropped, rewritten, or added sentences divided by the total sentences in a.

TABLE II

% of Totals by Edition	Sentences Dropped	Rewritten	Added
b	3	9	2
c	10	11	18
d	11	19	29
e	56	32	15
f	20	30	38

The number of dropped, rewritten, or added sentences in each edition divided by the total number of each type of variant.

TABLE III

Net gain: 31%
% by edition of net variations

	Sentences Dropped	Rewritten	Added
b	0.2	12	0.8
c	0.8	16	7
d	0.9	26	11
e	3.9	40	5
f	1.3	34	11

The number of dropped, rewritten, or added sentences in each edition divided by the total number of sentences in the previous edition.

TABLE IV

% of gain per edition

b	.8
c	6.4
d	9.4
e	4.7
f	10.6

The added sentences in each addition divided by the total number of sentences in that edition.

TABLE V

% of total variants

b	7
c	14
d	21
e	29
f	31

The total variants per edition divided by the total variants for all editions.

TABLE VI

	% of net variants
b	13
c	26
d	37
e	48
f	46

The total variants per edition divided by the total number of sentences in the previous edition.

APPENDIX II

TRANSCRIPT OF PUBLISHER'S RECORDS

The ledgers of John Murray, Publishers, are preserved at the publisher's office, 50 Albemarle Street, London. In the following edited transcript of the *Origin* entries I have omitted one column of occasionally mysterious figures which are in part probably cross-references to other files and in part unit prices. I have also omitted various other figures and unit prices, often inserted in pencil. In the original the debit and credit columns face each other on the verso and recto leaves of the open ledger.

Ledger E. Folio 158
Darwin On the Origin of Species

Dr

1859 Dec. 31	To Printing 1250 No Clowes	153	3	6
	" 53 ¾ Rms. Sht ½ Post Spalding	60	9	6
	" Author for 1st Edition	180	—	—
	" West Engraving & Printing Diagram	4	—	—
	" Entering Stationers Hall	—	5	—
	" Binding 1250 Copies	41	13	4
	" Advertising	45	—	—
	" Commn. alld to Agents	—	—	—
	" Registration at Paris & Leipsic	2	2	—
	" Balance Profit	57	4	2
		543	17	6

Cr

1859 Dec. 31	By 1250 Copies			
	5 Stationers' Hall			
	12 Allowed Author			
	58 41 Presented Reviews			
	1192 Sold 25 as 24	543	17	6
		543	17	6

Dr   Second Edition

1859 Dec. 31	To Printing 3000 No. Clowes	109	2	—
	" 129 Rms Sht ½ Post Spalding	140	17	—
	" Author for 2nd Edition	636	13	4
1860 June 30	" West Printing Diagram	7	6	3
	" Advertising	60	4	—
	" Commn. alld Agents on additional	14	3	—
	299 copies	14	19	—
	do———————153 copies	7	13	—
	" Binding 3000 Copies	100	—	—
	" Balance Profit	313	12	5
		1405	—	—

775

Cr
1859 Dec. 31   By 3000 Copies

	5 Stationers' Hall		
23	18 Allowed Author		
2977	Sold Viz.		

1020	Trade 25 as 24	465	10	—
1957	Do " "	939	10	—
2977				

1405	—	—

Ledger F. Folio 118
Darwin On the Origin of Species Third Edition

Dr

1861 Mar. 31	To Printing 2000 No. Clowes	108	7	8
	" 94 Rms. Sht ½ Post Spalding	105	15	—
	" West Printing Diagram	2	5	—
	" Author for the Edition	372	—	—
1865 Dec. 31	" Binding 2000 Copies	66	13	4
	" Advertising	29	3	6
	" Commn. alld Agents on 290 Copies	14	10	—
	" Balance Profit	239	19	6
		938	14	—

Cr

1861 Mar. 31   By 2000 copies

	5 Stationers' Hall		
	2 Presented		
13	6 Allowed Author		
1987	Sold Viz.		

637	Trade 25 as 24	290	14	—
1350	Do " "	648	—	—
1987				

938	14	—

4th Edn.

Dr

1866 Dec 31	To Printing 1500 No. Clowes	137	13	8
	" do Index Clowes	—	19	10
	" 77 Rms. Sht. ½ Post Spalding	80	18	—
	" West printing diagram	4	13	3
1867 June 30	" Binding 1500 Copies	53	2	6
	" Commn. to Agents 202 Copies	10	17	2
Septr. 19	" C. Darwin for the Edition	250	—	—
	" Advertizing	35	—	—
	" Balance Profit	160	6	10
		733	11	3

Cr

1866 Dec. 31   By 1500 Copies

           5 Statrs Hall
           6 Author
          10 Reviews
       21 — on hand June 30 1867
     1479 Sold viz

839 Trade 25 as 24		403	—	—
640 do		330	11	3
1479				
		733	11	3

5th Edition

Dr

1869 June 30	To Printing 2000 No. Clowes	133	1	6
	" 78 Rms D. Crown Dickinson	112	2	6
	" West Printing Diagrams	4	15	6
	" C. Darwin for Edition	315	—	—
	" Binding 2000 Copies	62	10	—
1870 June 30	" Advertising	30	—	—
	do	10	—	—
	" Comn to agents 255 Copies	13	14	3
	" Balance Profit	283	6	1
		954	9	9

Cr

1869 June 30   By 2000 Copies

           5 Strs Hall
           6 Author
           2 Presented
       70   57 on hand June 1872
     1930 Sold viz

875 Trade 25 as 24		420	—	—
1055 do		544	9	9
1930				
		964	9	9

6th Edition

Dr

1872 Mar 31	To Printing 3000 No Clowes	187	5	6
	" Clowes. set Stereo. Plates	35	15	—
	" 90 Rms D Crown Dick	112	10	—
	" W.T.Dallas for Glossary	10	—	—
	" West 3000 facsimiles	6	19	6
	" Binding 3000 copies	60	—	—
	" Author for the Edition	210	—	—
	" Comn to Agents 334 Copies	8	18	5
	" Advertizing	40	—	—
	" Balance Profit	114	6	11
		785	15	4

Cr

1872 June 30	By <u>57</u> on hand	carried below		
1872 Mar. 31	By 3000 Copies			
	5 Statrs Hall			
	12 Author			
	<u>34</u> 17 Reviews			
	2966 Sold viz			
	1481 Trade 25 as 24	355	10	—
	<u>1485</u> do " "	380	5	4
	2966			
1872 March	By Appleton Co. Set Stereoplates	50	—	—
		785	15	4
1872 June 30	By 57 on Hand 5th edition			
	40 Hodgson	13	5	3

Ledger G. Folio 337
Darwin On the Origin of Species Sixth Edition

Dr

1873 Mar 31	To Printing 2000 No. Clowes	39	15	—
	" 60 Rms Sm D. Post Dickinson	75	—	—
	" West 2000 Diagrams	4	12	3
	" Author for the Edition	180	—	—
	" Binding 2000 Copies	42	—	—
	" Advertizing	34	8	4
	" Comn to Agents 567 Copies	15	2	5
	" Balance Profit	102	2	—
		493	—	—

Cr

1873 Mar 31	By 2000 Copies			
	<u>5</u> Statn Hall			
	1995 Sold			
	1120 Trade 25 as 24	269	—	—
	<u>875</u> do " "	224	—	—
	1995			
		493	—	—

Dr

1875 Mar 31	To Printing 1500 No. Clowes	30	—	6
	" 30 Rms Dlled Dickn	57	—	—
	" West 1500 Diagrams	3	7	6
	" Author for the Edition	130	—	—
	" Binding 1500 Copies	31	10	—
	" Advertising	35	—	—
	" Comn to Agents 345 Copies	9	4	—
	" Balance Profit	75	19	4
		372	1	4

Cr

1875 Mar 31	By 1500 Copies (Making 16,250)				
	Sold viz				
	762 Trade 25 as 24		183	—	—
	738 do " "		189	1	4
	1500				
			372	1	4

Dr

1876 June 30	To Printing 1250 No Clowes		29	17	6
	" 25 Rms D Med Dickn		47	10	—
	" West 1250 Diagrams		2	16	3
	" Author for the edition		120	—	—
	" Binding 1250 Copies		26	5	—
	" Advertising		15	—	—
	" Com to Agents 227 Copies		6	1	1
	" Balance Profit		62	13	10
			310	3	8

Cr

1876 June 30	By 1250 Copies				
	Sold viz				
	630 Trade 25 as 24		151	5	—
	620 do " "		158	18	8
	1250				
			310	3	8

Dr

1878 Mar 31	To Printing 2000 No Clowes		45	2	6
	" 40 Rms D Medm Dickn		76	—	—
	" West 2000 Diagrams		4	11	2
	" Binding 800 Copies		16	16	—
	" Advertizing		5	—	—
	" Comn to Agents 86 Copies		2	5	10
			149	15	6

Cr

1876 June 30	By 2000 Copies				
	1 author				
	1406 1405 on hand June 1878				
	594 sold viz				
	507 Trade 25 as 24		121	15	—
	87 do " "		22	8	—
June 30	By Balance deficiency		5	12	6
			149	15	6

Dr

1878 June 30	To Balance deficiency		5	12	6
	" Binding 700 Copies		14	14	—
	" Advertizing		9	—	—
	" Comn to Agents 120 Copies		3	4	—
	" Author's 2/3 profit		109	17	11
	" Mr. M's 1/3 do		57	18	11
			197	7	4

1878 June 30   By 1405 on hand
              604 on hand June 1879

801 Sold viz			
497 Trade 25 as 24	119	10	—
304 " 25 as 24	77	17	4
801			
	197	7	7

Ledger H. Folio 46

Dr

1880 June 30 To Binding 500 Copies	10	10	—
" Advertising	8	5	—
" Comn to Agents 76 Copies	2	—	6
" Author's $\frac{2}{3}$ Profit	84	18	4
" Mr. M's $\frac{1}{3}$ "	42	9	2
	148	3	—

Cr

1879 June 30   By 604 on hand
                      Sold viz

423 Trade 25 as 24	101	15	—
181 do " "	46	8	—
604			
	148	3	—

Dr

1880 June 30 To Printing 2000 No Clowes	45	5	—
" 40 $\frac{1}{8}$ Rms D. Medium Dickn	68	4	3
" West 2000 Diagrams	4	10	—
" Binding 1100 Copies	23	2	—
" Advertizing	7	—	—
" Comn to Agents 211 Copies	5	12	6
" Author's $\frac{2}{3}$ Profit	55	15	9
" Mr M's $\frac{1}{3}$ do	27	17	10
	237	7	4

Cr

1880 June 30   By 2000 Copies
                    1039 on Hand, June 1881
                    961 sold

564 Trade 25 as 24	135	10	—
397 do " "	101	17	4
961			
	237	7	4

Dr

1882 Mar. 31	To Binding 900 Copies	18	18	—
	" Advertising	4	—	—
	" Comn to agens 101 Copies	2	13	10
	" Authors 2/3 Profit	153	17	11
	" Mr. M's 1/3 do	76	18	11
		256	8	8

Cr

1881 June 30	By 1039 on hand			
	Sold viz			
	622 Trade	149	10	—
	417 do	106	18	8
	1039			
		256	8	8

Dr

1882 June 30	To Printing 2000 No Clowes	43	16	6
	" 65 1/2 Rms D Crown Dickn	59	15	4
	" West 2000 Diagrams	4	11	2
	" Binding 1400 Copies	33	10	10
	" Advertising	4	6	11
	" Author's 2/3 Profit	99	12	11
	" Mr M's 1/3 "	49	16	6
		297	6	8

Cr

1882 June 30	By 2000 Copies			
	1 Presented			
	839 858 on Hand June 1882			
	1161 Sold 25 as 24	297	6	8
		297	6	8

Dr

1882 Dec. 30	To Printing 2000 Clowes	36	13	—
	" 60 Rms D Crown Dickn	55	—	—
	" West 2000 Diagrams	4	11	—
	" Binding 2150 Copies	51	10	2
	" Advertising	10	—	—
	" Comn to agents 410 Copies	10	18	8
	" Author's 2/3 Profit	265	11	—
	" Mr M's 1/3 do	132	15	6
		566	19	4

Cr

1882 June 30	By 838 on Hand			
	" 2000 copies			
	2838			
	1 Presented			
	567 566 on Hand June 1883			
	2271 Sold viz			
	914 Trade 25 as 24	219	10	—
	1357 do " "	347	9	4
	2271	566	19	4

Dr

1883 Dec 31	To Printing 2000 No Clowes	34	13	—
	" 60 Rms D Crown Dick	55	—	—
	" West 2000 Diagrams	4	10	—
	" Binding 1800 Copies	43	2	6
	" Advertising	4	10	—
	" Com'n to agents 354 Copies	9	8	10
	" Author's ⅔ Profit	194	18	—
	" Mr M's ⅓ do	97	9	—
		446	11	4

Cr

1883 June 30	By 566 on hand			
	" 2000 Copies			
	2566			
	763 on Hand June 1884			
	1803 Sold viz			
	956 Trade 25 as 24	229	10	—
	847 do " "	217	1	4
	1803			
		446	11	4

Ledger H. Folio 229

Dr

1885 Mar. 31	To Printing 2000 No Clowes	38	3	6
	" 60 Rms Crown Dick	55	—	—
	" West 2000 Diagrams	4	11	—
	" Binding 1250 Copies	29	19	—
	" Advertising	5	10	—
	" Comn to agents 285 Copies	7	12	—
	" Author ⅔ Profit	112	8	7
	" Mr M's ⅓ do	56	4	3
		309	8	4

Cr

1884 June 30	By 763 on Hand			
1885 Mar. 31	" 2000 Copies (29,500)			
	2763			
	1506 on Hand June 1885			
	1257 Sold viz			
	778 Trade 25 as 24	186	15	—
	479 do " "	122	13	4
	1257			
		309	8	4

1886 June 30	To Binding 1400 Copies	33	10	10
	" Advertising	6	10	—
	" Comn to agents 282 Copies	7	10	5
	" Author's $2/3$ Profit	215	17	5
	" Mr M's $1/3$ do	107	18	8
		371	7	4

Cr

1885 June 30	By 1506 on Hand			
	1 Presented			
	1505			
	889 Trade 25 as 24	213	10	—
	616 do " "	157	17	4
	1505	371	7	4

Dr

1886 Sep. 30	To Printing 2000 No Clowes	37	1	—
	" 60 Rms D Crown	55	—	—
	" West 2000 Diagrams	4	10	—
	" Binding 1600 Copies	38	6	8
	" Advertising	6	5	—
	" Comn to Agents 229 Copies	6	2	2
	" Author's $2/3$ Profit	136	5	6
	" Mr M's $1/3$ do	68	2	8
		351	13	—

Cr

1886 Sep. 30	By 2000 Copies			
	570 on Hand June 1887			
	1430 Sold viz			
	905 Trade 25 as 24	217	5	—
	525 do " "	134	8	—
	1430			
		351	13	—

Dr

1887 Dec. 31	To Printing 3500 No Clowes	67	12	—
	" 105 Rms D Crown Dickn	86	12	6
	" West Diagrams	10	4	9
	" Binding 2900 Copies	69	9	7
	" Advertising	9	10	—
	" Comn agents 112 Copies			
	1291 do	33	14	6
	" Cancelling Titles	—	5	—
	" Trade for 4 Copies Bound	—	4	—
	" Author's $2/3$ Profit	160	9	1
	" Mr M's $1/3$ do	80	4	7
		518	6	—

1887 June 30	By 570 on Hand				
	372 on Hand Novr 1887				
	198 Sold 24 as 24	50	18	8	
1887 Nov. 4	By 372 on Hand reduced to 6/—				
"	1500 Copies				
"	2000 do				
	3872				
	1504 on Hand June 1888				
	2368 Sold viz				
	798 Trade 25 as 24	153	4	—	
	1570 do " "	314	3	4	
	2368	518	6	—	

Dr

1889 Mar. 31	To Printing 2000 No Clowes	36	13	—
	" 60 Rms D Crown Dickn	49	10	—
	" West Diagrams	4	11	2
June 30	" Binding 2000 Copies	47	18	4
	" Advertising	6	5	—
	" Comn to agents 1818 Copies	36	7	6
	" Author's 2/3 Profit	169	16	9
	" Mr M's 1/3 do	84	18	5
		436	—	2

Cr

1888 June 30	By 1504 on Hand			
1889 Mar. 31	" 2000 Copies			
	3504			
	2 Presented			
	1311 1309 on Hand June 1888			
	2193 Sold viz			
	342 Trade 25 as 24	65	16	—
	1851 do " "	370	4	2
	2193	436	—	2

Ledger H. Folio 230

Dr

1890 Mar. 31	To Printing 2000 No Clowes	36	13	—
	" 60 Rms D Crown Dick	44	—	—
	" West Diagrams	6	3	9
	" Binding 1800 Copies	43	2	6
	" Advertising	3	—	—
	" Comn to agents 1451 Copies	29	6	3
	" Author's 2/3 Profit	136	14	8
	" Mr M's 1/3 do	67	17	4
		365	17	6

Cr

1889 June 30   By 1309 on Hand
                    2000 Copies

                    <u>3309</u>

                    <u>1469</u> on Hand June 1890

                    <u>1840</u> Sold viz

281 Trade 25 as 24	54	—	—
<u>1559</u> do " "	<u>311</u>	<u>17</u>	<u>6</u>
<u>1840</u>			
	<u>365</u>	<u>17</u>	<u>6</u>

Note: Subsequent editions were printed from plates originally used for the expensive two-volume edition. (See Appendix III) The next credit entry indicates that 635 copies were on hand in June, 1892.

APPENDIX III

BIBLIOGRAPHICAL DESCRIPTION

FIRST EDITION:

ON | THE ORIGIN OF SPECIES | BY MEANS OF NATURAL SELEC-
TION, | OR THE | PRESERVATION OF FAVOURED RACES IN
THE STRUGGLE | FOR LIFE. | BY CHARLES DARWIN, M.A., |
FELLOW OF THE ROYAL, GEOLOGICAL, LINNAEAN, ETC., SO-
CIETIES; | AUTHOR OF 'JOURNAL OF RESEARCHES DURING
H.M.S. BEAGLE'S VOYAGE | ROUND THE WORLD.' | LONDON: |
JOHN MURRAY, ALBEMARLE STREET. | 1859. | *The right of Trans-
lation is reserved.*

Coll: 12° (Uncut: 7 7/8 × 4 15/16), a^4, b^1, B-X^{12}, Y^{12} (—Y$_{12}$), 256 leaves,
   pp. [i-v] vi-ix [x] [1] 2-502; plate [1] is folded in between 116 and
   117.
Contents: p. [i]: half-title 'ON THE ORIGIN OF SPECIES.' p. [ii]:
   [wavy rule (2 3/16)] | epigraphs (from Whewell and Bacon) | [wavy
   rule (2 3/16)] | '*Down, Bromley, Kent,* | October 1st 1859.'. p. [iii]:
   title. p. [iv]: [rule (2 1/16)] 'LONDON: PRINTED BY W. CLOWES
   AND SONS, STAMFORD STREET, | AND CHARING CROSS.'. p.
   [v] vi-ix: 'CONTENTS' [short wavy rule (7/16)]. p. [x]: 'INSTRUC-
   TION TO BINDER.' p. [1]: 'ON THE ORIGIN OF SPECIES |
   [double short rule (3/4)] | INTRODUCTION.'. p. 490: text ends. p.
   491-502: 'INDEX.' [rule (3 17/32)]; [rule between columns]; [both
   rules on all pages of Index]. p. 502: Index ends. | 'THE END.' | [rule
   (2 1/16)] | printer's statement as on p. [iv].
RT] According to subject discussed on page. Occasionally RT runs across
   the double page. RT frequently correspond with sub-titles within
   chapters. In small caps.
Typography and paper: $b signed; $B-Y signed on 1st (B), 2nd (B 2), and
   5th (B 3) recto of each gathering.
   Text: 35 ll. (leaded) 5 3/4 (6 1/8) × 3 3/8. Type corresponds in all
   but a few minute particulars with face in *Specimens of some of the
   Printing Types for Book-Work, used in the Office of Charles Reed and
   Benjamin Pardon, at Paternoster Row, Saint Paul's, MDCCCL.* Text
   in Small Pica, chapter headings in Bourgeois, Index in Brevier. This
   face in turn corresponds in all but a few minute particulars (especially
   the 'f') with English Modern in the Reed and Fox specimen book of
   1856. (Both specimen books in the Columbia University Library.) This
   company had been R. Beazly & Co., late Thorowgood. The face used
   appears to be a new cutting, slightly changed, of the Thorowgood face
   illustrated (scale changed) in Charles Rosner, *Printer's Progress,* Cambridge, 1951. Printer was W. Clowes and Sons. Unfortunately, all of
   this firm's records were destroyed during the Second World War. Text,
   10-pt. with 2-pt. leading; RT and head-titles 8-pt.; index, 6-pt. with
   2-pt. leading.
   The paper (sheet and half post) purchased from Spalding (publisher's records), probably Spalding and Hodge, Wholesale Stationers,
   145 Drury Lane, founded 1812. (*The London Commercial List,* By

Estell & Co., London, 1871). White wove unwatermarked. Sheets bulk 1 5/16".

Plate: 1-leaf folded pasted on p. 116. Signed 'W. West lith. Hatton Garden.' in lower left-hand corner. Plate represents diagram of descent.

Binding: Dark-green zig-zag ribbed cloth. Front and back covers: identical blind-stamped panel with elaborate border. Spine: gilt-stamped [three rules width of spine] | 'ON THE | ORIGIN OF | SPECIES | [short rule] | DARWIN. | [two rules width of spine] | [elaborate triangular ornament point down] | [identical ornament point up] | [two rules width of spine] | LONDON | JOHN MURRAY.' | [three rules width of spine].

Light brown end-papers in front and back of wove unwatermarked paper lighter than sheets.

SECOND EDITION:

ON | THE ORIGIN OF SPECIES | BY MEANS OF NATURAL SELECTION, | OR THE | PRESERVATION OF FAVOURED RACES IN THE STRUGGLE | FOR LIFE. | BY CHARLES DARWIN, M.A., | FELLOW OF THE ROYAL, GEOLOGICAL, LINNEAN, ETC., SOCIETIES; | AUTHOR OF 'JOURNAL OF RESEARCHES DURING H.M.S. BEAGLE'S VOYAGE | ROUND THE WORLD.' | *FIFTH THOUSAND.* | LONDON: | JOHN MURRAY, ALBEMARLE STREET. | 1860. | *The right of Translation is reserved.*

Coll: 12° (uncut: 7 7/8 × 4 15/16), a^4, b^1, B-X^{12}, Y^{12} (—Y12), 256 leaves, pp. [i-v] vi-ix [x] [1] 2-502; plate [1] in fold between 116 and 117.

Contents: p. [i]: as in 1st. p. [ii]: epigraphs (quotations from Whell, Butler, Bacon. p. [iii] etc.: as in 1st.

RT] As in 1st, except for p. 433.

Typography and paper: as in 1st.

Plate: As in 1st.

Binding: As in 1st.

THIRD EDITION:

ON | THE ORIGIN OF SPECIES | BY MEANS OF NATURAL SELECTION, | OR THE | PRESERVATION OF FAVOURED RACES IN THE STRUGGLE | FOR LIFE. | BY CHARLES DARWIN, M.A., | FELLOW OF THE ROYAL, GEOLOGICAL, LINNEAN, ETC., SOCIETIES: | AUTHOR OF 'JOURNAL OF RESEARCHES DURING H.M.S. BEAGLE'S VOYAGE ROUND THE WORLD.' | THIRD EDITION, WITH ADDITIONS AND CORRECTIONS. | *(SEVENTH THOUSAND.)* | LONDON: | JOHN MURRAY, ALBEMARLE STREET. | 1861. | *The right of Translation is reserved.*

Coll.: 12° (uncut: 7 1/8 × 4 15/16), a^6, b^4, B-Z^{12}, 2A^6, 280 leaves, pp. [i-v] vi-ix [x] xii-xix [xx] [1] 2-538 [539-540]; plate [i] in fold between 146 and 147.

Contents: p. [i]: as in 1st. p. [ii]: as in 2nd with additional date '*Third Edit., March 1861.*'. p. [iii]: title. p. [iv]: advertisements 'BY THE SAME AUTHOR.' | (four titles) [rule (2 1/16)] | 'LONDON: PRINTED W. CLOWES AND SONS, STAMFORD STREET AND CHARING CROSS.'. p. [v] vi-ix: as in 1st. p. [x]: as in 1st, except 'page 123' for 'page 130'. p. [xi] xii. 'ADDITIONS AND CORRECTIONS TO THE SECOND AND THIRD EDITIONS' [short rule (1 1/16)]. p. xii: [rule (4 5/16)] | short rule (1 3/16) 'POSTSCRIPT.'. p. xiii-xix: 'AN HISTORICAL SKETCH OF THE RECENT PROG-

RESS OF OPINION ON THE ORIGIN OF SPECIES.' [short rule (29/32)]. p. [xx]: blank. p. [1]: as in 1st. p. 525: text ends. p. [526]: blank. p. 527: index begins as in 1st. p. 538: index ends as in 1st. p. [539]: 'WORKS ON | SCIENCE, NATURAL HISTORY, &c., | Published by Mr. Murray.' [short wavy rule (7/8)]. p. [539]: advt. ends.

RT] As in 2nd with adjustments for new pagination.

Typography and paper: As in 1st. Sheets bulk 1 7/16".

Plate: As in 1st. Pasted on p. 146 in error for p. 123 (in copy examined, University of Pennsylvania Library).

Binding: As in 1st.

FOURTH EDITION:

ON | THE ORIGIN OF SPECIES | BY MEANS OF NATURAL SELEC-TION, | OR THE | PRESERVATION OF FAVOURED RACES IN THE STRUGGLE | FOR LIFE. | BY CHARLES DARWIN, M.A., F.R.S. &C. | FOURTH EDITION, WITH ADDITIONS AND CORRECTIONS. | *(EIGHTH THOUSAND.)* | LONDON: | JOHN MURRAY, ALBE-MARLE STREET. | 1866. | *The right of Translation is reserved.*

Coll.: 12° (uncut: as in 1st), a^6, B-Z^{12}, 2A-2B^{12}, 2C^8, 2D^1, 308 leaves, pp. [i-v] vi-ix [x-xi] xii-xxi [xxii] [1] 2-594; plate [i] in fold between 130 and 131.

Contents: p. [i]: as in 1st. p. [ii]: as in first with new original and additional dates 'November 24th, 1859. (1st Edition.) | 'Fourth Edition, June, 1866.'. p. [iii]: title. p. [iv]: as in 3rd with eight titles. p. [v] vi-ix: as in 1st. p. [x]: as in 3rd. p. [xi] xii: 'ADDITIONS AND COR-RECTIONS, TO THE FOURTH EDITION.'. p. xiii-xxi: Historical Sketch as in 3rd. p. [xxii]: blank. p. [1]: as in 1st. p. 577: text ends. p. [578]: blank. p. 579-593: index as in 1st. p. [594]: blank.

RT] Through Chapter XIII to conform to chapter sub-titles in text, now rewritten, made more frequent, and centered on page. Chapter XIV, as in previous edns. Occasionally, where there are sub-sub-titles in text, not centered, RT subtitles on verso, sub-sub-titles on facing recto.

Typography and paper: As in 1st. A different, whiter, paper. Sheets bulk 1 7/16".

Binding: Dark-green pebble-grained cloth. Front and back blind-stamped with new design: larger, plain panel with simply decorated border. Spine as in 1st, but with new design for title and author. 'ON THE | ORIGIN | OF | SPECIES. | [short decorative rule (1/2)] | DARWIN.'. Roman letters are sans-serif. Dark-green end papers, weight as in 1st.

FIFTH EDITION:

ON | THE ORIGIN OF SPECIES | BY MEANS OF NATURAL SELEC-TION, | OR THE | PRESERVATION OF FAVOURED RACES IN THE STRUGGLE FOR LIFE. | BY CHARLES DARWIN, M.A., F.R.S., &C. | FIFTH EDITION, WITH ADDITIONS AND CORRECTIONS. | *(TENTH THOUSAND.)* | LONDON: | JOHN MURRAY, ALBE-MARLE STREET. | 1869. | *The Right of Translation is reserved.*

Coll.: 8° (uncut: 7 7/16 $\times$ 4 15/16), a^8, b^4, B-Z^8, 2A-2P^8, 2Q^2, 310 leaves, pp. [i-v] vi-xiii [xiv] xv-xxiii [xxiv] [1] 2-596; plate [i] in fold between 132 and 133.

Contents: p. [i]: as in 1st. p. [ii]: as in 2nd but with new place and dates, 'Down, Beckenham, Kent, | First Edition, November 24th, 1859. | Fifth Edition, May, 1869.'. p. [iii]: title. p. [iv]: as in 4th with nine titles in

different order, of which one by Fritz Müller. Printer's name reads
'. . . SONS, DUKE STREET, STAMFORD . . .'. p. [v] vi-x: Contents.
p. xi-xiii: 'ADDITIONS AND CORRECTIONS | TO THE FIFTH
EDITION.' [short decorated rule (15/32)]. p. [xiv]: 'INSTRUCTION
TO BINDER.', 'page 132' for 'page 130' in 4th. p. xv-xxiii: Historical
Sketch. p. [xxiv]: blank. p. [1] as in 1st. p. 579: text ends. p. 581-595:
Index as in 1st. p. 596: Index ends. [rule (2 1/16)] | 'P.S.' with three
lines of text. | 'THE END' | Printer's name as on p. [iv].

RT] As in 4th.

Typography and paper: Paper (double crown) bought from Dickinson (publisher's records), probably John Dickinson and Co., Wholesale Paper
Manufacturers, 65 Old Bailey, founded 1825, (*The London Commercial List, cf. supra* under 1st), one of the most important paper
manufacturers in England. (See Lewis Evans, *The Firm of John Dickinson & Co., Ltd.,* London, 1896). Like the preceding papers, white
wove unwatermarked, less white than in 4th. Sheets bulk 1 5/8″.

Plate: As in 1st, pasted to p. 133.

Binding: Dark-green pebble-grained cloth. Front and back: blind-stamped
with new design, four lines crossing to form a panel with leaf-deco-
rated line-points, in a three-line plain rectangle. Spine: gold-stamped
[decorated three-line rule width of spine] title and author as in 4th
with first line omitted | [rule] | undecorated spine | [rule width of
spine] | 'LONDON, JOHN MURRAY' | decorated three-line rule,
identical inverse of rule at top.

SIXTH EDITION:

THE | ORIGIN OF SPECIES | BY MEANS OF NATURAL SELEC-
TION, | OR THE | PRESERVATION OF FAVOURED RACES IN THE
STRUGGLE FOR LIFE. | BY CHARLES DARWIN, M.A., F.R.S., &C. |
SIXTH EDITION, WITH ADDITIONS AND CORRECTIONS. |
(*ELEVENTH THOUSAND.*) | LONDON: | JOHN MURRAY, ALBE-
MARLE STREET. | 1872. | *The right of Translation is reserved.*

Coll: 8° (7 3/8 × 4 5/8), a^8, b^4 (—b^4), B-Z^8, 2A-2F^8, 2G^4, 2H^1, 237 leaves,
pp. [i-v] vi-ix [x-xi] xii [xiii] xiv-xxi [xxii] [1] 2-458: plate [i] in
fold between 90 and 91.

Contents: p. [i]: 'THE ORIGIN OF SPECIES.'. p. [ii]: as in 5th with
'*Sixth Edition, Jan. 1872.*' for '*Fifth Edition, May, 1869.*'. p. [iii]:
title. p. [iv]: '*By the same Author.*' with ten titles, one by Müller.
Printer's name reads '. . . SONS, STAMFORD . . .' p. [v] vi-ix.
'CONTENTS.' | [short wavy rule (9/16)]. p. [x]: 'INSTRUCTION
TO THE BINDER' | [decorated rule (15/32)] | 'page 90' for 'page
132'. p. [xi]: 'ADDITIONS AND CORRECTIONS TO THE SIXTH
EDITION. | [short decorated rule as in 5th]. p. [xiii]: 'AN HIS-
TORICAL SKETCH OF THE PROGRESS OF OPINION ON THE
ORIGIN OF SPECIES, | PREVIOUSLY TO THE PUBLICATION
OF THE FIRST EDITION OF THIS WORK.' | [rule (1 1/8)]. p.
[1]: as in 1st. p. 429: text ends. p. 430-441: 'GLOSSARY | OF THE |
PRINCIPAL SCIENTIFIC TERMS USED IN THE | PRESENT
VOLUME.' | [short decorated rule (3/4)]. p. [442]: blank. p. 443-457:
Index as in 1st but in smaller type. p. 458. index ends. | 'THE END.' |
Printer's name reads as on p. [iv].

RT] As in 5th, with rule (3 1/4) underneath.

Typography and paper. Entirely reset, as 2nd through 5th were not. $b
signed; b 2 on recto of leaf 2. $B-2G signed on 1st (B) and 2nd (B 2)

recto of each gathering. Text, 8-pt. with 2-pt. leading; RT, 10-pt. italics; head-titles, 8-pt.; index, 6-pt. with 1-pt. leading.

Text: 43 ll. (leaded) 5 3/4 (6 1/16) × 3 1/16. Type is same or nearly same, but chapter headings and text are now set in a smaller face, Brevier, and the Glossary and Index in Pearl.

Paper from Dickinson. Whiter and less expensive (publisher's records) than in previous edns. Sheets bulk 1 1/4".

Binding: Dark-green cross-hatched cloth. Front and back: identical blind-stamped disk centered in simple four-ruled frame. Spine: [three blind-stamped rules width of spine] | gold-stamped 'ORIGIN | OF | SPE-CIES | [short rule (11/16)] | DARWIN. | LONDON. | JOHN MURRAY.' | [three blind-stamped rules width of spine]. Dark-brown end-papers on paper as in previous edns.

Second Impression.

Identical with 1st impression, except (1) '(*THIRTEENTH THOU-SAND.*)' and (2) '1873' on title and (3) differences in titles by Darwin on p. [iv].

Third Impression.

Identical with 2nd impression, except (1) '(*FIFTEENTH THOUSAND.*)' and (2) '1875' on title, (3) differences in titles by Darwin on p. [iv], (4) new binding design: Dark-green pebble-grained cloth. Front and back: blank-stamped with identical ornamental frame. Spine: gilt-stamped [ornamental rule across binding] | 'THE | ORIGIN | OF | SPECIES | [short rule] | DARWIN | [ornament] | LONDON | JOHN MURRAY.' | [ornamental rule identical with that at top].

Fourth Impression.

Not examined. 1876 (publisher's records).

Fifth Impression.

Identical with 3rd, except (1) '(*TWENTIETH THOUSAND.*)' and (2) '1878' on title, (3) differences in titles by Darwin on p. [iv], and (4) chocolate-purple end-papers. In either the fourth or fifth impression were made a few minor textual changes. (4) RT in 8-pt. roman.

Sixth Impression.

Identical with 5th, except (1) '(*TWENTY-SECOND THOUSAND.*)' and (2) '1880' on title.

Seventh Impression.

Identical with 6th, except (1) '(*TWENTY-FOURTH THOUSAND.*)' and (2) '1882' on title, (3) differences in titles by Darwin on p. [iv], (4) sheets bulk 1 5/16", (5) new binding design: Cloth as before. Front and back: blind-stamped with simple rectangular design with crossed corners within simple frame. Spine: gilt-stamped [ornamental band] | ORIGIN | OF | SPECIES | [ornamental rule] | DARWIN | [ornament] | LON-DON | JOHN MURRAY | [ornamental band identical with that at top], (6) slate-gray end-papers, and (7) uncut.

Eighth Impression.

Not examined. 1882 (publisher's records).

Ninth Impression.

Identical with 7th, except (1) '(*TWENTY-SIXTH THOUSAND.*)' and (2) '1884' on title, (3) differences in titles by Darwin on p. [iv], and (4) sheets bulk 1 3/16".

Tenth Impression.

Identical with 9th, except (1) '(*TWENTY-EIGHTH THOUSAND.*)' and (2) '1885' on title, and (3) slight variations in titles by Darwin on p. [iv].

Eleventh Impression.

Not examined. 1886 (publisher's records).

Twelfth Impression.

Identical with 10th, except (1) '(*THIRTY-FIFTH THOUSAND.*)' and (2) '1888' on title, (3) differences in titles by Darwin on p. [iv], (4) sheets bulk 1 1/4", and (5) new binding-cloth: Dark-green cloth lightly-grained.

Thirteenth Impression.

Identical with 12th, except (1) '(*THIRTY-SEVENTH THOUSAND.*)' and (2) '1889' on title, (3) slight differences on p. [iv], (4) sheets bulk 1 1/4", and (5) white end-papers printed with pale-blue all-over pattern.

Fourteenth Impression.

Identical with 13th, except (1) '(*THIRTY-NINTH THOUSAND.*)' and (2) '1890' on title, and (3) addition of F. Darwin's life of C. Darwin to titles on p. [iv].

Note: The printing totals on the titles are incorrect in the 6th ed., 1st impn., when an error of 2,000 was made, in the 2nd, 3rd, (?4th), impns., correct in the 5th, 6th, 7th, (?8th) impns., incorrect by 2,000 in the 9th (when the fact that there were two impressions in 1882 was overlooked), 10th, (?11th) impns., and correct in the 12th, 13th, and 14th impns.

Note: Index, 4th-6th edns., 8 pt. solid.

APPENDIX IV

Illustrations of the
Principal Binding Variants

First, Second, and Third Editions (Height: 8 1/16″)

Fourth Edition (Height: 8 1/32″)

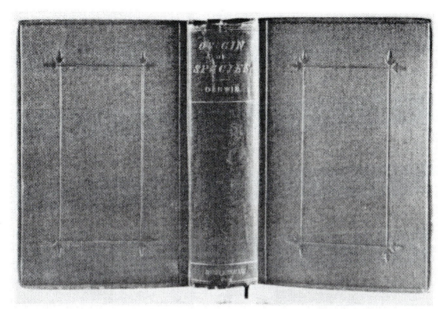

Fifth Edition (Height: 7 3/4″)

Sixth Edition, 1st and 2nd Impressions (Height: 7 7/16″)

Sixth Edition, 3rd through 6th Impressions (Height: 7 1/2″)

Sixth Edition, 7th through 14th Impressions (Height: 7 5/8″)

ALPHABETICAL LIST OF
CHAPTER SUB-TITLES

Minor varia are not included.

INDEX

[*Note:* This index is that of *f*, corrected, and supplemented with entries from previous editions, principally *a* and *d*, not found in *f*. Such entries are in italics.]

B

Babington, Mr., on British plants, 128.
Baer, Von, standard of Highness, 221.
————, comparison of bee and fish, 550.
————, embryonic similarity of the Vertebrata, 686.
Baker, Sir S., on the giraffe, 244.
Balancement of growth, 295.
Baleen, 247.
Bamboo with hooks, 365.
Barberry, flowers of, 188.
Barrande, M., on Silurian colonies, 522.
————, on the succession of species, 534.
————, on parallelism of palæozoic formations, 538.
————, on affinities of ancient species, 540.
Barriers, importance of, 563.
Bates, Mr., on glacial period in South America, 603, 607.
————, on mimetic butterflies, 666.
Batrachians on islands, 623.
Bats, how structure acquired, 331.
————, distribution of, 625.
Bear, catching water-insects, 333.
Beauty, how acquired, 367, 736.
Bee, Australian, extermination of, 160.
————, sting of, 374.
————, queen, killing rivals, 374.
Bees fertilising flowers, 156.
————, hive, not sucking the red clover, 183.
————, Ligurian, 184.
————, variation in habits, 385.
————, parasitic, 396.
————, hive, cell-making instinct, 402.
————, humble, cells of, 403.
Beetles, with deficient tarsi, 281.
————, wingless, in Madeira, 282.
Bentham, Mr., on British plants, 128.
————, on classification, 655.
Berkeley, Mr., on seeds in salt water, 575.
Bermuda, birds of, 620.
Birds, song of males, 175.
————, colour of, on continents, 277.
————, wingless, 280, 331.
————, beauty of, 371.
———— acquiring fear, 385.
————, footsteps, and remains of, in secondary rocks, 509.
————, fossil, in caves of Brazil, 554.
———— transporting seeds, 580.
———— annually cross the Atlantic, 582.

————, waders, 615.
————, of Madeira, Bermuda, and Galapagos, 620.
————, *with traces of embryonic teeth, 706.*
Bizcacha, 565.
————, affinities of, 671.
Bladder for swimming, in fish, 346.
Blindness of cave animals, 283.
Blyth, Mr., on distinctness of Indian cattle, 93.
————, on striped hemionus, 313.
————, on crossed geese, 433.
Boar, shoulder-pad of, 174.
Borrow, Mr., on the Spanish pointer, 110.
Bory St. Vincent, on Batrachians, 623.
Bosquet, M., on fossil Chthamalus, 510.
Boulders, erratic, on the Azores, 581.
Branchiae, 347.
———— of crustaceans, 354.
Branchiostoma, 224, 342.
Braun, Prof., on the seeds of Fumariaceæ, 238.
Brazil, glaciers in, 595.
Brent, Mr., on house-tumblers, 387.
————, *on hawks killing pigeons, 579.*
Brewer, Dr., on American cuckoo, 390.
Britain, mammals of, 626.
Broca, Prof., on Natural Selection, 232.
Bronn, various objections by, 232.
————, on duration of specific forms, 495.
Brown, Robert, on classification, 651.
————, Séquard, on inherited mutilations, 281.
Buckman, on variation in plants, 82.
Busk, Mr., on the Polyzoa, 257.
Butterflies, mimetic, 666.
Buzareingues, on sterility of varieties, 464.

C

Cabbage, varieties of, crossed, 189.
Calceolaria, 431.
Canary-birds, sterility of hybrids, 432.
Cape Verde islands, productions of, 631.
————, plants of, on mountains, 596.
Carabus, species of, in South America, 607.
Carpenter, Dr., on eozoon, 515.
————, on foraminifera, 549.
Carrier-pigeons killed by hawks, 579.
Carthamus, 238.
Cassini on flowers of compositæ, 293.

Extinction, as bearing on natural selection, 219.
———— of domestic varieties, 215.
————, 527.
Eye, structure of, 339.
————, correction for aberration, 373.
Eyes reduced in moles, 283.

F

Fabre, M., on hymenoptera fighting, 174.
————, on parasitic sphex, 397.
————, on Sitaris, 701.
Falconer, Dr., on naturalisation of plants in India, 148.
————, *on fossil crocodile, 522.*
————, on elephants and mastodons, 545.
———— and Cautley on mammals of sub-Himalayan beds, 555.
Falkland Islands, wolf of, 624.
Faults, 483.
Faunas, marine, 564.
Fear, instinctive, in birds, 389.
Feet of birds, young molluscs adhering to, 614.
Fertilisation variously effected, 357, 370.
Fertility of hybrids, 429.
————, from slight changes in conditions, 452.
Fertility of crossed varieties, 459.
Fir-trees destroyed by cattle, 155.
————, pollen of, 375.
Fish, flying, 331.
————, teleostean, sudden appearance of, 510.
————, eating seeds, 579, 616.
————, fresh-water, distribution of, 612.
Fishes, ganoid, now confined to fresh water, 199.
————, electric organs of, 350.
————, ganoid, living in fresh water, 531.
————, of southern hemisphere, 597.
Flat-fish, their structure, 250.
Flight, powers of, how acquired, 331.
Flint-tools, proving antiquity of man, 91.
Florida, pigs of, 170.
Flower, Prof., on the Larynx, 255.
————, on Halitherium, 540.
————, on the resemblance between the jaws of the dog and Thylacinus, 665.

————, on the homology of the feet of certain marsupials, 677.
Flowers, structure of, in relation to crossing, 180.
————, of compositæ and umbelliferæ, 237, 293.
————, beauty of, 370.
————, double, 416.
Flysch formation, destitute of organic remains, 488.
Forbes, Mr. D., on glacial action in the Andes, 593.
————, E., on colours of shells, 277.
————, on abrupt range of shells in depth, 324.
————, on poorness of palæontological collections, 487.
————, on continuous succession of genera, 525.
————, on continental extensions, 573.
————, on distribution during Glacial period, 584.
————, on parallelism in time and space, 644.
Forests, changes in, in America, 158.
Formation, Devonian, 545.
————, Cambrian, 514.
Formations, thickness of, in Britain, 481.
————, intermittent, 498.
Formica rufescens, 397.
————, sanguinea, 398.
————, flava, neuter of, 419.
Forms, lowly organised, long enduring, 225.
Frena, ovigerous, of cirripedes, 348.
Fresh-water productions, dispersal of, 646.
Fries, on species in large genera being closely allied to other species, 141.
Frigate-bird, 335.
Frogs on islands, 624.
Fruit-trees, gradual improvement of, 111, 289.
———— in United States, 170.
————, varieties of, acclimatised in United States, 289.
Fuci, crossed, 438, 448.
Fur, thicker in cold climates, 278.
Furze, 687.

G

Galapagos Archipelago, birds of, 619.
————, productions of, 630, 632.
Galaxias, its wide range, 612.
Galeopithecus, 330.

Game, increase of, checked by vermin, 153.

Gärtner, on sterility of hybrids, 427, 435.

———, on reciprocal crosses, 438.

———, on crossed maize and verbascum, 464.

———, on comparison of hybrids and mongrels, 467, 468.

Gaudry, Prof., on intermediate genera of fossil mammals in Attica, 539.

Geese, fertility when crossed, 433.

———, upland, 335.

Geikie, Mr. on subaerial denudation, 479.

Genealogy, important in classification, 656.

Generations, alternate, 684.

Geoffroy St. Hilaire, on balancement, 295.

———, on homologous organs, 677.

———, Isidore, on variability of repeated parts, 138.

———, on correlation, in monstrosities, 84.

———, on correlation, 291.

———, on variable parts being often monstrous, 304.

Geographical distribution, 562.

Geography, ancient, 756.

Geology, future progress of, 756.

———, imperfection of the record, 756.

Gervais, Prof., on Typotherium, 540.

Giraffe, tail of, 362.

———, structure of, 242.

Glacial period, 583.

———, affecting the North and South, 591.

Glands, mammary, 253.

Gmelin, on distribution, 583.

Gnathodon, fossil, 586.

Godwin-Austen, Mr., on the Malay Archipelago, 503.

Goethe, on compensation of growth, 295.

Gomphia, 239.

Gooseberry, grafts of, 442.

Gould, Dr. Aug. A., on land-shells, 628.

———, Mr., on colours of birds, 277.

———, on instincts of cuckoo, 393.

———, *on birds of the Galapagos, 630.*

———, on distribution of genera of birds, 637.

Gourds, crossed, 464.

Graba, on the Uria lacrymans, 179.

Grafting, capacity of, 441.

Granite, areas of denuded, 493.

Grasses, varieties of, 207.

Gray, Dr. Asa, on the variability of oaks, 133.

———, on man not causing variability, 164.

———, on sexes of the holly, 183.

———, on trees of the United States, 190.

———, on naturalised plants in the United States, 209.

———, on æstivation, 239.

———, on rarity of intermediate varieties, 326.

———, on Alpine plants, 583.

———, Dr. J. E., on striped mule, 313.

Grebe, 335.

Grimm, on asexual reproduction, 685.

Groups, aberrant, 671.

Grouse, colours of, 170.

———, red, a doubtful species, 129.

Growth, compensation of, 295.

Growth, correlation of, in domestic products, 84.

———, *correlation of, 97.*

Günther, Dr., on flat-fish, 252.

———, on prehensile tails, 253.

———, on the fishes of Panama, 564.

———, on the range of fresh-water fishes, 612.

———, on the limbs of Lepidosiren, 710.

H

Haast, Dr., on glaciers of New Zealand, 592.

Habit, effect of, under domestication, 83.

———, effect of, under nature, 280.

———, diversified, of same species, 332.

Häckel, Prof., on classification and the lines of descent, 676.

Hair and teeth, correlated, 292.

Halitherium, 540.

Harcourt, Mr. E. V., on the birds of Madeira, 620.

Hartung, M., on boulders in the Azores, 581.

Hazel-nuts, 576.

Hearne, on habits of bears, 333.

Hector, Dr., on glaciers of New Zealand, 592.

Heath, changes in vegetation, 154.

Heer, Oswald, on ancient cultivated plants, 91.

———, on plants of Madeira, 199.

Helianthemum, 239.

Helix pomatia, 629.

Helix, resisting salt water, 629.

Helmholtz, M., on the imperfection of the human eye, 373.

Helosciadium, 576.

Hemionus, striped, 313.

Hensen, Dr., on the eyes of Cephalopods, 353.

Herbert, W., on struggle for existence, 146.

——, on sterility of hybrids, 429.

Hermaphrodites crossing, 185.

Heron eating seed, 616.

Heron, Sir R., on peacocks, 175.

Heusinger, on white animals poisoned by certain plants, 84.

Hewitt, Mr., on sterility of first crosses, 448.

Hildebrand, Prof., on the self-sterility of Corydalis, 430.

Hilgendorf, on intermediate varieties, 495.

Himalaya, glaciers of, 592.

——, plants of, 596.

Hippeastrum, 430.

Hippocampus, 254.

Hofmeister, Prof., on the movements of plants, 261.

Hollyhock, varieties of, crossed, 465.

Holly-trees, sexes of, 182.

Hooker, Dr., on trees of New Zealand, 190.

——, on acclimatisation of Himalayan trees, 287.

——, on flowers of umbelliferæ, 293.

——, on the position of ovules, 235.

——, on glaciers of Himalaya, 592.

——, on algæ of New Zealand, 597.

——, on vegetation at the base of the Himalaya, 595.

——, on plants of Tierra del Fuego, 598.

——, on Australian plants, 597, 632.

——, on relations of flora of America, 602.

——, on flora of the Antarctic lands, 608, 631.

——, on the plants of the Galapagos, 68, 120.

——, on glaciers of the Lebanon, 591.

——, on man not causing variability, 164.

——, on plants of mountains of Fernando Po, 596.

Hooks on bamboos, 365.

Hooks on palms, 365.

—— on seeds, on islands, 622.

Hopkins, Mr., on denudation, 492.

Hornbill, remarkable instinct of, 423.

Horner, Mr., on the antiquity of Egyptians, 91.

Horns, rudimentary, 713.

Horse, fossil, in La Plata, 528.

——, proportions of, when young, 694.

Horses destroyed by flies in Paraguay, 155.

——, striped, 314.

Horticulturists, selection applied by, 107.

Huber, on cells of bees, 407.

——, P., on reason blended with, instinct, 381.

——, on habitual nature of instincts, 381.

——, on slave-making ants, 397.

——, on Melipona domestica, 403.

Hudson, Mr., on the Ground-Woodpecker of La Plata, 334.

——, on the Molothrus, 395.

Humble-bees, cells, of, 403.

Hunter, J., on secondary sexual characters, 299.

Hutton, Captain, on crossed geese, 433.

Huxley, Prof., on structure of hermaphrodites, 190.

——, on the affinities of the Sirenia, 540.

——, *on embryological succession, 553.*

——, on forms connecting birds and reptiles, 540.

——, on homologous organs, 683.

——, on the development of aphis, 690.

Hybrids and mongrels compared, 467.

Hybridism, 424.

Hydra, structure of, 345.

Hymenoptera, fighting, 174.

Hymenopterous insect, diving, 335.

Hyoseris, 238.

I

Ibla, 296.

Icebergs transporting seeds, 581.

Increase, rate of, 147.

Individuals, numbers favourable to selection, 192.

Individuals, many, whether simultaneously created, 572.

Inheritance, laws of, 86.

——, at corresponding ages, 86, 171.

———, principle not of recent origin, 111.

———, unconscious, 111.

———, natural, 163.

———, sexual, 173.

———, objections to term, 165.

——— *natural, circumstances favourable to, 191.*

——— natural, has not induced sterility, 443.

Sexes, relations of, 174.

Sexual characters variable, 305.

——— selection, 173.

Sheep, Merino, their selection, 106.

———, two sub-breeds, unintentionally produced, 110.

———, mountain varieties of, 159.

Shells, colours of, 277.

———, hinges of, 357.

———, littoral, seldom embedded, 487.

———, fresh-water, long retain the same forms, 549.

———, fresh-water, dispersal of, 613.

———, of Madeira, 621.

———, land, distribution of, 621.

———, land, resisting salt water, 628.

Shrew-mouse, 663.

Silene, infertility of crosses, 437.

Silliman, Prof., on blind rat, 283.

Sirenia, their affinities, 540.

Sitaris, metamorphosis of, 701.

Skulls of young mammals, 366, 680.

Slave-making instinct, 397.

Smith, Col. Hamilton, on striped horses, 315.

———, Mr. Fred., on slave-making ants, 398.

———, on neuter ants, 418.

———, *Mr., of Jordan Hill, on the degradation of coast-rocks, 480.*

Smitt, Dr., on the Polyzoa, 257.

Snake with tooth for cutting through egg-shell, 395.

Snap-dragon, 310.

Somerville, Lord, on selection of sheep, 106.

Sorbus, grafts of, 442.

Sorex, 663.

Spaniel, King Charles's breed, 109.

Specialisation of organs, 222.

Species, polymorphic, 124.

———, dominant, 137.

———, common, variable, 137.

——— in large genera variable, 138.

———, groups of, suddenly appearing, 506, 512.

——— beneath Silurian formations, 514.

——— successively appearing, 521.

——— changing simultaneously throughout the world, 532.

Spencer, Lord, on increase in size of cattle, 110.

———, Herbert, Mr., on the first steps in differentiation, 225.

———, on the tendency to an equilibrium in all forces, 453.

Sphex, parasitic, 397.

Spiders, development of, 690.

Spitz-dog crossed with fox, 460.

Sports in plants, 81.

Sprengel, C. C., on crossing, 186.

———, on ray-florets, 294.

Squalodon, 540.

Squirrels, gradations in structure, 329.

Staffordshire, heath, changes in, 154.

Stag-beetles, fighting, 174.

Star-fishes, eyes of, 339.

———, their pedicellariæ, 256.

Sterility from changed conditions of life, 80.

——— of hybrids, 426.

——— ———, laws of, 435.

——— ———, causes of, 443.

——— from unfavourable conditions, 450.

——— not induced through natural selection, 443.

———, *of certain varieties, 463.*

St. Helena, productions of, 619.

St. Hilaire, Aug., on variability of certain plants, 239.

———, on classification, 653.

St. John, Mr., on habits of cats, 386.

Sting of bee, 374.

Stocks, aboriginal, of domestic animals, 93.

Strata, thickness of, in Britain, 480.

Stripes on horses, 314.

Structure, degrees of utility of, 367.

Struggle for existence, 144.

Succession, geological, 521.

——— of types in same areas, 554.

Swallow, one species supplanting another, 160.

Swaysland, Mr., on earth adhering to the feet of migratory birds, 580.

Swifts, nests of, 413.

Swim-bladder, 347.

Switzerland, lake habitations of, 91.

System, natural, 648.

T

Tail of giraffe, 362.
———— of aquatic animals, 363.
————, prehensile, 253.
————, rudimentary, 713.
Tanais, dimorphic, 125.
Tarsi, deficient, 281.
Tausch, Dr., on umbelliferæ, 238.
Teeth and hair correlated, 292.
————, rudimentary, in embryonic, calf, 706, 746.
————, *embryonic, traces of, in birds, 706.*
Tegetmeier, Mr., on cells of bees, 405, 410.
Temminck, on distribution aiding classification, 655.
Tendrils, their development, 260.
Thompson, Sir W., on the age of the habitable world, 513.
————, on the consolidation of the crust of the earth, 728.
Thouin, on grafts, 442.
Thrush, aquatic species of, 335.
————, mocking, of the Galapagos, 635.
————, young of, spotted, 686.
————, nest of, 423.
Thuret, M., on crossed fuci, 438.
Thwaites, Mr., on acclimatisation, 287.
Thylacinus, 665.
Tierra del Fuego, dogs of, 388.
————, plants of, 608.
Timber-drift, 577.
Time, lapse of, 478.
———— by itself not causing modification, 197.
Titmouse, 333.
Toads on islands, 623.
Tobacco, crossed varieties of, 465.
Tomes, Mr., on the distribution of bats, 625.
Torell, on a monocotyledonous Cambrian plant, 514.
Transitions in varieties rare, 322.
Traquair, Dr., on flat-fish, 252.
Trautschold, on intermediate varieties, 495.
Trees on islands belong to peculiar orders, 623.
———— with separated sexes, 189.
Trifolium pratense, 157, 183.
———— incarnatum, 183.
Trigonia, 531.
Trilobites, 512.
————, sudden extinction of, 531.

Trimen, Mr., on imitating-insects, 669.
Trimorphism in plants, 125, 453.
Troglodytes, 423.
Tuco-tuco, blind, 283.
Tumbler pigeons, habits of, hereditary, 387.
————, young of, 694.
Turkey-cock, tuft of hair on breast, 176.
————, naked skin on head, 365.
————, young of, instinctively wild, 389.
Turnip and cabbage, analogous variations of, 308.
Type, unity of, 378.
Types, succession of, in same areas, 554.
Typotherium, 540.

U

Udders enlarged by use, 83.
————, rudimentary, 707.
Ulex, young leaves of, 687.
Unbelliferæ, flowers and seeds of, 293.
————, outer and inner florets of, 237.
Unity of type, 378.
Uria lacrymans, 179.
Use, effects of, under domestication, 83.
————, effects of, in a state of nature, 280.
Utility, how far important in the construction of each part, 367.

V

Valenciennes, on fresh-water fish, 613.
Variability of mongrels and hybrids, 467.
Variation under domestication, 77.
———— caused by reproductive system being affected by conditions of life, 79.
———— under nature, 120.
————, laws of, 275.
————, correlated, 84, 291, 366.
Variations appear at corresponding ages, 86, 171.
———— analogous in distinct species, 308.
Varieties, natural, 118.
————, struggle between, 159.
————, domestic, extinction of, 204.
————, transitional, rarity of, 322.
————, when crossed, fertile, 462.
————, when crossed, sterile, 460.
————, classification of, 659.

P.S. [*in e only*]—The statement given at page [514], that a monocotyledonous plant has been discovered in the Cambrian formation in Sweden, I have been assured is probably erroneous.

THE END

Printed in the United Kingdom
by Lightning Source UK Ltd.
114135UKS00001B/1